Vehicular Technologies: Increasing Connectivity
Edited by Miguel Almeida

Published by InTech
Janeza Trdine 9, 51000 Rijeka, Croatia

Publishing Process Manager Katarina Lovrecic

Technical Editor Teodora Smiljanic

Cover Designer InTech Design Team

Additional hard copies can be obtained from orders@intechopen.com

Vehicular Technologies: Increasing Connectivity, Edited by Miguel Almeida
p. cm.
ISBN 978-953-307-223-4

For EU product safety concerns:
IN TECH d.o.o., Prolaz Marije Krucifikse Kozulić 3, 51000 Rijeka, Croatia,
info@intechopen.com or visit our website at intechopen.com.

Meet the editor

Miguel Almeida is both experienced in publishing papers for relevant conferences, and in working at the Telecommunication industry. His research background allowed him to participate in several European research projects, while his industry related experience has allowed him to manage advanced network performance solutions and software architecture definitions.
With an Engineering Degree and a Master Degree in Telecommunications, he is currently pursuing his PhD under the topics of performance management of remote devices. His main areas of interest are: Cross Layer Performance Analysis, Vehicular and Mobile Networks, 4G Architectures, End User Behavior Tracking, the Internet of Things Paradigm and Remote Device Management solutions.

Contents

Preface **XI**

Chapter 1 **A Non-Stationary MIMO Vehicle-to-Vehicle Channel Model Derived From the Geometrical T-Junction Model** **1**
Ali Chelli and Matthias Pätzold

Chapter 2 **Simulation of SISO and MIMO Multipath Fading Channels** **11**
Antonio Petrolino and Gonçalo Tavares

Chapter 3 **User Scheduling and Partner Selection for Multiplexing-based Distributed MIMO Uplink Transmission** **35**
Ping-Heng Kuo and Pang-An Ting

Chapter 4 **Resource Allocation for Multi-User OFDMA-Based Wireless Cellular Networks** **51**
Dimitri Kténas and Emilio Calvanese Strinati

Chapter 5 **From Linear Equalization to Lattice-Reduction-Aided Sphere-Detector as an Answer to the MIMO Detection Problematic in Spatial Multiplexing Systems** **71**
Sébastien Aubert and Manar Mohaisen

Chapter 6 **DFT Based Channel Estimation Methods for MIMO-OFDM Systems** **97**
Moussa Diallo,Maryline Hélard, Laurent Cariou and Rodrigue Rabineau

Chapter 7 **Channels and Parameters Acquisition in Cooperative OFDM Systems** **115**
D. Neves, C. Ribeiro, A. Silva and A. Gameiro

Chapter 8 **Fast Power and Channel Adaptation for Mobile Users in OFDMA Multi-Cell Scenarios** **137**
L. Reggiani, L. Galati Giordano and L. Dossi

Chapter 9 **Statistical Properties of the Capacity of Double Nakagami-*m* Channels for Applications in V2V Dualhop Communication Systems 153**
Gulzaib Rafiq, Bjørn Olav Hogstad and Matthias Pätzoldt

Chapter 10 **Resource Allocation and User Scheduling in Coordinated Multicell MIMO Systems 165**
Edgar Souza, Robson Vieira, Mari Kobayashi and Mérouane Debbah

Chapter 11 **Hybrid Evolutionary Algorithm-based Schemes for Subcarrier, Bit, and Power Allocation in Multiuser OFDM Systems 185**
Wei-Cheng Pao, Yung-Fang Chen and Yun-Teng Lu

Chapter 12 **Reduced-Complexity PAPR Minimization Schemes for MC-CDMA Systems 205**
Mariano García Otero and Luis A. Paredes Hernández

Chapter 13 **Cognitive Radio Communications for Vehicular Technology – Wavelet Applications 223**
Murroni Maurizio and Popescu Vlad

Chapter 14 **Multiple Antenna-Aided Spectrum Sensing Using Energy Detectors for Cognitive Radio 239**
Seung-Hoon Hwang and Jun-Ho Baek

Chapter 15 **New Method to Generate Balanced 2^n-PSK STTCs 261**
P. Viland, G. Zaharia and J.-F. Hélard

Chapter 16 **Correlation Coefficients of Received Signal I and Q Components in a Domain with Time and Frequency Axes under Multipath Mobile Channel with LOS and NLOS 281**
Shigeru Kozono, Kenji Ookubo, Takeshi Kozima and Tomohiro Hamashima

Chapter 17 **Multimodulus Blind Equalization Algorithm Using Oblong QAM Constellations for Fast Carrier Phase Recovery 299**
Jenq-Tay Yuan and Tzu-Chao Lin

Chapter 18 **Peak-to-Average Power Ratio Reduction for Wavelet Packet Modulation Schemes via Basis Function Design 315**
Ngon Thanh Le, Siva D. Muruganathan and Abu B. Sesay

Chapter 19 **Outage Performance and Symbol Error Rate Analysis of L-Branch Maximal-Ratio Combiner for κ-μ and η-μ Fading 333**
Mirza Milišić, Mirza Hamza and Mesud Hadžialić

Chapter 20 **Technological Issues in the Design of Cost-Efficient Electronic Toll Collection Systems 359**
José Santa, Rafael Toledo-Moreo, Benito Úbeda, Miguel A. Zamora-Izquierdo and Antonio F. Gómez-Skarmeta

Chapter 21 **Propagation Aspects in Vehicular Networks 375**
Lorenzo Rubio, Juan Reig and Herman Fernández

Chapter 22 **Propagation Path Loss Modelling in Container Terminal Environment 415**
Slawomir J. Ambroziak, Ryszard J. Katulski, Jaroslaw Sadowski and Jacek Stefanski

Chapter 23 **Link Budgets: How Much Energy is Really Received 433**
Aarne Mämmelä, Adrian Kotelba, Marko Höyhtyä and Desmond P. Taylor

Preface

This book covers the most recent advances concerning the ability to overcome connectivity limitations and extend the link capacity of vehicular systems. Ranging from the advances on radio access technologies to intelligent mechanisms deployed to enhance cooperative communications, cognitive radio and multiple antenna systems have been given particular highlight.

While some contributions do not offer an immediate response to the challenges that appear in some vehicular scenarios, they provide insight and research conclusions, from which Vehicle Networking Design can greatly benefit. Finding new ways to overcome the limitations of these systems will increase network reachability, service delivery, from infrastructure to vehicles, and the inter-vehicle connectivity. Having this in mind, particular attention was paid to the propagation issues and channel characterization models. To overcome the current limitations over these systems, this book is mainly comprised of the following topics:

1. Multiple Antenna Systems, Cognitive Radio and Cooperative Communications: focusing on multiple smart antenna systems, MIMO, OFDM, MC-CDMA systems, cognitive radio advances.
2. Transmission and Propagation: evaluating the propagation aspects of these systems, link layer coding techniques, mobile/radio oriented technologies, channel characterization, channel coding.

It is our understanding that advances on vehicular networking technologies can greatly benefit from the research studies presented herein. In this book, we tried to summarize the areas concerning physical and link layers with contributions that expose the state of the art for vehicular networks. We are thankful to all of those who contributed to this book and who made it possible.

Miguel Almeida
University of Aveiro
Portugal

1

A Non-Stationary MIMO Vehicle-to-Vehicle Channel Model Derived From the Geometrical T-Junction Model

Ali Chelli and Matthias Pätzold
University of Agder
Norway

1. Introduction

According to the European commission (*Road Safety Evolution in EU*, 2009), 1.2 million road accidents took place in the European Union in 2007. These road accidents have resulted in 1.7 million injuries and more than 40 thousand deaths. It turned out that human errors were involved in 93% of these accidents. V2V communication is a key element in reducing road casualties. For the development of future V2V communication systems, the exact knowledge of the statistics of the underlying fading channel is necessary. Several channel models for V2V communications can be found in the literature. For example, the two-ring channel model for V2V communications has been presented in (Pätzold et al., 2008). There, a reference and a simulation model have been derived starting from the geometrical two-ring model. In (Zajić et al., 2009), a three-dimensional reference model for wideband MIMO V2V channels has been proposed. The model takes into account single-bounce and double-bounce scattering in vehicular environments. The geometrical street model (Chelli & Pätzold, 2008) captures the propagation effects if the communicating vehicles are moving along a straight street with local roadside obstructions (buildings, trees, etc.). In (Acosta et al., 2004), a statistical frequency-selective channel model for small-scale fading is presented for a V2V communication links.
The majority of channel models that can be found in the literature rely on the stationarity assumption. However, measurement results for V2V channels in (Paier et al., 2008) have shown that the stationarity assumption is valid only for very short time intervals. This fact arises the need for non-stationary channel models. Actually, if the communicating cars are moving with a relatively high speed, the AoD and the AoA become time-variant resulting in a non-stationary channel model. The traditional framework invoked in case of stationary stochastic processes cannot be used to study the statistical properties of non-stationary channels. In the literature, quite a few time-frequency distributions have been proposed to study non-stationary deterministic signals (Cohen, 1989). A review of these distributions can be found in (Cohen, 1989). Many commonly used time-frequency distributions are members of the Cohen class (O'Neill & Williams, 1999). It has been stated in (Sayeed & Jones, 1995) that the Cohen class, although introduced for deterministic signals, can be applied on non-stationary stochastic processes.

In this chapter, we present a non-stationary MIMO V2V channel model. The AoD and the AoA are supposed to be time dependent. This assumption makes our channel model non-stationary. The correlation properties of a non-stationary channel model can be obtained using a multi-window spectrogram (Paier et al., 2008). For rapidly changing spectral content however, finding an appropriate time window size is a rather complicated task. The problem is that a decrease in the time window size improves the time resolution, but reduces the frequency resolution. To overcome this problem, we make use of the Choi-Williams distribution proposed in (Choi & Williams, 1989). The extremely non-isotropic propagation environment is modelled using the T-junction scattering model (Zhiyi et al., 2009). In contrast to the original multi-cluster T-model, we assume to simplify matters that each cluster consists of only one scatterer. Under this assumption, the reference and the simulation model are identical. The main contribution of this chapter is that it presents a non-stationary channel model with time-variant AoD and AoA. Moreover, analytical expressions for the correlation properties of the non-stationary channel model are provided, evaluated numerically, and then illustrated.

The rest of the chapter is organized as follows. In Section 2, the geometrical T-model is presented. Based on this geometrical model, we derive a reference (simulation) model in Section 3. In Section 4, the correlation properties of the proposed channel model are studied. Numerical results of the correlation functions are presented in Section 5. Finally, we draw the conclusions in Section 6.

2. The Geometrical T-junction Model

A typical propagation scenario for V2V communications at a T-junction is presented in Fig. 1. Fixed scatterers are located on both sides of the T-junction. In order to derive the statistical properties of the corresponding MIMO V2V channel, we first need to find a geometrical model that describes properly the vehicular T-junction propagation environment. This geometrical model is illustrated in Fig. 2. It takes into account double-bounce scattering under non-line-of sight conditions. Each building is modelled by one scatterer which makes our model extremely non-isotropic. The scatterers in the neighborhood of the transmitter MS_T are denoted by S_m^T $(m = 1, 2, \ldots, M)$, whereas the scatterers close to the receiver MS_R are designated by S_n^R $(n = 1, 2, \ldots, N)$. The total number of scatterers near to the transmitter is denoted by N, while the total number of scatterers near to the receiver is designated by M. The transmitter and the receiver are moving towards the intersection point with the velocities v_T and v_R, respectively. The direction of motions of the transmitter and the receiver w.r.t. the x-axis are referred to as ϕ_T and ϕ_R, respectively. The AoD are time-variant and are denoted by $\alpha_m^T(t)$, while the symbol $\beta_n^R(t)$ stands for the AoA. The AoD and the AoA are independent since double-bounce scattering is assumed. The transmitter and the receiver are equipped with an antenna array encompassing M_T and M_R antenna elements, respectively. The antenna element spacing at the transmitter side is denoted by δ_T. Analogously, the antenna element spacing at the receiver side is referred to as δ_R. The tilt angle of the transmit antenna array is denoted by γ_T, while γ_R stands for the tilt angle for the receive antenna array. The transmitter (receiver) is located at a distance h_1^T (h_1^R) from the left-hand side of the street and at a distance h_2^T (h_2^R) from the right-hand side seen in moving direction.

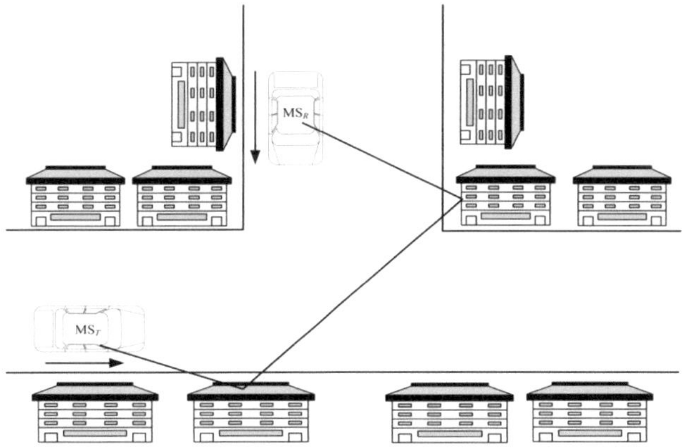

Fig. 1. Typical propagation scenario for V2V communications at a T-junction.

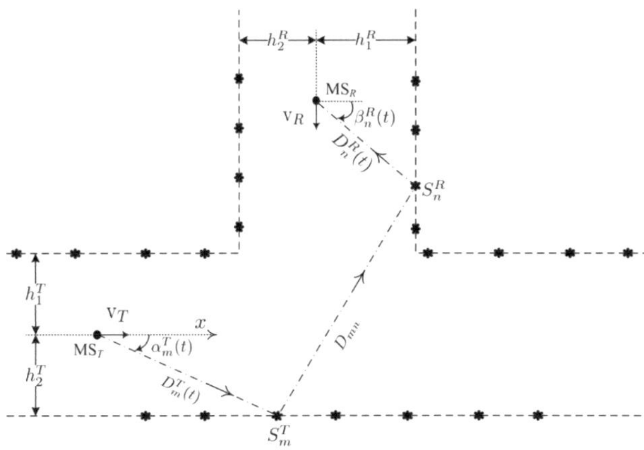

Fig. 2. The geometrical T-Junction model for V2V communications.

3. The Reference Model

The starting point for the derivation of the reference model for the MIMO V2V channel is the geometrical T-junction model presented in Fig. 2. For the reference model, we assume double-bounce scattering from fixed scatterers. We distinguish between the scatterers near to the transmitter and the scatterers close to the receiver. It can be seen from Fig. 2 that a wave emitted from the lth transmit antenna element A_l^T $(l = 1, 2, \ldots, M_T)$ travels over the scatterers S_m^T and S_n^R before impinging on the kth receive antenna element A_k^R $(k = 1, 2, \ldots, M_R)$. Using the wave propagation model in (Pätzold et al., 2008), the complex channel gain $g_{kl}(\vec{r}_T, \vec{r}_R)$ describing the link A_l^T–A_k^R of the underlying $M_T \times M_R$ MIMO V2V channel model can be expressed in the present case as

$$g_{kl}(\vec{r}_T, \vec{r}_R) = \sum_{m=1,n=1}^{M,N} c_{mn}\, e^{j\left(\theta_{mn}(t) + \vec{k}_m^T \cdot \vec{r}_T - \vec{k}_n^R \cdot \vec{r}_R - k_0 d_{mn}(t)\right)}. \quad (1)$$

The symbols c_{mn} and $\theta_{mn}(t)$ stand for the the joint gain and the joint phase shift caused by the scatterers S_m^T and S_n^R. The joint channel gain can be written as $c_{mn} = 1/\sqrt{MN}$ (Pätzold et al., 2008). The phase shift $\theta_{mn}(t)$ is a stochastic process, as the AoD $\alpha_m^T(t)$ and the AoA $\beta_n^R(t)$ are time-variant. This is in contrast to the models proposed in (Pätzold et al., 2008) and (Zhiyi et al., 2009), where the phase shift is a random variable. The joint phase shift can be expressed as $\theta_{mn}(t) = (\theta_m(t) + \theta'_n(t)) \bmod 2\pi$, where mod stands for the modulo operation. The terms $\theta_m(t)$ and $\theta'_n(t)$ are the phase shifts associated with the scatterers S_m^T and S_n^R, respectively.
The second phase term in (1), $\vec{k}_m^T \cdot \vec{r}_T$, is caused by the movement of the transmitter. The wave vector pointing in the propagation direction of the mth transmitted plane wave is denoted by $\vec{k}_m^T$, while $\vec{r}_T$ stands for the spatial translation vector of the transmitter. The scalar product $\vec{k}_m^T \cdot \vec{r}_T$ can be expanded as

$$\vec{k}_m^T \cdot \vec{r}_T = 2\pi f_{\max}^T \cos(\alpha_m^T(t) - \phi_T)t \tag{2}$$

where $f_{\max}^T = \mathrm{v}_T/\lambda$ denotes the maximum Doppler frequency associated with the mobility of the transmitter. The symbol λ refers to the wavelength. The time-variant AoD $\alpha_m^T(t)$ can be expressed as

$$\alpha_m^T(t) = \begin{cases} -\pi + g_2(t) & \text{if} \quad -\pi \le \alpha_m^T(t) \le -\frac{\pi}{2} \\ g_2(t) & \text{if} \quad -\frac{\pi}{2} < \alpha_m^T(t) \le 0 \\ g_1(t) & \text{if} \quad 0 < \alpha_m^T(t) \le \frac{\pi}{2} \\ \pi + g_1(t) & \text{if} \quad \frac{\pi}{2} < \alpha_m^T(t) \le \pi \end{cases} \tag{3}$$

where

$$g_1(t) = \arctan\left(\frac{h_1^T \tan(\alpha_m^T(t_0))}{h_1^T - \mathrm{v}_T(t-t_0)\tan(\alpha_m^T(t_0))}\right) \tag{4}$$

$$g_2(t) = \arctan\left(\frac{h_2^T \tan(\alpha_m^T(t_0))}{h_2^T - \mathrm{v}_T(t-t_0)\tan(\alpha_m^T(t_0))}\right). \tag{5}$$

We assume that the AoD seen from the transmitter side can be considered as constant for a given time interval if the angular deviation does not exceed a certain threshold. For instance, the AoD $\alpha_m^T(t_1)$ at time instant t_1 and the AoD $\alpha_m^T(t_2)$ at time instant t_2 are equal if the angle difference $|\alpha_m^T(t_1) - \alpha_m^T(t_2)| \le \epsilon_\alpha$, with ϵ_α is a very small positive value. In this way, the AoD $\alpha_m^T(t)$ can be written as

$$\alpha_m^T(t) = \alpha_{m,i-1}^T \text{ if } t_{i-1} \le t < t_i \text{ for } i = 1, 2, \ldots \tag{6}$$

The term $\alpha_{m,i-1}^T$ is a constant that can be obtained from (3) by setting the time t to t_{i-1}. The length of the intervals $[t_{i-1}, t_i)$ and $[t_i, t_{i+1})$ can be quite different for $i = 1, 2, \ldots$. The phase shift introduced by a scatterer is generally dependent on the direction of the outgoing wave. Hence, a change in the AoD $\alpha_m^T(t)$ results in a new random phase shift. Since the AoD $\alpha_m^T(t)$ is defined piecewise, the phase shift $\theta_m(t)$ is also defined piecewise as follows

$$\theta_m(t) = \theta_{m,i-1} \text{ if } t_{i-1} \le t < t_i \text{ for } i = 1, 2, \ldots \tag{7}$$

where $\theta_{m,0}$, $\theta_{m,1}$, ... are independent identically distributed (i.i.d.) random variables uniformly distributed over $[0, 2\pi)$.
The third phase term in (1), $\vec{k}_n^R \cdot \vec{r}_R$, is associated with the movement of the receiver. The symbol $\vec{k}_n^R$ stands for the wave vector pointing in the propagation direction of the nth received plane wave, while $\vec{r}_R$ represents the spatial translation vector of the receiver. The scalar product $\vec{k}_n^R \cdot \vec{r}_R$ can be expanded as

$$\vec{k}_n^R \cdot \vec{r}_R = -2\pi f_{\max}^R \cos(\beta_n^R(t) - \phi_R)t \tag{8}$$

where $f_{\max}^R = \mathrm{v}_R/\lambda$ denotes the maximum Doppler frequency caused by the receiver movement. Using the geometrical T-junction model shown in Fig. 2, the time-variant AoA $\beta_n^R(t)$ can be expressed as

$$\beta_n^R(t) = \begin{cases} -\pi + g_3(t) & \text{if} \quad -\pi \le \beta_n^R(t) \le -\frac{\pi}{2} \\ g_4(t) & \text{if} \quad -\frac{\pi}{2} < \beta_n^R(t) \le \frac{\pi}{2} \\ \pi + g_3(t) & \text{if} \quad \frac{\pi}{2} < \beta_n^R(t) \le \pi \end{cases} \tag{9}$$

where

$$g_3(t) = \arctan\left(\frac{h_2^R \tan(\beta_n^R(t_0)) - \mathrm{v}_R(t - t_0)}{h_2^R}\right) \tag{10}$$

$$g_4(t) = \arctan\left(\frac{h_1^R \tan(\beta_n^R(t_0)) - \mathrm{v}_R(t - t_0)}{h_1^R}\right). \tag{11}$$

We assume that the AoA seen from the receiver side can be considered as constant for a given time interval if the angular deviation does not exceed a certain threshold. For instance, the AoA $\beta_n^R(t_1')$ at time instant t_1' and the AoA $\beta_n^R(t_2')$ at time instant t_2' are equal if the angle difference $|\beta_n^R(t_1') - \beta_n^R(t_2')| \le \epsilon_\alpha$. In this way, the AoA $\beta_n^R(t)$ can be written as

$$\beta_n^R(t) = \beta_{n,j-1}^R \text{ if } t_{j-1}' \le t < t_j' \text{ for } j = 1, 2, \ldots \tag{12}$$

The term $\beta_{n,j-1}^R$ is a constant that can be obtained from (9) by setting the time t to t_{j-1}'. The length of the intervals $[t_{j-1}', t_j')$ and $[t_j', t_{j+1}')$ can be quite different for $j = 1, 2, \ldots$. The phase shift introduced by a scatterer is generally dependent on the direction of the incoming wave. Hence, a change in the AoA $\beta_n^R(t)$ results in a new random phase shift. Since the AoA $\beta_n^R(t)$ is defined piecewise, the phase shift $\theta_n'(t)$ is also defined piecewise as follows

$$\theta_n'(t) = \theta_{n,j-1}' \text{ if } t_{j-1}' \le t < t_j' \text{ for } j = 1, 2, \ldots \tag{13}$$

where $\theta_{n,0}'$, $\theta_{n,1}'$, ... are i.i.d. random variables uniformly distributed over $[0, 2\pi)$.
After substituting (2) and (8) in (1), the complex channel gain $g_{kl}(t)$ can be expressed as

$$g_{kl}(t) = \sum_{m,n=1}^{M,N} \frac{a_m^T\, b_n^R\, c_{mn}^{TR}}{\sqrt{MN}} e^{j\left(2\pi(f_m^T + f_n^R)t + \theta_{mn}(t)\right)} \tag{14}$$

where

$$a_m^T = e^{j\pi \frac{\delta_T}{\lambda}(M_T - 2l + 1)\cos(\alpha_m^T(t) - \gamma_T)} \tag{15}$$

$$b_n^R = e^{j\pi \frac{\delta_R}{\lambda}(M_R - 2k + 1)\cos(\beta_n^R(t) - \gamma_R)} \tag{16}$$

$$c_{mn}^{TR} = e^{-j\frac{2\pi}{\lambda}\left(D_m^T(t) + D_{mn} + D_n^R(t)\right)} \tag{17}$$

$$f_m^T = f_{\max}^T \cos(\alpha_m^T(t) - \phi^T) \tag{18}$$

$$f_n^R = f_{\max}^R \cos(\beta_n^R(t) - \phi^R). \tag{19}$$

with $D_m^T(t)$ denoting the distance from the transmitter to the scatterer S_m^T. The term D_{mn} represents the distance between the scatterers S_m^T and S_n^R,while $D_n^R(t)$ corresponds to the distance from the receiver to the scatterer S_n^R, as shown in Fig. 2.

4. Correlation Properties

For wide-sense stationary processes, the temporal ACF depends only on the time difference τ. However, for non-stationary processes, the temporal ACF does not only depend on the time difference τ, but also on the time t. Due to its time dependance, the ACF of non-stationary processes is called local ACF (Cohen, 1989). Several definitions for the local ACF have been proposed in literature. In this paper, we utilize the definition of the local ACF proposed by Wigner (Cohen, 1989), which is given by

$$r_{g_{kl}}(t,\tau) := E\{g_{kl}(t+\tau/2)g_{kl}^*(t-\tau/2)\} \tag{20}$$

where $(\cdot)^*$ denotes the complex conjugation and $E\{\cdot\}$ stands for the expectation operator. By applying the expectation operator on the i.i.d. random variables $\theta_{m,i}$ $(i = 0,1,\ldots)$ and $\theta'_{n,j}$ $(j = 0,1,\ldots)$ and exploiting their independence, we can express the local ACF as

$$r_{g_{kl}}(t,\tau) = r_{g_{kl}}^T(t,\tau) \cdot r_{g_{kl}}^R(t,\tau). \tag{21}$$

where

$$r_{g_{kl}}^T(t,\tau) = \frac{1}{M}\sum_{m=1}^{M} e^{j2\pi\left(f_m^T\left(t+\frac{\tau}{2}\right)\left(t+\frac{\tau}{2}\right) - f_m^T\left(t-\frac{\tau}{2}\right)\left(t-\frac{\tau}{2}\right)\right)} \tag{22}$$

$$r_{g_{kl}}^R(t,\tau) = \frac{1}{N}\sum_{n=1}^{N} e^{j2\pi\left(f_n^R\left(t+\frac{\tau}{2}\right)\left(t+\frac{\tau}{2}\right) - f_n^R\left(t-\frac{\tau}{2}\right)\left(t-\frac{\tau}{2}\right)\right)}. \tag{23}$$

Note that the local ACF $r_{g_{kl}}(t,\tau)$ is written as a product of the local transmit ACF $r_{g_{kl}}^T(t,\tau)$ and the local receive ACF $r_{g_{kl}}^T(t,\tau)$ since we assume a limited number of scatterers in the proposed model. The expression of the local ACF is derived using the Wigner method (Cohen, 1989). By applying the Fourier transformation on the local ACF in (21), we obtain the Wigner time-frequency distribution. The former even though a member of the Cohen class of distributions, suffers from the cross-term problem (Cohen, 1989). To deal with this issue, a kernel function aiming to reduce the cross-terms need to be introduced. One of the effective distributions in diminishing the effect of cross-terms is the Choi-Williams distribution (Choi & Williams, 1989). Choi and Williams devised their kernel function in such a way that a relatively large weight is given to $g_{kl}(u+\tau/2)g_{kl}^*(u-\tau/2)$ if u is close to t. In this way, they emphasis the local behaviour of the channel and guarantee that the non-stationarities will not be smeared in time and frequency. The kernel function for the Choi-Williams distribution is given by $\phi(\xi,\tau) = e^{-\xi^2\tau^2/\sigma}$. It follows that the generalized local ACF can be expressed as (Choi & Williams, 1989)

$$
\begin{aligned}
K(t,\tau;\phi) &= \int_{-\infty}^{\infty}\int_{-\infty}^{\infty} e^{j2\pi\xi(u-t)}\phi(\xi,\tau) r_{g_{kl}}(u,\tau)\, du\, d\xi \\
&= \int_{-\infty}^{\infty} \frac{r_{g_{kl}}(u,\tau)}{\sqrt{4\pi\tau^2/\sigma}} \exp\Big(-\frac{(u-t)^2}{4\tau^2/\sigma}\Big)\, du. \qquad (24)
\end{aligned}
$$

The generalized local ACF presented above can be used for both stationary and non-stationary processes. Actually, if the process is stationary, the local ACF $r_{g_{kl}}(u,\tau)$ equals $r_{g_{kl}}(\tau)$. Using (24), it turns out that the generalized local ACF for stationary processes equals the classical ACF, i.e., $K(t,\tau;\phi) = r_{g_{kl}}(\tau)$.

For stationary processes, the power spectral density can be obtained from the Fourier transformation of the temporal ACF. Analogously, for non-stationary processes, the time-frequency distribution can be obtained from the generalized local ACF by applying the Fourier transformation. The time-frequency distribution gives an insight into how the power spectrum varies with time t. The time-frequency distribution $W(t,f;\phi)$ can be written as

$$
\begin{aligned}
W(t,f;\phi) &= \int_{-\infty}^{\infty} K(t,\tau;\phi)\, e^{-j2\pi f\tau}\, d\tau \\
&= \int_{-\infty}^{\infty}\int_{-\infty}^{\infty} A(\xi,\tau;\phi)\, e^{-j2\pi(f\tau - t\xi)}\, d\tau d\xi \qquad (25)
\end{aligned}
$$

where $A(\xi,\tau;\phi)$ is the ambiguity function.

The local space CCF can be expressed as

$$
\begin{aligned}
\rho_{kl,k'l'}(t,\delta_T,\delta_R) &= E\{g_{kl}(t) g^*_{k'l'}(t)\} \\
&= \frac{1}{M}\sum_{m=1}^{M} e^{-j2\pi\frac{\delta_T}{\lambda}(l-l')\cos(\alpha_m^T(t)-\gamma_T)} \\
&\quad \cdot \frac{1}{N}\sum_{n=1}^{N} e^{-j2\pi\frac{\delta_R}{\lambda}(k-k')\cos(\beta_n^R(t)-\gamma_R)} \\
&= \rho_{ll'}^T(t,\delta_T)\cdot\rho_{kk'}^R(t,\delta_R). \qquad (26)
\end{aligned}
$$

In (26), the AoD $\alpha_m^T(t)$ and the AoA $\beta_n^R(t)$ are given by (6) and (12), respectively. Note that the local space CCF is written as a product of the local transmit space correlation function (CF) $\rho_{ll'}^T(t,\delta_T)$ and the local receive space CF $\rho_{kk'}^R(t,\delta_R)$.

5. Numerical Results

In this section, the analytical expressions presented in the previous section are evaluated numerically and then illustrated. The propagation environment encompasses twelve scatterers around the transmitter. Six scatterers are located on the left side of the transmitter and the remaining scatterers are on the right side. The distance between two successive scatterers is set to 20 m. We consider the same number of scatterers around the receiver. The transmitter and the receiver have a velocity of 70 km/h and a direction of motion determined

by $\phi_T = 0$ and $\phi_R = -\pi/2$, respectively. The transmitter and the receiver antenna tilt angles γ_T and γ_R are equal to $\pi/2$. The street parameters are chosen as $h_1^T = h_2^T = 50$ m and $h_1^R = h_2^R = 50$ m. The parameter ϵ_α is set to 0.1.
The absolute value of the resulting generalized local ACF $K(t, \tau; \phi)$ is illustrated in Fig. 3. From this figure, we can see that the shape of the local ACF changes for different values of t, which is due to the non-stationarity of the channel model. If the channel model is stationary, we would observe the same shape of the local ACF at different time instants t. The absolute value of the time-frequency distribution, $|W(t, f; \phi)|$, shown in Fig. 4, is obtained from the generalized local ACF by applying the Fourier transform w.r.t. the time lag τ. It can be seen from this figure how the Doppler spectrum of the channel varies with time t. For the chosen propagation scenario, it can be observed from Fig. 4 that the zero Doppler frequency has the highest power for all time instants t. The power of the non-zero Doppler frequencies decays for a certain period of time before increasing again. The absolute value of the local transmit space CF $|\rho_{12}^T(t, \delta_T)|$ is presented in Fig. 5. It can be seen from this figure that the amplitude of this function is more sensitive to the transmit antenna spacing δ_T than to the time t. The absolute value of the local receive space CF $|\rho_{12}^R(t, \delta_R)|$ is illustrated in Fig. 6. For the chosen scenario, this function decays faster than the local transmit space CF w.r.t. the antenna spacing.

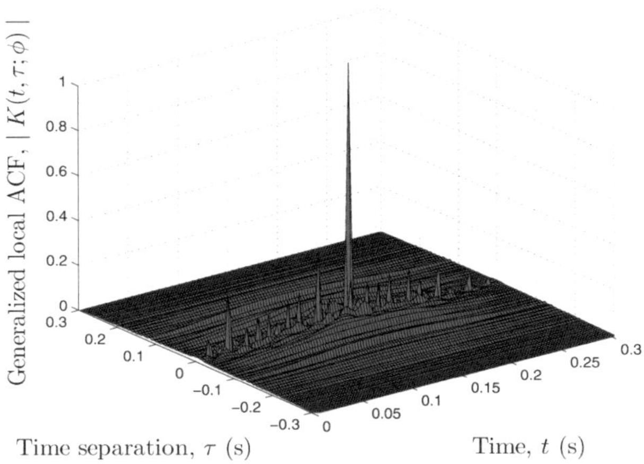

Fig. 3. The absolute value of the generalized local ACF $|K(t, \tau; \phi)|$.

6. Conclusion

In this chapter, we have presented a non-stationary MIMO V2V channel model. Based on the geometrical T-junction model, we have derived an expression for the time-variant channel gain taking into account double-bounce scattering from fixed scatterers. We have assumed a limited number of scatterers. Under this assumption, the reference model equals the simulation model. In vehicular environments, the high speed of the communicating vehicles results in time-variant AoD and AoA. This property is taken into account in our channel model, which makes the model non-stationary. To study the statistical properties of the proposed channel model, we utilized the Choi-Williams distribution. We have provided analytical expressions of the generalized local ACF, the time-frequency distribution, and the local space CCF. The latter can be written as a product of the local transmit space CF and

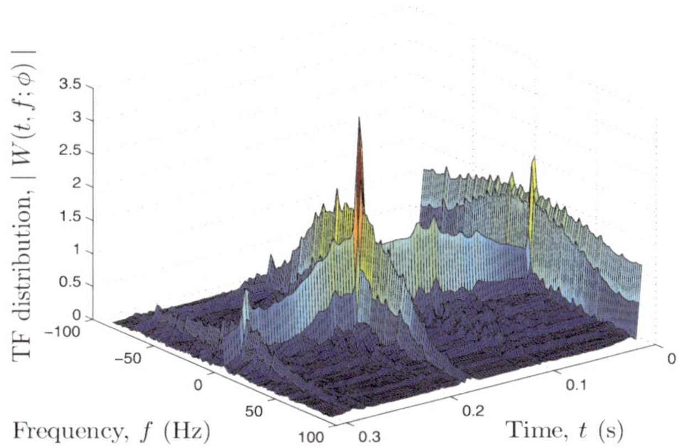

Fig. 4. The absolute value of the time-frequency distribution $|W(t, f; \phi)|$.

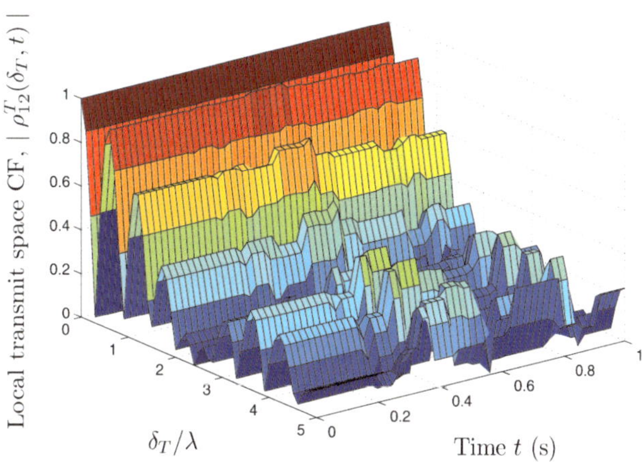

Fig. 5. Absolute value of the local transmit space CF $|\rho_{12}^{T}(t, \delta_T)|$.

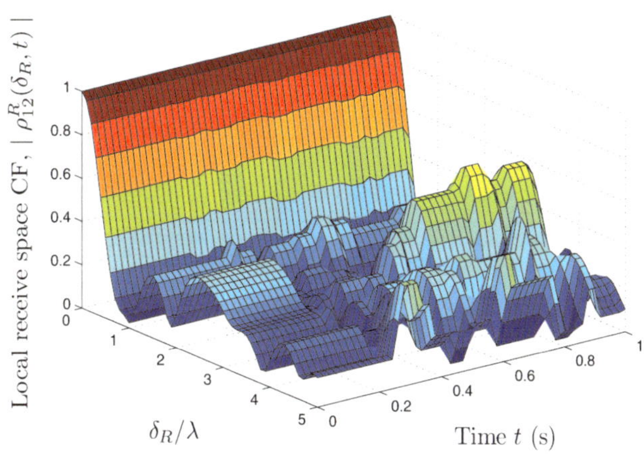

Fig. 6. Absolute value of the local receive space CF $|\rho_{12}^{R}(t, \delta_R)|$.

the local receive space CF. Supported by our analysis, we can conclude that the stationarity assumption is violated for V2V channels, especially if the mobile speed is high and the observation interval is large. Non-stationary channel models are needed as a tool for designing future V2V communication systems. In future work, the effect of moving scatterers on the channel statistics will be studied.

7. References

Acosta, G., Tokuda, K. & Ingram, M. (2004). Measured joint doppler-delay power profiles for vehicle-to-vehicle communications at 2.4 GHz, *Proc. IEEE Global Telecommunications Conference, 2004. GLOBECOM '04.*, Vol. 6, Dallas, TX, pp. 3813–3817.

Chelli, A. & Pätzold, M. (2008). A wideband multiple-cluster MIMO mobile-to-mobile channel model based on the geometrical street model, *Proc. 19th IEEE International Symposium on Personal, Indoor and Mobile Radio Communications, PIMRC 2008*, Cannes, France, pp. 1–6.

Choi, H.-I. & Williams, W. (1989). Improved time-frequency representation of multicomponent signals using exponential kernels, *IEEE Transactions on Acoustics, Speech, and Signal Processing* 37(6): 862–871.

Cohen, L. (1989). Time-frequency distributions-a review, *Proceedings of the IEEE* 77(7): 941–981.

O'Neill, J. & Williams, W. (1999). Shift covariant time-frequency distributions of discrete signals, *IEEE Transactions on Signal Processing* 47(1): 133–146.

Paier, A., Zemen, T., Bernado, L., Matz, G., Karedal, J., Czink, N., Dumard, C., Tufvesson, F., Molisch, A. F. & Mecklenbräuker, C. F. (2008). Non-WSSUS vehicular channel characterization in highway and urban scenarios at 5.2 GHz using the local scattering function, *Proc. 2008 International ITG Workshop on Smart Antennas (WSA 2008)*, Darmstadt, Germany, pp. 9–15.

Pätzold, M., Hogstad, B. & Youssef, N. (2008). Modeling, analysis, and simulation of MIMO mobile-to-mobile fading channels, *IEEE Transactions on Wireless Communications* 7(2): 510–520.

Road Safety Evolution in EU (2009). *Technical report*, European Commision/ Directorate General Energy Transport.

Sayeed, A. & Jones, D. (1995). Optimal kernels for nonstationary spectral estimation, *IEEE Transactions on Signal Processing* 43(2): 478–491.

Zajić, A. G., Stüber, G. L., Pratt, T. G. & Nguyen, S. T. (2009). Wideband MIMO mobile-to-mobile channels: geometry-based statistical modeling with experimental verification, *IEEE Transactions on Vehicular Technology* 58(2): 517–534.

Zhiyi, H., Wei, C., Wei, Z., Pätzold, M. & Chelli, A. (2009). Modelling of MIMO vehicle-to-vehicle fading channels in T-junction scattering environments, *3rd European Conference on Antennas and Propagation, 2009. EuCAP 2009.*, Berlin, Germany, pp. 652–656.

2

Simulation of SISO and MIMO Multipath Fading Channels

Antonio Petrolino[1] and Gonçalo Tavares[2]
[1]*Instituto de Engenharia de Sistemas e Computadores - Investigação e Desenvolvimento, INESC-ID, Lisbon*
[2]*INESC-ID and Instituto Superior Técnico, IST, Lisbon*
Portugal

1. Introduction

Accurate software simulation is an important step for testing mobile communication systems. It allows the evaluation of the system performance without the need for physical experiments. At an early testing stage, simulations are usually done considering transmissions over additive white Gaussian noise (AWGN) channels, which do not account for obstacles or motion between the transmitter (Tx) and the receiver (Rx). According to the AWGN channel model, only the thermal noise which comes from the receiver amplifiers and from many natural sources is considered and added to the transmitted signal. This means that the received signal is a copy of the transmitted signal embedded in Gaussian noise, whose intensity depends on the propagation scenario (including the amplifiers of the Tx and the Rx) and can be controlled in simulation by tuning the Gaussian process variance.

It is easy to understand that AWGN channel models do not capture the conditions of a real radio channel where the presence of obstacles between the Tx and the Rx causes multipath propagation and may lead to *frequency selective* channels. In this situation, several replicas of the transmitted signal get to the Rx antenna with different propagation delays. If these delays are not negligible compared to the transmission interval T_s, the attenuation caused by the channel in the frequency-domain is not constant and a distortion is introduced[1]. Moreover, if there is motion between the Tx and the Rx, the received signal may be affected by a time-domain multiplicative distortion called *fading*. This happens because, being the Rx in motion relative to the Tx, the propagation channel characteristics vary with time and their effect can be modeled as a time-domain distortion affecting the transmitted signal. By combining both the effects of multipath propagation and fading on a wireless transmission, the received signal baseband equivalent $r(t)$ may be expressed as

$$r(t) = \sum_{i=1}^{N(t)} \rho_i(t) e^{j\theta_i(t)} s(t - \tau_i(t)) + n(t) \tag{1}$$

where $s(t)$ is the transmitted signal baseband equivalent, $\rho_i(t)$ and $\theta_i(t)$ are the time-varying attenuation and phase shift respectively of the i-th path and $n(t)$ is a white Gaussian process

[1] Otherwise, when the propagation delays are negligible compared to the signal interval, the channel is said to be *flat*.

modeling thermal noise and noise added by the receiver amplifiers. Although in the case of time varying channels $N(t)$ and $\tau_i(t)$ are functions of the time, for the sake of simplicity, it is common in the literature to ignore the dependence on time of the number of significant paths and the propagation delays, that is to set $N(t) = N$ and $\tau_i(t) = \tau_i$. Eq. (1) can be rewritten as

$$\begin{aligned} r(t) &= \sum_{i=1}^{N} \rho_i(t) e^{j\theta_i(t)} s(t - \tau_i) + n(t) \\ &= \sum_{i=1}^{N} h_i(t) s(t - \tau_i) + n(t) \end{aligned} \tag{2}$$

where $h_i(t) \triangleq \rho_i(t) e^{j\theta_i(t)}$. Eq. (2) represents the received signal in the case of transmission over a *frequency-selective* channel.
In the case of *flat* multipath channels a further simplification is possible since $\tau_i \approxeq \overline{\tau} \ll T_s \; \forall i$. Eq. (2) can be rewritten as

$$r(t) = \sum_{i=1}^{N} h_i(t) \cdot s(t - \overline{\tau}) + n(t). \tag{3}$$

Moreover, if a large number of paths is considered i.e., $N \gg 1$, the central limit theorem can be applied to the sum in (3) leading to

$$r(t) = h(t) \cdot s(t - \overline{\tau}) + n(t) \tag{4}$$

where $h(t) = h_R(t) + jh_I(t)$ is a complex Gaussian random process.
By comparing equations (2) and (4), it is clear that the *selectivity* in the frequency-domain requires the contribution of each path to be considered and that the fading process, being the combination of N complex processes, each modeling the fading affecting the corresponding propagation path, can not be approximated by a Gaussian random process.
As far as flat fading channels are concerned, a detailed description of the process $h(t)$ was derived by Clarke in 1968. According to the *model of Clarke* (Clarke, 1968), for a mobile-to-fixed channel, the fading quadrature components are independent each with the same autocorrelation function (ACF) given by (Pätzold, 2002)

$$\mathcal{R}_{h_R}(\tau) = \mathcal{R}_{h_I}(\tau) = \mathcal{R}_h(\tau) = \frac{\sigma_h^2}{2} \cdot J_0(2\pi f_D \tau) \tag{5}$$

where σ_h^2 is the fading power, $J_0(\cdot)$ is the zero-order Bessel function of the first kind and f_D is the maximum Doppler frequency shift. Two other widely used statistics in channel modeling for the statistical characterization of a fading process are the level crossing rate (LCR) and the average fade duration (AFD) of the process. The LCR L_R is defined as the expected number of times per second that the envelope $|h(t)|$ crosses the threshold R in the positive direction. For Rayleigh fading, it is given as (Patel et al., 2005)

$$L_R = \sqrt{2\pi} f_D \rho e^{-\rho^2}, \tag{6}$$

where $\rho = R/\sqrt{\mathbb{E}[|h(t)|^2]}$ (here, $\mathbb{E}[\cdot]$ denotes expectation). The AFD T_R is defined as the expected value for the length of the time intervals in which the stochastic process $h(t)$ is below a given level R. For Rayleigh fading, it is given as (Patel et al., 2005)

$$T_R = \frac{e^{\rho^2} - 1}{\sqrt{2\pi} f_D \rho}. \tag{7}$$

It is important to note that some characteristics of the fading process, like the amplitude probability density function (PDF), can change considerably depending on the propagation scenario. For example, when a line-of-sight path is not available it can be shown that the complex Gaussian process $h(t)$ in (4) has zero-mean. Consequently, the fading process envelope PDF follows a Rayleigh distribution. Otherwise, when a line-of-sight path is available, the Gaussian process $h(t)$ has non-zero-mean leading to a Rician distributed envelope. However, experience shows that often the hypotheses of the model of Clarke are not verified and the fading severity turns out to be different from Rayleigh or Rician. In such cases, distributions like Nakagami-m (Nakagami, 1960), Weibull (Weibull, 1951) or Hoyt (Hoyt, 1947) match more closely with real fadings (Vehicular Technology Society Committee on Radio Propagation, 1988).

Simulation of fading processes with Rayleigh and Rician fading envelopes has been extensively studied e.g., the sum-of-sinusoids (SoS) method proposed in (Jakes, 1974), the simulator based on the inverse fast Fourier Transform (IFFT) proposed in (Smith, 1975), the filtering method presented in (Komninakis, 2003) and the simulator based on the Karhunen-Loève (KL) expansion presented in (Petrolino et al., 2008a).

The first approach, proposed by Jakes in (Jakes, 1974) is limited to the simulation of the Clarke's model for which it has been developed. It is based on the representation of the received signal as a superposition of a finite number of sinusoids having equal amplitude and uniformly spaced angle of arrivals. Jakes' simulator has to be considered as *deterministic* (Pätzold et al., 1998), since the generated waves depend on parameters which, once chosen, remain fixed for the duration of the simulation run. The SoS method was improved in (Pop & Beaulieu, 2001), with the addition of random phases, with the goal of making the generated process wide sense stationary (WSS).

The second approach, named IFFT-based method, was proposed in (Smith, 1975) and extended in (Young & Beaulieu, 2000). With this method, in order to achieve a fading output with the desired ACF, two real sequences of zero-mean independent Gaussian random variables (rvs) are multiplied in the frequency-domain by the same frequency mask equal to the square root of the power spectral density (PSD) corresponding to the desired ACF. Then, the obtained sequences are added in quadrature and a complex IFFT is taken. Due to the linearity of the IFFT, the output is a time-domain Gaussian sequence with the desired ACF.

The third approach is known as filtering method and resorts to digital filtering of computer-generated independent Gaussian rvs by a filter whose magnitude frequency response is given by the square root of the desired PSD. Due to the linearity of the filtering operation, the filter output sequence is still Gaussian and has the desired ACF. A solution based on the combination of an infinite impulse response (IIR) filter with a nearly ideal interpolator is proposed in (Komninakis, 2003).

The last approach is based on the idea of expanding the complex random fading process as a modified KL expansion. Let $h(t)$ be a correlated zero-mean complex Gaussian random process in $0 < t < T_0$ whose quadrature components are independent each showing the same ACF. The process $h(t)$ can be represented by the series expansion (Van Trees, 1968)

$$h(t) = \lim_{L\to\infty} \sum_{l=0}^{L-1} h_l \phi_l(t) \tag{8}$$

where the set of functions $\{\phi_l(t)\}_{l=0}^{L-1}$ is the basis of the expansion and the $\{h_l\}_{l=0}^{L-1}$ are the expansion coefficients, that is the projections of the process $h(t)$ on the basis functions. The representation in (8) is known as the Karhunen-Loève expansion for the process $h(t)$

(Van Trees, 1968). Since $h(t)$ is a zero-mean complex Gaussian random process and the $\{\phi_l(t)\}_{l=0}^{L-1}$ are orthogonal, the $\{h_l\}_{l=0}^{L-1}$ are zero-mean, independent complex Gaussian rvs with variances $\{\sigma_l^2\}_{l=0}^{L-1}$. A possible expansion for the process $h(t)$ is achieved by setting the process $h(t)$ correlation matrix eigenvectors and eigenvalues as the basis functions $\{\phi_l(t)\}_{l=0}^{L-1}$ and the variances $\{\sigma_l^2\}_{l=0}^{L-1}$, respectively (Petrolino et al., 2008b). Inversely, in fading channel simulation we need to generate a correlated random process starting from the *a priori* knowledge of its ACF which is determined by the channel model being used. This can be achieved by following the approach proposed in (Petrolino et al., 2008a), where the $\{\phi_l(t)\}_{l=0}^{L-1}$ is imposed as a complete set of exponential functions, leading to a *modified* KL expansion. Due to the particular choice of the basis functions, the eigenvalues can be calculated for $l = 0, \ldots, L-1$, as (Petrolino et al., 2008b, eq. (9))

$$\sigma_l^2 = \int_{-T_0/2}^{T_0/2} \mathcal{R}_h(\tau) \cos\left(\frac{2\pi l \tau}{T_0}\right) d\tau. \tag{9}$$

This means that if we are able to calculate the integral in (9) for every l, then the process $h(t)$ may be synthesized by (8) using a set of L computer-generated independent zero-mean Gaussian rvs, with variances $\{\sigma_l^2\}_{l=0}^{L-1}$.
At this point a N-samples version $h[n]$ of the fading process $h(t)$ may be obtained by sampling the interval T_0 with sampling frequency $f_s = N/T_0$

$$h[n] = \sum_{l=0}^{L-1} h_l \cdot e^{j\frac{2\pi}{N}ln} \quad n = 0, \ldots, N-1 \tag{10}$$

where the $\{h_l\}_{l=0}^{L-1} \sim N(0, \{\sigma_l^2\}_{l=0}^{L-1})$ are independent Gaussian rvs. The dimension of the expansion basis is (Van Trees, 1968) $L = 2BT_0$ where B is the unilateral bandwidth of the process and T_0 its duration. Since $T_0 = NT_s$, the KL expansion dimension reduces to $L = 2BNT_s = \frac{2B}{f_s}N \leq N$, where the inequality follows from the sampling theorem. This means that $h_l = 0$ for $l \geq N$, so it is possible to rewrite (10) as

$$h[n] = \sum_{l=0}^{N-1} h_l \cdot e^{j\frac{2\pi}{N}ln} \quad n = 0, \ldots, N-1 \tag{11}$$

that is, $h[n]$ is the IDFT of the sequence $\{h_l\}_{l=0}^{N-1}$ and may be computed efficiently using the fast Fourier transform (FFT) algorithm. Note that simulators like (Young & Beaulieu, 2000), (Komninakis, 2003) and (Petrolino et al., 2008a) impose the desired ACF by linear transformations of Gaussian processes. By the properties of this class of processes, the Gaussian PDF is preserved. However, this is not the case with non-Gaussian processes. As a consequence, such simulators can not be directly used to generate autocorrelated non-Gaussian sequences like Nakagami-*m*, Weibull or Hoyt. Regarding Nakagami-*m* fading, simulators have been proposed in (Beaulieu & Cheng, 2001) and (Filho et al., 2007), both of which require already-correlated Rayleigh sequences which must be generated by existing simulators. In (Beaulieu & Cheng, 2001), the correlated Nakagami sequence is obtained by applying a double transformation to the Rayleigh sequence, while in (Filho et al., 2007), an uncorrelated Nakagami sequence is rearranged according to the Rayleigh sequence ranking in order to match the desired ACF.

An important class of systems which have gained success in the last years are multiple-input multiple-output (MIMO) systems. Consequently, MIMO fading channel simulators are increasingly required by MIMO designers for system performance evaluation and assessment. In MIMO systems, multi-element antenna arrays are used at both sides of a radio link and the communication occurs over N_{min} parallel channels, where N_{min} is the minimum of the number of Tx or Rx antennas (Gesbert et al., 2000). Due to finite antenna separation, a spatial correlation between the channels always exists and greatly affects the link capacity of MIMO systems (Wang, 2006). The MIMO channel spatial correlation can be described by a $MN \times MN$ correlation matrix. This matrix can be obtained in several ways depending on the reference model under consideration. In this chapter we will use the Kronecker Based Stochastic Model (KBSM) (Wang, 2006). According to the KBSM model, the spatial correlation matrix $\mathbf{S}_{MIMO}$ is obtained as

$$\mathbf{S}_{MIMO} = \mathbf{S}_{BS} \otimes \mathbf{S}_{MS} \tag{12}$$

where $\mathbf{S}_{BS}$ is the $M \times M$ spatial correlation matrix at the BS, $\mathbf{S}_{MS}$ is the $N \times N$ spatial correlation matrix at the MS and $\otimes$ stands for the Kronecker product. For the referred reasons, a MIMO fading channel simulator must contemplate the possibility of generating random processes cross-correlated according to a spatial correlation structure determined by the used model.
In general, the fading channel simulation can be divided in three tasks:

1. Choice of the most convenient fading channel model.
2. Generation of uncorrelated rvs with the desired PDF. This point will be discussed in Sections 2 and 3.
3. Imposition of the desired ACF to the generated sequence without affecting the initial PDF. This point will be analyzed in Section 4.

This chapter is organized as follows. Section 2 presents a technique for the generation of uncorrelated arbitrary jointly distributed phase and envelope processes. The proposed method is based on the two-dimensional extension of the Metropolis-Hastings (MH) algorithm (Hastings, 1970; Metropolis et al., 1953). Starting from jointly Gaussian random variate pairs, this method achieves the desired PDF by an acceptance-rejection (AR) algorithm whose probability of acceptance is set in a convenient way.
In Section 3, the two-dimensional MH algorithm is applied to the generation of Nakagami-*m* and Hoyt envelope fading processes.
Section 4 presents a method for inducing the desired correlation structure on wireless channel simulation processes. It will be shown that it can be used for simulating both the spatial correlation between the propagation channels for MIMO systems and the time-ACF in case of single-input single-output (SISO) systems. The procedure is based on the method proposed by Iman & Conover (IC) in 1982 (Iman & Conover, 1982). This method is able to induce a desired correlation matrix on an input matrix whose columns are random vectors with independent arbitrary marginal distributions (Petrolino & Tavares, 2010a), (Petrolino & Tavares, 2010b). The IC method is distribution-free, which means that it preserves the marginal PDF of the input sequences while inducing the desired correlation. This important feature enables the generation of sequences with arbitrarily distributed PDFs and arbitrary correlation matrix.
Section 5 presents simulation results obtained by applying the IC method to the simulation of MIMO and SISO channels. In the MIMO case the simulated spatial correlation matrices are compared with the targets, while in the SISO case the comparison is made between the target and the simulated time ACFs. In both cases, perfect preservation of the initial PDF is achieved.
Finally, Section 6 concludes the chapter.

2. The Metropolis-Hastings algorithm

In this Section, we present a computational efficient and accurate method for simulating every type of fading (e.g. Nakagami-m, Weibull, Hoyt, etc.) starting from PDFs (e.g. Gaussian or Uniform distribution) which can be easily generated by known and efficient methods. This method is based on the Metropolis-Hastings algorithm (Hastings, 1970; Metropolis et al., 1953) with an AR sampling proposed by Tierney in (Tierney, 1994).

Consider now the problem of simulating a fading channel characterized by arbitrary jointly distributed envelope and phase. For this purpose, let $h(t) = x(t) + jy(t)$ be a complex fading process and let $h(t) = \rho(t)e^{j\theta(t)}$ be its representation in polar coordinates.

Two possible ways exist for generating $h(t)$: the first is the direct generation of the envelope and phase processes $\rho(t)$ and $\theta(t)$ respectively, the second is generating the IQ component processes $x(t)$ and $y(t)$. The first solution is attractive in the sense that the characterization of fading channels is usually done in terms of envelope and phase PDFs. However, in telecommunications, system testing is usually carried out by considering the transmission of the signal IQ components over the fading channel under consideration.

Recalling that the IQ components joint PDF $f_{XY}(x,y)$ and the phase-envelope joint PDF $f_{R,\theta}(\rho,\theta)$ are related by

$$f_{XY}(x,y) = \frac{f_{R,\Theta}(\rho,\theta)}{|J|} \tag{13}$$

where $|J|$ is the Jacobian of the transformation

$$x = \rho\cos\theta, \quad y = \rho\sin\theta \tag{14}$$

we conclude that if we are able to generate pairs of rvs (x_i, y_i) following the joint PDF $f_{XY}(x,y)$ in (13), then the phase-amplitude joint PDF will be the desired $f_{R,\Theta}(\rho,\theta)$. The main issue in simulating fading processes is that usually, the distribution $f_{XY}(x,y)$ does not belong to any family of well-known PDFs and can not be directly generated. Here, we propose a method which allows the efficient generation of rvs with arbitrary distribution.

Suppose we want to generate a set of N pairs $(x_1^{(i)}, x_2^{(i)})|_{i=1,\dots,N}$ following an *arbitrary target joint density* $f(\cdot)$ whose analytical expression is known, and that we have pairs (y_1, y_2) from a *candidate-generating joint density* $h(\cdot)$ that can be simulated by some known, inexpensive method. After setting an arbitrary initial pair $(x_1^{(0)}, x_2^{(0)})$ for initializing the algorithm, the MH method can be summarized as follows (Chib & Greenberg, 1995):

- Repeat the following steps for $i = 0, \dots, N-1$
- Generate (y_1, y_2) from $h(\cdot)$ and u from a uniform distribution $u \sim \mathcal{U}(0,1)$
- **IF** $u \le \alpha$
  Set $(x_1^{(i+1)}, x_2^{(i+1)}) = (y_1, y_2)$
- **ELSE**
  Set $(x_1^{(i+1)}, x_2^{(i+1)}) = (x_1^{(i)}, x_2^{(i)})$

where $\alpha = \alpha[(x_1^{(n)}, x_2^{(n)}), (y_1, y_2)]$ is the (conditional) probability of accepting the candidate pair (y_1, y_2) given that the current pair is $(x_1^{(n)}, x_2^{(n)})$. This probability is usually called *probability of move* (Chib & Greenberg, 1995), because it represents the probability of the process moving to a new state. When the MH algorithm was proposed, this probability was set to (Tierney, 1994)

$$\alpha = \min\left\{\frac{f(y_1, y_2)}{f(x_1^{(i)}, x_2^{(i)})}, 1\right\} \tag{15}$$

which means that the higher is the target PDF at the candidate pair, the higher is the probability to accept it. Here we use a modified version of the MH algorithm.
This modification has been discussed in (Tierney, 1994) and it is based on the idea of introducing an acceptance-rejection step for generating candidates for a MH algorithm. Pairs (y_1, y_2) are generated from $Kh(\cdot)$ until a pair satisfying $u < f(y_1, y_2)$, where $u \sim \mathcal{U}\{0, Kh(y_1, y_2)\}$, in obtained. The constant K is set such that $f(x_1, x_2) \leq Kh(x_1, x_2)$ for all (x_1, x_2) and controls the acceptance rate. According to this method, only those pairs for which the target density $f(\cdot)$ is high are accepted as candidates for the MH algorithm. Moreover, the probability of move α is set as follows. Let

$$\mathbf{C} = \{(a, b) : f(a, b) < Kh(a, b)\} \tag{16}$$

be the set where $Kh(a, b)$ *dominates* $f(a, b)$. Four possible cases occur and the probability α is set as follows (Chib & Greenberg, 1995):

- *Case (a)*: $(x_1^{(n)}, x_2^{(n)}) \in \mathbf{C}$ and $(y_1, y_2) \in \mathbf{C}$.
  In this case $Kh(\cdot)$ dominates $f(\cdot)$ for both the current and the candidate pair. This means that with higher probability both pairs belong to $Kh(\cdot)$ rather than to the target PDF $f(\cdot)$. In this case the algorithm always accepts the new pair i.e., $\alpha = 1$;
- *Case (b)*: $(x_1^{(n)}, x_2^{(n)}) \in \mathbf{C}$ and $(y_1, y_2) \notin \mathbf{C}$.
  In this case the current pair belongs with higher probability to $Kh(\cdot)$ rather than to $f(\cdot)$. Moreover, for the candidate pair, $f(\cdot)$ is higher than $Kh(\cdot)$. This means that the probability of belonging to the target density is higher for the candidate pair than for the current pair. Also in this case the algorithm always accepts the new one i.e., $\alpha = 1$;
- *Case (c)*: $(x_1^{(n)}, x_2^{(n)}) \notin \mathbf{C}$ and $(y_1, y_2) \in \mathbf{C}$.
  The probability of belonging to $f(\cdot)$ is higher for the current pair than for the candidate pair. This means that the transition to other states must be controlled. In this case the methods sets

  $$\alpha = \frac{Kh(x_1^{(n)}, x_2^{(n)})}{f(x_1^{(n)}, x_2^{(n)})}.$$

  Note that the higher the value of the target density $f(\cdot)$ at the current pair, the lower the probability to accept the candidate pair;
- *Case (d)*: $(x_1^{(n)}, x_2^{(n)}) \notin \mathbf{C}$ and $(y_1, y_2) \notin \mathbf{C}$.
  Both the current and the candidate pair belong with higher probability to the target PDF $f(\cdot)$ rather than to the candidate $Kh(\cdot)$. Also in this case the probability of move depends on the target and the candidate densities. It is easy to understand that it must be set such it is in general higher than in the previous case. The method in this case sets

  $$\alpha = \min\left\{\frac{f(y_1, y_2) \cdot h(x_1^{(n)}, x_2^{(n)})}{f(x_1^{(n)}, x_2^{(n)}) \cdot h(y_1, y_2)}, 1\right\}.$$

  Note that the higher the product $f(x_1^{(n)}, x_2^{(n)}) \cdot h(y_1, y_2)$, the lower the probability of move.

Compared with the standard MH algorithm, this version of the algorithm provides a PDF with a better match with the target distribution. This is achieved by observing that the initial distribution support can be divided in two kind of zones, one producing good values with higher probability than the other. It is important to note that, contrary to the standard MH algorithm, this method will occasionally reject candidates if the process is at a pair $(x_1^{(n)}, x_2^{(n)}) \notin \mathbf{C}$ (Tierney, 1994). By keeping the same pair for a certain number of consecutive steps, Metropolis-based methods introduce in general correlation along the sampled series. This can be harmful if uncorrelated processes are required. Our goal in this paper is to generate uncorrelated fading processes with desired envelope and phase PDFs. In this way, distribution-free algorithms can be applied to the generated pairs for imposing the desired correlation if second order statistics specification is required. We got round this difficulty by applying a random shuffling to the generated sequence at the end of the algorithm.

3. Simulation of uncorrelated arbitrary jointly distributed phase and envelope processes with the MH algorithm

In this Section we will show some results obtained by applying the previously described MH algorithm to two different propagation scenarios. We simulated the IQ components of fading channels characterized by arbitrarily distributed envelope and phase processes. As required by the algorithm we deduced analytical expressions of the target IQ components joint PDF $f(\cdot)$ from results available in the literature. In both the following examples the candidate pairs (y_1, y_2) follow the bivariate joint Gaussian PDF. They are independent Gaussian rvs with zero-mean and the same variance σ^2, so that their joint PDF reduces to the product of their Gaussian marginal PDFs and is given by

$$h(y_1, y_2) = \frac{1}{2\pi\sigma^2} \exp\left(-\frac{y_1^2 + y_2^2}{2\sigma^2}\right). \tag{17}$$

The constant K introduced in equation (16) has been adjusted by experimentation to achieve an acceptance rate of about 50%.
A rv representing a realization of a fading process at a fixed instant of time is

$$C = X + jY \tag{18}$$

where X and Y are rvs representing the fading process IQ components and

$$R = \sqrt{X^2 + Y^2} \quad \Theta = \tan^{-1}(Y/X) \tag{19}$$

are the envelope and the phase of C respectively. From (14), since the Jacobian is $|J| = \rho$, we have that

$$f_{XY}(x, y) = \frac{f_{R,\Theta}(\rho, \theta)}{\rho}. \tag{20}$$

3.1 Nakagami-m fading

For Nakagami-m fading, the envelope PDF is given by the well-known formula (Nakagami, 1960)

$$f_R(\rho) = \frac{2m^m \rho^{2m-1}}{\Omega^m \Gamma(m)} \exp\left(-\frac{m\rho^2}{\Omega}\right), \rho \geq 0 \; m > \frac{1}{2} \tag{21}$$

where $\Omega = \mathbb{E}[R^2]$ is the mean power, $m = \Omega^2/V[R^2]$ is the Nakagami fading parameter ($V[\cdot]$ denotes variance) and $\Gamma(\cdot)$ is the Gamma function.
For independent I and Q components, the corresponding phase PDF has been obtained in (Yacoub et al., 2005) and is given by

$$f_\Theta(\theta) = \frac{\Gamma(m)|\sin(2\theta)|^{m-1}}{2^m\Gamma^2(m/2)}, \; -\pi \le \theta < \pi. \tag{22}$$

Considering the envelope and the phase as independent rvs, we have

$$f_{R,\Theta}(\rho,\theta) = f_R(\rho) \times f_\Theta(\theta). \tag{23}$$

The substitution of (23) into (20) leads to

$$f_{XY}(x,y) = \frac{m^m|\sin(2\theta)|^{m-1}\rho^{2m-2}}{2^{m-1}\Omega^m\Gamma^2(m/2)} \exp\left(-\frac{m\rho^2}{\Omega}\right) \tag{24}$$

where $\rho = \sqrt{x^2+y^2}$ and $\theta = \tan^{-1}(y/x)$. We applied the proposed method for the generation of 2^{20} pairs of rvs following (24) starting from independent jointly Gaussian rvs with zero mean and variance $\sigma^2 = \Omega/2m$. The target and the simulated joint PDF are plotted in Fig. 1(a) and 1(b) respectively. In Fig. 2(a) and 2(b) the simulated envelope and phase PDFs are plotted

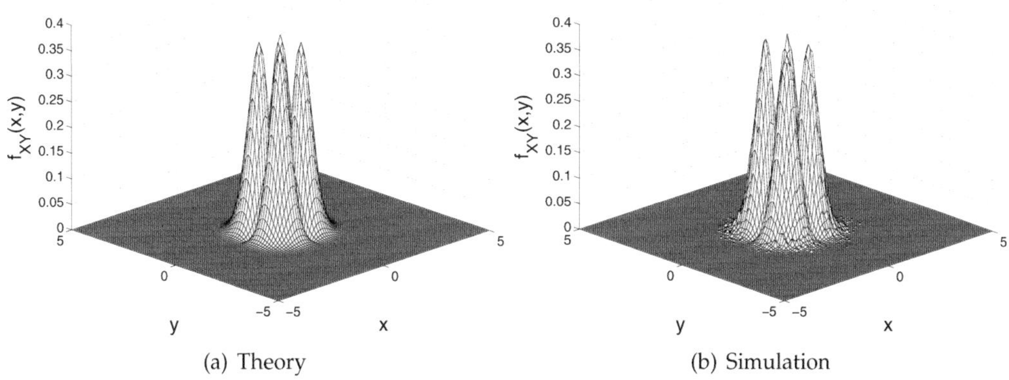

Fig. 1. Theoretical (a) and simulated (b) IQ components joint PDF for Nakagami-m fading with $m = 2$ and $\Omega = 1$.

respectively against the theoretical references. In both cases the agreement with the theory is very good. We now show how the application of a Metropolis-based method may introduce a (possibly not desired) correlation in the generated sequence and how this problem may be solved by random shuffling the sampled pairs. The correlation is due to the fact that until accepting a new candidate pair, the MH algorithm keeps the last pair as a good pair for sampling the target density $f(\cdot)$. However, the comparison of figures 3(a) and 3(b) shows that the (low) correlation introduced by the algorithm is completely canceled by the random shuffling.

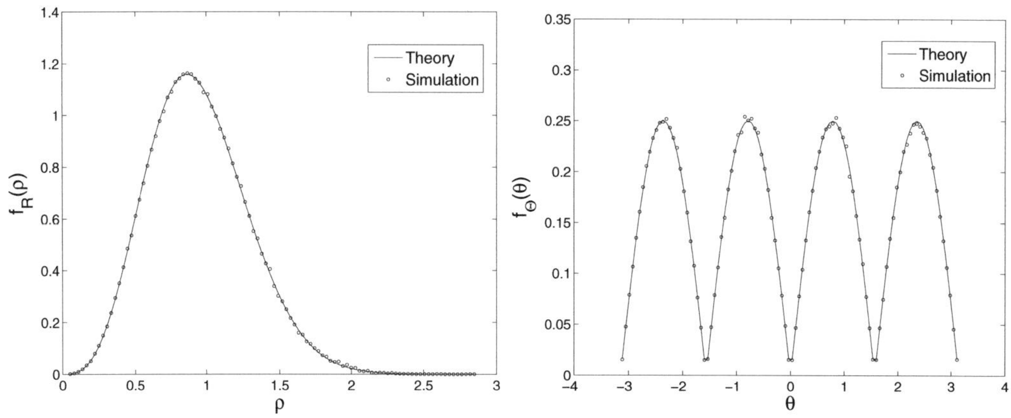

(a) Envelope PDF (b) Phase PDF

Fig. 2. Simulated Nakagami-m envelope PDF (a) and phase PDF (b) plotted against theory for 2^{20} generated pairs, $m = 2$ and $\Omega = 1$.

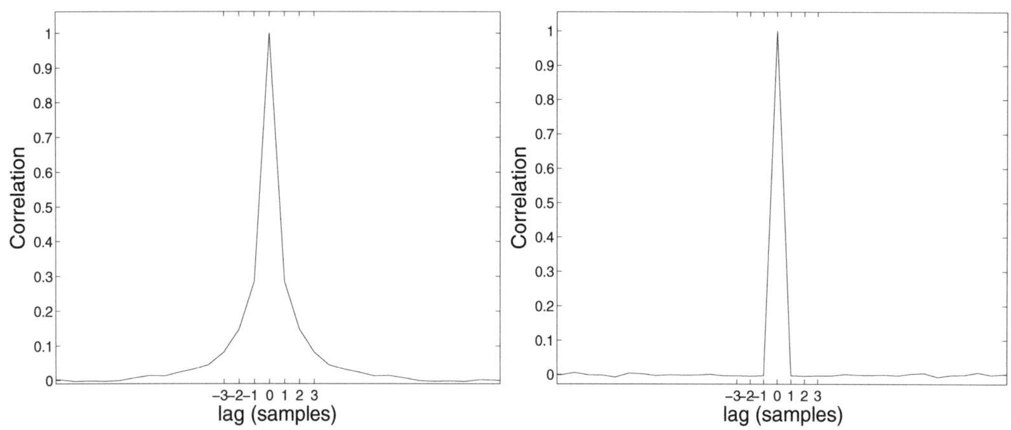

(a) Autocorrelation before shuffling (b) Autocorrelation after random shuffling

Fig. 3. Autocorrelation of the generated process in-phase component before (a) and after (b) the random shuffling.

3.2 Hoyt fading

For Hoyt fading, the envelope PDF is given by (Hoyt, 1947)

$$f_R(\rho) = \frac{2\rho}{\Omega\sqrt{1-b^2}} \exp\left(-\frac{\rho^2}{\Omega(1-b^2)}\right) \times I_0\left(\frac{b\rho^2}{\Omega(1-b^2)}\right), \quad \rho \geq 0 \tag{25}$$

where $\Omega = \mathbb{E}[R^2]$, $b \in [-1,1]$ is the Hoyt fading parameter and $I_0(\cdot)$ is the modified Bessel function of the first kind and zero-th order. The corresponding phase PDF is (Hoyt, 1947)

$$f_\Theta(\theta) = \frac{\sqrt{1-b^2}}{2\pi(1-b\cos(2\theta))}, -\pi \leq \theta < \pi. \tag{26}$$

Considering also in this case independent envelope and phase processes and applying relation (20), we obtain the IQ components target joint PDF

$$f_{XY}(x,y) = \frac{1}{\Omega\pi(1-b\cos(2\theta))} \exp\left(-\frac{\rho^2}{\Omega(1-b^2)}\right) \times I_0\left(\frac{b\rho^2}{\Omega(1-b^2)}\right) \tag{27}$$

where $\rho = \sqrt{x^2+y^2}$ and $\theta = \tan^{-1}(y/x)$. We applied the proposed method for the generation of 2^{20} pairs of rvs following (27) starting from independent jointly Gaussian rvs with zero mean and variance $\sigma^2 = \Omega(1-b^2)/2$. The target and the simulated joint PDF are plotted in Fig. 4(a) and 4(b) respectively.

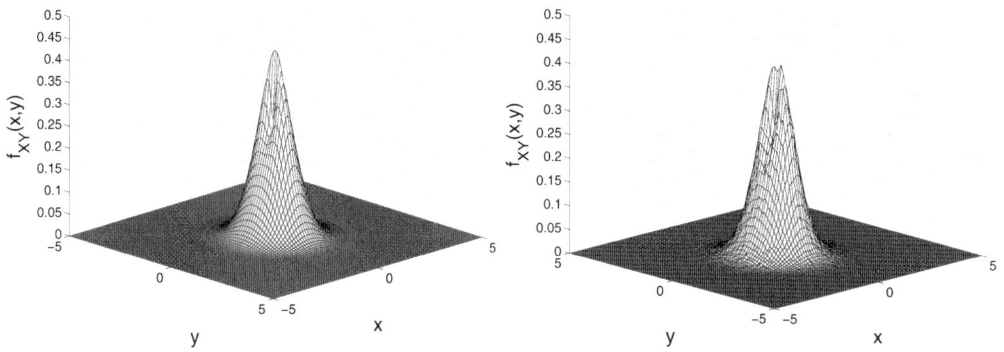

(a) Theory

(b) Simulation

Fig. 4. Theoretical (a) and simulated (b) IQ components joint PDF for Hoyt fading with b=0.25 and $\Omega = 1$.

In Fig. 5(a) and 5(b) the simulated envelope and phase PDFs are plotted respectively against the theoretical references. Also in this case the agreement with the theory is very good.

4. The Iman-Conover method

Let us consider a $n \times P$ matrix $\mathbf{X}_{in}$, with each column of $\mathbf{X}_{in}$ containing n uncorrelated arbitrarily distributed realizations of rvs. Simply stated, the Iman-Conover method is a procedure to induce a desired correlation between the columns of $\mathbf{X}_{in}$ by rearranging the samples in each column. Let $\mathbf{S}$ be the $P \times P$ symmetric positive definite target correlation matrix. By assumption, $\mathbf{S}$ allows a Cholesky decomposition

$$\mathbf{S} = \mathbf{C}^T\mathbf{C} \tag{28}$$

where $\mathbf{C}^T$ is the transpose of some $P \times P$ upper triangular matrix $\mathbf{C}$. Let $\mathbf{K}$ be a $n \times P$ matrix with independent, zero-mean and unitary standard deviation columns. Its linear correlation matrix $\mathbf{R}_{lin}(\mathbf{K})$ is given by

$$\mathbf{R}_{lin}(\mathbf{K}) = \frac{1}{n}\mathbf{K}^T\mathbf{K} = \mathbf{I} \tag{29}$$

where $\mathbf{I}$ is the $P \times P$ identity matrix. As suggested by the authors in (Iman & Conover, 1982), the so-called "scores matrix" $\mathbf{K}$ may be constructed as follows: let $\mathbf{u} = [u_1 \dots u_n]^{\mathrm{T}}$ denote the

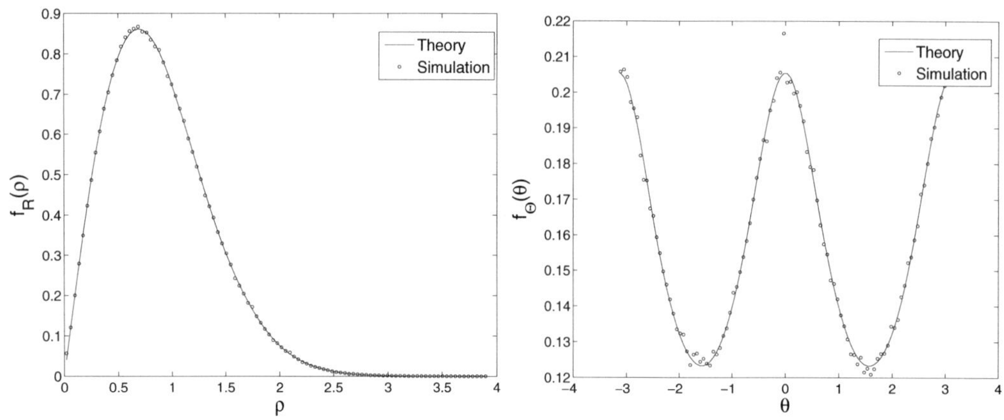

(a) Envelope PDF (b) Phase PDF

Fig. 5. Simulated Hoyt envelope PDF (a) and phase PDF (b) plotted against theory for 2^{20} generated pairs, $b = 0.25$ and $\Omega = 1$.

column vector with elements $u_i = \Phi^{-1}(\frac{i}{n+1})$, where $\Phi^{-1}(\cdot)$ is the inverse function of the standard normal distribution function, then $\mathbf{K}$ is formed as the concatenation of P vectors

$$\mathbf{K} = \frac{1}{\sigma_\mathbf{u}}[\mathbf{u}\ \mathbf{v}_1\ \ldots\ \mathbf{v}_{P-1}] \tag{30}$$

where $\sigma_\mathbf{u}$ is the standard deviation of $\mathbf{u}$ and the $\mathbf{v}_i$ are random permutations of $\mathbf{u}$. The columns of $\mathbf{K}$ have zero-mean and unitary standard deviation by construction and the random permutation makes them independent. Consider now the matrix

$$\mathbf{T} = \mathbf{KC} \tag{31}$$

and note that, according to equations (28) and (29), $\mathbf{T}$ has a linear correlation matrix equal to the target correlation matrix $\mathbf{S}$:

$$\mathbf{R}_{lin}(\mathbf{T}) = n^{-1}\mathbf{T}^T\mathbf{T} = \mathbf{C}^T \underbrace{n^{-1}\mathbf{K}^T\mathbf{K}}_{\mathbf{I}} \mathbf{C} = \mathbf{S}. \tag{32}$$

The basis of the IC method is as follows: generate the matrix $\mathbf{T}$ as explained above and then produce a new matrix $\mathbf{X}$ which is a rearranged version of $\mathbf{X}_{in}$ so that its samples have the same ranking position of the corresponding samples of $\mathbf{T}$. With reference to Fig. 6, note that the matrix $\mathbf{T}$ depends on the correlation matrix we want to induce and on the "scores matrix" $\mathbf{K}$ which is deterministic. So, it does not have to be evaluated during simulation runs, i.e., it may be computed *offline*. The only operation which must be executed in real time is the generation and the rearrangement of $\mathbf{X}$, which is not computationally expensive.
A scheme explaining how the rank matching between matrices $\mathbf{T}$ and $\mathbf{X}$ is carried out column by column is shown in Fig. 7.
At this point two observations are necessary:

- The *rank* correlation matrix of the rearranged data $\mathbf{R}_{rank}(\mathbf{X})$ will match exactly $\mathbf{R}_{rank}(\mathbf{T})$ and though $\mathbf{T}$ is constructed so that $\mathbf{R}_{lin}(\mathbf{T}) = \mathbf{S}$, the rank correlation matrix $\mathbf{R}_{rank}(\mathbf{T})$ is also close to $\mathbf{S}$. This happens because when there are no prominent outliers, as is the

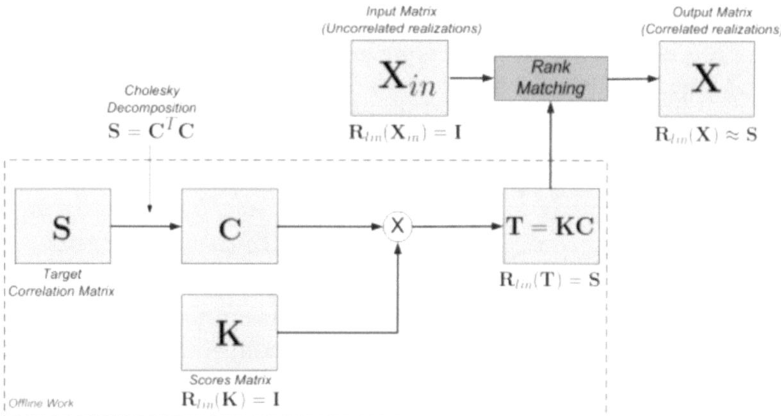

Fig. 6. Block diagram of the IC method.

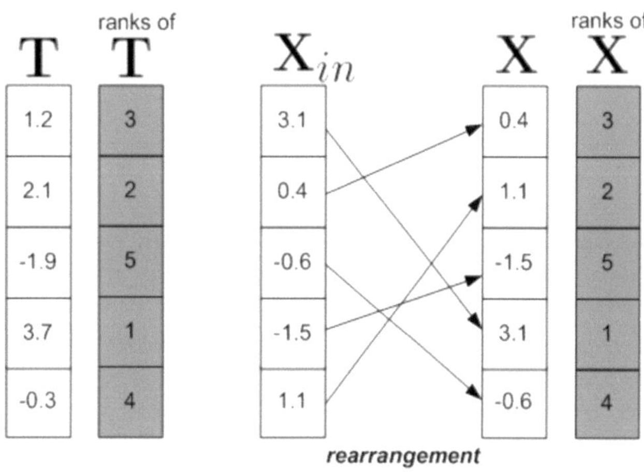

Fig. 7. Rank matching.

case of **T** whose column elements are normally distributed, linear and rank correlation coefficients are very close to each other and so $\mathbf{R}_{rank}(\mathbf{T}) \approx \mathbf{R}_{lin}(\mathbf{T}) = \mathbf{S}$. Moreover, for WSS processes we may expect $\mathbf{R}_{lin}(\mathbf{X}) \approx \mathbf{R}_{rank}(\mathbf{X})$ (Lehmann & D'Abrera, 1975) and, since $\mathbf{R}_{rank}(\mathbf{X}) = \mathbf{R}_{rank}(\mathbf{T}) \approx \mathbf{R}_{lin}(\mathbf{T}) = \mathbf{S}$, then $\mathbf{R}_{lin}(\mathbf{X}) \approx \mathbf{S}$.

- Row-wise, the permutation resulting from the rearrangement of **X** appears random and thus the n samples in any given column remain uncorrelated.

The rearrangement of the input matrix **X** is carried out according to a ranking matrix which is *directly* derived from the desired correlation matrix and this operation does not affect the input marginal distributions. This means that it is possible to induce arbitrary (rank) correlations between vectors with arbitrary distributions, a fundamental task in fading channel simulation. In the sequel we analyze the effects of the finite sample size error and its compensation.

As introduced above, the linear correlation matrix of $\mathbf{T} = \mathbf{KC}$ is, by construction, equal to **S**. However, due to finite sample size, a small error can be introduced in the simulation by the non-perfect independence of the columns of **K**. This error can be corrected with an adjustment proposed by the authors of this method in (Iman & Conover, 1982). Consider that $\mathbf{E} = \mathbf{R}_{lin}(\mathbf{K})$

is the correlation matrix of $\mathbf{K}$ and that it allows a Cholesky decomposition $\mathbf{E} = \mathbf{F}^T\mathbf{F}$. We have verified that for $\mathbf{K}$ constructed as in (30), $\mathbf{E}$ is generally very close to positive-definite. In some cases however, a regularization step may be required, by adding a small $\delta > 0$ to each element in the main diagonal of $\mathbf{E}$. This does not sacrifice accuracy because δ is very small. Following the rationale above, if the matrix $\mathbf{T}$ is now constructed such that $\mathbf{T} = \mathbf{K}\mathbf{F}^{-1}\mathbf{C}$, then it has covariance matrix

$$\frac{1}{n}\mathbf{T}^T\mathbf{T} = \mathbf{C}^T\mathbf{F}^{-T}\underbrace{\mathbf{K}^T\mathbf{K}}_{\mathbf{E}=\mathbf{F}^T\mathbf{F}}\mathbf{F}^{-1}\mathbf{C} = \mathbf{C}^T\underbrace{\mathbf{F}^{-T}\mathbf{F}^T}_{\mathbf{I}}\underbrace{\mathbf{F}\mathbf{F}^{-1}}_{\mathbf{I}}\mathbf{C} = \mathbf{C}^T\mathbf{C} = \mathbf{S} \tag{33}$$

as desired. In (33), the notation $\mathbf{F}^{-T}$ stands for the inverse of the transpose of $\mathbf{F}$. With this adjustment, a possible error introduced by the non-perfect independence of the columns of $\mathbf{K}$ is completely canceled.

With the goal of quantifying the gain obtained with this adjustment it is possible to calculate the MSE between the output linear correlation matrix $\mathbf{R}_{lin}(\mathbf{X})$ and the target $\mathbf{S}$ with and without the error compensation. We define the matrix $\mathbf{D} \triangleq \mathbf{R}_{lin}(\mathbf{X}) - \mathbf{S}$ and the MSE as

$$\mathrm{MSE} = \frac{1}{MN}\sum_{i=1}^{M}\sum_{j=1}^{N} d_{ij}^2 \tag{34}$$

where d_{ij} is the element of $\mathbf{D}$ with row index i and column index j.

4.1 Using the Iman-Conover method for the simulation of MIMO and SISO channels

Let us consider the MIMO propagation scenario depicted in Fig. 8, with M transmitting antennas at the Base Station (BS) and N receiving antennas at the Mobile Station (MS). Considering all possible combinations, the MIMO channel can be subdivided into MN subchannels. Moreover, due to multipath propagation, each subchannel is composed by L uncorrelated paths. In general these paths are complex-valued. For simplicity we will consider that these paths are real-valued (in-phase component only) but the generalization of the method for complex paths is straightforward. The IC method can be used for MIMO channels simulation by simply storing the MIMO channel samples in a $n \times MNL$ matrix $\mathbf{X}_{in}$, where MNL is the number of subchannels considering multipath and n is the number of samples generated for each subchannel. The target (desired) spatial correlation coefficients of the MIMO channel are stored in a $MNL \times MNL$ symmetric positive definite matrix $\mathbf{S}_{MIMO}$ (Petrolino & Tavares, 2010a). After this, the application of the IC method to matrix $\mathbf{X}_{in}$, will produce a new matrix $\mathbf{X}$. The columns of $\mathbf{X}$ contain the subchannels realizations which are spatially correlated according to the desired correlation coefficients in $\mathbf{S}_{MIMO}$ and whose PDFs are preserved.

We have also used the IC method for the simulation of SISO channels. This idea comes from the following observation. As the final result of the IC method we have a rearranged version of the input matrix $\mathbf{X}_{in}$, so that its columns are correlated as desired. This means that if we extract any row from this matrix, the samples therein are correlated according to the ACF contained in the $P \times P$ target correlation matrix $\mathbf{S}_{SISO}$ (Petrolino & Tavares, 2010b). Note also that for stationary processes, $\mathbf{S}_{SISO}$ is Toeplitz so all rows will exhibit the same ACF.

If we consider P as the number of correlated samples to be generated with the IC method and n as the number of uncorrelated fading realizations (often required when Monte Carlo simulations are necessary), after the application of the IC method we obtain a $n \times P$ matrix $\mathbf{X}$ whose correlation matrix is equal to the target $\mathbf{S}_{SISO}$. Its n rows are uncorrelated and each contains a P-samples fading sequence with the desired ACF. These n columns can be extracted and used for Monte Carlo simulation of SISO channels.

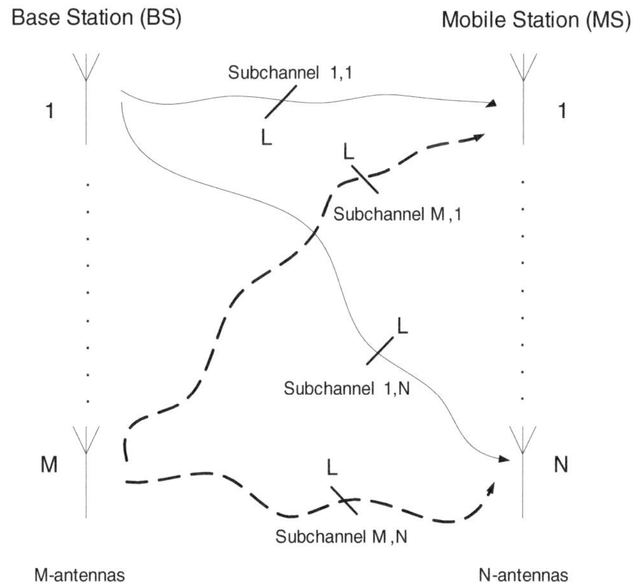

Fig. 8. Propagation scenario for multipath MIMO channels.

5. Simulation of MIMO and SISO channels with the IC method

In this Section we present results obtained by applying the IC method to the simulation of different MIMO and SISO channels [2]. The MH candidate pairs are jointly Gaussian distributed with zero mean and a variance which is set depending on the target density. The constant K introduced in equation (16) has been adjusted by experimentation to achieve an acceptance rate of about 50%.

5.1 Simulation of a 2×1 MIMO channel affected by Rayleigh fading

Let us first consider a scenario with $M = 2$ antennas at the BS and $N = 1$ antennas at the MS. For both subchannels, multipath propagation exists but, for simplicity, the presence of only 2 paths per subchannel ($L = 2$) is considered. This means that the channel matrix $\mathbf{X}$ is $n \times 4$, where n is the number of generated samples. Equi-spaced antenna elements are considered at both ends with half a wavelength element spacing. We consider the case of no local scatterers close to the BS as usually happens in typical urban environments and the power azimuth spectrum (PAS) following a Laplacian distribution with a mean azimuth spread (AS) of 10°. Given these conditions, an analytical expression for the spatial correlation at the BS is given in (Pedersen et al., 1998, eq.12), which leads to

$$\mathbf{S}_{BS} = \begin{bmatrix} 1 & 0.874 \\ 0.874 & 1 \end{bmatrix}. \tag{35}$$

Since $N = 1$, the correlation matrix collapses into an autocorrelation coefficient at the MS, i.e.,

$$\mathbf{S}_{MS} = 1. \tag{36}$$

[2] In the following examples, the initial uncorrelated Nakagami-m, Weibull and Hoyt rvs have been generated using the MH algorithm described in Sections 2 and 3.

Remembering that $\mathbf{S}_{MIMO} = \mathbf{S}_{BS} \otimes \mathbf{S}_{MS}$, we have an initial correlation matrix

$$\mathbf{S}_{MIMO} = \begin{bmatrix} 1 & 0.874 \\ 0.874 & 1 \end{bmatrix}. \tag{37}$$

Equation (37) represents the spatial correlation existing between the subchannel paths at any time delay. If we remember that in this example we are considering $L = 2$ paths for each subchannel, recalling that non-zero correlation coefficients exist only between paths which are referred to the same path delay, we obtain the final zero-padded spatial correlation matrix $\mathbf{S}_{MIMO}$ as

$$\mathbf{S}_{MIMO} = \begin{bmatrix} 1 & 0.874 & 0 & 0 \\ 0.874 & 1 & 0 & 0 \\ 0 & 0 & 1 & 0.874 \\ 0 & 0 & 0.874 & 1 \end{bmatrix}. \tag{38}$$

As is done in most studies (Gesbert et al., 2000), the subchannel samples are considered as realizations of uncorrelated Gaussian rvs. In the following examples a sample length of $n =$ 100000 samples has been chosen. Since very low values of the MSE in (34) are achieved for large n, simulations have been run without implementing the error compensation discussed in the previous Section. The simulated spatial correlation matrix $\hat{\mathbf{S}}_{MIMO}$ is

$$\hat{\mathbf{S}}_{MIMO} = \begin{bmatrix} 1.0000 & 0.8740 & -0.0000 & -0.0001 \\ 0.8740 & 1.0000 & -0.0000 & -0.0001 \\ -0.0000 & -0.0000 & 1.0000 & 0.8740 \\ -0.0001 & -0.0001 & 0.8740 & 1.0000 \end{bmatrix}. \tag{39}$$

The comparison between (38) and (39) demonstrates the accuracy of the proposed method. The element-wise absolute difference between $\hat{\mathbf{S}}_{MIMO}$ and $\mathbf{S}_{MIMO}$ is of the order of 10^{-4}.

5.2 Simulation of a 2×3 MIMO channel affected by Rayleigh, Weibull and Nakagami-m fading

We now present a set of results to show that the IC method is distribution-free, which means that the subchannels marginal PDFs are preserved. For this reason we simulate a single-path ($L = 1$) 2×3 MIMO system and consider that the subchannel envelopes arriving at the first Rx antenna are Rayleigh, those getting the second antenna are Weibull distributed with a fading severity parameter $\beta = 1.5$ (fading more severe than Rayleigh) and finally those arriving at the third one are Nakagami-m distributed with $m = 3$ (fading less severe than Rayleigh). In the first two cases the phases are considered uniform in $[0, 2\pi)$, while in the case of Nakagami envelope, the corresponding phase distribution (Yacoub et al., 2005, eq. 3) is also considered. For the sake of simplicity, from now on only the real part of the processes are considered. Being the real and imaginary parts of the considered processes equally distributed, the extension of this example to the imaginary components is straightforward. The Rayleigh envelope and uniform phase leads to a Gaussian distributed in-phase component[3] $f_{X_r}(x_r)$ for subchannels $\mathbf{H}_{1,1}$ and $\mathbf{H}_{2,1}$. The in-phase component PDF $f_{X_w}(x_w)$ of subchannels $\mathbf{H}_{1,2}$ and $\mathbf{H}_{2,2}$ has been obtained by transformation of the envelope and phase rvs into their corresponding in-phase and quadrature (IQ) components rvs. It is given by

$$f_{X_w}(x_w) = \int_{-\infty}^{\infty} f_{X_w Y_w}(x_w, y_w)\, dy_w \tag{40}$$

[3] In the following, the subscripts r, w and n refer to the fading distribution (Rayleigh, Weibull and Nakagami-m) in the respective subchannels.

where

$$f_{X_wY_w}(x_w, y_w) = \frac{\beta}{2\pi\Omega_w}(x_w^2 + y_w^2)^{\frac{\beta-2}{2}} \exp\left(-\frac{(x_w^2 + y_w^2)^{\frac{\beta}{2}}}{\Omega_w}\right) \tag{41}$$

is the IQ components joint PDF obtained considering a Weibull distributed envelope and uniform phase in $[0, 2\pi]$ and where Ω_w is the average fading power. Finally, the in-phase PDF $f_{X_n}(x_n)$ for subchannels $\mathbf{H}_{1,3}$ and $\mathbf{H}_{2,3}$ is obtained by integrating the IQ joint PDF corresponding to the Nakagami-m fading. It has been deduced by transformation of the envelope and phase rvs, whose PDFs are known (Yacoub et al., 2005), into their corresponding IQ components rvs. It is given by

$$f_{X_nY_n}(x_n, y_n) = \frac{m^m |\sin(2\theta)|^{m-1}\rho^{2m-2}}{2^{m-1}\Omega_n^m \Gamma^2(m/2)} \exp\left(-\frac{m\rho^2}{\Omega_n}\right) \tag{42}$$

where $\rho = \sqrt{x_n^2 + y_n^2}$, $\theta = \tan^{-1}(y_n/x_n)$, $\Gamma(\cdot)$ is the Gamma function and Ω_n is the average fading power. At the BS, the same conditions of the previous examples are considered, except for the antenna separation which in this case is chosen to be one wavelength. Given these conditions, the spatial correlation matrix at the BS is

$$\mathbf{S}_{BS} = \begin{bmatrix} 1 & 0.626 \\ 0.626 & 1 \end{bmatrix}. \tag{43}$$

Note that, due to the larger antenna separation, the spatial correlation between the antennas at the BS is lower than in the previous examples, where a separation of half a wavelength has been considered. Also at the MS, low correlation coefficients between the antennas are considered. This choice is made with the goal of making this scenario (with three different distributions) more realistic. Hence, we assume

$$\mathbf{S}_{MS} = \begin{bmatrix} 1 & 0.20 & 0.10 \\ 0.20 & 1 & 0.20 \\ 0.10 & 0.20 & 1 \end{bmatrix}. \tag{44}$$

It follows that

$$\mathbf{S}_{MIMO} = \begin{bmatrix} 1 & 0.2000 & 0.1000 & 0.6268 & 0.1254 & 0.0627 \\ 0.2000 & 10.2000 & 0.1254 & 0.6268 & 0.1254 & \\ 0.1000 & 0.2000 & 1 & 0.0627 & 0.1254 & 0.6268 \\ 0.6268 & 0.1254 & 0.0627 & 1 & 0.2000 & 0.1000 \\ 0.1254 & 0.6268 & 0.1254 & 0.2000 & 1 & 0.2000 \\ 0.0627 & 0.1254 & 0.6268 & 0.1000 & 0.2000 & 1 \end{bmatrix}. \tag{45}$$

The simulated correlation matrix is

$$\hat{\mathbf{S}}_{MIMO} = \begin{bmatrix} 1.0000 & 0.1974 & 0.0995 & 0.6268 & 0.1238 & 0.0623 \\ 0.1974 & 1.0000 & 0.1961 & 0.1240 & 0.6144 & 0.1229 \\ 0.0995 & 0.1961 & 1.0000 & 0.0623 & 0.1227 & 0.6209 \\ 0.6268 & 0.1240 & 0.0623 & 1.0000 & 0.1973 & 0.0994 \\ 0.1238 & 0.6144 & 0.1227 & 0.1973 & 1.0000 & 0.1956 \\ 0.0623 & 0.1229 & 0.6209 & 0.0994 & 0.1956 & 1.0000 \end{bmatrix}. \tag{46}$$

Also in this case the simulated spatial correlation matrix provides an excellent agreement with the target correlation as is seen from the comparison between (45) and (46). Figs. 9, 10 and 11 demonstrate that the application of the proposed method does not affect the subchannels PDF. After the simulation run the agreement between the theoretical marginal distribution (evaluated analytically) and the generated samples histogram is excellent.

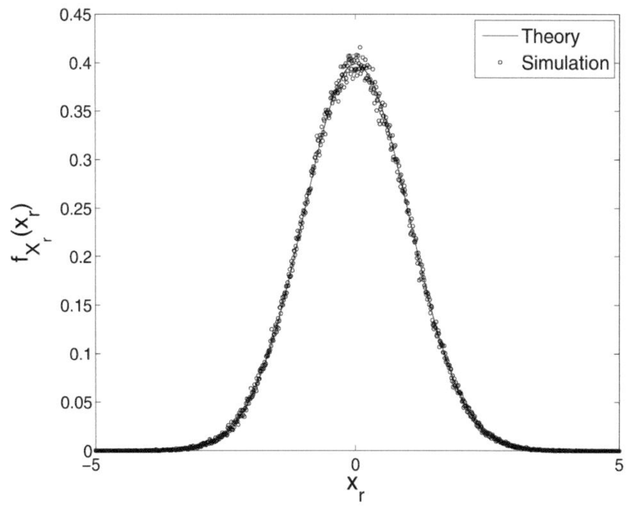

Fig. 9. PDF of the in-phase component for subchannels $\mathbf{H}_{1,1}$ and $\mathbf{H}_{2,1}$ affected by Rayleigh fading after the application of the method, plotted against the theoretical reference.

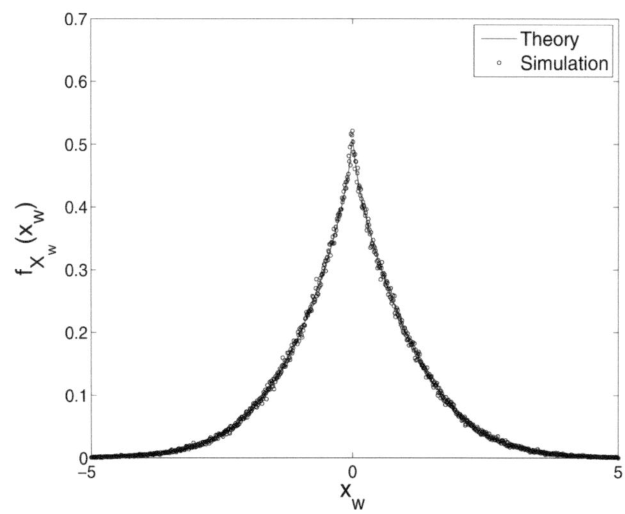

Fig. 10. PDF of the in-phase component for subchannels $\mathbf{H}_{1,2}$ and $\mathbf{H}_{2,2}$ affected by Weibull ($\beta = 1.5$) fading after the application of the method, plotted against the theoretical reference.

5.3 SISO fading channel simulation with the IC method

The use of non-Rayleigh envelope fading PDFs has been gaining considerable success and acceptance in the last years. Experience has indeed shown that the Rayleigh distribution

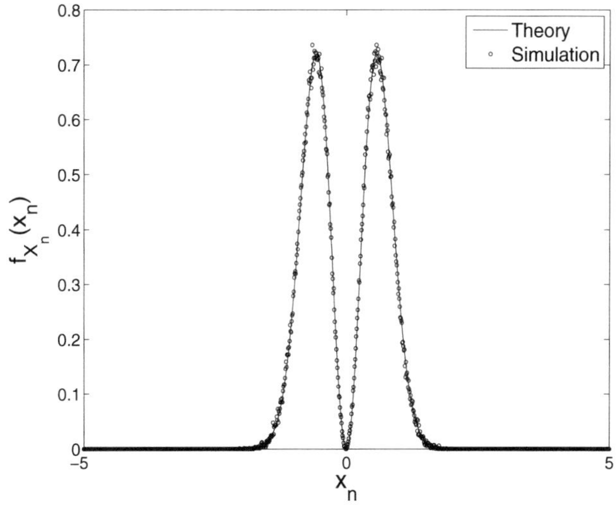

Fig. 11. PDF of the in-phase component for subchannels $\mathbf{H}_{1,3}$ and $\mathbf{H}_{2,3}$ affected by Nakagami-m ($m = 3$) fading after the application of the method, plotted against the theoretical reference.

shows a good matching with real fadings only in particular cases of fading severity. More general distributions (e.g., the Nakagami-m and Weibull) allow a convenient channel severity to be set simply by tuning one parameter and are advisable for modeling a larger class of fading envelopes. As reported in the Introduction, in the case of Rayleigh fading, according to the model of Clarke, the fading process $h(t)$ is modeled as a complex Gaussian process with independent real and imaginary parts $h_R(t)$ and $h_I(t)$ respectively. In the case of isotropic scattering and omni-directional receiving antennas, the normalized ACF of the quadrature components is given by (5). An expression for the normalized ACF of the Rayleigh envelope process $r(t) = \sqrt{h_R^2(t) + h_I^2(t)}$ can be found in (Jakes, 1974). It is given by

$$\mathcal{R}_r(\tau) = {}_2F_1\left[-\frac{1}{2}, -\frac{1}{2}; 1; J_0^2(2\pi f_D \tau)\right] \tag{47}$$

where ${}_2F_1(\cdot,\cdot;\cdot;\cdot)$ is a Gauss hypergeometric function.

In this Section we present the results obtained by using the IC method for the direct simulation of correlated Nakagami-m and Weibull envelope sequences (Petrolino & Tavares, 2010b). In particular, the simulation problem has been approached as follows: uncorrelated envelope fading sequences have been generated with the MH algorithm and the IC method has been afterwards applied to the uncorrelated sequences with the goal of imposing the desired correlation structure, derived from existing channel models. Since the offline part of the IC method has been discussed in the previous Section, here we present the online part for the simulation of SISO channels. So, from now on, we consider that a matrix $\mathbf{T}_{rank}$, containing the ranking positions of the matrix $\mathbf{T}$ has been previously computed and is available. Recalling the results of Section 4, there are only two operations that must be executed during the online simulation run:

1. Generate the $n \times P$ input matrix $\mathbf{X}_{in}$ containing nP uncorrelated rvs with the desired PDF.

2. Create matrix $\mathbf{X}$ which is a rearranged version of $\mathbf{X}_{in}$ according to the ranks contained in $\mathbf{T}_{rank}$ to get the desired ACF.

Each of the n rows of $\mathbf{X}$ represents an P-samples fading sequence which can be used for channel simulation.

5.3.1 Simulation of a SISO channel affected by Nakagami-m fading

In 1960, M. Nakagami proposed a PDF which models well the signal amplitude fading in a large range of propagation scenarios. A rv R_N is Nakagami-m distributed if its PDF follows the distribution (Nakagami, 1960)

$$f_{R_N}(r) = \frac{2m^m r^{2m-1}}{\Gamma(m)\Omega^m} \exp\left(-\frac{mr^2}{\Omega}\right), \quad r \geq 0, \ m > \frac{1}{2} \tag{48}$$

where $\Gamma(\cdot)$ is the gamma function, $\Omega = \mathbb{E}[R_N^2]$ is the mean power and $m = \Omega^2/(R_N^2 - \Omega)^2 \geq \frac{1}{2}$ is the Nakagami-m fading parameter which controls the depth of the fading amplitude.
For $m = 1$ the Nakagami model coincides with the Rayleigh model, while values of $m < 1$ correspond to more severe fading than Rayleigh and values of $m > 1$ to less severe fading than Rayleigh. The Nakagami-m envelope ACF is found as (Filho et al., 2007)

$$R_{R_N}(\tau) = \frac{\Omega\Gamma^2(m + \frac{1}{2})}{m\Gamma^2(m)} {}_2F_1\left[-\frac{1}{2}, -\frac{1}{2}; m; \rho(\tau)\right] \tag{49}$$

where $\rho(\tau)$ is an autocorrelation coefficient (ACC) which depends on the propagation scenario. In this example and without loss of generality, we consider the isotropic scenario with uniform distributed waves angles of arrival for which $\rho(\tau) = J_0^2(2\pi f_D \tau)$. With the goal of reducing the simulation error induced by the finiteness of the sample size n, the simulated ACF has been evaluated on all the n rows of the rearranged matrix $\mathbf{X}$ and finally the arithmetic mean has been taken. Fig. 12 shows the results of the application of the proposed method to the generation of correlated Nakagami-m envelope sequences. As is seen, the simulated ACF matches the theoretical reference very well.

5.3.2 Simulation of a SISO channel affected by Weibull fading

The Weibull distribution is one of the most used PDFs for modeling the amplitude variations of the fading processes. Indeed, field trials show that when the number of radio wave paths is limited the variation in received signal amplitude frequently follows the Weibull distribution (Shepherd, 1977). The Weibull PDF for a rv R_W is given by (Sagias et al., 2004)

$$f_{R_W}(r) = \frac{\beta r^{\beta-1}}{\Omega} \exp\left(-\frac{r^\beta}{\Omega}\right) \quad r, \beta \geq 0 \tag{50}$$

where $\mathbb{E}[r^\beta] = \Omega$ and β is the fading severity parameter. As the value of β increases, the severity of the fading decreases, while for the special case of $\beta = 2$ the Weibull PDF reduces to the Rayleigh PDF (Sagias et al., 2004).
The Weibull envelope ACF has been obtained in (Yacoub et al., 2005) and validated by field trials in (Dias et al., 2005). Considering again an isotropic scenario as was done in the case of Nakagami fading, it is given by

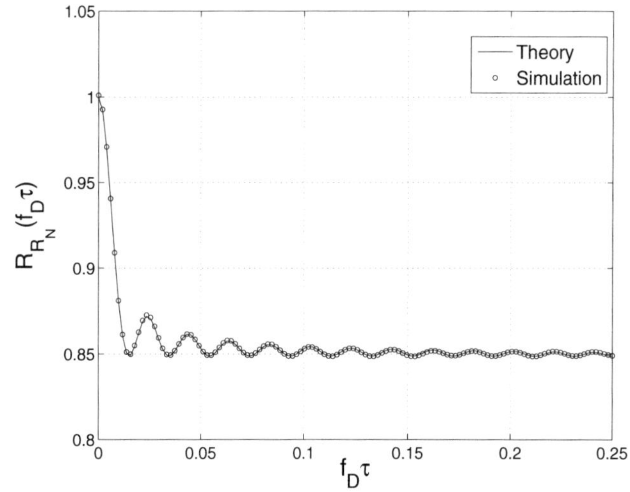

Fig. 12. Normalized linear ACF of the simulated Nakagami-m fading process plotted against the theoretical normalized ACF for $m = 1.5$, normalized Doppler frequency $f_D = 0.05$, $n = 10000$ realizations and $P = 2^{10}$ samples.

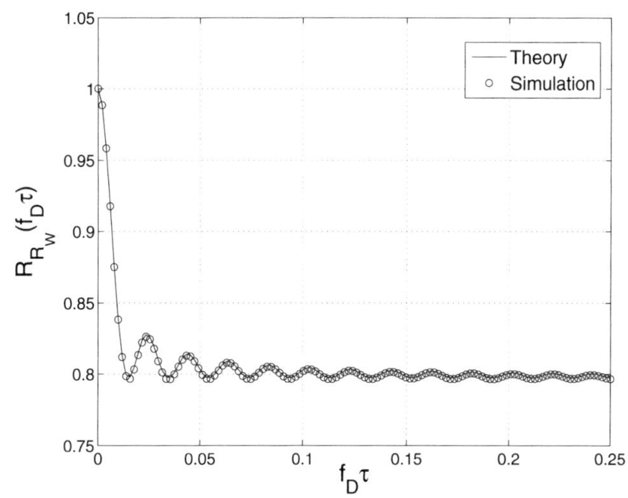

Fig. 13. Normalized linear ACF of the simulated Weibull fading process plotted against the theoretical normalized ACF for $\beta = 2.5$, normalized Doppler frequency $f_D = 0.05$, $n = 10000$ realizations and $P = 2^{10}$ samples.

$$R_{R_W}(\tau) = \Omega^{2/\beta}\Gamma^2\left(1+\frac{1}{\beta}\right){}_2F_1\left[-\frac{1}{\beta},-\frac{1}{\beta};1;J_0^2(2\pi f_D\tau)\right]. \quad (51)$$

Also in this example an isotropic scenario with uniform distributed waves angles of arrival is considered. The simulated ACF has been evaluated on all the n rows of the rearranged matrix $\mathbf{X}$ and finally the arithmetic mean has been taken. Fig. 13 shows the results of the application of the proposed method to the generation of correlated Weibull envelope

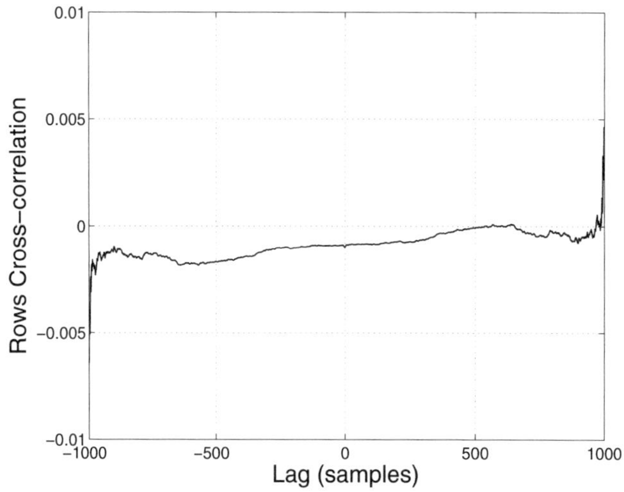

Fig. 14. Average row crosscorrelation of the simulated Weibull fading process for $\beta = 2.5$, normalized Doppler frequency $f_D = 0.05$, $n = 1000$ realizations and $P = 2^{10}$ samples.

sequences. The simulated ACF matches the theoretical reference quite well. Fig. 14 plots the average row crosscorrelation obtained from a simulation run of a Weibull fading process with autocorrelation (51), for $n = 1000$ realizations and $P = 2^{10}$ samples. As is seen, the crosscorrelation between rows is quite small (note the scale). This means that the n realizations produced by the method are statistically uncorrelated, as is required for channel simulation.

6. Conclusions

In this chapter, we presented a SISO and MIMO fading channel simulator based on the Iman-Conover (IC) method (Iman & Conover, 1982). The method allows the simulation of radio channels affected by arbitrarily distributed fadings. The method is distribution-free and is able to induce any desired spatial correlation matrix in case of MIMO simulations (Petrolino & Tavares, 2010a) and any time ACF in case of SISO channels (Petrolino & Tavares, 2010b), while preserving the initial PDF of the samples. The proposed method has been applied in different MIMO and SISO scenarios and has been shown to provide excellent results. We have also presented a simulator for Gaussian-based fading processes, like Rayleigh and Rician fading. These simulators are based on the Karhunen-Loève expansion and have been applied to the simulation of mobile-to-fixed (Petrolino et al., 2008a) and mobile-to-mobile (Petrolino et al., 2008b) fading channels. Moreover a technique for generating uncorrelated arbitrarily distributed sequences has been developed with the goal of combining it with the IC method. This technique is based on the Metropolis-Hastings algorithm (Hastings, 1970; Metropolis et al., 1953) and allows the application of the IC method to a larger class of fadings.

7. References

Beaulieu, N. & Cheng, C. (2001). An efficient procedure for Nakagami-*m* fading simulation, *Global Telecommunications Conf., 2001. GLOBECOM '01. IEEE*, Vol. 6, pp. 3336–3342.

Chib, S. & Greenberg, E. (1995). Understanding the Metropolis-Hastings algorithm, *The American Statistician* 49(4): 327–335.

Clarke, R. H. (1968). A statistical theory of mobile-radio reception, *Bell Systems Tech. Jou.* 47(6): 957–1000.

Dias, U., Fraidenraich, G. & Yacoub, M. (2005). On the Weibull autocorrelation function: field trials and validation, *Microwave and Optoelectronics, 2005 SBMO/IEEE MTT-S International Conference on*, pp. 505–507.

Filho, J., Yacoub, M. & Fraidenraich, G. (2007). A simple accurate method for generating autocorrelated Nakagami-*m* envelope sequences, *IEEE Comm. Lett.* 11(3): 231–233.

Gesbert, D., Bolcskei, H., Gore, D. & Paulraj, A. (2000). MIMO wireless channels: capacity and performance prediction, *Global Telecommunications Conference, 2000. GLOBECOM '00. IEEE*, Vol. 2, pp. 1083–1088.

Hastings, W. K. (1970). Monte Carlo sampling methods using Markov chains and their applications, *Biometrika* 57(1): 97–109.

Hoyt, R. (1947). Probability functions for the modulus and angle of the normal complex variate, *Bell System Technical Journal* 26: 318–359.

Iman, R. & Conover, W. (1982). A distribution-free approach to inducing rank correlation among input variables, *Communications in Statistics-Simulation and Computation* 11: 311–334.

Jakes, W. C. (1974). *Microwave Mobile Communications*, John Wiley & Sons, New York.

Komninakis, C. (2003). A fast and accurate Rayleigh fading simulator, *Global Telecommunications Conference, 2003. GLOBECOM '03. IEEE* 6: 3306–3310.

Lehmann, E. L. & D'Abrera, H. J. M. (1975). *Nonparametrics: Statistical Methods Based on Ranks*, McGraw-Hill, New York.

Metropolis, N., Rosenbluth, A. W., Rosenbluth, M. N., Teller, A. H. & Teller, E. (1953). Equation of state calculations by fast computing machines, *The Journal of Chemical Physics* 21(6): 1087–1092.

Nakagami, N. (1960). The *m*-distribution, a general formula for intensity distribution of rapid fading, *in* W. G. Hoffman (ed.), *Statistical Methods in Radio Wave Propagation*, Oxford, England: Pergamon.

Patel, C., Stuber, G. & Pratt, T. (2005). Simulation of Rayleigh-faded mobile-to-mobile communication channels, *IEEE Trans. on Comm.* 53(11): 1876–1884.

Pätzold, M. (2002). *Mobile Fading Channels*, John Wiley & Sons, New York.

Pätzold, M., Killat, U., Laue, F. & Li, Y. (1998). On the statistical properties of deterministic simulation models for mobile fading channels, *IEEE Trans. on Veh. Tech.* 47(1): 254–269.

Pedersen, K., Mogensen, P. & Fleury, B. (1998). Spatial channel characteristics in outdoor environments and their impact on BS antenna system performance, *IEEE 48th Vehicular Technology Conference, 1998. VTC 98*, Vol. 2, pp. 719–723.

Petrolino, A., Gomes, J. & Tavares, G. (2008a). A fading channel simulator based on a modified Karhunen-Loève expansion, *Radio and Wireless Symposium, 2008 IEEE*, Orlando (FL), USA, pp. 101 –104.

Petrolino, A., Gomes, J. & Tavares, G. (2008b). A mobile-to-mobile fading channel simulator based on an orthogonal expansion, *IEEE 67th Vehicular Technology Conference: VTC2008-Spring*, Marina Bay, Singapore, pp. 366 –370.

Petrolino, A. & Tavares, G. (2010a). Inducing spatial correlation on MIMO channels: a distribution-free efficient technique, *IEEE 71st Vehicular Technology Conference: VTC2010-Spring*, Taipei, Taiwan, pp. 1–5.

Petrolino, A. & Tavares, G. (2010b). A simple method for the simulation of autocorrelated and arbitrarily distributed envelope fading processes, *IEEE International Workshop on Signal Processing Advances in Wireless Communications.*, Marrakech, Morocco.

Pop, M. & Beaulieu, N. (2001). Limitations of sum-of-sinusoids fading channel simulators, *IEEE Trans. on Comm.* 49(4): 699–708.

Sagias, N., Zogas, D., Karagiannidis, G. & Tombras, G. (2004). Channel capacity and second-order statistics in Weibull fading, *IEEE Comm. Lett.* 8(6): 377–379.

Shepherd, N. (1977). Radio wave loss deviation and shadow loss at 900 MHz, *IEEE Trans. on Veh. Tech.* 26(4): 309–313.

Smith, J. I. (1975). A computer generated multipath fading simulation for mobile radio, *IEEE Trans. on Veh. Tech.* 24(3): 39–40.

Tierney, L. (1994). Markov chains for exploring posterior distributions, *Annals of Statistics* 22(4): 1701–1728.

Van Trees, H. L. (1968). *Detection, Estimation, and Modulation Theory - Part I*, Wiley, New York.

Vehicular Technology Society Committee on Radio Propagation (1988). Coverage prediction for mobile radio systems operating in the 800/900 MHz frequency range, *IEEE Trans. on Veh. Tech.* 37(1): 3–72.

Wang, C. (2006). Modeling MIMO fading channels: Background, comparison and some progress, *Communications, Circuits and Systems Proceedings, 2006 International Conference on*, Vol. 2, pp. 664–669.

Weibull, W. (1951). A statistical distribution function of wide applicability, *ASME Jou. of Applied Mechanics, Trans. of the American Society of Mechanical Engineers* 18(3): 293–297.

Yacoub, M., Fraidenraich, G. & Santos Filho, J. (2005). Nakagami-*m* phase-envelope joint distribution, *Electr. Lett.* 41(5): 259–261.

Young, D. & Beaulieu, N. (2000). The generation of correlated Rayleigh random variates by inverse discrete Fourier transform, *IEEE Trans. on Comm.* 48(7): 1114–1127.

3

User Scheduling and Partner Selection for Multiplexing-based Distributed MIMO Uplink Transmission

Ping-Heng Kuo and Pang-An Ting
Information and Communication Laboratories,
Industrial Technology Research Institute (ITRI), Hsinchu
Taiwan

1. Introduction

During the last decade, multiple-input multiple-output (MIMO) systems, where both the transmitter and receiver are equipped with multiple antennas, have been verified to be a very promising technique to break the throughput bottleneck in future wireless communication networks. Based on countless results of analysis and field measurements that have been undertaken by many researchers around the world, it is indubitable that MIMO transceiver schemes can significantly improve the system performance. Some well-known standards for next-generation wireless broadband, including 3GPP Long Term Evolution (LTE) and IEEE 802.16 (WiMAX), adopt MIMO as a key feature of the physical layer. For instance, the standard of IEEE 802.16e supports three possible options of *advanced antenna systems* (AAS), namely transmit diversity (TD), beamforming (BF), and spatial multiplexing (SM). The extensions of these multi-antenna techniques are also envisaged in future standards such as IEEE 802.16m and LTE-A.

The potential benefits of MIMO systems are mainly attributed to the multiplexing/diversity gains provided by multiple antenna elements. With spatial multiplexing, multiple independent data streams are transmitted in a single time/frequency resource allocation, so spectral efficiency is thereby increased. In practice, it may be difficult, if not impossible, to equip multiple antenna elements at mobile stations (MS) due to their small physical sizes. In such cases, concurrent transmission of multiple data streams is not feasible, as the maximum number of data streams is limited by $m = min(M_t, M_r)$, where M_t and M_r are the number of antennas at the transmitter and receiver, respectively. Furthermore, even if multiple antenna elements can be installed on a small mobile device, multiple closely-packed antennas may result in high spatial fading correlation, which theoretically leads to a reduced-rank channel and therefore degrades the performance of spatial multiplexing [Shiu et al., 2000].

In correspondence, the concept of *distributed MIMO communications* has emerged, which has received considerable attention from both academia and industry in recent years. By treating multiple distributed nodes as a single entity, each node can emulate a portion of a virtual antenna array, and the advantages of MIMO techniques can therefore be exploited with appropriate protocols. A common example of distributed MIMO is collaborative spatial multiplexing (CSM), which enhances system capacity by allowing multiple separate users within a cell to send their uplink signals in the same time-frequency resource unit [Balachandran et al., 2009]. In this case, each user is virtually tantamount to a portion of the transmit antenna array of a point-to-point SM-based MIMO system. On the other hand, some other distributed MIMO topologies may involve the operations of data relaying; for instance, in cooperative relaying, each of the nodes may need to receive and then forward the messages on behalf of other users. The baseline configuration of a cooperative scheme encompasses three main roles, namely source (S), relay (R), and destination (D). Furthermore, the conventional half-duplex (all nodes are unable to transmit and receive simultaneously) cooperative transmission procedure is consisted of two phases. In the first phase (Phase 1), the relay obtains messages from the source. In the second phase (Phase 2), the relay forwards the data it has received during Phase 1 to the destination to complete the transmission.

This chapter considers user scheduling and partner selection problems for CSM and cooperative relaying spatial multiplexing (CRSM) respectively. For CSM-based uplink schemes, we assume it allows $N_u = 2$ users to transmit data simultaneously, so the base station (BS) needs to select two out of U users in each signalling interval. If the BS possesses the channel state information (CSI) of all users, it can perform *channel-dependent scheduling* by assessing certain quality metrics of the prevailing channel condition. Since CSM creates a virtual MIMO channel in the uplink, some intrinsic MIMO channel metrics may be employed for scheduling. One of the conventional scheduling algorithms is to choose the user pair with the highest capacity in order to maximize the system throughput [Wang et al., 2008]. This, however, may not be the best solution for CSM, because the resultant MIMO channel structure may not be *spatial multiplexing-preferred* in terms of the error performance even if it maximizes the channel capacity. Additionally, a scheduler that always selects the user pair with the highest capacity may not be able to give an acceptable fairness performance, especially in low mobility scenarios (such as indoor applications) where the channels of most users do not change rapidly.

Not many papers have appeared to examine scheduling strategies for CSM in spite of their importance. The authors of [Lee & Lee, 2008] have developed a scheduler for CSM based on antenna correlations at the BS. In this chapter, the problem is studied from a different perspective. We reasonably presume that the antenna spacings at the BS are large enough so the spatial correlation is negligible. An objective of this work is to scrutinize different scheduling modes and selection metrics for CSM, in terms of their search complexity, error performance, and fairness. Recently, we became aware that [Wang et al., 2008] has the similar goal. Nonetheless, [Wang et al., 2008] only considers *Semi Round Robin* user pairing mode (which will also be discussed subsequently in this chapter), while our study further examines some other user pairing modes of MIMO channel metric-dependent scheduling for CSM.

Most of the early developments in the context of cooperative relaying communications are aimed to provide higher spatial diversity gain [Laneman et al., 2004]. Due to the fact that the destination node in an uplink scenario (the base station) usually possesses multiple antennas, the extensions to cooperative spatial multiplexing have also appeared in the literature to achieve higher spectral efficiency [Kim & Cherukuri, 2005]. In CRSM, the source node first shares a portion of the data with the relay node in Phase 1, and then distinct data portions are jointly transmitted by the source and the relays to the destination in Phase 2. In general, the number of available spatial links in the MIMO channel is limited by the number of antennas at transmitter. Analogously, it refers to the number of relays that cooperates with the source in CRSM. When the user population is dense, the number of potential relay candidates is likely to exceed the number of relays required for cooperation. Additionally, cooperation with too many relays may jeopardize the availability of system resources. Hence, a recruitment process is needed to select active relays from all potential candidates. Apparently, how the active relays are chosen would affect the overall system performance. The issues of relay selection have been studied extensively (for examples, [Norsratinia & Hunter, 2007] and references therein) for cooperative diversity schemes, and the relevant research on CRSM is relatively sparse in existing literature.

From the discussions above, it is apparent that both user-scheduling for CSM and partner recruitment for CRSM need a MIMO channel-related selection metric. Thus, the main objective of this chapter is to develop a selection metric that can be applied to both CSM and CRSM. To be specific, the proposed metric is developed based on the condition number of the virtual MIMO channel, the magnitude of which could be employed to determine whether the MIMO channel is suitable for spatial multiplexing operation.

This chapter is organized as following. The channel model of a CSM system is elaborated in the next section. We review, as well as propose, several different user pairing modes and selection metrics in Section 3 and 4. Then, in Section 5, the proposed user selection criterion is further applied to design a partner recruitment algorithm for CRSM. The simulation-based investigation results are presented in Section 6, and a concise conclusion is drawn in Section 7.

2. System model for collaborative spatial multiplexing

This chapter firstly consider a CSM uplink transmission scenario, where $M_r \geq 2$ antennas are installed at the BS for reception, while each of the U users is equipped with $M_t = 1$ antenna. For sake of convenience, we postulate the number of users that intend to transmit with CSM, U, is always an even number. In each signalling interval, $N_u = 2$ out of U users are scheduled to transmit data simultaneously in the same time-frequency resource unit via CSM, as illustrated in Fig. 1. The channel vector of the uth user can be written as:

$$\mathbf{h}_u = \begin{bmatrix} h_{1,u} \\ h_{2,u} \\ \vdots \\ h_{M_r,u} \end{bmatrix}, \tag{1}$$

where $h_{i,u}$ is the fading coefficient between the ith antenna of the BS and the uth user. All fading coefficients are modelled as complex Gaussian random variables with zero-mean (Rayleigh fading) and unit variance. We assume that the BS has acquired the full channel state information (CSI) of every user through pilot signals. As mentioned previously, it is reasonable to assume that the distances among antennas at the BS are sufficiently large, so all entries in the channel vector (1) are independent random variables. Note that the channel vectors of any two users are also independent as they are spatially dispersed. Thus, the overall channel for CSM formed by a pair of users is virtually equivalent to a $M_r \times 2$ MIMO system matrix with independent, identical distributed (i.i.d) entries:

$$\mathbf{H} = \begin{bmatrix} \mathbf{h}_u & \mathbf{h}_v \end{bmatrix} = \begin{bmatrix} h_{1,u} & h_{1,v} \\ h_{2,u} & h_{2,v} \\ \vdots & \vdots \\ h_{M_r,u} & h_{M_r,v} \end{bmatrix}, \tag{2}$$

where are u and v are user indices. Moreover, due to user mobility, each column of $\mathbf{H}$ (the channel vector of each user) evolves with time in accordance with some time-correlation function $\rho(\tau)$. Hence,

$$\mathbf{h}(t+\tau) = \rho\mathbf{h}(\tau) + \sqrt{1-\rho^2}\mathbf{a}, \tag{3}$$

where τ represents time displacement, and $\mathbf{a}$ is a random Gaussian vector with identical dimension and statistics as $\mathbf{h}$. The value of ρ depends on τ and the Doppler frequency, f_D, of the associating user. The impacts of f_D on scheduling performance will be discussed in subsequent sections.

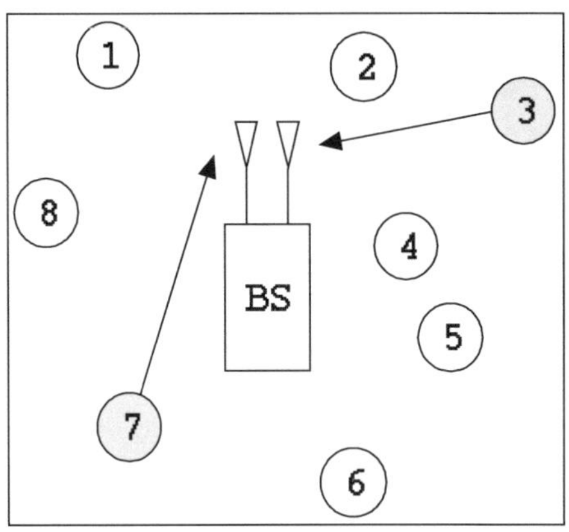

Fig. 1. The illustration of a collaborative spatial multiplexing (CSM) system with $M_t = 1$, $M_r = 2$ and $U = 8$. In this example, both user 3 and user 7 are scheduled for uplink transmission in the same resource unit, thereby creates a 2×2 virtual MIMO channel

3. User pairing modes for CSM

It is assumed that the BS schedules a pair of users in each of the signalling intervals for CSM transmission. Here we define three different modes of user pairing, namely *Pre-Defined Pairing, Instantaneous Pairing*, and *Semi Round Robin*. Each of them renders different search complexity. Note that the term *search complexity* (denoted as W) in this chapter is defined as the number of possible paring choices that the BS has to examine before making the final decision. In other words, W represents the number of required iterations.

3.1 Mode 1: pre-defined pairing

In the initial stage of this mode, the BS arbitrarily divides all users into multiple pairs. That is, the BS first define a list of $U/2$ pairs in a random manner. Note that each user pair forms a virtual MIMO channel, and is therefore associated with a certain MIMO channel metric. Then, in each of the subsequent signaling intervals, the BS chooses a pair from this pre-defined list based on the channel metric. When the number of users in the cell changes (the occurrence of handover), the BS initiates the pairing again to renew the candidate list. In fact, since the BS has full channel knowledge, the users could be paired base on their channel status (such as user channel orthogonality). However, the pair list may need to be updated more frequently due to channel variation in time, which potentially increases system complexity. Hence, we focus on random pre-defined pairing schemes in this chapter. In each of the signaling interval, the BS has to search over $U/2$ pair candidates and schedule one of these pairs for transmission. Thus, the search complexity of mode 1, W_1, is

$$W_1 = \frac{U}{2} \tag{4}$$

Another advantage of using this mode is that each pair can be regarded as a single entity with two antennas. Thus, many well-known single-user scheduling strategies for conventional multiple access scenarios, such as proportional fair algorithms, can also be applied. Nonetheless, this is beyond the scope of this chapter.

3.2 Mode 2: instantaneous pairing

In this mode, exhaustive search is performed in every signaling interval to find the best user pair. Thus, the BS has to inspect the associated MIMO channel metrics of all possible user pairs. Intuitively, the system performance can be optimized with this pairing mode. The search complexity this pairing mode can be expressed as:

$$W_2 = \frac{U!}{(U-2)!2!} \tag{5}$$

3.3 Mode 3: semi round robin

When the user mobility levels are relatively low, the users with bad channels may be starved for a long time, and the overall fairness performance of the scheduler is therefore degraded. As there are two user vacancies for CSM scheduling, the BS may simply reserve the first vacancy for scheduling in a *round robin* manner. That is, the users occupy the first vacancy in a signaling interval by taking turns, without considering its channel status. The second vacancy, on the other hand, is given to one of the remaining users that can realize the most appropriate MIMO channel $\mathbf{H}$ with the first reserved user. Thus, this mode attempts to strike a balance between scheduling fairness and system performance. Since a user is reserved, the BS only has to find one of the remaining candidates to pair up with the first user, the search complexity is therefore:

$$W_3 = U - 1 \tag{6}$$

By comparing these three modes, it is apparent that mode 2 should offer the optimum resultant system performance. However, mode 2 may be prohibited in practice due to its high computational complexity, especially when U is large. The search complexities for these three modes are compared as the functions of U in Fig. 2.

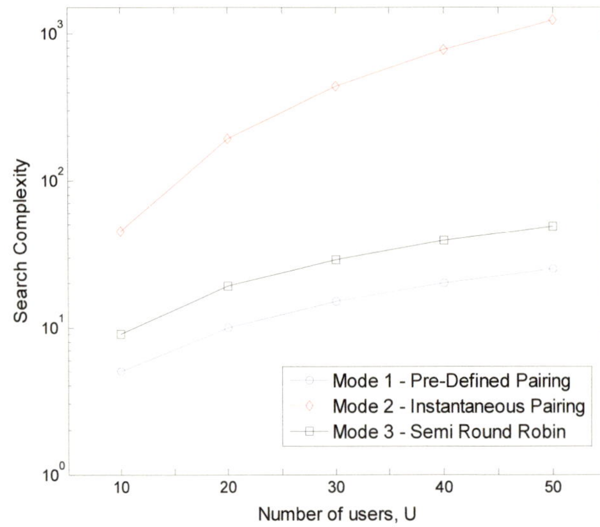

Fig. 2. Search complexities of three user pairing modes with different number of users in the system

4. Selection metrics

All pairing modes described in the preceding section schedule users in accordance to a certain MIMO channel metric. In this section, we elaborate some possible selection criteria that could be coupled with these pairing modes, including a newly proposed metric based on MIMO channel condition number.

4.1 Random Selection (RS)

This is the simplest selection method. The BS can choose the user pair without taking instantaneous channel condition into account. In pairing mode 1, the BS randomly selects a pair of users from the pre-defined list. In mode 2, on the other hand, two users are arbitrarily chosen and paired instantaneously. Finally, in pairing mode 3, the first scheduled user is scheduled in round robin fashion, while another user is randomly picked up by the BS.

4.2 Maximum Capacity (MC)

An intuitive way of scheduling is to select the pair of users that can realize the highest capacity. In CSM, a virtual MIMO channel is formed by cascading the channel vectors of two users. The BS can calculate the corresponding MIMO capacity by the following well-known formula:

$$C = \sum_{k=1}^{2} \log_2(1 + P\lambda_k) \tag{7}$$

where P is signal to noise ratio (SNR) on each of the MIMO spatial links, and λ_k is the kth eigenvalue of the channel correlation matrix $\mathbf{HH}^{\dagger}$. Note that we denote eigenvalues in descent order: $\lambda_1 > \lambda_2$. In each signaling interval, the BS finds the user pair that has the maximum channel capacity among all the available choices. That is, the pair index is chosen based on

$$j^* = \arg\max_j C_j, j = 1 \ldots W, \quad (8)$$

where C_j is the MIMO channel capacity realized by the jth choice among W options.

4.3 The proposed metric: Maximum Capacity with Spatial Multiplexing Preferred Channel (MC-SMPC)

In this chapter, we wish to propose a new selection metric that can improve the reception error rates. Although selecting the users based on the channel capacity can optimize the spectral efficiency, the resultant MIMO link may not be a spatial multiplexing-preferred channel, and the receiver is therefore more likely to decode the message erroneously. Additionally, the channel capacity is merely a theoretical bound and cannot be directly translated to the system throughput. In [Heath & Love, 2005], the authors have derived a method to determine whether a MIMO system with linear receiver is more suitable for multiplexing or diversity. In particular, it can be shown that a $M_r \times 2$ MIMO channel is spatial multiplexing-preferred if its condition number, $\kappa = \lambda_1 / \lambda_2$, satisfies the following criterion:

$$\kappa \le \gamma = \frac{2^B - 1}{(2^{B/2} - 1)2}, \quad (9)$$

where B is the total number of bits that will be transmitted in one signaling interval. If both scheduled users in CSM send the data with identical modulation formats in the uplink transmission, B is simply the double of the bit number from the individual user. For examples, both users in the scheduled pair send their data with QPSK and 16-QAM when $B = 8$ and 4 respectively. Note that the condition number characterizes the spatial selectivity of a MIMO channel, and spatial multiplexing schemes are more appropriate for MIMO channels with spatial streams that are closer in magnitude. Thus, we may utilize the criterion of (9) to identify the user pairs that are capable to realize a spatial multiplexing-preferred MIMO channel. Then, among all these identified pairs, the BS chooses the pair with the highest capacity for scheduling. In short, we propose a novel metric:

$$\Phi_j = \max\left[0, C_j\left(\frac{\gamma - \kappa_j}{|\gamma - \kappa_j|}\right)\right], j = 1 \ldots W \quad (10)$$

and the pair of user is scheduled by choosing the option index:

$$j^* = \arg\max_j \Phi_j, \quad (11)$$

If none of W options has a SM-preferred channel, i.e. all $\kappa_j > \gamma$ and hence all $\Phi_j = 0$, the BS simply choose the pair based on (8). This leads to an adaptive scheduling scheme which switches its selection metric between (11) and (8) depending on if all Φ_j are zero. Although the mean channel capacity of this method may be lower than the one in Section 4.2, but it should provide better error performance, assuming that linear receiver is used at the BS. Along with the advantages in error rates, using this novel metric may also improve the system fairness. In low mobility scenarios, the channels of the users do not change significantly over time. Hence, scheduling based on capacity may "starve" the users who have low quality channels for a long period of time. Nevertheless, the newly proposed selection metric takes the channel condition number into account, which reduces the chance of consecutively scheduling specific users. It can be explained by using the nature of condition number and capacity in terms of auto-correlation functions (ACF). As shown in Fig. 3, the ACF for $\kappa(t)$ decays faster than the one for $C(t)$, which means the variation of κ is more sensitive to channel fluctuations. Hence, in low mobility cases, scheduling based on the proposed metric can provide a better fairness performance, as the BS is less likely to schedule specific users for too long, due to the faster variation of the condition number.

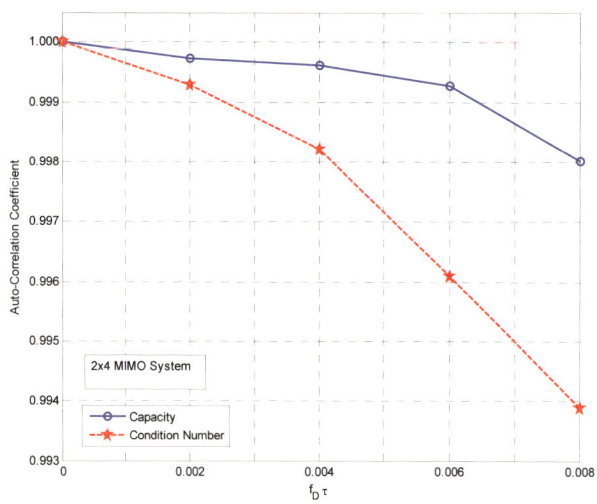

Fig. 3. Autocorrelation functions for MIMO channel capacity and condition number

5. Cooperative Relaying Spatial Multiplexing (CRSM)

In the last section, a novel metric for CSM user scheduling has been proposed. Here we further apply this metric to develop a relay recruitment algorithm for CRSM. In particular, the single-relay CRSM topology delineated in [Kim et al., 2007] is considered. At the beginning, the source node evenly splits the whole data stream, $\mathbf{x}$, into two segments (x_1 and x_2). Then, one of its N neighbouring nodes is chosen to play the role of relay. Phase 1 transmission is commenced once the active relay is chosen. During Phase 1, the source shares the data segment, x_2, with the relay. In Phase 2, the relay forwards the data segment that it has received and decoded in Phase 1 ($\hat{x}_2$) to the destination, and the source also concurrently participates the transmission by sending x_1 to the destination. The topology of this CRSM scheme is illustrated in Fig. 4.

For CRSM schemes with decode-and-forward protocols, it is essential for the relay to correctly decode the messages from the source during Phase 1. Otherwise Phase 2 transmission would become pointless as the data forward by the relay is erroneous at the first place. This implies that the link strength between the source and the relay must be strong enough to ensure reliable transmission in Phase 1. Thus, it is more appropriate for CRSM to operate in a clustered system [Yuksel & Erkip, 2007], in which multiple closely-placed user nodes establish a cluster. In the cluster, inter-user links can be modelled as simple AWGN channels [Ng et al., 2007], and the information exchange between the source and the relays is certainly easier. In practice, a clustered system could be found in indoor applications, where the user nodes are relatively closer to each other. Also, since mobility levels of users in indoor wireless networks are generally low, cooperative schemes (many of which require quasi-static channels) are more feasible.

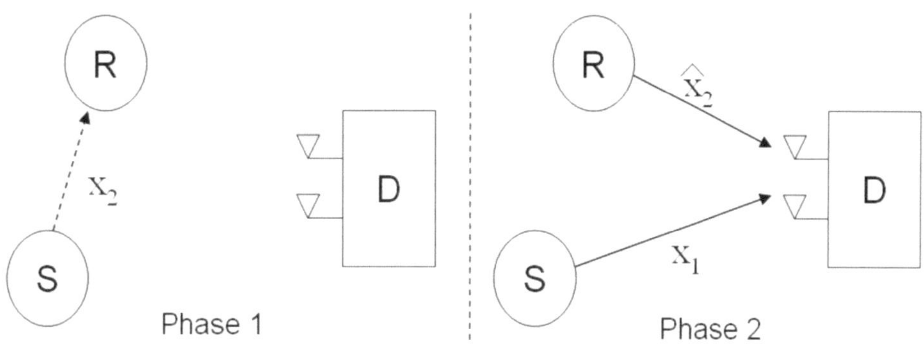

Fig. 4. The illustration of a decode-and-forward CRSM scheme: in Phase 1, the source node (S) shares a portion of data with the relay node (R). Both S and R concurrently send different portions of data in Phase 2 transmission

5.1 Channel model and assumptions

A single-cell, single-source uplink scenario is considered. The base station (destination) has $M_r > 1$ antennas, and each of the user nodes (including the source and all potential relays) is equipped with single antenna. The nodes are assumed to lie in either *AWGN Zone* or *Rayleigh Zone* of the source depending on their locations [Yuksel & Erkip, 2007]. The link from the source to a node is modelled as an AWGN channel with a scalar gain if the distance between these two terminals is smaller than the threshold q. Otherwise, a Rayleigh fading channel is used to model this inter-node link [Ng et al., 2007]. In this chapter, we assume that the source has established a cluster with some other user nodes in proximity based on certain protocols and techniques. It is further assumed that the user cluster and the destination are separated by a distance larger than q. Thus, while all intra-cluster links are AWGN, the link between the destination and every node in the cluster is modelled as a Rayleigh flat-fading channel. The model is depicted in Fig. 5, where we have termed the cluster as *S-R Cluster* since only the nodes within the cluster are considered as relay candidates R_C.

Virtually, during Phase 2, each of the user nodes (including the relay and the source itself) emulates a transmit antenna of a spatial multiplexing system. Hence, the effective virtual

MIMO channel matrix realized in Phase 2 is formed by cascading channel vectors of the source and relay:

$$\mathbf{H}_{eff} = [\mathbf{h}_S \quad \mathbf{h}_R] = \begin{bmatrix} h_{1,S} & h_{1,R} \\ h_{2,S} & h_{2,R} \\ \vdots & \vdots \\ h_{M_r,S} & h_{M_r,R} \end{bmatrix}, \tag{12}$$

where $\mathbf{h}_S$ and $\mathbf{h}_R$ represent the channel vector from the source and the relay to the destination (similar to (1)), respectively. Similarly, the entries of the fading channel vectors are modelled as a complex Gaussian with zero-mean and unit variance in this work. Furthermore, we assume that the source has full knowledge regarding the quality (such as channel state information and battery power) of the other nodes within the user cluster.

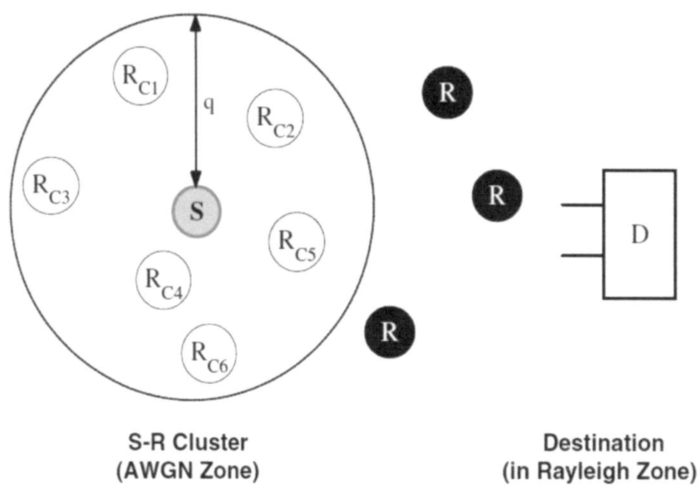

Fig. 5. A source establishes a cluster with some other user nodes that are within its AWGN Zone, and the destination is located in its Rayleigh Zone. The user nodes outside the cluster are not considered as candidates of relays

6. Relay recruitment algorithm for CRSM

6.1 Initial elimination

Once the source has obtained a list of N potential candidates in the cluster, it should first identify unqualified candidates and remove them from the list. As aforementioned, if errors occur in Phase 1 transmission, Phase 2 becomes pointless since the data forwards by the relay is erroneous in the first place. To make sure that the relays can decode the message correctly in Phase 1, the nodes that have weak intra-cluster links to the source should be discarded, as their link gains to the source are not strong enough to support reliable Phase 1 transmission. Additionally, the nodes that have insufficient battery power to participate the cooperation should notify the source and quit from the list. After withdrawing u nodes in this initial elimination, the number of remaining nodes on the candidate list is $Q = N - u$.

The next task is to find a suitable $\mathbf{h}_R$ by selecting a relay out of the remaining Q candidates. In the special cases where $N=0$ (no other nodes are found in the cluster) or $N=u$ (all candidates in the cluster are unqualified), the source transmits the data to the destination directly without cooperation, and simple combining techniques (such as maximum ratio combining) can be leveraged at the receiver.

6.2 Decision metric

Intuitively, in order to maximize the system throughput, the relay could be chosen based on the maximum capacity criterion described in Section 4.2. However, as aforementioned, although the channel capacity can be optimized by choosing the relays based on the maximum value of (7), the characteristic of the resultant effective channel may not be appropriate for spatial multiplexing. To be specific, the error performance of a spatial multiplexing system is degraded if the channel matrix is *ill-conditioned*. As mentioned in Section 4.3, the condition number could be employed to judge if a MIMO channel is multiplexing-preferred. In particular, if the condition number is smaller than the threshold level given in (9), then the MIMO channel is more suitable for the operation of spatial multiplexing. Thus, the criterion proposed in Section 4.3 can be applied as the decision metric for relay selection of CRSM:

$$j^* = \arg\max_j \Phi_j, j = 1 \ldots Q \tag{13}$$

where

$$\Phi_j = \max\left[0, C_j\left(\frac{\gamma - \kappa_j}{|\gamma - \kappa_j|}\right)\right], j = 1 \ldots Q \tag{14}$$

The definitions of C, κ and γ are available in Section 4. In circumstances when the values of $\Phi_1, \ldots \Phi_N$ are all zero (none of the relay candidates can realize a spatial multiplexing-preferred effective channel), the source computes the values of Φ_j for (13) again with a different formula:

$$\Phi_j == C_j, j = 1 \ldots Q \tag{15}$$

This is simply tantamount to directly choosing a relay that maximizes the capacity.

6.3 Procedural summary

To recapitulate, the steps of the proposed recruitment algorithm is summarized as following:

1. Withdraw u unqualified candidates in the cluster from consideration. If $N=0$ or $N=u$ ($Q=0$), the source transmits the data to the destination directly without cooperation, and the algorithm terminates here. Otherwise, proceed to step 2.
2. Find the index of the relay $j^* = \arg\max_j \Phi_j$, where Φ_j is computed using (14). The algorithm finishes here if $\Phi_{j^*} > 0$. Otherwise, proceed to step 3.
3. If $\Phi_{j^*} = 0$, repeat step 2 with the computation of Φ_j using (15) in lieu of (14).

7. Simulation results

In order to illustrate the performance of the proposed selection metric in both CSM and CRSM schemes, several simulations have been carried out. In particular, the performance of the proposal is evaluated via observations on the resultant error rates. For all simulations, we assume a zero-forcing (ZF) MIMO detector is employed at the receiver (BS). Furthermore, in all cases, we assume that B = 8 or 4 bits are transmitted in one signaling interval, so each of the two nodes involved in cooperation sends 16-QAM (4 bits per node) or QPSK (2 bits per node) message in the multiplexing phase. We first examine the performance of CSM. By setting $U = 40$, vector symbol error rates (VSER) of CSM schemes with B = 8 and 4 under different pairing modes and selection metrics are shown in Fig. 6 and 7 respectively. Apparently, the proposed selection criterion (MC-SMPC) can achieve the best error performance regardless the pairing mode. As mentioned previously, the condition number-based metric also has the advantage of scheduling fairness as compared to maximum capacity, as the value of condition number changes more rapidly in time-varying channels. To show this, fairness performance of different pairing modes and selection metrics are shown in Fig. 8, and the fairness is measured by the Coefficient of Variance (CoV), which is defined as the ratio of standard deviation to mean number of times that each user is scheduled, so a lower CoV generally indicates better fairness. Remarkably, as shown in Fig. 8, user selections based on MC-SMPC are fairer than the ones based on MC in all three pairing modes, regardless the total number of users.

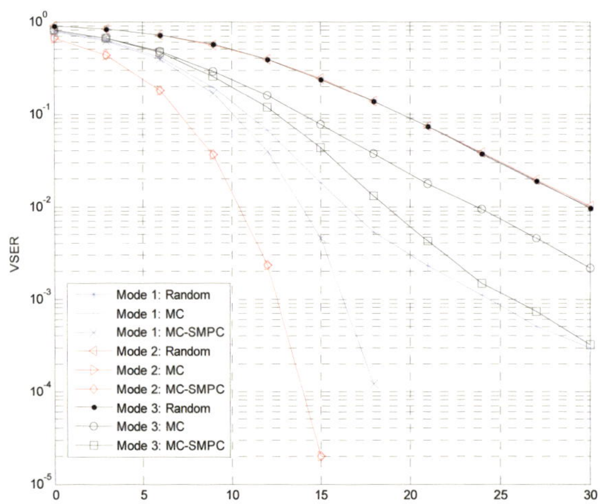

Fig. 6. The VSER performance of different scheduling strategies for CSM schemes with B = 8 and U = 40. Generally MC-SPMC gives better performance in most cases

Similarly, for CRSM, we compare VSER of three different relay selection criteria, including random selection, maximum capacity, and the proposed method based on condition number. In both Fig. 9 and 10, it is apparent that our proposed selection method can give the best error performance, since it first filters out the candidates that are not able to realize a spatial multiplexing-preferred MIMO channel. The gain of the proposed metric over the other selection criteria is more obvious in high SNR regime, Note that, due to *multi-candidate diversity*, we expect the performance to improve with the number of relay candidates. As the number of candidates increases, it's more likely for the algorithm to find a more appropriate node to play the role of relay. As compared to the selection based on maximum capacity, the achievable capacity of the proposed method is only slightly lower. In order to inspect the loss in capacity, we further compared the average capacity of these relay selection metrics under different SNR in Fig. 11. Additionally, the cumulative distribution functions (CDFs) of the capacity are also compared in Fig. 12.

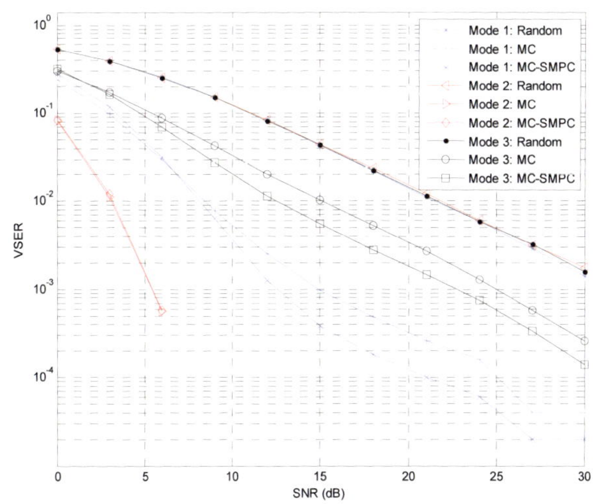

Fig. 7. The VSER performance of different scheduling strategies for CSM schemes with $B = 4$ and $U = 40$. Generally MC-SPMC gives better performance in most cases

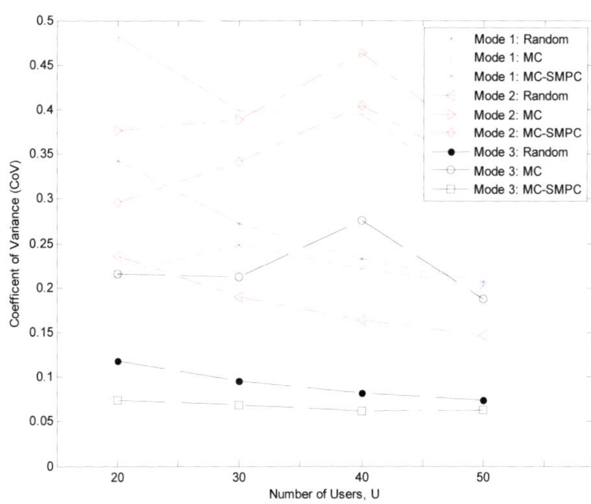

Fig. 8. The comparison of fairness of different user scheduling strategies for CSM schemes

8. Conclusions

This chapter mainly discussed two distributed MIMO uplink transmission schemes, including CSM and CRSM.

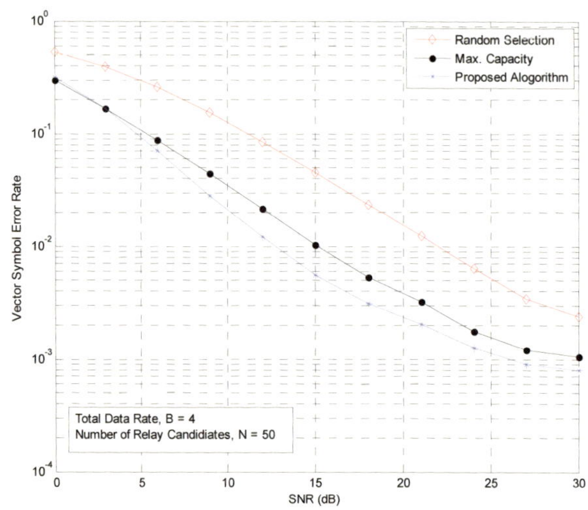

Fig. 9. The VSER performance of different relay selection strategies for CRSM schemes with $B = 4$ and $N = 50$

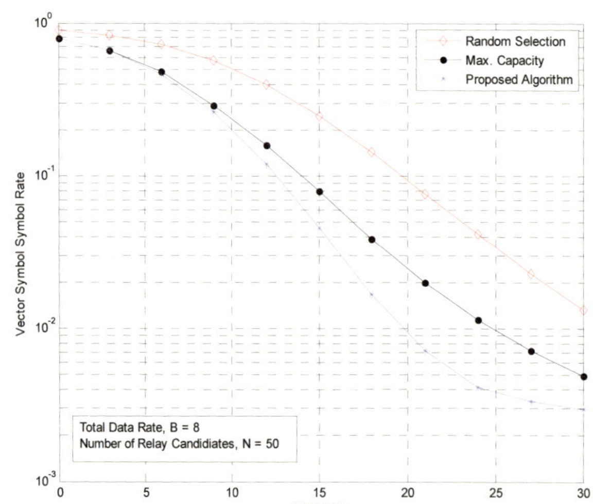

Fig. 10. The VSER performance of different relay selection strategies for CRSM schemes with $B = 8$ and $N = 50$

Both of these schemes allow multiple user nodes to form a virtual antenna array, thereby increases the system throughput via spatial multiplexing. In CSM, the BS is required to schedule a pair of users to perform concurrent transmission. In CRSM, on the other hand, the source node needs to select a neighbour user as the relay to cooperate with. This chapter proposed a novel selection criterion based on MIMO channel condition number, which aims to provide a spatial multiplexing-preferred virtual MIMO channel for collaborative uplink transmission. In accordance to computer simulations, it is clear that the proposed selection criterion outperforms some other metrics in terms of the error performance, regardless it is being invoked for user scheduling in CSM or relay selection in CRSM.

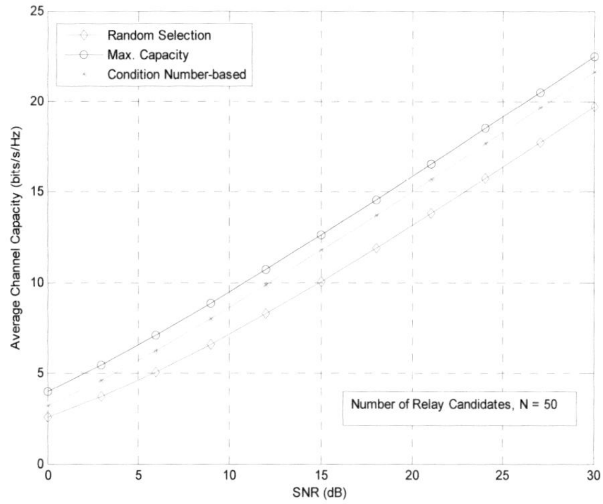

Fig. 11. The comparison of mean channel capacity for different relay selection algorithms with $N = 50$

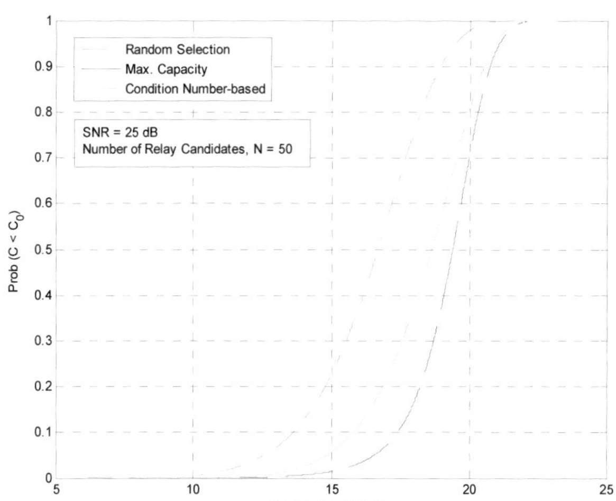

Fig. 12. The comparison of capacity cumulative density functions (cdf) for different relay selection algorithms with a SNR of 25dB and $N = 50$

9. References

Shiu, D.; Foschini, G.J.; Gans, M.J. & Kahn, J.M. (2000). Fading correlation and its effect on the capacity of multielement antenna systems, *IEEE Trans. on Commun.*, Vol. 48, No. 3, pp. 502-513, Mar. 2000.

Laneman, J. N.; Tse, D. & Wornell, G. W. (2004). Cooperative diversity in wireless networks: efficient protocols and outage behavior, *IEEE Trans. Inform. Theory*, Vol. 50, pp. 3062–3080, Dec. 2004.

Kim, S. W. & Cherukuri, R. (2005). Cooperative spatial multiplexing for highrate wireless communications, *Proceeding of IEEE SPWAC*, pp. 181–185, New York City, USA, June 2005.

Heath, R. W. & Love, D. J. (2005). Multimode antenna selection for spatial multiplexing systems with linear receivers, *IEEE Trans. Signal Processing*, Vol. 53, pp. 3042–3056, Aug. 2005.

Norsratinia, A. & Hunter, T. E. (2007). Grouping and partner selection in cooperative wireless networks, *IEEE J. Select. Areas Commun.*, vol. 25, pp. 369–378, Feb. 2007.

Yuksel, M. & Erkip, E. (2007). Multiple-antenna cooperative wireless systems: a diversity-multiplexing tradeoff perspective, *IEEE Trans. Inform. Theory*, Vol. 53, pp. 3371–3393, Oct. 2007.

Ng, C.; Jindal, N.; Goldsmith, A. J. & Mitra, U. (2007). Capacity gain from two-transmitter and two-receiver cooperation, *IEEE Trans. Inform. Theory*, Vol. 53, pp. 3822–3827, Oct. 2007.

Lee, J. & Lee, Y. (2008). Receive correlation-based user scheduling for collaborative spatial multiplexing systems, *Proceeding of IEEE Int'l. Conf. on Commun.* (ICC), pp. 3447–3451, Beijing, China, May 2008.

Wang, X.; Wang, W.; Zhu, T. & Zhao, Z. (2008). Spatial multiuser pairing scheduling strategies for virtual MIMO systems, *Proceeding of Int'l. Conf. on Commun. Systems* (ICCS), pp. 823–827, Guangzhou, China, Nov. 2008.

Balachandran, K.; Calin, D.; Gopalakrishnan, N.; Kang, J.; Kogiantis, A.; Li, S.; Ozarow, L.; Ramakrishna, S.; Rudrapatna, A. & Sun, L. (2009). Design and performance analysis of collaborative spatial multiplexing for IEEE 802.16e-based systems, *Bell Labs Technical Journal*, Vol. 13, Issue 4, pp. 97-117, Feb. 2009.

4

Resource Allocation for Multi-User OFDMA-Based Wireless Cellular Networks

Dimitri Kténas and Emilio Calvanese Strinati
CEA, LETI, MINATEC, F-38054 Grenoble
France

1. Introduction

Modern wideband communication systems present a very challenging multi-user communication problem: many users in the same geographic area will require high on-demand data rates in a finite bandwidth with a variety of heterogeneous services such as voice (VoIP), video, gaming, web browsing and others. Emerging broadband wireless systems such as WiMAX and 3GPP/LTE employ Orthogonal Frequency Division Multiple Access (OFDMA) as the basic multiple access scheme. Indeed, OFDMA is a flexible multiple access technique that can accommodate many users with widely varying applications, data rates, and Quality of Service (QoS) requirements. Because the multiple access is performed in the digital domain (before the IFFT operation), dynamic and efficient bandwidth allocation is possible. Therefore, this additional scheduling flexibility helps to best serve the user population. Diversity is a key source of performance gain in OFDMA systems. In particular, OFDMA exploits multiuser diversity amongst the different users, frequency diversity across the sub-carriers, and time diversity by allowing latency. One important observation is that these sources of diversity will generally compete with each other. Therefore, efficient and robust allocation of resources among multiple heterogeneous data users sharing the same resources over a wireless channel is a challenging problem to solve.
The scientific content of this chapter is based on some innovative results presented recently in two conference papers (Calvanese Strinati et al., VTC 2009)(Calvanese Strinati et al., WCNC 2009).
The goals of this chapter are for the reader to have a basic understanding of resource allocation problem in OFDMA-based systems and, to have an in-depth insight of the state-of-the-art research on that subject. Eventually, the chapter will present what we have done to improve the performance of currently proposed resource allocation algorithms, comparing performance of our approaches with state-of-the-art ones. A critical discussion on advantages and weaknesses of the proposed approaches, including future research axes, will conclude the chapter.

2. Basic principles of resource allocation for OFDMA-based wireless cellular networks

The core topic investigated in this chapter is the performance improvement of Resource Allocation for Multi-User OFDMA-based wireless cellular networks. In this section we present the basic principles of resource allocation for multiple users to efficiently share the limited resources in OFDMA-based wireless mobile communication systems while meeting the QoS constraints. In OFDMA-based wireless cellular networks the resource allocation process is split in three families of allocation mechanisms: *priority scheduling*, *frequency scheduling* and *retransmission scheduling* techniques. While the merging of those two first scheduling mechanisms is a well investigated subject and it is called Time/Frequency dependent packet scheduling (TFDPS), smart design of re-schedulers present still some challenging open issues. TFDPS scheduling techniques are designed to enable the scheduler to exploit both time and frequency diversity across the set of time slots and sub-carriers offered by OFDMA technology. To this end, in order to fully exploit multi-user diversity in OFDMA systems, frequency scheduling algorithms besides try to select the momentary best set of sub-carriers for each user aiming at optimizing a overall criterion. In real commercial communication systems such as WiMAX and 3GPP/LTE, the frequency scheduler allocates chunks of sub-carriers rather than individual sub-carriers. The advantage of such chunk allocation is twofold: first, the allocation algorithm complexity is notably reduced; second, the signalling information required is shorten. In the literature several TFDPS scheduling algorithms have been proposed. The general scope of such scheduling algorithms is to grant access to resources to a subset of users which at a given scheduling moment positively satisfy a given cost function. Some algorithms were designed for OFDMA based systems to profit of the multi-user diversity of a wireless system and attempt to instantaneously achieve an objective (such as the total sum throughput, maximum throughput fairness, or pre-set proportional rates for each user) regardless to QoS constraints of the active users in the system. On the other hand, some scheduling algorithms were designed to support specific QoS constraints, either taking into account channel state information or not. Alternatively, one could attempt to maximize the scheduler objective (such as maximization of the overall system throughput, and/or fairness among users) over a time window, which provides significant additional flexibility to the scheduling algorithms. In this case, in addition to throughput and fairness, a third element enters the tradeoff, which is latency. In an extreme case of latency tolerance, the scheduler could simply just wait for the user to get close to the base station before transmitting. Since latencies even on the order of seconds are generally unacceptable, recent scheduling algorithms that balance latency and throughput and achieve some degree of fairness have been investigated. In (Ryu et al., 2005), urgency and efficient based packet scheduling (UEPS) was proposed to support both RT (Real Time) and NRT (Non Real Time) traffics, trying to provide throughput maximization for NRT traffic and meeting QoS constraints for RT traffics. However, UEPS bases its scheduling rule on a set of utility functions which depend on the traffic type characteristics and the specific momentary set of active users in the network. The correct choice of these utility functions have a strong impacts on the effectiveness of the UEPS algorithm. In (Yuen et al., 2007), a packet discard policy for real-time traffic only (CAPEL) was proposed. This paper stresses the issue of varying transmission delay and proposes to sacrifice some packets that have small probability to be successfully delivered and save the system resources for more useful packets. Again in section 4, we will present and comment some of the most known priority scheduling algorithms in the specific context of OFDMA-based wireless cellular networks, while our proposal will be extensively described in section 5.

Nevertheless, even with well designed TFDPS schedulers, the resource allocation process has to deal with error at destination. As a consequence, additional resource has to be allocated for accidental occurrences of request of retransmission. Nowadays, smart design of re-schedulers is still an open issue. A re-scheduler copes with negative acknowledge (NACK) packets which can be quite frequent in mobile wireless communications. Therefore, a re-scheduler must reallocate resources for NACK packets in a efficient and robust manner. Efficient, since it might reduce the average number of retransmission associated to NACK packets. Robust re-scheduling, in the way of minimizing the residual PER (PER_{res}). Thus, adaptive mechanisms such as Adaptive Modulation and Coding (AMC) can achieve a target PER_{res} with less stringent physical layer requirement, but with higher throughput, power saving, latency improvement and reduction of MAC signalling. In section 4, we will present and comment the most known retransmission scheduling algorithms while our proposal will be extensively described in section 5.

3. System model

The system model is mainly based on the 3GPP/LTE downlink specifications (TR25.814, 2006)(TS36.211, 2007), where both components of the cellular wireless network, i.e. base stations (BS) and mobile terminals (UE), implement an OFDMA air interface. Using the terminology defined in (TSG-RAN1#48, 2007), OFDM symbols are organized into a number of physical resource blocks (PRB or chunk) consisting of 12 contiguous sub-carriers for 7 consecutive OFDM symbols (one slot). Each user is allocated one or several chunks in two consecutive slots, i.e. the time transmission interval (TTI) or sub-frame is equal to two slots and its duration is 1ms. With a bandwidth of 10MHz, this leads to 50 chunks available for data transmission. The network has 19 hexagonal three-sectored cells where each BS transmits continuously and with maximum power. We mimic the traffic of the central cell, while others BSs are used for down-link interference generation only. Fast fading is generated using a Jakes model for modeling a 6-tap delay line based on the Typical Urban scenario (TSG-RAN1#48, 2007), with a mobile speed equal to 3km/h. Flat fading is assumed for the neighboring cells. A link-to-system (L2S) interface is used in order to accurately model the physical layer at the system level. This L2S interface is based on EESM (Effective Exponential SINR Mapping) as proposed in (Brueninghaus et al., 2005).

In the central cell, the BS has a multiuser packet scheduler which determines the resource allocations, AMC (Adaptive Modulation and Coding) parameters and Hybrid Automatic Repeat reQuest (HARQ) policy within the next slot. While the scheduler sends downlink control messages that specify the resource allocation and the link adaptation parameters adopted in the next time slot, UEs send positive or negative acknowledgment (ACK/NACK) to inform the scheduler of correct/incorrect decoding of the received data. Perfect channel state information (CSI) is assumed for all links. Nevertheless, a feedback delay is introduced between the time when CSI is available at the destination and the time when the packet scheduler performs the resource allocation.

In this model the possible presence of mixed traffic flows which present different and competing Quality of Service (QoS) requirements is studied. Two traffic classes are considered: real-time traffic (RT) and non real-time traffic (NRT). As RT traffic, we consider Voice over IP traffic (VoIP) which is modeled according to (TSG-RAN1#48, 2007). This is equivalent to a 2-state voice activity model with a source rate of 12.2kbps, an encoder frame length of 20ms and a total voice payload on air interface of 40 bytes. For RT traffic, we also consider near real-time video source (NRTV), which we model according to (TR25.892, 2004) as a source video with rate of 64 kbps and a deterministic inter-arrival time between the beginning of each frame equal to 100ms. The mean and maximum packet sizes are respectively equal to 50 and 250 bytes. As NRT traffic we consider an HyperText Transfer Protocol (HTTP), as specified in (TR25.892, 2004), that is divided into ON/OFF periods representing respectively web-page downloads and the intermediate reading times. More details on the adopted system model are summarized on table 1.

Parameter	Value
Network	
Carrier frequency	2.0 GHz
Bandwidth	10 MHz
Inter-site distance	500 m
Minimum distance	35 m
TTI duration	1 ms
Cell layout	Hexagonal grid, 19 three-sectored cells
Link to System interface	EESM
Traffic model	VoIP, NRTV, HTTP
Nb of antennas (Tx, Rx)	(1,1)
Access Technique	OFDMA
Total Number of sub-carriers	600
Nb of sub-carriers per chunk (PRB)	12
Total Nb of Chunks	50
Propagation Channel	
Parameter	Value
Fast fading	Typical urban 6-tap model, 3 km/h
Interference	White
UE	
Parameter	Value
Channel estimation	ideal
CQI reporting	ideal
Turbo decoder	max Log-MAP (8 iterations)
Dynamic Resource Allocation	
Parameter	Value
Nb of MCS	12 (from QPSK 1/3 to 64-QAM 3/4)
AMC PER_{target}	10 %
CQI report	Each TTI, with 2 ms delay
Packet Scheduling	MCI, PF, EDF, MLWDF, HYGIENE
Sub-carriers Allocation Strategy	Chunk based allocation
Number of control channels per TTI	16
HARQ	
Parameter	Value
Stop and Wait	synchronous adaptive
Number of processes	6
Retransmission Interval	6 ms
Maximum Nb of retransmissions	up to 3
Combining technique	Chase

Table 1. Main system model parameters

A limited number of control channels per TTI is considered, as the control channel capacity is always limited in realistic systems. In this study, that number, which corresponds to the maximum number of scheduled users in a TTI, is equal to 16, that is the double of the number given in (Henttonen et al., 2008) for a 3GPP/LTE system with a bandwidth of 5 MHz. For the first transmission attempt, the MCS (Modulation and Coding Scheme) selection is based on the EESM link quality metric. As suggested in the 3GPP LTE standard, AMC algorithm selects the same MCS for all chunks allocated to one UE. This solution has the advantage of make both signaling and AMC algorithm easier to be implemented on real equipment. Concerning adaptive HARQ, as done in (Pokhariyal et al., 2006), all the time a retransmission is scheduled, the scheduler re-computes the set of frequency chunks previously allocated to the negative acknowledged packets, depending on the re-scheduling policy.

4. Survey on resource allocation mechanisms

In this section we will focus on three main families of resource allocation techniques for packet based transmissions. The first one is related to packet scheduling algorithms that decide in which priority order resources are allocated to the different competing flows. We will consider some of the most esteemed priority schedulers, namely the maximum channel to interference ratio (MCI) (Pokhariyal et al., 2006), the proportional fair (PF) (Norlund et al., 2004), the earliest deadline first (EDF) (Chiusssi et al., 1998) and the Modified Largest Weighted Deadline First (MLWDF) (Andrews et al., 2001) schedulers. The second technique deals with frequency scheduling: the frequency dependent packet scheduler (FDPS) allocates frequency resources (hereafter chunks) to the population of users that will be served in the next transmission intervals. FDPS maps best chunks to best users, where the notion of best users depends on the priority rule of the scheduler. Any priority based selection methods such as MCI per chunk or PF per chunk selection methods (Pokhariyal et al., 2006) can be adopted. Eventually, the third technique is related to packet retransmissions and aims at deciding how chunks are allocated or reallocated to packets which require a retransmission. It could be either persistent or hyperactive methods (Pokhariyal et al., 2006), depending wether the chunk allocation for all NACK packets is kept or recomputed.
In the following, each of these techniques has a dedicated subsection to discuss in detail their limitations and advantages.

4.1 Priority scheduling

Many researchers address the problem of defining an efficient and robust resource allocation strategy for multiple heterogeneous data users sharing the same resources over a wireless channel. Priority scheduler can deal with both allocation of time and frequency resources, in order to exploit multi-user diversity in both domains. This is often referred as time/frequency domain packet scheduling (TFDPS). In this sub-section, priority scheduling is related to the time domain dimension.
Four of these well known priority scheduling algorithms are investigated in this work: max C/I (MCI) scheduler, proportional fair (PF) scheduler, Earliest Deadline First (EDF) scheduler, and Modified Largest Weighted Delay First (MLWDF) scheduler. These priority scheduling algorithms have been proposed aiming at satisfying either delay, throughput, fairness constraints of all active users or as many as possible users. While some scheduling algorithms take into account only the time constraints of the traffic flows (e.g. EDF), others take into account the momentary channel state to optimize the overall cell throughput (e.g. MCI), or, a compensation model to improve fairness among UEs (e.g. PF), or a compound of all these goals (e.g. MLWDF). The key features and drawbacks of such schedulers are the following:
MCI: Its goal is to maximize the instantaneous system throughput regardless to any traffic QoS constraints. Therefore, MCI always chooses the set of users whose momentary link quality is the highest. Even if maximum system throughput can be achieved with MCI, users whose momentary channels are not good for a relatively long period may starve and consequently release their connections. MCI is indeed inadequate for real-time traffic.
PF: Its goal is to maximize the long-term throughput of the users relative to their average

channel conditions. Thus, its goal is to trade-off fairness and capacity maximization by allocating resources to users having best instantaneous rate (over one or several chunks) relative to their mean served rate calculated using a smoothed average over an observation time window (TW_i) (Pokhariyal et al., 2006)(TSG-RAN1#44bis, 2006). While PF is a good scheduler for best effort traffic, it is less efficient for real-time traffics.

EDF: It allocates resources first to packets with smaller remaining TTLs (Time To Live) thus each packet is prioritized according to its remaining TTL (R_{TTL}). As a consequence, by serving users in order to match everyone's deadline, EDF is designed for RT traffics. The drawback of this scheduler is that multiuser diversity is not exploited since any momentary channel state information is taken into account in the scheduling rule.

MLWDF: It aims at keeping queues stable (fairness) while trying to serve users with momentary better channel conditions (throughput maximization). Contrary to EDF and MCI scheduling algorithms, MLWDF is designed to cope with mixed traffic scenarios. The major drawback of this scheduler is that its performance depends on the design of three parameters, the maximum probability for a packet to exceed TTL (for RT traffic), the requested rate (for NRT traffic) and the averaging window for rate computation. Thus correct choice of the adequate set of parameters can be system state dependent, especially in heterogenous mixed traffic scenarios.

4.2 Frequency scheduling

FDPS maps *'best'* chunks to *'best'* users. The notion of *'best'* users depends on the priority rule of the scheduler. At time i, UE k has a metric $P_{k,n}(i)$ for chunk n, which is given for instance by $P_{k,n}(i) = R_{k,n}(i)/T_k(i)$ or by $P_{k,n}(i) = R_{k,n}(i)$, respectively for PF per chunk and MCI per chunk schedulers. $R_{k,n}(i)$ is the instantaneous supportable rate for UE k at chunk n, depending on each UE's channel quality indicator (CQI) while $T_k(i)$ is the previously mean served rate. For each time i, the *'best'* UE of each chunk n is scheduled. That is the scheduled UE at chunk n is $U_n(i) = \underset{k}{\operatorname{argmax}}\, P_{k,n}(i)$.

The adoption of realistic traffic models provides different performance if compared to non realistic full buffer models. The chunk allocation process is indeed strongly influenced by the amount of data present in users' queues: with the use of non-full buffer models, resources are only allocated to users that effectively have data to send. Thus, to find the *'best'* chunk(s) for each user, several solutions may be considered. In this section, we consider two common chunk allocation algorithms whose principles are derived from (Ramachandran et al., 2008):

Matrix-based chunk allocation: it iteratively picks the *'best'* user-chunk pair in the two dimensional matrix of chunks and users. The matrix contains the metrics $P_{k,n}(i)$ of all possible user-chunk pairs.

Sequential chunk allocation: it does only the first iteration of the *matrix-based chunk allocation*. Therefore, when a user that has been selected at the first chunk-pick has not unscheduled packet in its queue, the next user with unscheduled packets in the same matrix-row will be selected. Only when the system is forced to have full-queue traffic, both chunk allocation algorithms perform the same. Otherwise, *sequential chunk allocation* may perform sub-optimally.

Note that with EDF scheduling for OFDMA based transmission, allocation is decoupled. In a first step, each packet is prioritized according to its remaining TTL (R_{TTL}) and then chunks are allocated to the ordered packets in order to maximize spectral efficiency. This approach is more efficient than the previous one, at the expense of an increase complexity at the transmitter.

4.3 Retransmissions

The *re-scheduler* allocates chunks for retransmission according to one of the common following chunk reallocation policies:

Persistent: the *re-scheduler* persists in allocating the same set of chunks previously allocated to NACK packets. The idea is to reduce both control signaling, complexity at the BS and latency. This approach used in (TSG-RAN1#Adhoc, 2007) is typically adopted for real-time traffic such as VoIP associated to small payloads.

Hyperactive: as done in (Pokhariyal et al., 2006), each time a retransmission is scheduled, the scheduler re-computes the set of *best* frequency chunks previously allocated to NACK packets.

5. Improving RRM effectiveness

As seen in section 4, TFDPS algorithms such as the maximum channel to interference ratio (MCI) per chunk or the proportional fair (PF) per chunk were designed for OFDMA based systems to profit of the multi-user diversity of a wireless system and attempt to instantaneously achieve an objective (such as the total sum throughput, maximum throughput fairness, or pre-set proportional rates for each user) regardless to QoS constraints of the active users in the system. More precisely, MCI scheduler allocates resources to users with the highest momentary instantaneous capacity; PF scheduler tries to balance the resource allocation and serve momentary good users (not necessarily the best) while providing long term throughput fairness (equal data rates amongst all users). On the other hand, some scheduling algorithms were designed to support specific QoS constraints. For instance, Earliest Deadline First (EDF) is designed to deal with real-time QoS constraints regardless to the momentary user's channel quality. Other schedulers are designed to cope with the coexistence of RT and NRT traffics (mixed traffic), as the Modified Largest Weighted Deadline First (MLWDF) algorithm. Its design objective is to maintain delay (or throughput) of each traffic smaller (or greater) that a predefined threshold value with a given probability, at the expense of an adequate set of parameters that is system state dependent.

With our first proposal, the goal is to design efficient Time/Frequency domain packet scheduling algorithms in order to maximize the overall system capacity while supporting QoS for mixed traffic flows considering either homogeneous and heterogeneous traffics. We propose to split the resource allocation process into three steps, as defined in (Calvanese Strinati et al., VTC 2009). In a first step we identify which entities (packets for RT traffics and users for NRT ones) are *rushing*. Then in step two we deal with urgencies: we allocate resources only to entities that have an high probability of missing their QoS requirements regardless to their momentary link quality. Then, if any resources (here chunks) are still unscheduled, in a third step of the proposed scheduling algorithm, we allocate resources to users with highest momentary link quality, regardless to their QoS constraints. We call the proposed algorithm *HurrY-Guided-Irrelevant-Eminent-NEeds* (HYGIENE) scheduling.

With our second proposal we tried to tackle the issue of frequency scheduling combined with retransmissions. Indeed, as pointed out in previous section, while FDPS is a well investigated subject, smart design of *re-schedulers* is still an open issue. The re-scheduler must reallocate resources for NACK packets in a efficient and robust manner.

Decoding errors are classically attributed to insufficient instantaneous signal-to-noise-ratio (SNR) level, as it is for gaussian channels. Therefore, when a packet is not correctly decoded, its retransmission is traditionally scheduled as soon as possible and on the same frequency resource until either it is successfully transmitted or retry limit is reached. Nevertheless, the mobile wireless channel is not gaussian. A more appropriate model for such channel is the non-ergodic block fading channel for which information theory helps us to define a novel approach for re-scheduling. Actually, in non-ergodic channels decoding errors are mainly caused by adverse momentary channel instance and unreliable PER predictions (Lampe et al., 2002)(Emilio Calvanese Strinati, 2005) adopted for the AMC mechanism. As a consequence, a smart *re-scheduler* should permit to forecast, given the momentary chunks instance related to the unsuccessful transmission, if correct packet decoding is impossible even after a large number of retransmissions. To this end, in our second investigation, we present a novel *re-scheduler* which exploits both information associated to a NACK as proposed in (Emilio Calvanese Strinati, 2007) (i.e. channel outage instances and CRC) to allocate the set of *'best'* suited chunks for NACK packets. In other words, we recompute the chunk allocation only if the previously selected chunks do not permit correct decoding for the selected Modulation And Coding Scheme (MCS). We call the proposed on-demand *re-scheduler* criterion as *2-bit lazy*.

5.1 Proposed HYGIENE scheduling algorithm

EDF-like schedulers do not profit of time diversity as much as they should do. MCI and PF like schedulers aiming at maximizing the cell throughput regardless of the user QoS, are totally insensitive to any time constraints of the data traffic. Based on these observations, we propose to split the resource allocation process into three steps. First a *Rushing Entity Classifier* (REC) identifies *rushing entities* that must be treated with higher priority. Depending on the nature of the traffic, entities are UEs (NRT traffic) or packets (RT). Therefore, rushing entity classification is traffic-dependent. Second the proposed scheduler deals with urgencies: we schedule the transmission of rushing entities regardless to their momentary link quality. If any resources (here chunks) are still unscheduled, in a third step, HYGIENE allocates resources to those users with better momentary link quality, regardless to their time constraints. The proposed scheduling algorithm is summarized as follows:

Step 1: The REC classifies *entities* (packets or UEs) waiting to be scheduled as *rushing* or *non-rushing*. With RT traffic, packets are classified as *rushing* if $Th_{rush} \cdot TTL + \eta \geq R_{TTL}$. Where Th_{rush} is a threshold on the QoS deadline which depends on the traffic type, η is a constant which takes into account both retransmission interval and maximum allowed number of retransmissions. With NRT traffic, UEs and not packets are classified by the REC. Therefore, the ith UE (UE_i) is classified as *rushing* if it has been under-served during TW_i. More precisely, every TTI the REC checks for each UE_i if $(TW_i - t_{now,i}) \leq (QoS_i - tx_{data,i})/R_{min}$. Where $t_{now,i}$ is the elapsed time since the beginning of TW_i, QoS_i the QoS requirements of the UE class of traffic, $tx_{data,i}$ the total data transmitted by user i during $(TW_i - t_{now,i})$ and R_{min} the minimum transmission rate of the system. Note that Th_{rush}, η and TW_i are scheduler design parameters.
Step 2: Resources (chunks) are allocated to *rushing entities* with an EDF-like scheduler which allocates *best* chunk(s) to entities with higher deadline priority. Deadline priority metrics differ between RT and NRT traffics: while with RT traffic deadline priority depends on R_{TTL}, with NRT traffic it depends on the lack of data transmitted in TW_i. Again, chunks are selected in order to maximize the spectral efficiency.
Step 3: All unscheduled resources (chunks) are allocated to users which maximize the cell throughput regardless to any QoS constraints of active UEs. Thus, the allocation is done according to MCI per chunk, following the 'matrix-based chunk allocation' described previously with $P_{k,n}(i) = R_{k,n}(i)$.

5.2 Proposed 2-bit lazy frequency re-scheduling algorithm

Many delay-constrained communication systems, such as OFDM systems, can be characterized as instances of block fading channel (Ozarow et al., 1994). Since the momentary instance of the wireless channel has a finite number of states n_c the channel is non-ergodic, and it admits a null Shannon capacity (Ozarow et al., 1994). The information theoretical limit for such channels is established by defining an outage probability. The outage probability is then defined as the probability that the instantaneous mutual information for a given fading instance is smaller than the information rate R associated to the transmitted packet:

$$\mathrm{P_{out}} = \Pr(I(\gamma, \alpha) < R) \tag{1}$$

where $I(\gamma, \alpha)$ is a random variable representing the instantaneous mutual information for a given fading instance α and γ is the instantaneous SNR.
For an infinitely large block length, the outage probability is the lowest error probability that can be achieved by a channel encoder and decoder pair. Therefore, when an information outage occurs, correct packet decoding is not possible. The outage probability is an information theoretic bound on the packet error rate (PER) in block fading, and thus no system can have a PER that is better than the outage probability.
For a generic code $\mathcal{C}$, assuming Maximum Likelihood decoding, we can express the packet error probability of the code $\mathcal{C}$ as:

$$P_e^{\mathcal{C}}(\gamma) = P_{e|out}^{\mathcal{C}}(\gamma)\mathrm{P}_{\mathrm{out}}^{\mathcal{C}}(\gamma) + P_{e|\overline{out}}^{\mathcal{C}}(\gamma)(1 - \mathrm{P}_{\mathrm{out}}^{\mathcal{C}}(\gamma)) \tag{2}$$

where $P_{e|out}^{\mathcal{C}}$ and $P_{e|\overline{out}}^{\mathcal{C}}(\gamma)$ are respectively the packet error probability when transmission is in outage and when it is not. For capacity achieving codes Eq. (2) can be tightly upper bounded by:

$$P_e^{\mathcal{C}}(\gamma) \lessapprox \mathrm{P}_{\mathrm{out}}^{\mathcal{C}}(\gamma) + \underbrace{P_{e|\overline{out}}^{\mathcal{C}}(\gamma)(1 - \mathrm{P}_{\mathrm{out}}^{\mathcal{C}}(\gamma))}_{\mathrm{P}_{\mathrm{noise}}^{\mathcal{C}}(\gamma)} \tag{3}$$

Considering capacity approaching codes an analytical expression of $\mathrm{P}_{\mathrm{noise}}^{\mathcal{C}}(\gamma)$ is not trivial, but the inequality (3) still holds. We can indeed distinguish two components of the packet error probability: the code outages due to fading instance and noise respectively.
In our work we propose to exploit at the transmitter side the knowledge on both components of the PER: the code outages due to fading instance and noise respectively. As proposed in (Calvanese Strinati et al., WCNC 2009), the receiver can send a 2-bit ACK/NACK to feedback such information: one bit informs on successful/unsuccesfull decoding (CRC), the other on code outages due to fading instance. Alternatively, the classic 1-bit feedback (CRC) can be computed at the receiver and, code outages due to fading instance can be directly estimated at the transmitter side if the channel coefficients are known at the transmitter. Based on these assumptions, we propose the *2-bit lazy frequency re-scheduler*. The goal of *2-bit lazy frequency re-scheduler* is to strongly limit unsuccessful retransmissions attempts. To this end, when retransmissions are scheduled, the proposed *re-scheduler* checks both components of the packet error probability outlined by equation (3). The *2-bit lazy frequency re-scheduler* works as follows:

Step 1: When a retransmission is required (NACK on CRC), the receiver or the transmitter (depending on the system implementation) checks if decoding failure is associated to a

channel outage.
Step 2: If $I(\gamma, \alpha) < R$, transmission is in outage and *best* chunk allocation is recomputed only for $NACK_{out}$ packets.
Step 3: Otherwise, if $I(\gamma, \alpha) \geq R$, retransmission is due only to a unfavorable noise instance and the *2-bit lazy frequency re-scheduler* reallocates the same set of chunks for the packet retransmission.

To detect a channel outage it is necessary to compute the instantaneous mutual information associated to previous transmission(s) of the NACK packet. Such instantaneous mutual information can be computed as follows:

$$I(\gamma, \alpha) = \frac{1}{n_c} \sum_{i=1}^{n_c} I_i \left(\sum_{k=1}^{K} \frac{|\alpha_{i,k}|^2}{\sigma_k{}^2} \right)$$

where

$$I_i = \log_2(M) - \frac{1}{M} \sum_{k=1}^{M} E_z \left[\log_2 \left(\sum_{q=1}^{M} A_{i,k,q} \right) \right] \tag{4}$$

and

$$A_{i,k,q} = \exp[-\frac{|\alpha_i a_k + z - \alpha_i a_q|^2 - |z|^2}{2\sigma^2}]$$

Note that equation (4) is derived from (Ungerboeck, 1982) where **a** is the real or complex discrete signal transmitted vector. Moreover, all information required can be directly available at the receiver: M (size of the M-QAM modulation alphabet) and R are known since the MCS is known at the receiver; both α_i and the noise variance σ^2 are known at the receiver using training pilots based channel estimation; **a** is known from the demapper. z are the Gaussian noise samples, with zero-mean and variance equal to σ^2. Mutual information is computed over the n_c sub-carriers and the K current transmissions on which the packet is transmitted.
While *hyperactive* re-scheduler recomputes chunk allocation for all NACK packets, *lazy* does it only for $NACK_{out}$ packets. Both re-schedulers can adopt any FDPS such as MCI per chunk, PF per chunk or others. Complexity added by packet outage detection is low because the mutual information can be computed easily thanks to Look-Up Tables (LUT) or polynomial expansion. Thus, the overall complexity of the proposed *lazy* re-scheduler is in between the two classical *1-bit persistent* and *1-bit hyperactive* methods.
It is possible to further improve the effectiveness of chunk re-allocation algorithms. First, banning some chunks during a given period for a sub-set of user at step 2, may prevent from repetitive errors in the chunk allocation process. Second, $NACK_{out}$ packet detection can also be based on *accumulative* mutual information of both current and future packet transmission attempts in a given set of chunks. In this case, the instantaneous mutual information is computed as in (4) except that the summation is done over K+1 transmissions, and under the assumption that $\frac{|\alpha_{i,K+1}|^2}{\sigma_{K+1}{}^2} = \frac{|\alpha_{i,K}|^2}{\sigma_K{}^2}$.

6. Numerical results

In this section the effectiveness of the two proposed approaches, HYGIENE scheduling and Lazy frequency (re)scheduling, is evaluated comparing it with the classical resource allocation techniques presented in section 4. Schedulers are compared in terms of maximum achievable cell traffic load in different traffic scenarios, considering either single traffic, mixed real-time traffic and heterogeneous mixed traffic scenarios, following the metrics defined in (TR25.814,

2006)(TSG-RAN1#48, 2007). Performance are also assessed in terms of residual Packet Error Rate (through its cumulative density function) and chunk re-allocation cost, while varying the number of maximum retransmissions $rxtx_{max}$.
Simulation results are given for the system and traffic models presented in section 3. Results are averaged over 100 independent dynamic runs, where at the beginning of each run UEs are randomly uniformly located in the central cell. Positions, bi-dimensional log-normal shadowing and path loss values are kept constant for the duration of each run. Each run simulates 100 seconds of network activity and at each TTI channel realizations are updated.

6.1 Packet scheduling

In this first subsection, we assess the effectiveness of our proposed HYGIENE scheduling algorithm comparing it to four scheduling algorithms often investigated in the literature: MCI, PF, MLWDF and EDF. For this performance evaluation, the following assumption holds: all the time a retransmission is scheduled, the scheduler re-computes the set of frequency chunks previously allocated to the negative acknowledged packets. Furthermore, for MLWDF scheduling, we adopt the same parameters as the ones suggested in (Andrews et al., 2001).
Schedulers are compared in terms of maximum achievable cell traffic load in three different traffic scenarios:

Scenario A (*single traffic* scenario): unique traffic type in the cell for all UEs.

Scenario B (*mixed real-time traffic* scenario): coexistence of VoIP and NRTV traffic in the same cell.

Scenario C (*heterogeneous mixed traffic* scenario): coexistence of VoIP and HTTP traffic in the same cell.

To evaluate the maximum achievable cell traffic load we use the metrics defined in (TR25.814, 2006)(TSG-RAN1#48, 2007). The maximum achievable cell traffic load for real-time traffics is defined as the number of users in the cell when more than 95% of the users are satisfied. VoIP and NRTV users are considered satisfied if their residual BLER is below 2% and their transfer delay is respectively below 50ms and 100ms. HTTP users are considered satisfied if their average bit rate is at least 128 Kbps.
On figure 1 we show our simulation results for VoIP, NRTV and HTTP traffics considering *scenario A*. Under single VoIP traffic, the highest system load is achieved with EDF and HYGIENE (up to 540 VoIP UEs). MCI, PF and MLWDF achieve respectively up to 445, 440, and 360 satisfied VoIP UEs. Performance gap between EDF or HYGIENE and MCI or PF is not surprising. Actually, since both PF and MCI aim at maximizing the cell throughput regardless of the user time QoS constraints, with the increasing number of real-time flows, many users may face momentary service starvation and consequently, exceed the maximum delivery delay (50 ms). This is not the case with EDF or HYGIENE since both schedulers allocate best chunk(s) to entities with higher QoS deadline priority. What can look surprising is the poorer performance of MLWDF scheduling. Classical performance evaluations for MLWDF show that MLWDF is a good scheduler with both RT and NRT traffics. However, in such studies an unlimited number of control channels per TTI is assumed. We compare performance in a more realistic scenario where the number of control channels per TTI, and thus the maximum number of scheduled users per TTI (UE/TTI), is limited to 16. Thus, we observe by simulation that such limitation has significant impact only on MLWDF capacity performance.
For single NRTV traffic, maximum cell capacity performance obtained with any of the investigated schedulers is very similar, ranging from up to 95 satisfied UEs with MLWDF

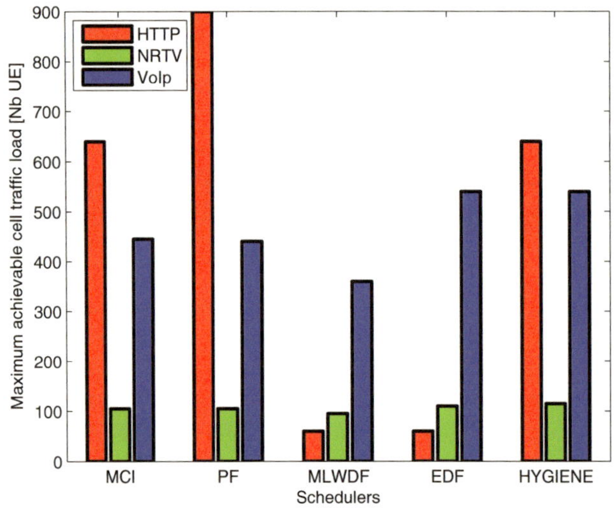

Fig. 1. Scenario A (*single traffic*): maximum achievable cell capacity with PF, MCI, MLWDF, EDF and HYGIENE schedulers.

(worst case), to up to 115 satisfied UEs with HYGIENE. With single HTTP traffic, best performance is obtained as expected with PF, having up to 900 HTTP UEs satisfied. MCI and HYGIENE perform the same (640 UEs each) while both EDF and MLWDF can satisfy very few UEs (up to 60 UEs).

On figure 2 we show our results for coexistent VoIP and NRTV traffics (*scenario B*). In our simulations we fix the number of NRTV traffic to 75 and we vary the number of VoIP. Best performance is obtained with HYGIENE, having up to 250 VoIP UEs while 75 NRTV UEs are satisfied too. Other schedulers perform as follows. EDF scheduler serves more VoIP UEs (up to 220 VoIP) than PF (up to 140 VoIP) and MLWDF (up to 70 VoIP). Worst performance is obtained with the non QoS aware MCI scheduler, having no VoIP UEs satisfied when 75 NRTV UEs are satisfied. When considering the coexistence of 75 NRTV UEs and 425 VoIP UEs, we obtained by simulation that limiting respectively to 16, 32 and 50 UE/TTI, MLWDF achieves a user satisfaction equals to 41.4%, 99.6% and 100%. In the last two cases, MLWDF performs even better than the other schedulers subject to the same restriction, except the HYGIENE one. Again, we can see that the number of control channels has significant impact on MLWDF capacity performance.

On figure 3 we mimic a heterogeneous network traffic. We fix the number of HTTP flows to 200 UEs while we evaluate the maximum VoIP UEs capacity. When scheduling is based on EDF or MLWDF ordering rules, any UE (HTTP and VoIP) can be satisfied. As expected, we observe that EDF scheduler results totaly inadequate since it cannot efficiently deal with NRT traffic. Furthermore, we observe again how MLWDF is deeply penalized by the UE/TTI limitation. Besides, MCI serves up to 180 satisfied VoIP UEs, PF up to 370 VoIP UEs. Best performance is obtained with HYGIENE scheduler, which serves up to 390 satisfied VoIP UEs. Contrarily to (*scenario A* with HTTP only), HYGIENE scheduler performs better than PF in this mixed scenario, showing the supremacy of the rushing approach. The above results were obtained with empirically optimized rushing thresholds optimized.

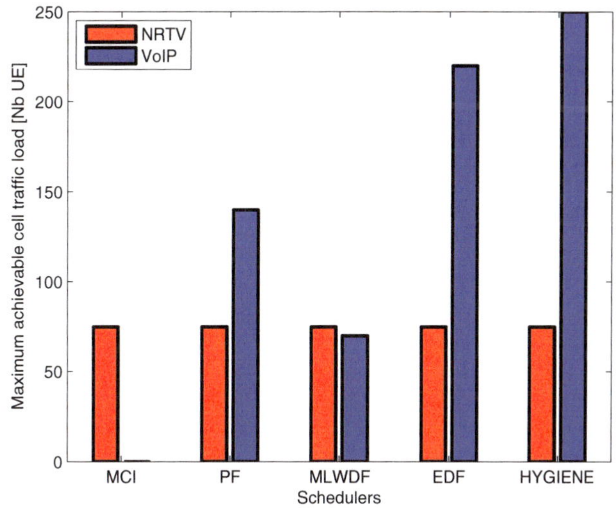

Fig. 2. Scenario B (*mixed real-time traffic*): maximum achievable cell capacity with PF, MCI, MLWDF, EDF and HYGIENE schedulers imposing 75 active NRTV flows.

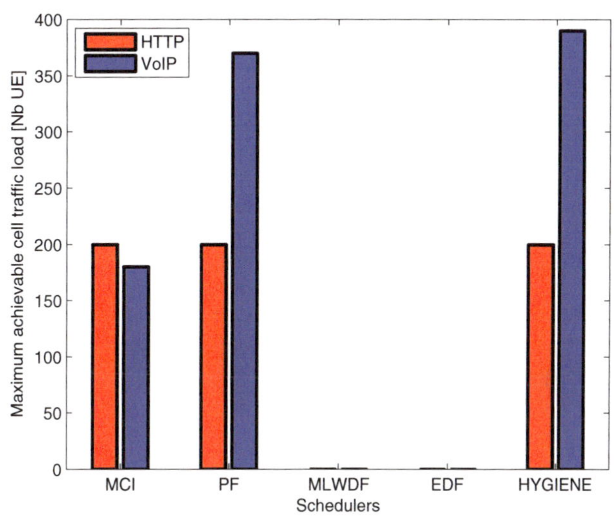

Fig. 3. Scenario C (*mixed heterogeneous traffic*): maximum achievable cell capacity with PF, MCI, MLWDF, EDF and HYGIENE schedulers imposing 200 active HTTP flows.

On figure 4 we mimic coexistent activity of 225 VoIP and 75 NRTV UEs testing different rushing thresholds for both VoIP and NRTV: $Th_{rush,VoIP}$ and $Th_{rush,NRTV}$. Our goal is to determine whether HYGIENE performance depends on an optimal combination of $(Th_{rush,VoIP}, Th_{rush,NRTV})$. Simulations show that a large range of $(Th_{rush,VoIP}, Th_{rush,NRTV})$ slightly affects user satisfaction ($Th_{rush,VoIP} \leq 40\%$ and $Th_{rush,NRTV} \leq 90\%$).

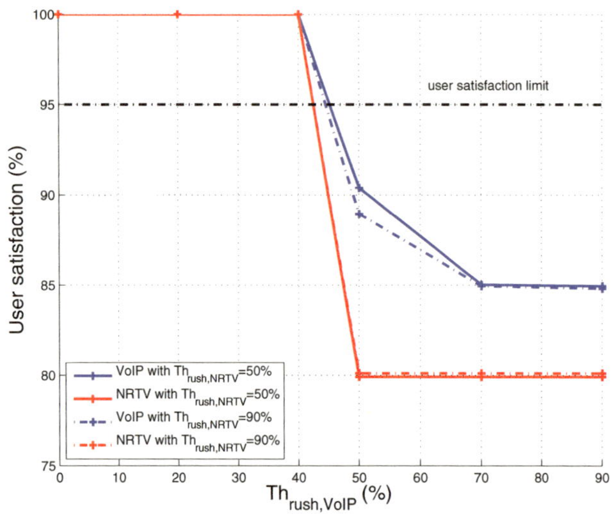

Fig. 4. Scenario B (*mixed real-time traffic*): sensitivity of HYGIENE performance on rushing threshold design.

We also looked for the quasi-optimal range of $Th_{rush,VoIP}$ and $Th_{rush,NRTV}$ in the single traffic scenario. We observed that user satisfaction for VoIP UEs is not affected if $Th_{rush,VoIP} \geq 20\%$ and, for NRTV UEs is constant for any $Th_{rush,NRTV}$ value.

6.2 Coupling of priority scheduling with multi-user re-scheduler

In this section we investigate the effectiveness of coupling a priority packet scheduler with a well designed multi-user *re-scheduler*. To this aim, we compare the performance of three classical priority packet scheduling algorithms (MCI, PF and EDF) coupled with *1-bit persistent*, *1-bit hyperactive* and *2-bit lazy* frequency *re-schedulers*. Performance is compared in terms of maximum achievable system capacity, PER_{res} cumulative density function (CDF) and chunk re-allocation cost for the system and traffic models presented in section 3. Results obtained for *1-bit persistent*, *1-bit hyperactive* and *2-bit lazy* re-schedulers are respectively plotted with orange, red and blue colors.

On figures 5, 6 and 7 we compare the pairs of priority and re-schedulers in terms of maximum achievable system capacity respectively with $rxtx_{max} = 1$ and $rxtx_{max} = 2$. To evaluate the maximum achievable cell traffic load we use the metrics defined in (TR25.814, 2006) and updated in (TSG-RAN1#48, 2007). On figure 5 we show our simulation results for VoIP traffic with $rxtx_{max} = 1$ and *matrix-based chunk allocation*. With this scheduling configuration *1-bit hyperactive* or *2-bit lazy* performs the same, outperforming *persistent* re-scheduling respectively of 120%, 135% and 150% with PF, EDF and MCI packet schedulers. Best performance is obtained coupling EDF with *1-bit hyperactive* or *2-bit lazy*, having a cell capacity of 400 UE. When using the HYGIENE scheduler (not plotted here), we observed the same conclusions: HYGIENE with *1-bit hyperactive* and *2-bit lazy* reached a cell capacity of 420 UE while HYGIENE with persistent rescheduling only achieved a cell capacity of 170 UEs. We also investigated two other scheduling scenarios when $rxtx_{max} = 1$: VoIP traffic with *sequential chunk allocation* and, NRTV traffic with both chunk allocation scheduling. We did not plot our simulation results for these scenarios because in both cases QoS constraints are not met.

On figure 6 we show our simulation results for VoIP traffic with $rxtx_{max} = 2$ and *sequential chunk allocation*. With this scheduling configuration system capacity improvement obtained with *2-bit lazy* instead of the other two re-schedulers is significant: capacity is multiplied by 2.6 even with respect to the hyperactive scheme. Again, best performance is obtained for the pair EDF and *2-bit lazy*, having the maximum system capacity of 540 UEs. *2-bit lazy* outperforms *1-bit hyperactive* when chunk allocation is sequential since in this case chunk allocation is less effective (chunk search is not exhaustive) and can even introduces additional errors. It can happen that when a retransmission is scheduled, the new pair user-chunk(s) can be associated to a higher error probability. *2-bit lazy* is more robust to such error since chunks are not reallocated when outage does not occur. On the contrary, with matrix-based chunk allocation, an exhaustive search of the *best* user-chunk(s) pair is done. As a consequence, this phenomenon disappears and *1-bit hyperactive* performs as *2-bit lazy*. Furthermore our results show how performance of non real-time QoS based schedulers (e.g. MCI) can be significantly improved with *2-bit lazy* re-scheduler.

On figure 7 we show our simulation results for NRTV traffic with $rxtx_{max} = 2$ and *matrix-based chunk allocation*. Gains between persistent and lazy retransmission schedulers are respectively equal to 5%, 6.3% and 7% with PF, EDF and MCI packet schedulers. As for the above scenarios, best performance is obtained with EDF priority scheduling, having the maximum system capacity of 120 NRTV UEs when retransmissions are rescheduled with *2-bit lazy* or *1-bit hyperactive*. Note that even when same performance is obtained with *1-bit hyperactive* and *2-bit lazy* re-schedulers, complexity is significantly reduced by *2-bit lazy* as it will be discussed later. Dealing with our HYGIENE scheduler (not plotted here), it achieves quite the same performance as the ones obtained with EDF, with a slight gain for *1-bit hyperactive* and *1-bit persistent* (cell capacity of 120 UEs instead of 117 UEs).

On figure 8 the three re-schedulers coupled with *sequential chunk allocation* are compared in terms of PER_{res} CDF for 180 VoIP traffic activity. The priority scheduler is the MCI and $rxtx_{max} = 3$. VoIP traffic QoS constraints impose a target of $PER_{res} < 0.02$ for at least 95% of users. Simulation results show how, while *1-bit persistent* re-scheduler cannot guarantee such QoS requirements, both *1-bit hyperactive* and *2-bit lazy* re-schedulers do: 95% of users have respectively a PER_{res} of $2.6 \cdot 10^{-1}$, $6 \cdot 10^{-3}$ and $2.8 \cdot 10^{-3}$. Therefore *2-bit lazy* has best performance also in terms of PER_{res} CDF.

On table 2 we compare *1-bit hyperactive* and *2-bit lazy* re-schedulers in terms of chunk re-computation ratio (η), which is the percentage of chunk re-allocation per information packet. We compute η for the three re-schedulers as follows:

Persistent: chunk re-allocation is never done, $\eta = 0$;

Hyperactive: since chunk re-allocation is done for all NACK packets, η is the ratio between the sum of all NACKs and the sum of all transmitted information packets;

Lazy outage: since chunk re-allocation is done only for $NACK_{out}$ packets, η is the ratio between the sum of $NACK_{out}$ and the sum of all transmitted information packets.

Note that re-scheduling is activated only if the number of retransmissions does not exceed $rxtx_{max}$.

Numerical results on table 2 are reported for *matrix based chunk-allocation* and $rxtx_{max} = 2$. We verify that while PER_{res} and capacity are at least not degraded (often improved, see figures 5, 6, 7 and 8) by *2-bit lazy* re-scheduling, *1-bit hyperactive* does chunk re-computation more often. For instance, coupling MCI with *1-bit hyperactive* we observe respectively for VoIP and NRTV traffics $\eta = 7.3\%$ and $\eta = 9.5\%$. Coupling MCI with *2-bit lazy*, the re-computation ratio is

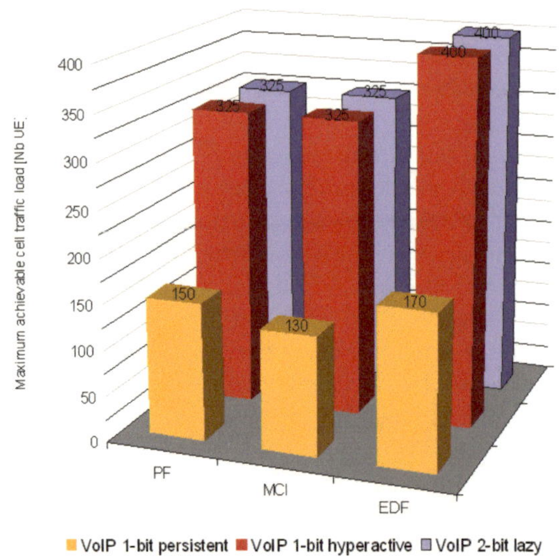

Fig. 5. $\mathbf{rxtx_{max}} = \mathbf{1}$: VoIP traffic. Comparison of 1-bit persistent, 1-bit hyperactive and 2-bit lazy frequency *re-schedulers* coupled with PF, MCI and EDF schedulers plus matrix-based chunk allocation

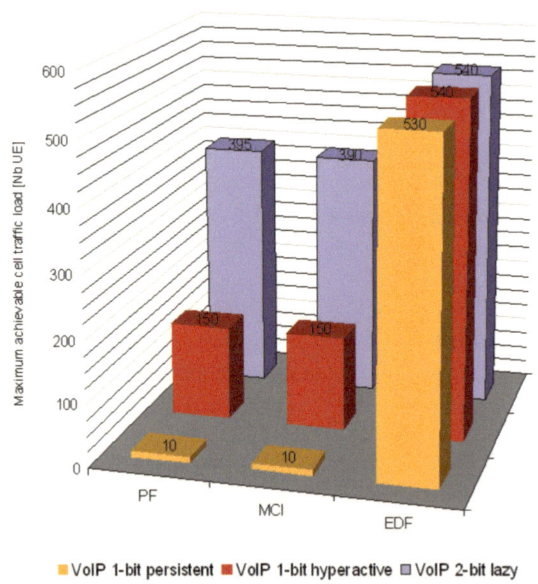

Fig. 6. $\mathbf{rxtx_{max}} = \mathbf{2}$: VoIP traffic. Comparison of 1-bit persistent, 1-bit hyperactive and 2-bit lazy frequency *re-schedulers* coupled with PF, MCI and EDF schedulers plus sequential chunk allocation

approximately divided by 10 for VoIP and by 5 for NRTV traffics. Indeed, *2-bit lazy* permits to notably reduce chunk re-allocation costs since it recomputes chunk allocation merely for $NACK_{out}$ packets. Comparing *2-bit lazy* with *1-bit persistent*, which is a very low complexity

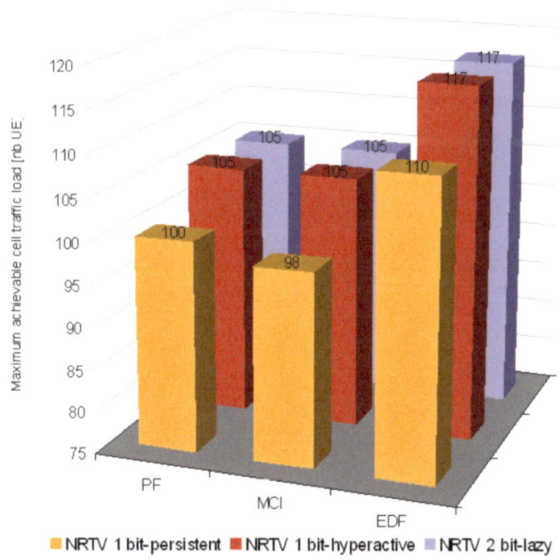

Fig. 7. $\mathbf{rxtx_{max}} = \mathbf{2}$: NRTV traffic. Comparison of 1-bit persistent, 1-bit hyperactive and 2-bit lazy frequency *re-schedulers* coupled with PF, MCI and EDF schedulers plus matrix-based chunk allocation

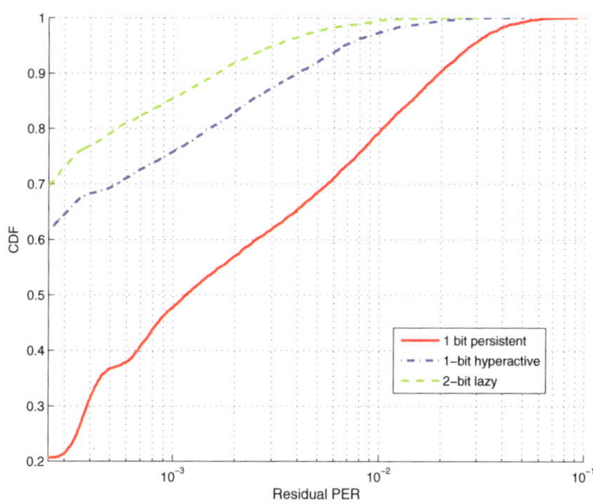

Fig. 8. $\mathbf{rxtx_{max}} = \mathbf{3}$: VoIP traffic. CDF of the residual PER for 180 UE/sector. Comparison of 1-bit persistent, 1-bit hyperactive and 2-bit lazy frequency *re-schedulers* coupled with MCI scheduler and sequential chunk allocation

chunk allocation re-scheduler, *2-bit lazy* notably performs best in terms of both maximum achievable system capacity and PER_{res} at the expense of very small additional complexity cost.

Chunk re-computation ratio can be further reduced by 4 when $NACK_{out}$ detection is based on *accumulative* mutual information as suggested at the end of section 5. For the same simulation scenario of table 2, the re-computation ratio is 0.12% for (MCI, VoIP) and 0.98% for (EDF, NRTV).

Scheduler	re-scheduler	VoIP	NRTV
PF	1-bit hyperactive	7.2%	9.1%
PF	2-bit lazy	0.6%	1.4%
MCI	1-bit hyperactive	7.3%	9.5%
MCI	2-bit lazy	0.5%	1.7%
EDF	1-bit hyperactive	7.6%	12.7%
EDF	2-bit lazy	0.8%	2.8%

Table 2. Chunk re-computation ratio

7. Conclusions and future research

In this chapter we have first presented an overview of currently investigated radio resource management solutions for OFDMA-based wireless cellular networks. We discussed advantages and weaknesses of main reference scheduling algorithms such as MCI, PF, MLWDF and EDF, in a system that implements a realistic OFDMA air interface based on the 3GPP/LTE downlink specifications where non full buffer traffic and a limited number of control channel per TTI were assumed. We focused our investigation on real-time, non real-time and coexisting real-time and non real-time traffic scenarios. We underlined that while EDF does not profit from multi-user diversity, MCI and PF schedulers target at maximizing the cell throughput regardless of the user's QoS constraints. Then we concentrated on the combining of FDPS and retransmission schedulers. We come out with two proposals.

First, we defined a novel scheduling algorithm, the HYGIENE scheduler. HYGIENE splits the resource allocation process in three steps: first, it identifies which *entities* (UE or packets) must be scheduled with high priority; second, it deals with *rushing entities*; third, remaining resources (if any) are allocated to users with highest momentary throughput. We evaluate the effectiveness of the proposed HYGIENE scheduler comparing it with the above reference schedulers. Our simulations substantiate how HYGIENE is a highly flexible and effective scheduler for a variety of traffic scenarios.

Second, we proposed a simple re-scheduling algorithm to efficiently deal with retransmissions. We propose that the *re-scheduler* checks, before reallocating chunks for NACK packet, if correct packet decoding is theoretically possible given the momentary channel instance and the pair (MCS, allocated chunk set). We base our re-scheduling strategy on the pair of information: channel outage instances and simple decoding errors (CRC). Thus, the *re-scheduler* recomputes the chunk allocation only for packets for which previous NACK transmission was in outage. The proposed method permits to reduce the residual PER while reducing the average number of retransmissions and increasing the overall cell capacity. Performance obtained is very favorable. We have better or equal performance than *1-bit hyperactive* re-scheduling while notably reducing the retransmission algorithm complexity. Furthermore, our results show how performance of non real-time QoS oriented schedulers (such as MCI and PF) can be significantly improved adopting the proposed *2-bit lazy re-scheduler*, especially with VoIP traffic flows. We have also shown by simulation that the

combination of both proposals (HYGIENE and smart re-scheduler) is an appealing solution to deal with real traffic and HARQ mechanisms.
Further work will focus on the combination of our proposals with CQI feedback schemes in order to assess the robustness of our proposals with respect to partial and inaccurate CQI reporting schemes.

8. References

Calvanese Strinati, E.; Kténas, D. (2009). HYGIENE Scheduling for OFDMA Wireless Cellular Networks, *Proceedings of IEEE VTC Spring*, Spain, April 2009, Barcelona.

Calvanese Strinati, E.; Kténas, D. (2009). Multi-User Dynamic (Re)transmission Scheduler for OFDMA Systems, *Proceedings of IEEE WCNC*, Hungary, April 2009, Budapest.

Ryu, S; Ryu, B; Seo, H.; Shin, M. (2005). Urgency and Efficiency based Packet Scheduling Algorithm for OFDMA wireless System, *Proceedings of IEEE ICC*, Seoul, 16-20 May 2005, Korea.

Yuen, C.W; Lau, W.C; Yue, O.C. (2007). CAPEL: A Packet Discard Policy for Real-time Traffic over Wireless Networks, *Proceedings of IEEE ICC*, Scotland, 24-28 June 2007, Glasgow.

3GPP TSG RAN (2006). 3GPP TR.25814, Physical Layer Aspects for Evolved UTRA (Release7)', v7.1.0 (2006-09).

3GPP TSG-RAN (2007). 3GPP TS36.211, Physical Channels and Modulation (Release 8), v8.1.0 (2007-11).

3GPP TSG-RAN1#48 (2007). Orange Labs, China Mobile, KPN, NTT DoCoMo, Sprint, T-Mobile, Vodafone and Telecom Italia, R1-070674, LTE physical layer framework for performance verification, 12th-16th February 2007, St Louis, USA.

K. Brueninghaus et al. (2005). Link Performance Models for System Level Simulations of Broadband radio Access Systems, *Proceedings of IEEE PIMRC*, Germany, September 2005, Berlin.

3GPP TSG RAN (2004). 3GPP TR 25.892, Feasibility Study for Orthogonal Frequency Division Multiplexing (OFDM) for UTRAN enhancement, v6.0.0 (2004-06).

T. Henttonen et al. (2008). Performance of VoIP with Mobility in UTRA Long Term Evolution, *Proceedings of IEEE VTC spring*, Singapore, May 2008, Singapore.

Pokhariyal, A.; Kolding, T.E.; Mogensen, P.E. (2006). Performance of Downlink Frequency Domain Packet Scheduling for the UTRAN Long Term Evolution, *Proceedings of IEEE PIMRC*, Finland, September 2006, Helsinki.

Norlund K.; Ottosson, T.; Brunstrom, A. (2004). Fairness measures for best effort traffic in wireless networks, *Proceedings of IEEE PIMRC*, Spain, September 2004, Barcelona.

Chiussi, F.M.; Sivaraman, V. (1998). Achieving High Utilization in Guaranteed Services Networks using Early-Deadline-First Scheduling, *Proceedings of 6th Int'l. Wksp. QoS*, May 1998.

Andrews, M.; Kumaran, K.; Ramanan, K.; Stolyar, A.; Whiting, P.; Vijayakumar, R. (2001). Providing Quality of Service over a Shared Wireless Link, *IEEE Communications Magazine*, February 2001, pp.150-154.

3GPP TSG-RAN1#44bis (2006). Motorola, R1-060877, Frequency Domain Scheduling for E-UTRA, 27th-31st March 2006, Athens, Greece.

V. Ramachandran et al. (2008). Frequency Selective OFDMA Scheduler with Limited Feedback, *Proceedings of IEEE WCNC*, USA, April 2008, Las Vegas.

3GPP TSG-RAN WG1 (2007). Samsung, R1-071971, E-UTRA Performance erification: VoIP, 22-23 April 2007.

Lampe, M.; Rohling, H.; Zirwas, W (2002). Misunderstandings about link adaptation for frequency selective fading channels, *Proceedings of IEEE International Symposium on Personal, Indoor, and Mobile Radio Communications*, Portugal, September 2002, Lisbon.

Emilio Calvanese Strinati (2005). Radio Link Control for Improving the QoS of Wireless Packet Transmission, *Ecole Nationale Supérieure des Télécomunications de Paris, PhD Thesis*, December 2005.

Emilio Calvanese Strinati, Système de télécommunication à adaptation de liaison et décodage conditionnel, *Technical Report of CEA-LETI (submitted for patent application)*, February 2007.

Ozarow, L.H.; Shamai, S.; Wyner, A.D., Information theoretic considerations for cellular mobile radio, *IEEE Trans. on Vehicular Tech.*, May 1994, vol. 43, no. 2, pp. 359-378.

Ungerboeck, G., Channel Coding with Multilevel/Phase Signals, *IEEE Transaction on Information Theory*, January 1982, vol. 28, pp. 55-67.

5

From Linear Equalization to Lattice-Reduction-Aided Sphere-Detector as an Answer to the MIMO Detection Problematic in Spatial Multiplexing Systems

Sébastien Aubert[1] and Manar Mohaisen[2]
[1]ST-Ericsson/INSA IETR
[2]Korea University of Technology and Education (KUT)
[1]France
[2]Republic of Korea

1. Introduction

Employing multiple antennas at both the transmitter and the receiver linearly boosts the channel capacity by $\min(n_T, n_R)$, where n_T and n_R are the number of transmit and receive antennas, respectively A. Telatar (1999). Multiple-Input Multiple-Output (MIMO) technologies are classified into three categories: (*i*) MIMO diversity, (*ii*) MIMO Spatial Multiplexing (MIMO-SM) and (*iii*) beamforming that will not be addressed here since it particularly deals with transmitter algorithms. MIMO diversity techniques are deployed to increase the reliability of communications by transmitting or receiving multiple copies of the same signal at different resource entities of the permissible dimensions, *i.e.*, time, frequency, or space. In contrast, the target of MIMO-SM is to increase the capacity of the communication channel. To this end, independent symbols are transmitted simultaneously from the different transmit antennas. Due to its attracting implementation advantages, Vertical Bell Laboratories Layered Space-Time (V-BLAST) transmitter structure is often used in the practical communication systems P. Wolniansky, G. Foschini, G. Golden, and R. Valenzuela (1998).

In 3GPP Long Term Evolution-Advanced (3GPP LTE-A) 3GPP (2009), the challenge of de-multiplexing the transmitted symbols via SM techniques, *i.e.* detection techniques, stands as one of the main limiting factors in linearly increasing system's throughput without requiring additional spectral resources. The design of detection schemes with high performance, low latency, and applicable computational complexity is being a challenging research topic due to the power and latency limitations of the mobile communication systems M. Mohaisen, H.S. An, and K.H.Chang (2009).

1.1 System model and problem statement

We consider a MIMO-SM system employing n_T transmit antennas and n_R receive antennas, where $n_T \leq n_R$ P. Wolniansky, G. Foschini, G. Golden, and R. Valenzuela (1998). The

This work was supported by ST-Ericsson and by research subsidy for newly-appointed professor of Korea University of Technology and Education for year 2011

simultaneously transmitted symbols given by the vector $\mathbf{x} \in \Omega_{\mathbb{C}}^{n_T}$ are drawn independently from a Quadrature Amplitude Modulation (QAM) constellation, where $\Omega_{\mathbb{C}}$ indicates the constellation set with size $|\Omega_{\mathbb{C}}|$.
Under the assumption of narrow-band flat-fading channel, the received vector $\mathbf{r} \in \mathbb{C}^{n_R}$ is given by:

$$\mathbf{r} = \mathbf{H}\mathbf{x} + \mathbf{n}, \tag{1}$$

where $\mathbf{n} \in \mathbb{C}^{n_R}$ is the Additive White Gaussian Noise (AWGN) vector whose elements are drawn from i.i.d. circularly symmetric complex Gaussian processes with mean and variance of zero and σ_n^2, respectively. $\mathbf{H} \in \mathbb{C}^{n_R \times n_T}$ denotes the complex channel matrix whose element $\mathbf{H}_{i,j} \sim \mathcal{CN}(0,1)$ is the channel coefficient between the j-th transmit antenna and the i-th receive antenna.
Working on the transmitted vector $\mathbf{x}$, $\mathbf{H}$ generates the complex lattice

$$\mathcal{L}(\mathbf{H}) = \left\{\mathbf{z} = \mathbf{H}\mathbf{x} | \mathbf{x} \in \Omega_{\mathbb{C}}^{n_T}\right\} = \left\{\mathbf{x}_1\mathbf{H}_1 + \mathbf{x}_2\mathbf{H}_2 + \cdots + \mathbf{x}_{n_T}\mathbf{H}_{n_T} | \mathbf{x}_i \in \Omega_{\mathbb{C}}^{n_T}\right\}, \tag{2}$$

where the columns of $\mathbf{H}$, $\{\mathbf{H}_1, \mathbf{H}_2, \cdots, \mathbf{H}_{n_T}\}$, are known as the basis vectors of the lattice $\mathcal{L} \in \mathbb{C}^{n_R}$ L. Lovász (1986). Also, n_T and n_R refer to the *rank* and *dimension* of the lattice $\mathcal{L}$, respectively, where the lattice is said to be *full-rank* if $n_T = n_R$.

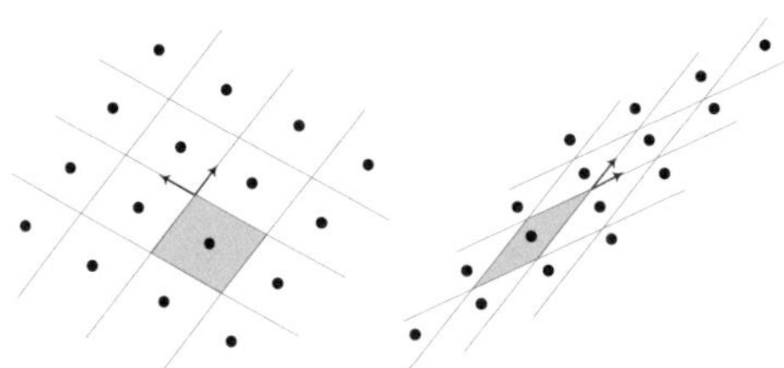

(a) (b)

Fig. 1. Examples of 2-dimensional real lattices with orthogonal bases (a) and correlated bases (b).

Figure 1(a) shows an example of a 2-dimensional real lattice whose basis vectors are $\mathbf{H}_1 = [0.39\ \ 0.59]^T$ and $\mathbf{H}_2 = [-0.59\ \ 0.39]^T$, and Figure 1(b) shows another example of a lattice with basis vectors $\mathbf{H}_1 = [0.39\ \ 0.60]^T$ and $\mathbf{H}_2 = [0.50\ \ 0.30]^T$. The elements of the transmitted vector $\mathbf{x}$ are withdrawn independently from the real constellation set $\{-3, -1, 1-, 3\}$. Herein we introduce the *orthogonality defect od* which is usually used as a measure of the orthogonality of the lattice basis:

$$od = \frac{\prod_{i=1}^{n_T} \|\mathbf{H}_i\|}{|\det\{\mathbf{H}\}|}, \tag{3}$$

where $\det\{\cdot\}$ refers to determinant and $od \geq 1$. The od of the lattices depicted in Figure 1(a) and Figure 1(b) are 1 and 2.28, respectively. This indicates that the first set of basis vectors is perfectly orthogonal while the second set is correlated, which implies inter-layer interferences and induces the advantage of joint detectors among others. The form of the resulting Voronoi regions of the different lattice points, an example is indicated in gray in Figure 1, also indicates the orthogonality of the basis; when the basis vectors are orthogonal with equal norms, the resulting Voronoi regions are squares, otherwise different shapes are obtained.
In light of the above and from a geometrical point of view, the signal detection problem is

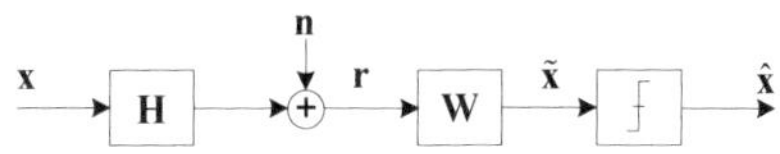

Fig. 2. Block diagram of the linear detection algorithms.

defined as finding the lattice point $\hat{\mathbf{z}} = \mathbf{H}\hat{\mathbf{x}}$, such that $\|\mathbf{r} - \hat{\mathbf{z}}\|^2$ is minimized, where $\|\cdot\|$ is the Euclidean norm and $\hat{\mathbf{x}}$ is the estimate of the transmitted vector $\mathbf{x}$.

1.2 Maximum-likelihood detection

The optimum detector for the transmit vector estimation is the well-known Maximum-Likelihood Detector (MLD) W. Van Etten (1976). MLD employs a brute-force search to find the vector $\mathbf{x}_k$ such that the a-posteriori probability $P\{\mathbf{x}_k \,|\, \mathbf{r}\},\ k = 1, 2, \cdots, |\Omega_{\mathbb{C}}|^{n_T}$, is maximized; that is,

$$\hat{\mathbf{x}}_{\text{ML}} = \arg\max_{\mathbf{x} \in \Omega_{\mathbb{C}}^{n_T}} \left(P\{\mathbf{x}_k | \mathbf{r}\} \right). \tag{4}$$

After some basic probability manipulations, the optimization problem in (4) is reduced to:

$$\hat{\mathbf{x}}_{\text{ML}} = \arg\max_{\mathbf{x} \in \Omega_{\mathbb{C}}^{n_T}} \left(p\left(\mathbf{r} | \mathbf{x}_k\right) \right), \tag{5}$$

where $p(\mathbf{r}|\mathbf{x}_k)$ is the probability density function of $\mathbf{r}$ given $\mathbf{x}_k$. By assuming that the elements of the noise vector $\mathbf{n}$ are i.i.d. and follow Gaussian distribution, the noise covariance matrix becomes $\Sigma_n = \sigma_n^2 \mathbf{I}_{n_R}$. As a consequence, the received vector is modelled as a multivariate Gaussian random variable whose mean is $(\mathbf{H}\mathbf{x}_k)$ and covariance matrix is Σ_n. The optimization problem in (5) is rewritten as follows:

$$\hat{\mathbf{x}}_{\text{ML}} = \arg\max_{\mathbf{x} \in \Omega_{\mathbb{C}}^{n_T}} \left(\frac{1}{\pi^{n_R} \det(\Sigma)} \exp^{-(\mathbf{r}-\mathbf{H}\mathbf{x}_k)^H \Sigma_n^{-1} (\mathbf{r}-\mathbf{H}\mathbf{x}_k)} \right) = \arg\min_{\mathbf{x} \in \Omega_{\mathbb{C}}^{n_T}} \left(\|\mathbf{r} - \mathbf{H}\mathbf{x}_k\|^2 \right). \tag{6}$$

This result coincides with the conjuncture based on the lattice theory given in section 1.1. The computational complexity of the MLD is known to be exponential in the modulation set size $|\Omega_{\mathbb{C}}|$ and the number of transmit antennas n_T. For mobile communications systems, which are computational complexity and latency limited, MLD becomes infeasible. In the following Sections, we review the conventional sub-optimal detection algorithms J. Wang, and B. Daneshrad (2005), and analyse their advantages and inconveniences.

2. Linear detection algorithms

The idea behind linear detection schemes is to treat the received vector by a filtering matrix $\mathbf{W}$, constructed using a performance-based criterion, as depicted in Figure 2 A. Paulraj, R. Nabar, and D. Gore (2003), C. Windpassenger (2004), B. Schubert (2006). The well known Zero-Forcing (ZF) and Minimum-Mean Square Error (MMSE) performance criteria are used in the Linear ZF (LZF) and MMSE (LMMSE) detectors.

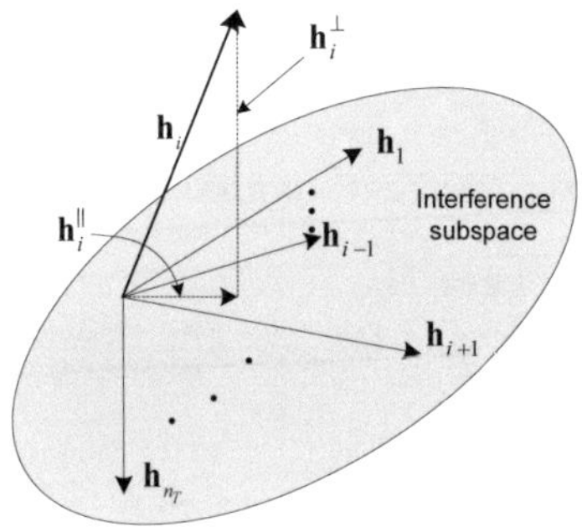

Fig. 3. Geometrical representation of the linear zero-forcing detection algorithm.

2.1 Linear Zero-Forcing detector

LZF detector treats the received vector by the pseudo-inverse of the channel matrix, resulting in full cancellation of the interference with colored noise. The detector in matrix form is given by:

$$\mathbf{W}_{\text{ZF}} = \left(\mathbf{H}\mathbf{H}^{H}\right)^{-1}\mathbf{H}^{H}, \tag{7}$$

where $(\cdot)^H$ is the Hermitian transpose.

Figure 3 depicts a geometrical representation of the LZF detector. Geometrically, to obtain the k-th detector's output, the received vector is processed as follows:

$$\tilde{\mathbf{x}}_k = \underline{\mathbf{w}}_k\mathbf{r} = \frac{\left(\mathbf{H}_k^{\perp}\right)^{H}}{\left\|\mathbf{H}_k^{\perp}\right\|^2}\mathbf{r} = \mathbf{x}_k + \mu_k, \tag{8}$$

where $\underline{\mathbf{w}}_k$ is the k-th row of $\mathbf{W}$, $\mathbf{H}_k^{\perp}$ is the perpendicular component of $\mathbf{H}_i$ on the interference space, and μ_k equals $\underline{\mathbf{w}}_k\mathbf{n}$. Note that $\mathbf{H}_k^{\perp}$ equals $\left(\mathbf{H}_k - \mathbf{H}_k^{||}\right)$, where $\mathbf{H}_k^{||}$ is the parallel component of $\mathbf{H}_k$ to the interference subspace. Then, the mean and variance of the noise μ_k at the output of the LZF detector are 0 and $\|\underline{\mathbf{w}}_k\|^2\sigma_n^2$, respectively. When the channel matrix is ill-conditioned, e.g., if a couple or more of columns of the channel matrix are correlated, $\mathbf{H}_k^{||}$ becomes large, and the noise is consequently amplified.

2.2 Linear minimum-mean square error detector

To alleviate the noise enhancement problem induced by the ZF equalization, the LMMSE can be used. The LMMSE algorithm optimally balances the residual interference and noise enhancement at the output of the detector. To accomplish that, the filtering matrix $\mathbf{W}_{\text{MMSE}}$ is given by:

$$\mathbf{W}_{\text{MMSE}} = \arg\min_{\mathbf{G}} \left(\mathrm{E}\left[\|\mathbf{G}\mathbf{r} - \mathbf{x}\|^2\right]\right), \tag{9}$$

where $\mathrm{E}[\cdot]$ denotes the expectation. Due to the orthogonality between the received vector and the error vector given in (9), we have:

$$\mathrm{E}\left[\left(\mathbf{W}_{\text{MMSE}}\mathbf{r} - \mathbf{x}\right)\mathbf{r}^{H}\right] = \mathbf{0}, \tag{10}$$

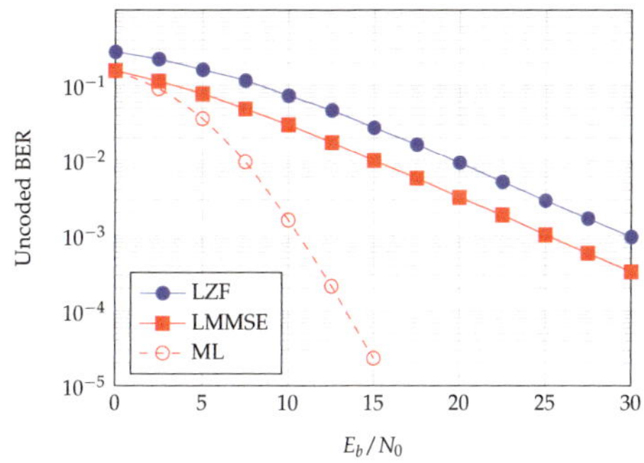

Fig. 4. Uncoded BER as a function of E_b/N_0, Complex Rayleigh 4×4 MIMO channel, LZF, LMMSE and ML detectors, QPSK modulations at each layer.

and by extending the left side of (10), it directly follows that:

$$\mathbf{W}_{\text{MMSE}} = \left(\Phi_{\mathbf{xx}}^{-1} + \mathbf{H}^H \Phi_{\mathbf{nn}}^{-1} \mathbf{H}\right)^{-1} \mathbf{H}^H \Phi_{\mathbf{nn}}^{-1} = \left(\mathbf{H}^H \mathbf{H} + \frac{\sigma_n^2}{\sigma_x^2} \mathbf{I}_{n_T}\right)^{-1} \mathbf{H}^H, \tag{11}$$

where $\Phi_{\mathbf{nn}}$ equals $\sigma_n^2 \mathbf{I}$ and $\Phi_{\mathbf{xx}}$ equals $\sigma_x^2 \mathbf{I}$ are the covariance matrices of the noise and the transmitted vectors, respectively. Theoretically, at high Signal-to-noise Ratio (SNR), the LMMSE optimum filtering converges to the LZF solution. However, we show in M. Mohaisen, and K.H. Chang (2009b) that the improvement by the LMMSE detector over the LZF detector is not only dependent on the plain value of the noise variance, but also on how close σ_n^2 is to the singualr values of the channel matrix. Mathematically, we showed that the ratio between the condition number of the filtering matrices of the linear MMSE and ZF detectors is approximated as follows:

$$\frac{cond(\mathbf{W}_{\text{MMSE}})}{cond(\mathbf{W}_{\text{ZF}})} \approx \frac{1 + \sigma_n^2/\sigma_1^2(\mathbf{H})}{1 + \sigma_n^2/\sigma_N^2(\mathbf{H})}, \tag{12}$$

where σ_1 and σ_N are the maximal and minimal singular values of the channel matrix $\mathbf{H}$, and σ_n^2 is the noise variance. Also, $cond(\mathbf{A}) = (\sigma_1(\mathbf{A})/\sigma_N(\mathbf{A}))$ is the condition number that attains a minimum value of one for orthogonal $\mathbf{A}$.

Figure 4 shows the Bit Error Rate (BER) of the linear detection algorithms in 4×4 MIMO multiplexing system, using 4-QAM signalling. Although the BER performance of LMMSE is close to that of MLD for low E_b/N_0 values, the error rate curves of the two linear detection algorithms have a slope of -1, viz., diversity order equals one, whereas the diversity order of the MLD equals $n_R = 4$.

3. Decision-feedback detection

3.1 Introduction

Although linear detection approaches are attractive in terms of computational complexity, they lead to degradation in the BER performance, due to independent detection of $\mathbf{x}$ components. Superior performance can be obtained if non-linear approaches are employed, as in the Decision-Feedback Detection (DFD) algorithms. In DFD approach, symbols are detected successively, where already-detected components of $\mathbf{x}$ are subtracted out from the

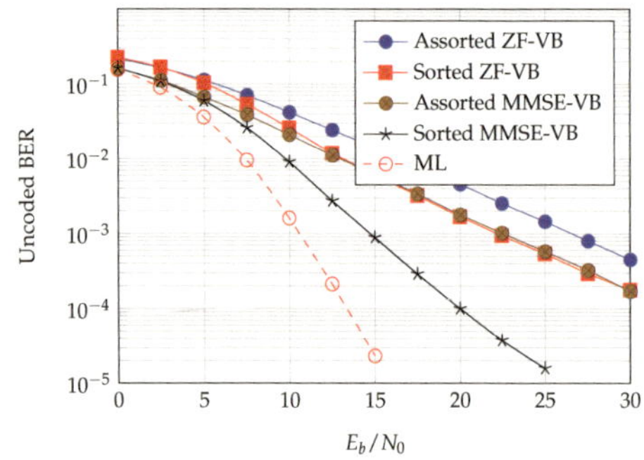

Fig. 5. Uncoded BER as a function of E_b/N_0, Complex Rayleigh 4×4 MIMO channel, Assorted ZF-VB, Sorted ZF-VB, Assorted MMSE-VB, Sorted MMSE-VB and ML detectors, QPSK modulations at each layer.

received vector. This leads to a system with fewer interferers. In the following two sections, we introduce two categories of DFD algorithms.

3.2 The V-BLAST detection algorithm

In V-BLAST, symbols are detected successively using the aforementioned linear detection approaches. At the end of each iteration, the already-detected component of $\mathbf{x}$ is subtracted out from the received vector P. Wolniansky, G. Foschini, G. Golden, and R. Valenzuela (1998), S. Haykin, and M. Moher (2005). Also, the corresponding column of the matrix $\mathbf{H}$ is removed. When decision-feedback approach is used, error propagation becomes a challenging issue. Therefore, the order in which symbols are detected has a great impact on the system performance.

The idea behind the ZF-based V-BLAST (ZF-VB) algorithm is to detect the components of $\mathbf{x}$ that suffer the least noise amplification at first. For the first decision, the pseudo-inverse, *i.e.*, $\mathbf{W}$ equals $\mathbf{H}^{\dagger}$, of the matrix $\mathbf{H}$ is obtained. By assuming that the noise components are i.i.d and that noise is independent of $\mathbf{x}$, then the row of $\mathbf{W}$ with the least Euclidean norm corresponds to the required component of $\mathbf{x}$. That is $k_1 = \arg\min_j (\|\underline{\mathbf{w}}_j\|^2)$ and $\tilde{\mathbf{x}}_{k_1} = \underline{\mathbf{w}}_{k_1}\mathbf{r}^{(1)}$, where $\mathbf{r}^{(1)}$ = $\mathbf{r}$, the superscript indicates the iteration number, and $\hat{\mathbf{x}}_{k_1} = \mathcal{Q}_{\Omega_{\mathbb{C}}}(\tilde{\mathbf{x}}_{k_1})$ is the decision for the k_1-th component of $\mathbf{x}$. The interference due to the k_1-th symbol is then cancelled out as follows: $\mathbf{r}^{(2)} = \mathbf{r}^{(1)} - \hat{\mathbf{x}}_{k_1}\mathbf{H}_{k_1}$ and $\mathbf{H}^{(2)} = \left[\cdots, \mathbf{H}_{k_1-1}, \mathbf{H}_{k_1+1}, \cdots\right]$. This strategy is repeated up to the last component of $\mathbf{x}$.

In analogy with the linear detection approaches, MMSE-based V-BLAST (MMSE-VB) improves the BER performance, by alleviating the noise enhancement problem. Therefore, the filtering matrix $\mathbf{G}^{(i)}$ at the i-th iteration is given by:

$$\mathbf{G}^{(i)} = \left(\mathbf{H}^{H^{(i)}}\mathbf{H}^{(i)} + \frac{\sigma_n^2}{\sigma_x^2}\mathbf{I}\right)^{-1}\mathbf{H}^{H^{(i)}}, \tag{13}$$

rather than $\mathbf{H}^{(i)\dagger}$ in the case of ZF-VB.

Figure 5 shows the BER performance of the V-BLAST for several detection algorithms. In the assorted V-BLAST schemes, the symbols are detected in an ascending order, *i.e.*, $\mathbf{x}_1, \mathbf{x}_2, \cdots, \mathbf{x}_{n_T}$, without considering their noise conditions. Obviously, signal ordering leads to improvements for both the ZF-VB and MMSE-VB algorithms, and the improvement is larger in the case of the MMSE-VB algorithm. We note also that the diversity order of the MMSE-VB

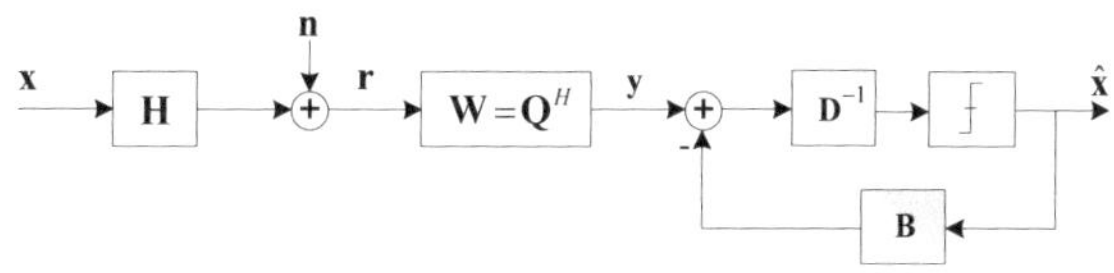

Fig. 6. QRD-DFD.

is larger than 1, unlike the other V-BLAST algorithms which have a diversity order close to 1. This indicates less error propagation in the case of MMSE-VB, compared to other V-BLAST detection algorithms.
The computational complexity of the BLAST detection algorithm, both ZF-VB and MMSE-VB, is $\mathcal{O}(n^4)$, which is infeasible due to power and latency limitations of the mobile communication systems. Although several techniques were proposed to reduce the complexity of the BLAST detection algorithm, it is still complex B. Hassibi (2000); H. Zhu, Z. Lei, and F. Chin (2004); J. Benesty, Y. Huang, and J. Chen (2003).

3.3 QR Decomposition-based detection-feedback detection

The DFD scheme, that lies in the QR Decomposition (QRD) of the channel matrix S. Aubert, M. Mohaisen, F. Nouvel, and K.H. Chang (2010), requires only a fraction of the computational efforts required by the V-BLAST detection algorithm D. Shiu, and J. Kahn (1999). This is why QRD-based DFD (QRD-DFD) is preferable for power and latency limited wireless communication systems.
In QRD-DFD, the channel matrix is decomposed into the multiplication of a unitary matrix $\mathbf{Q} \in \mathbb{C}^{n_R \times n_T}$, *i.e.*, $\mathbf{Q}^H\mathbf{Q} = \mathbf{I}_{n_T}$, and an upper triangular matrix $\mathbf{R} \in \mathbb{C}^{n_T \times n_T}$; that is $\mathbf{H} = \mathbf{QR}$. Then, the received vector is multiplied by the Hermitian transpose of $\mathbf{Q}$, we have

$$\mathbf{y} = \mathbf{Rx} + \mathbf{v} = (\mathbf{D} + \mathbf{B})\,\mathbf{x} + \mathbf{v}, \tag{14}$$

where $\mathbf{y} = \mathbf{Q}^H\mathbf{r}$ and $\mathbf{v} = \mathbf{Q}^H\mathbf{n}$. Note that the noise statistics does not changeable due to the orthogonality of $\mathbf{Q}$. The matrix $\mathbf{D}$ is a diagonal matrix whose diagonal elements are the diagonal elements of $\mathbf{R}$, and $\mathbf{B}$ is strictly upper triangular matrix. As a consequence, the MIMO system becomes spatially causal which implies that:

$$\mathbf{y}_k = \mathbf{R}_{k,k}\tilde{\mathbf{x}}_k + \sum_{i=k+1}^{n_T} \mathbf{R}_{k,j}\hat{\mathbf{x}}_i, \tag{15}$$

where $\tilde{\mathbf{x}}_k$ is a candidate symbol and $\hat{\mathbf{x}}_k$ is the estimate, both for the $k-$th component of $\mathbf{x}$. Therefore,

$$\hat{\mathbf{x}}_k = \mathcal{Q}_{\Omega_C}\left(\frac{\mathbf{y}_k - \sum_{i=k+1}^{n_T} \mathbf{R}_{k,j}\hat{\mathbf{x}}_i}{\mathbf{R}_{k,k}}\right). \tag{16}$$

Note that due to the structure of the matrix $\mathbf{R}$, the last component of $\mathbf{x}$, *i.e.*, x_{n_T}, is interference-free, hence, it can be detected first. Already-detected component of $\mathbf{x}$ is cancelled out to detect the following component. This technique is repeated up to the first component of $\mathbf{x}$, *i.e.*, $\mathbf{x}_1$ D. Wübben, R. Böhnke, J. Rinas, V. Kühn, and K.-D. Kammeyer (2001); D. Wübben, R. Böhnke, V. Kühn, and K.-D. Kammeyer (2003); M. Mohaisen, and K.H. Chang (2009a).
Fig. 6 depicts the detailed QRD-based detection algorithm (ZF-QRD). Note that the feedback loop is equivalent to $(\mathbf{D} + \mathbf{B})^{-1} = \mathbf{R}^{-1}$. Figure 7 depicts the BER performance of the the QRD-based DFD algorithms. The MMSE-SQRD algorithm has the best performance but its diversity order converges to unity for high E_b/N_0 values.

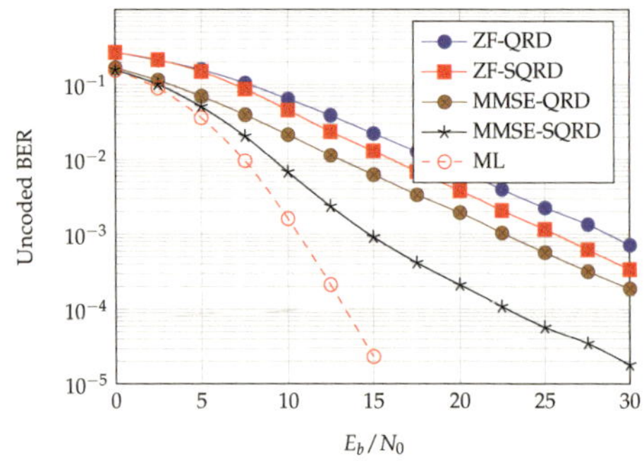

Fig. 7. Uncoded BER as a function of E_b/N_0, Complex Rayleigh 4×4 MIMO channel, ZF-QRD, ZF-SQRD, MMSE-QRD, MMSE-SQRD and ML detectors, QPSK modulations at each layer.

4. Tree-search detection

Several tree-search detection algorithms have been proposed in the literature that achieve quasi-ML performance while requiring lower computational complexity. In these techniques, the lattice search problem is presented as a tree where nodes represent the symbols' candidates. In the following, we introduce three tree-search algorithms and discuss their advantages and drawbacks.

4.1 Sphere Decoder

The Sphere Decoder (SD) was proposed in the literature to solve several lattice search problems J. Boutros, N. Gresset, L. Brunel, and M. Fossorier (2003). Based on Hassibi and Vikalo analysis, SD achieves quasi-ML performance with polynomial average computational complexity for large range of SNR B. Hassibi, and H. Vikalo (2001). Hence, instead of testing all the hypotheses of the transmitted vector, SD restricts the search in (6) to the lattice points that reside in the hypersphere of radius d and implicitly centred at the unconstrained ZF estimate. Therefore,

$$\hat{\mathbf{x}}_{\text{SD}} = \underset{\mathbf{x} \in \Omega_{\mathbb{C}}^{n_T}}{\arg\min} \left(\|\mathbf{r} - \mathbf{H}\mathbf{x}\|^2 \leq d^2 \right). \tag{17}$$

The order in which hypotheses are tested at each detection level is defined by the employed search strategy, namely Fincke-Pohst (FP) U. Fincke, and M. Pohst (1985) or Schnorr-Euchner (SE) C. Schnorr, and M. Euchner (1994) strategies. SE strategy orders, and then examines, the hypotheses based on their Euclidean distance from the unconstrained ZF solution; closer hypothesis is tested first. On the other hand, FP strategy tests the hypotheses at each layer without considering the distance from the unconstrained ZF solution. That is why FP strategy leads to higher computational complexity. Figure 8 shows the details of these two strategies for 6-Pulse Amplitude Modulation (PAM) constellation where $\hat{\mathbf{z}}_i$ is the unconstrained ZF solution of the $i-$th transmitted symbol and the numbers under the constellation points represent the order in which hypotheses are tested. It is clear that employing SE strategy leads to reduction in the complexity since the most probable hypothesis for any considered layer, independently of others, is tested first.

It could be shown that the BER performance of the SD coincides with that of the optimum detector. However, some drawbacks remain. In particular, the complexity of SD is variable

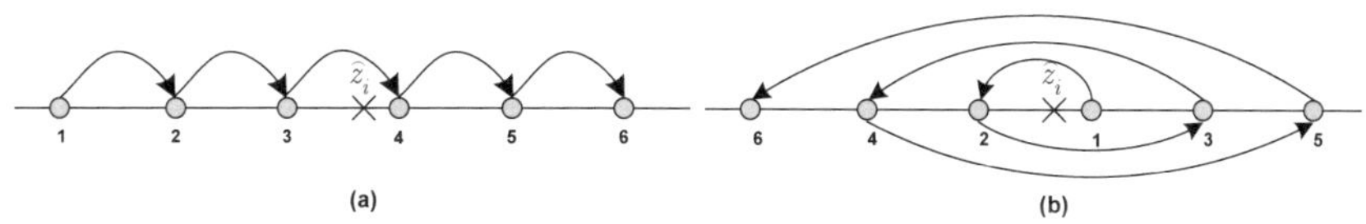

Fig. 8. Search strategies employed by the sphere decoder FP strategy (a) and SE strategy (b).

and depends on the conditionality of the channel matrix and the noise variance, where the worst-case complexity of SD is consequently comparable with that of MLD. That is, the worst-case complexity of SD is exponential. In fact, Jalden and Otterson have shown in J. Jaldén, and B. Ottersten (2005) that even the average complexity of SD is exponential for a fixed SNR value. Also, the SD has a sequential nature because it requires the update of the search radius at every time a new lattice point with smaller accumulative metric is found. This limits the possibility of parallel processing and hence reduces the detection throughput, *i.e.*, increases the detection latency.

To fix the complexity of the detection stage and allow a parallel processing, the QRD with M-algorithm (QRD-M) has been proposed. The QRD-M algorithm and several related complexity reduction techniques are introduced in the following sections. Also, it is equivalently denoted as the K-Best.

4.2 QRD-M detection algorithm

In the QRD-M detection algorithm, only a fixed number, M, of symbol candidates is retained at each detection level J. Anderson, and S. Mohan (1984); K.J. Kim, J. Yue, R.A. Iltis, and J.D. Gibson (2005). At the first detection level, the root node is extended to all the possible candidates of $\mathbf{x}_{n_T}$, the accumulative metrics of the resulting branches are calculated and the best M candidates with the smallest metrics are retained for the next detection level. At the second detection level, the retained M candidates at the previous level are extended to all possible candidates. The resulting $(|\Omega_C| \times M)$ branches are sorted according to their accumulative metrics where the M branches with the smallest accumulative metrics are retained for the next detection level. This strategy is repeated down to the last detection level. By employing the QRD-M strategy, near-ML BER performance are reached for a sufficient M value, as depicted in Figure 9, while the computational complexity of the detection algorithm becomes fixed and only dependent on the size of the modulation set $|\Omega_C|$ and the number of transmit antennas n_T. It also makes a parallel implementation possible. Note that the overall number of visited nodes by the QRD-M algorithm $= (|\Omega_C| + (n_T - 1) \times |\Omega_C| \times M)$.

Although the conventional QRD-M has a fixed complexity which is an advantage, it does not take into consideration the noise and channel conditions. Thus, unnecessary computations are usually done when the channel is well-conditioned and the signal to noise ratio is high. Also, for high $|\Omega_C|$ and n_T, the computational complexity of the QRD-M algorithm becomes high, where as a consequence the detection latency increases.

In order to solve this point, several algorithms have been proposed in the literature. A solution that lies on applying a variable M has been widely studied and is denoted as the dynamic QRD-M algorithm. It offers promising performance results in the case of a SQRD pre-processing step. These algorithms reduce the number of retained candidates at each detection level with tolerable degradation in the performance H. Kawai, K. Higuchi, N. Maeda, and M. Sawahashi (2006); K. Jeon, H. Kim, and H. Park (2006); M. Mohaisen, and K.H. Chang (2010) and references therein. In M. Mohaisen, K.H. Chang, and B.T. Koo (2009), two

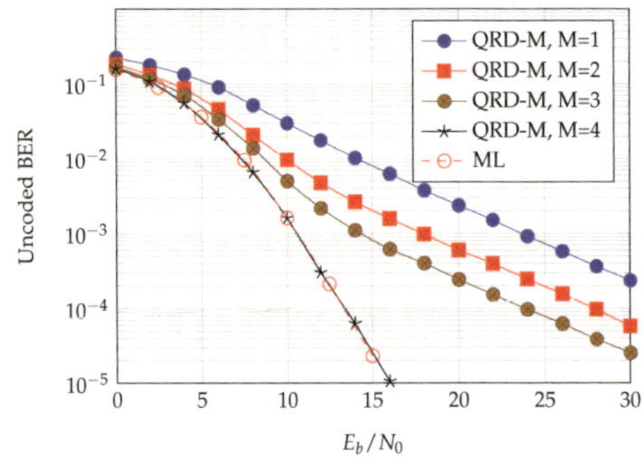

Fig. 9. Uncoded BER as a function of E_b/N_0, Complex Rayleigh 4×4 MIMO channel, QRD-M algorithm for multiple M values and ML detectors, QPSK modulations at each layer.

algorithms have been proposed, that in addition of reducing the computational complexity of the QRD-M algorithm, they either reduce the processing delay by parallelizing the detection stage or reduce the hardware requirements by iteratively processing the QRD-M algorithm. While a variable M also implies a variable computational complexity, a particular case has to be introduced.

4.3 Fixed-complexity sphere decoder

Fixed-complexity Sphere Decoder (FSD) was proposed by Barbero *et al.* to overcome the aforementioned drawbacks of SD. FSD achieves a quasi-ML performance by performing the following two-stage tree search L. Barbero, and J. Thompson (2008a;b):

- Full expansion: In the first p levels, a full expansion is performed, where all symbols replicas candidates are retained to the following levels;
- Single expansion: A single expansion of each retained branch is done in the remaining $(n_T - p)$ levels, where only the symbol replica candidate with the lowest accumulative metric is considered for next levels.

Because all possible symbols candidates are retained in the first p levels, the reliability of signals detected in these levels does not affect the final detection performance compared to MLD. Therefore, signals with the least robustness are detected in the full expansion stage. On the other hand, in the remaining $(n_T - p)$ levels, signals are sorted based on their reliability, where signals with the least noise amplification are detected first. In the conventional FSD, the V-BLAST algorithm is employed to obtain the required signal ordering by the FSD.
Figure 10 depicts the BER performance of the FSD for $p = 1$ in 4×4 MIMO-SM system using 4-QAM. Results show that the ordering has a crucial effect on the the performance of the FSD. For instance, both the performance and the attained diversity order are degraded when the ordering stage is skipped or when a non-optimal signal ordering is used. A low complexity FSD ordering scheme that requires a fraction of the computations of the V-BLAST scheme was proposed in M. Mohaisen, and K.H. Chang (2009a), where a close to optimum performance was achieved by embedding the sorting stage in the QR factorization of the channel matrix.

4.4 MMSE-centred sphere decoder

The SD principle may be extended. By defining the Babai point - only in the case of a depth-first search algorithm - as the first obtained solution by the algorithm, the induced Babai point in this case is implicitly a ZF-QRD. In the case of a QRD-M algorithm, this

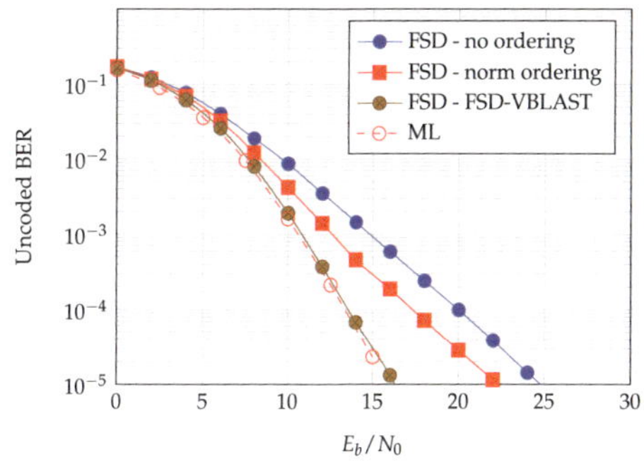

Fig. 10. Uncoded BER as a function of E_b/N_0, Complex Rayleigh 4×4 MIMO channel, FSD algorithms with $p = 1$ and ML detectors, QPSK modulations at each layer.

definition is extended and is considered as the solution that would be directly reached, without neighborhood study. Another useful notation that has to be introduced is the sphere search around the center search $\mathbf{x}_C$, namely the signal in any equation of the form $\|\mathbf{x}_C - \mathbf{x}\|^2 \leq d^2$, where $\mathbf{x}$ is any possible hypothesis of the transmitted vector $\mathbf{x}$, which is consistent with the equation of an $(n_T - 1)-$ sphere.
Classically, the SD formula is centred at the unconstrained ZF solution and the corresponding detector is denoted in the sequel as the naïve SD. Consequently, a fundamental optimization may be considered by introducing an efficient search center that results in an already close-to-optimal Babai point. In other words, to obtain a solution that is already close to the ML solution. This way, it is clear that the neighborhood study size can be decreased without affecting the outcome of the search process. In the case of the QRD-M algorithm, since the neighborhood size is fixed, it will induce a performance improvement for a given M or a reduction of M for a given target BER.
The classical SD expression may be re-arranged, leading to an exact formula that has been firstly proposed by Wong *et al.*, aiming at optimizations for a VLSI implementation through an efficient Partial Euclidean distance (PED) expression and early pruned nodes K.-W. Wong, C.-Y. Tsui, S.-K. Cheng, and W.-H. Mow (2002):

$$\mathbf{x}_{ZF-DFD} = \underset{\mathbf{x} \in \Omega_{\mathbb{C}}^{n_T}}{\operatorname{argmin}} \|\mathbf{R}\mathbf{e}_{ZF}\|^2, \tag{18}$$

where $\mathbf{e}_{ZF} = \mathbf{x}_{ZF} - \mathbf{x}$ and $\mathbf{x}_{ZF} = (\mathbf{H}^H\mathbf{H})^{-1}\mathbf{H}^H\mathbf{r}$. Equation (18) clearly exhibits the point that the naïve SD is unconstrained ZF-centred and implicitly corresponds to a ZF-QRD procedure with a neighborhood study at each layer.
The main idea proposed by B.M. Hochwald, and S. ten Brink (2003); L. Wang, L. Xu, S. Chen, and L. Hanzo (2008); T. Cui, and C. Tellambura (2005) is to choose a closer-to-ML Babai point than the ZF-QRD, which is the case of the MMSE-QRD solution. For sake of clearness with definitions, we say that two ML equations are equivalent if the lattice points argument outputs of the minimum distance are the same, even in the case of different metrics. Two ML equations are equivalent iff:

$$\underset{\mathbf{x} \in \Omega_{\mathbb{C}}^{n_T}}{\operatorname{argmin}} \{\|\mathbf{r} - \mathbf{H}\mathbf{x}\|^2\} = \underset{\mathbf{x} \in \Omega_{\mathbb{C}}^{n_T}}{\operatorname{argmin}} \{\|\mathbf{r} - \mathbf{H}\mathbf{x}\|^2 + c\}, \tag{19}$$

where c is a constant.
In particular, Cui *et al.* T. Cui, and C. Tellambura (2005) proposed a general equivalent

minimization problem: $\hat{\mathbf{x}}_{ML} = \underset{\mathbf{x}\in\Omega_{\mathbb{C}}^{n_T}}{\text{argmin}}\{\|\mathbf{r}-\mathbf{H}\mathbf{x}\|^2+\alpha\mathbf{x}^H\mathbf{x}\}$, by noticing that signals $\mathbf{x}$ have to be of constant modulus. This assumption is obeyed in the case of QPSK modulation and is not directly applicable to 16-QAM and 64-QAM modulations, even if this assumption is not limiting since a QAM constellation can be considered as a linear sum of QPSK points T. Cui, and C. Tellambura (2005).

This expression has been applied to the QRD-M algorithm by Wang *et al.* in the case of the unconstrained MMSE-center which leads to an MMSE-QRD procedure with a neighborhood study at each layer L. Wang, L. Xu, S. Chen, and L. Hanzo (2008). In this case, the equivalent ML equation is rewritten as:

$$\hat{\mathbf{x}}_{ML} = \underset{\mathbf{x}\in\Omega_{\mathbb{C}}^{n_T}}{\text{argmin}}\ (\mathbf{x}_C-\mathbf{x})^H\left(\mathbf{H}^H\mathbf{H}+\sigma^2\mathbf{I}\right)(\mathbf{x}_C-\mathbf{x})\,. \tag{20}$$

Through the use of the Cholesky Factorization (CF) of $\mathbf{H}^H\mathbf{H}+\sigma^2\mathbf{I}=\mathbf{U}^H\mathbf{U}$ in the MMSE case ($\mathbf{H}^H\mathbf{H}=\mathbf{U}^H\mathbf{U}$ in the ZF case), the ML expression equivalently rewrites:

$$\hat{\mathbf{x}}_{ML} = \underset{\mathbf{x}\in\Omega_{\mathbb{C}}^{n_T}}{\text{argmin}}\ (\tilde{\mathbf{x}}-\mathbf{x})^H\mathbf{U}^H\mathbf{U}(\tilde{\mathbf{x}}-\mathbf{x})\,, \tag{21}$$

where $\mathbf{U}$ is upper triangular with real elements on diagonal and $\tilde{\mathbf{x}}$ is any (ZF or MMSE) unconstrained linear estimate.

5. Lattice reduction

For higher dimensions, the ML estimate can be provided correctly with a reasonable complexity using a Lattice Reduction (LR)-aided detection technique.

5.1 Lattice reduction-aided detectors interest

As proposed in H. Yao, and G.W. Wornell (2002), LR-Aided (LRA) techniques are used to transform any MIMO channel into a better-conditioned (short basis vectors norms and roughly orthogonal) equivalent MIMO channel, namely generating the same lattice points. Although classical low-complexity linear, and even (O)DFD detectors, fail to achieve full diversity as depicted in D. Wübben, R. Böhnke, V. Kühn, and K.-D. Kammeyer (2004), they can be applied to this equivalent (the exact definition will be introduced in the sequel) channel and *significantly improve performance* C. Windpassinger, and R.F.H. Fischer (2003). In particular, it has been shown that LRA detectors achieve the full diversity C. Ling (2006); M. Taherzadeh, A. Mobasher, and A.K. Khandani (2005); Y.H. Gan, C. Ling, and W.H. Mow (2009). By assuming $i<j$, Figure 11 depicts the decision regions in a trivial two-dimensional case and demonstrates to the reader the reason why LRA detection algorithms offer better performance by approaching the optimal ML decision areas D. Wübben, R. Böhnke, V. Kühn, and K.-D. Kammeyer (2004). From a singular value theory point of view, when the lattice basis is reduced, its singular values becomes much closer to each other with equal singular values for orthogonal basis. Therefore, the power of the system will be distributed almost equally on the singular values and the system become more immune against the noise enhancement problem when the singular values are inverted during the equalization process.

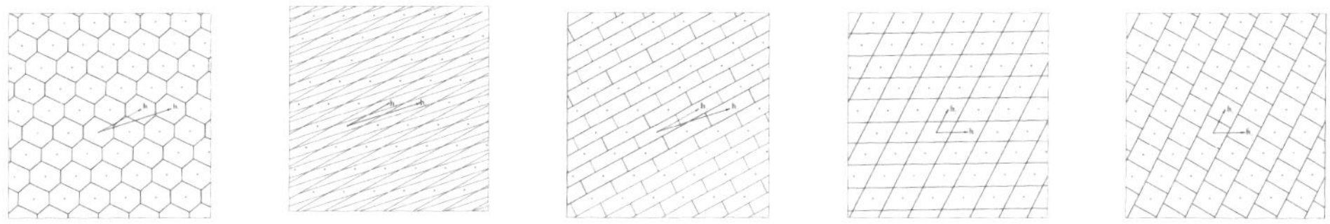

(a) ML (b) LD (c) DFD (d) LRA-LD (e) LRA-DFD

Fig. 11. Undisturbed received signals and decision areas of (a) ML, (b) LD, (c) DFD, (d) LRA-LD and (e) LRA-DFD D. Wübben, R. Böhnke, V. Kühn, and K.-D. Kammeyer (2004).

5.2 Summary of the lattice reduction algorithms

To this end, various reduction algorithms, namely the optimal (the orthogonality is maximized) but NP-hard Minkowski B.A. Lamacchia (1991), Korkine-Zolotareff B.A. Lamacchia (1991) algorithms E. Agrell , T. Eriksson, A. Vardy, and K. Zeger (2002), the well-known LLL reduction A.K. Lenstra, H.W. Lenstra, and L. Lovász (1982), and Seysen's B.A. Lamacchia (1991); M. Seysen (1993) LR algorithm have been proposed.

5.3 Lattice definition

By interpreting the columns $\mathbf{H}_i$ of $\mathbf{H}$ as a *generator basis* , note that $\mathbf{H}$ is also referred to as the lattice basis whose columns are referred to as "basis vectors", the lattice $\Lambda(\mathbf{H})$ is defined as all the complex integer combinations of $\mathbf{H}_i$, *i.e.*,

$$\Lambda(\mathbf{H}) \triangleq \left\{ \sum_{i=1}^{n_T} a_i \mathbf{H}_i \mid a_i \in \mathbb{Z}_{\mathbb{C}} \right\}, \tag{22}$$

where $\mathbb{Z}_{\mathbb{C}}$ is the set of complex integers which reads: $\mathbb{Z}_{\mathbb{C}} = \mathbb{Z} + j\mathbb{Z}$, $j^2 = -1$.
The lattice $\Lambda(\tilde{\mathbf{H}})$ generated by the matrix $\tilde{\mathbf{H}}$ and the lattice generated by the matrix $\mathbf{H}$ are *identical* iff all the lattice points are the same. The two aforementioned bases generate an identical lattices iff $\tilde{\mathbf{H}} = \mathbf{HT}$, where the $n_T \times n_T$ transformation matrix is *unimodular* E. Agrell , T. Eriksson, A. Vardy, and K. Zeger (2002), *i.e.*, $\mathbf{T} \in \mathbb{Z}_{\mathbb{C}}^{n_T \times n_T}$ and such that $|\det(\mathbf{T})| = 1$.
Using the reduced channel basis $\tilde{\mathbf{H}} = \mathbf{HT}$ and introducing $\mathbf{z} = \mathbf{T}^{-1}\mathbf{x}$, the system model given in (1) can be rewritten D. Wübben, R. Böhnke, V. Kühn, and K.-D. Kammeyer (2004):

$$\mathbf{r} = \tilde{\mathbf{H}}\mathbf{z} + \mathbf{n}. \tag{23}$$

The idea behind LRA equalizers or detectors is to consider the identical system model above. The detection is then performed with respect to the reduced channel matrix ($\tilde{\mathbf{H}}$), which is now roughly orthogonal by definition, and to the equivalent transmitted signal that still belongs to an integer lattice since $\mathbf{T}$ is unimodular D. Wübben, R. Böhnke, V. Kühn, and K.-D. Kammeyer (2004). Finally, the estimated $\hat{\mathbf{x}}$ in the original problem is computed with the relationship $\tilde{\mathbf{x}} = \mathbf{T}\hat{\mathbf{z}}$ and by hard-limiting $\tilde{\mathbf{x}}$ to a valid symbol vector. These steps are summarized in the block scheme in Figure 12.

The following Subsections briefly describe the main aspects of the LLL Algorithm (LA) and the Seysen's Algorithm (SA).

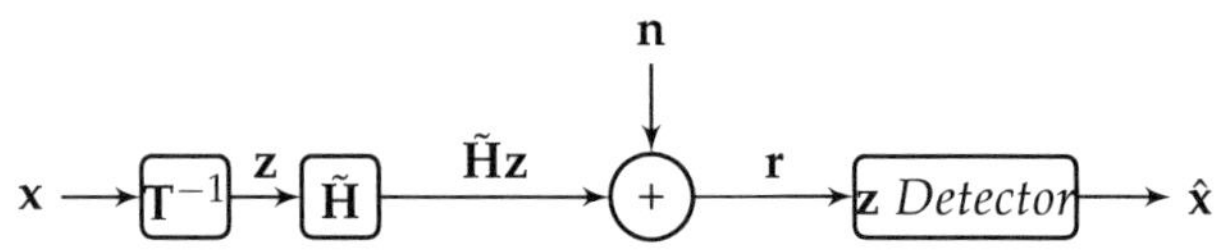

Fig. 12. LRA detector bloc scheme.

5.4 LLL algorithm

The LA is a local approach that transforms an input basis $\mathbf{H}$ into an LLL-reduced basis $\tilde{\mathbf{H}}$ that satisfies both of the orthogonality and norm reduction conditions, respectively:

$$|\Re\{\mu_{i,j}\}|,\ |\Im\{\mu_{i,j}\}| \leq \tfrac{1}{2},\ \forall\, 1 \leq j < i \leq n_T, \tag{24}$$

where $\mu_{i,j} \triangleq \frac{<\mathbf{H}_i, \tilde{\mathbf{H}}_j>}{\|\tilde{\mathbf{H}}_j\|^2}$, and:

$$\|\tilde{\mathbf{H}}_i\|^2 = (\delta - |\mu_{i,i-1}|^2)\|\tilde{\mathbf{H}}_{i-1}\|^2,\ \forall\, 1 < i \leq n_T, \tag{25}$$

where δ, with $\frac{1}{2} < \delta < 1$, is a factor selected to achieve a good quality-complexity trade-off A.K. Lenstra, H.W. Lenstra, and L. Lovász (1982). In this book chapter, δ is assumed to be $\delta = \frac{3}{4}$, as commonly suggested, and $\tilde{\mathbf{H}}_i = \tilde{\mathbf{H}}_i - \sum_{j=1}^{i-2}\{\lceil \mu_{i,j} \rfloor \mathbf{H}_j\}$. Another classical result consists of directly considering the Complex LA (CLA) that offers a saving in the average complexity of nearly 50% compared to the straightforward real model system extension with negligible performance degradation Y.H. Gan, C. Ling, and W.H. Mow (2009).

Let us introduce the QR Decomposition (QRD) of $\mathbf{H} \in \mathbb{C}^{n_R \times n_T}$ that reads $\mathbf{H} = \mathbf{QR}$, where the matrix $\mathbf{Q} \in \mathbb{C}^{n_R \times n_T}$ has orthonormal columns and $\mathbf{R} \in \mathbb{C}^{n_T \times n_T}$ is an upper-triangular matrix. It has been shown D. Wübben, R. Böhnke, V. Kühn, and K.-D. Kammeyer (2004) the QRD of $\mathbf{H} = \mathbf{QR}$ is a possible starting point for the LA, and it has been introduced L.G. Barbero, T. Ratnarajah, and C. Cowan (2008) that the Sorted QRD (SQRD) provides a better starting point since it finally leads to a significant reduction in the expected computational complexity D. Wübben, R. Böhnke, V. Kühn, and K.-D. Kammeyer (2004) and in the corresponding variance B. Gestner, W. Zhang, X. Ma, and D.V. Anderson (2008).

By denoting the latter algorithm as the SQRD-based LA (SLA), these two points are depicted in Figure 13 (a-c) under DSP implementation-oriented assumptions on computational complexities (see S. Aubert, M. Mohaisen, F. Nouvel, and K.H. Chang (2010) for details).

Instead of applying the LA to the only basis $\mathbf{H}$, a simultaneous reduction of the basis $\mathbf{H}$ and the dual basis $\mathbf{H}^{\#} = \mathbf{H}(\mathbf{H}^H\mathbf{H})^{-1}$ D. Wübben, and D. Seethaler (2007) may be processed.

5.5 Seysen's algorithm

At the beginning, let us introduce the *Seysen's orthogonality measure* M. Seysen (1993)

$$\mathcal{S}(\tilde{\mathbf{H}}) \triangleq \sum_{i=1}^{n_T} \left\|\tilde{\mathbf{H}}_i\right\|^2 \left\|\tilde{\mathbf{H}}_i^{\#}\right\|^2, \tag{26}$$

where $\tilde{\mathbf{H}}_i^{\#}$ is the i-th basis vector of the dual lattice, *i.e.*, $\tilde{\mathbf{H}}^{\#H}\tilde{\mathbf{H}} = \mathbf{I}$.

The SA is a global approach that transforms an input basis $\mathbf{H}$ (and its dual basis $\mathbf{H}^{\#}$) into a Seysen-reduced basis $\tilde{\mathbf{H}}$ that (locally) minimizes $\mathcal{S}$ and that satisfies, $\forall\, 1 \leq i \neq j \leq n_T$ D. Seethaler, G. Matz, and F. Hlawatsch (2007)

$$\lambda_{i,\,j} \triangleq \left\lfloor \frac{1}{2}\left(\frac{\tilde{\mathbf{H}}_j^{\#H}\tilde{\mathbf{H}}_i^{\#}}{\|\tilde{\mathbf{H}}_i^{\#}\|^2} - \frac{\tilde{\mathbf{H}}_j^{H}\tilde{\mathbf{H}}_i}{\|\tilde{\mathbf{H}}_j^{\#}\|^2}\right)\right\rceil = 0. \tag{27}$$

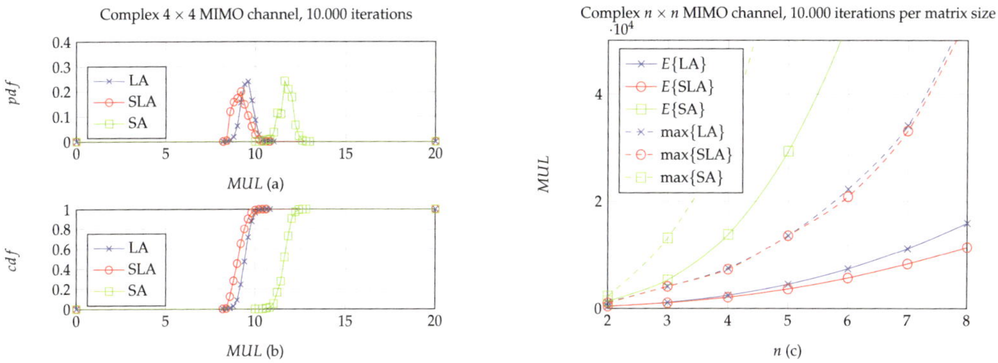

Fig. 13. PDF (a) and CDF (b) of the number of equivalent MUL of the LA, SLA and SA, and average and maximum total number of equivalent MUL of the LA, SLA and SA as a function of the number of antennas n (c).

SA computational complexity is depicted in Figure 13 (a-c) as a function of the number of equivalent *real multiplication* MUL, which allow for some discussion.

5.5.1 Concluding remarks

The aforementioned LR techniques have been presented and both their performances (orthogonality of the obtained lattice) D. Wübben, and D. Seethaler (2007) and computational complexities L.G. Barbero, T. Ratnarajah, and C. Cowan (2008) have been compared when applied to MIMO detection in the Open Loop (OL) case. In Figure 14 (a-f), the od, $cond$, and S of the reduced basis provided by the SA compared to the LA and SLA are depicted. These measurements are known to be popular measures of the quality of a basis for data detection C. Windpassinger, and R.F.H. Fischer (2003). However, this orthogonality gain is obtained at the expense of a higher computational complexity, in particular compared to the SLA. Moreover, it has been shown that a very tiny uncoded BER performance improvement is offered in the case of LRA-LD only D. Wübben, and D. Seethaler (2007). In particular, in the case of LRA-DFD detectors, both LA and SA yield almost the same performance L.G. Barbero, T. Ratnarajah, and C. Cowan (2008).

According to the curves depicted in Figure 13 (a), the mean computational complexities of LA, SLA and SA are $1,6.10^4$, $1,1.10^4$ and $1,4.10^5$ respectively in the case of a 4×4 complex matrix. The variance of the computational complexities of LA, SLA and SA are 3.10^7, $2,3.10^7$ and $2,4.10^9$ respectively, which illustrates the aforementioned reduction in the mean computational complexity and in the corresponding variance and consequently highlights the SLA advantage over other LR techniques.

In Figure 14, the Probability Density Function (PDF) and Cumulative Density Function (CDF) of $ln(cond)$, $ln(od)$ and $ln(S)$ for LA, SLA and SA are depicted and compared to the performance without lattice reduction. It can be observed that both LA and SLA offer exactly the same performance, with the only difference in their computational complexities. Also, there is a tiny improvement in the *od* when SA is used as compared to (S)LA. This point will be discussed in the sequel.

The LRA algorithm preprocessing step has been exposed and implies some minor modifications in the detection step.

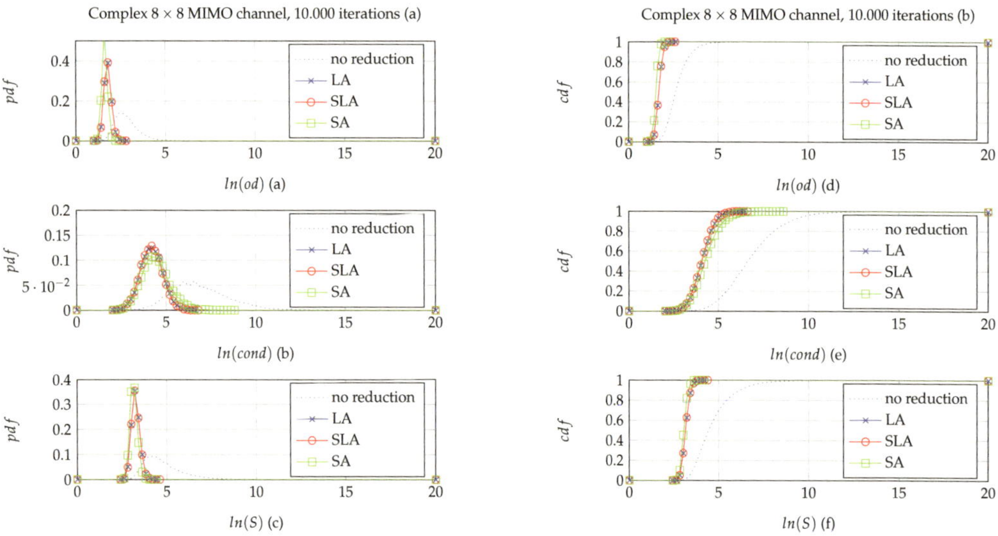

Fig. 14. PDF (a-c) and CDF (d-f) of $ln(cond)$ (a, d), $ln(od)$ (b, e) and $ln(S)$ (c, f) by application of the LA, SLA and SA and compared to the original basis.

5.6 Lattice reduction-aided detection principle

The key idea of the LR-aided detection schemes is to understand that the finite set of transmitted symbols $\Omega_{\mathbb{C}}^{n_T}$ can be interpreted as the De-normalized, Shifted then Scaled (DSS) version of the infinite integer subset $\mathcal{Z}_{\mathbb{C}}^{n_T} \subset \mathbb{Z}_{\mathbb{C}}^{n_T}$ C. Windpassinger, and R.F.H. Fischer (2003), where $\mathbb{Z}_{\mathbb{C}}^{n_T}$ is the infinite set of complex integers, *i.e.,*:

$$\Omega_{\mathbb{C}}^{n_T} = 2a\mathcal{Z}_{\mathbb{C}}^{n_T} + \tfrac{1}{2}\mathbf{T}^{-1}\mathbf{1}_{\mathbb{C}}^{n_T}), \tag{28}$$

and reciprocally

$$\mathcal{Z}_{\mathbb{C}}^{n_T} = \tfrac{1}{2a}\Omega_{\mathbb{C}}^{n_T} - \tfrac{1}{2}\mathbf{T}^{-1}\mathbf{1}_{\mathbb{C}}^{n_T}, \tag{29}$$

where a is a power normalization coefficient (*i.e.,* $1/\sqrt{2}, 1/\sqrt{10}$ and $1/\sqrt{42}$ for QPSK, 16QAM and 64QAM constellations, respectively) and $\mathbf{1}_{\mathbb{C}}^{n_T} \in \mathbb{Z}_{\mathbb{C}}^{n_T}$ is a complex displacement vector (*i.e.,* $\mathbf{1}_{\mathbb{C}}^{n_T} = [1+j, \cdots, 1+j]^T$ in the complex case).
At this step, a general notation is introduced. Starting from the system equation, it can be rewritten equivalently in the following form, by de-normalizing, by dividing by two and subtracting $\mathbf{H}\mathbf{1}_{\mathbb{C}}^{n_T}/2$ from both sides:

$$\frac{\mathbf{r}}{2a} - \frac{\mathbf{H}\mathbf{1}_{\mathbb{C}}^{n_T}}{2} = \frac{\mathbf{H}\mathbf{x}}{2a} + \frac{\mathbf{n}}{2a} - \frac{\mathbf{H}\mathbf{1}_{\mathbb{C}}^{n_T}}{2} \Leftrightarrow \frac{1}{2}\left(\frac{\mathbf{r}}{a} - \mathbf{H}\mathbf{1}_{\mathbb{C}}^{n_T}\right) = \mathbf{H}\frac{1}{2}\left(\frac{\mathbf{x}}{a} - \mathbf{1}_{\mathbb{C}}^{n_T}\right) + \frac{1}{2a}\mathbf{n}, \tag{30}$$

where $\mathbf{H}\mathbf{1}_{\mathbb{C}}^{n_T}$ is a simple matrix-vector product to be done at each channel realization.
By introducing the DSS signal $\mathbf{r}_{\mathbb{Z}} = \frac{1}{2}\left(\frac{\mathbf{r}}{a} - \mathbf{H}\mathbf{1}_{\mathbb{C}}^{n_T}\right) = dss\left\{\mathbf{r}\right\}$ and the re-Scaled, re-Shifted then Normalized (SSN) signal $\mathbf{x}_{\mathbb{Z}} = \frac{1}{2}\left(\frac{\mathbf{x}}{a} - \mathbf{1}_{\mathbb{C}}^{n_T}\right) = ssn\left\{\mathbf{x}\right\}$, which makes both belonging to $\mathbf{H}\mathcal{Z}_{\mathbb{C}}^{n_T}$ and $\mathcal{Z}_{\mathbb{C}}^{n_T}$, respectively, the expression reads:

$$\mathbf{r}_{\mathbb{Z}} = \mathbf{H}\mathbf{x}_{\mathbb{Z}} + \frac{\mathbf{n}}{2a}. \tag{31}$$

This intermediate step allows to define the symbols vector in the reduced transformed constellation through the relation $\mathbf{z}_{\mathbb{Z}} = \mathbf{T}^{-1}\mathbf{x}_{\mathbb{Z}} \in \mathbf{T}^{-1}\mathcal{Z}^{n_T} \subset \mathbb{Z}^{n_T}$. Finally, the lattice-reduced channel and reduced constellation expression can be introduced:

$$\mathbf{r}_{\mathbb{Z}} = \tilde{\mathbf{H}}\mathbf{z}_{\mathbb{Z}} + \frac{\mathbf{n}}{2a}. \tag{32}$$

The LRA detection steps comprise the $\hat{\mathbf{z}}_{\mathbb{Z}}$ estimation of $\mathbf{z}_{\mathbb{Z}}$ with respect to $\mathbf{r}_{\mathbb{Z}}$ and the mapping of these estimates onto the corresponding symbols belonging to the $\Omega_{\mathbb{C}}^{n_T}$ constellation through the $\mathbf{T}$ matrix. In order to finally obtain the $\hat{\mathbf{x}}$ estimation of $\mathbf{x}$, the DSS $\tilde{\mathbf{x}}_{\mathbb{Z}}$ signal is obtained following the $\tilde{\mathbf{z}}_{\mathbb{Z}}$ quantization with respect to $\mathbb{Z}_{\mathbb{C}}^{n_T}$ and re-scaled, re-shifted, then normalized again.
The estimation for the transmit signal is $\hat{\mathbf{x}} = \mathcal{Q}_{\Omega_{\mathbb{C}}^{n_T}}\{\tilde{\mathbf{x}}\}$, as described in the block scheme in Figure 15 in the case of the LRA-ZF solution, and can be globally rewritten as

$$\hat{\mathbf{x}} = \mathcal{Q}_{\Omega_{\mathbb{C}}^{n_T}}\left\{a\left(2\mathbf{T}\mathcal{Q}_{\mathbb{Z}_{\mathbb{C}}^{n_T}}\{\tilde{\mathbf{z}}_{\mathbb{Z}}\} + \mathbf{1}_{\mathbb{C}}^{n_T}\right)\right\}, \tag{33}$$

where $\mathcal{Q}_{\mathbb{Z}_{\mathbb{C}}^{n_T}}\{\cdot\}$ denotes the quantization operation of the n_T-th dimensional integer lattice, for which per-component quantization is such as $\mathcal{Q}_{\mathbb{Z}_{\mathbb{C}}^{n_T}}\{\mathbf{x}\} = [\lfloor \mathbf{x}_1 \rceil, \cdots, \lfloor \mathbf{x}_{n_T} \rceil]^T$, where $\lfloor \cdot \rceil$ denotes the rounding to the nearest integer.

Due to its performance versus complexity, the LA is a widely used reduction algorithm.

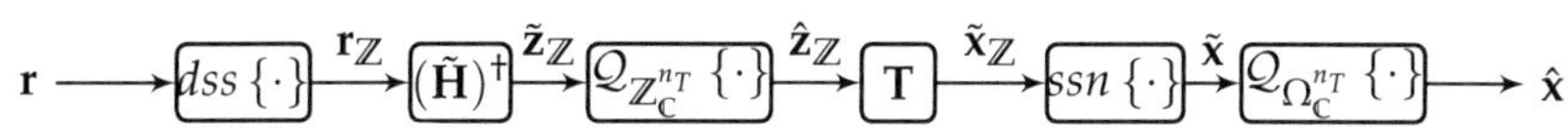

Fig. 15. LRA-ZF detector block scheme.

This is because SA requires a high additional computations compared to the LA to achieve a small, even negligible, gain in the BER performance L.G. Barbero, T. Ratnarajah, and C. Cowan (2008), as depicted in Figure 14. Based on this conjecture, LA will be considered as the LR technique in the remaining part of the chapter.
Subsequently to the aforementioned points, the SLA computational complexity has been shown J. Jaldén, D. Seethaler, and G. Matz (2008) to be unbounded through distinguishing the SQRD pre-processing step and the LA related two conditions. In particular, while the SQRD offers a polynomial complexity, the key point of the SLA computational complexity estimation lies in the knowledge of the number of iterations of both conditions. Since the number of iterations depends on the condition number of the channel matrix, it is consequently unbounded J. Jaldén, D. Seethaler, and G. Matz (2008), which leads to the conclusion that the worst-case computational complexity of the LA in the Open Loop (OL) case is exponential in the number of antennas. Nevertheless, the mean number of iterations (and consequently the mean total computational complexity) has been shown to be polynomial J. Jaldén, D. Seethaler, and G. Matz (2008) and, therefore, a thresholded-based version of the algorithm offers convenient results. That is, the algorithm is terminated when the number of iterations exceeds a pre-defined number of iterations.

5.7 Simulation results

In the case of LRA-LD, the quantization is performed on $\mathbf{z}$ instead of $\mathbf{x}$. The unconstrained LRA-ZF equalized signal $\tilde{\mathbf{z}}_{LRA-ZF}$ are denoted $(\tilde{\mathbf{H}}^H\tilde{\mathbf{H}})^{-1}\tilde{\mathbf{H}}^H\mathbf{r}$ and $\mathbf{T}^{-1}\tilde{\mathbf{x}}_{ZF}$, simultaneously D.

Wübben, R. Böhnke, V. Kühn, and K.-D. Kammeyer (2004). Consequently, the LRA-ZF estimate is $\hat{\mathbf{x}} = \mathcal{Q}_{\Omega_C^{n_T}}\{\mathbf{T}\mathcal{Q}_{\mathbb{Z}^{n_T}}\{\tilde{\mathbf{z}}_{LRA-ZF}\}\}$. Identically, the LRA-MMSE estimate is given as $\hat{\mathbf{x}} = \mathcal{Q}_{\Omega_C^{n_T}}\{\mathbf{T}\mathcal{Q}_{\mathbb{Z}^{n_T}}\{\tilde{\mathbf{z}}_{LRA-MMSE}\}\}$, considering the unconstrained LRA-MMSE equalized signal $\tilde{\mathbf{z}}_{LRA-MMSE} = (\tilde{\mathbf{H}}^H\tilde{\mathbf{H}} + \sigma^2\mathbf{T}^H\mathbf{T})^{-1}\tilde{\mathbf{H}}^H\mathbf{r}$.
It has been shown D. Wübben, R. Böhnke, V. Kühn, and K.-D. Kammeyer (2004) that the consideration of the MMSE criterion by reducing the extended channel matrix $\mathbf{H}_{ext} = [\mathbf{H};\, \sigma\mathbf{I}_{n_T}]$, leading to $\tilde{\mathbf{H}}_{ext}$, and the corresponding extended receive vector $\mathbf{r}_{ext}$ leads to both an important performance improvement and while reducing the computational complexity compared to the straightforward solution. In this case, not only the $\tilde{\mathbf{H}}$ conditioning is considered but also the noise amplification, which is particularly of interest in the case of the LRA-MMSE linear detector. In the sequel, this LR-Aided linear detector is denoted as LRA-MMSE Extended (LRA-MMSE-Ext) detector.
The imperfect orthogonality of the reduced channel matrix induces the advantageous use of DFD techniques D. Wübben, R. Böhnke, V. Kühn, and K.-D. Kammeyer (2004). By considering the QRD outputs of the SLA, namely $\tilde{\mathbf{Q}}$ and $\tilde{\mathbf{R}}$, the system model rewrites $\tilde{\mathbf{z}}_{LR-ZF-QRD} = \tilde{\mathbf{Q}}^H\mathbf{r}$ and reads simultaneously $\tilde{\mathbf{R}}\mathbf{z} + \tilde{\mathbf{Q}}^H\mathbf{n}$. The DFD procedure can than then be performed in order to iteratively obtain the $\hat{\mathbf{z}}$ estimate. In analogy with the LRA-LD, the extended system model can be considered. As a consequence, it leads to the LRA-MMSE-QRD estimate that can be obtained via rewriting the system model as $\tilde{\mathbf{z}}_{LR-MMSE-QRD} = \tilde{\mathbf{Q}}_{ext}^H\mathbf{r}_{ext}$ and reads simultaneously $\tilde{\mathbf{R}}_{ext}\mathbf{z} + \tilde{\mathbf{n}}$, where $\tilde{\mathbf{n}}$ is a noise term that also includes residual interferences.
Figure 16 shows the uncoded BER performance versus E_b/N_0 (in dB) of some well-established LRA-(pseudo) LDs, for a 4×4 complex MIMO Rayleigh system, using QPSK modulation (a, c) and 16QAM (b, d) at each layer. The aforementioned results are compared to some reference results; namely, ZF, MMSE, ZF-QRD, MMSE-QRD and ML detectors. It has been shown that the (S)LA-based LRA-LDs achieve the full diversity M. Taherzadeh, A. Mobasher, and A.K. Khandani (2005) and consequently offer a strong improvement compared to the common LDs. The advantages in the LRA-(Pseudo)LDs are numerous. First, they constitute efficient detectors in the sense of the high quality of their hard outputs, namely the ML diversity is reached within a constant offset, while offering a low overall computational complexity.Also, by noticing that the LR preprocessing step is independent of the SNR, a promising aspect concerns the Orthogonal Frequency-Division Multiplexing (OFDM) extension that would offer a significant computational complexity reduction over a whole OFDM symbol, due to both the time and coherence band. However, there remains some important drawbacks. In particular, the aforementioned SNR offset is important in the case of high order modulations, namely 16-QAM and 64-QAM, despite some aforementioned optimizations. Another point is the LR algorithm's sequential nature because of its iterative running, which consequently limits the possibility of parallel processing. The association of both LR and a neighborhood study is a promising, although intricate, solution for solving this issue. For a reasonable K, a dramatic performance loss is observed with classical K-Best detectors in Figure 9. For a low complexity solution such as LRA-(Pseudo) LDs, a SNR offset is observed in Figure 16. Consequently, the idea that consists in reducing the SNR offset by exploring a neighborhood around a correct although suboptimal solution becomes obvious.

6. Lattice reduction-aided sphere decoding

While it seems to be computationally expensive to cascade two NP-hard algorithms, the promising perspective of combining both the algorithms relies on achieving the ML diversity

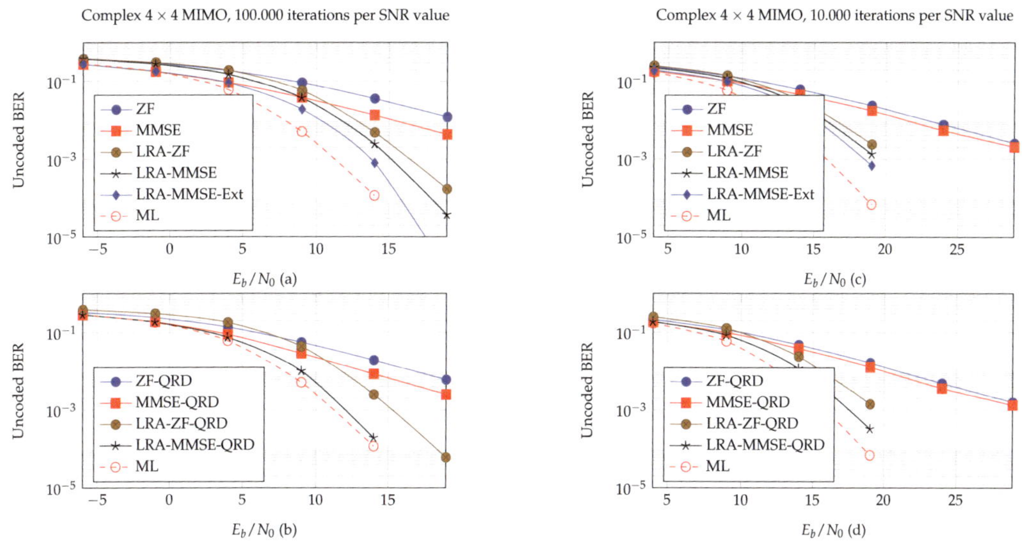

Fig. 16. Uncoded BER as a function of E_b/N_0, Complex Rayleigh 4×4 MIMO channel, ZF, MMSE, LRA-ZF, LRA-MMSE, LRA-MMSE-Ext and ML detectors (a, c), ZF-QRD, MMSE-QRD, LRA-ZF-QRD, LRA-MMSE-QRD and ML detectors (b, d), QPSK modulations at each layer (a-b) and 16QAM modulations at each layer (c-d).

through a LRA-(Pseudo)LD and on reducing the observed SNR offset thanks to an additional neighborhood study. This idea senses the neighborhood size would be significantly reduced while near-ML results would still be reached.

6.1 Lattice reduction-aided neighborhood study interest

Contrary to LRA-(O)DFD receivers, the application of the LR technique followed by the K-Best detector is not straightforward. The main problematic lies in the consideration of the possibly transmit symbols vector in the reduced constellation, namely $\mathbf{z}$. Unfortunately, the set of all possibly transmit symbols vectors can not be predetermined since it does not only depend on the employed constellation, but also on the $\mathbf{T}^{-1}$ matrix. Consequently, the number of children in the tree search and their values are not known in advance. A brute-force solution to determine the set of all possibly transmit vectors in the reduced constellation, $\mathbf{Z}_{all}$, is to get first the set of all possibly transmit vectors in the original constellation, $\mathbf{X}_{all}$, and then to apply the relation $\mathbf{Z}_{all} = \mathbf{T}^{-1}\mathbf{X}_{all}$ for each channel realization. Clearly, this possibility is not feasible since it corresponds to the computational complexity of the ML detector. To avoid this problem, some feasible solutions, more or less efficient, have been proposed in the literature.

6.2 Summary of the lattice reduction-aided neighbourhood study algorithms

While the first idea of combining both the LR and a neighborhood study has been proposed by Zhao *et al.* W. Zhao, and G.B. Giannakis (2006), Qi *et al.* X.-F. Qi, and K. Holt (2007) introduced in detail a novel scheme-Namely LRA-SD algorithm-where a particular attention to neighborhood exploration has been paid. This algorithm has been enhanced by Roger *et al.* S. Roger, A. Gonzalez, V. Almenar, and A.M. Vidal (2009) by, among others, associating LR and K-Best. This offers the advantages of the K-Best concerning its complexity and parallel nature, and consequently its implementation. The hot topic of the neighborhood study size reduction is being widely studied M. Shabany, and P.G. Gulak (2008); S. Roger, A. Gonzalez, V. Almenar, and A.M. Vidal (2009). In a first time, let us introduce the basic idea that makes the LR theory appropriate for application in complexity - and latency - limited communication systems. Note that the normalize-shift-scale steps that have been previously introduced, will not be addressed again.

6.3 The problem of the reduced neighborhood study

Starting from Equation (32), both the sides of the lattice-reduced channel and reduced constellation can be left-multiplied by $\tilde{\mathbf{Q}}^H$, where $[\tilde{\mathbf{Q}}, \tilde{\mathbf{R}}] = QRD\{\tilde{\mathbf{H}}\}$. Therefore, a new relation is obtained:

$$\tilde{\mathbf{Q}}^H \mathbf{r}_{\mathbb{Z}} = \tilde{\mathbf{R}}\mathbf{z}_{\mathbb{Z}} + \tilde{\mathbf{n}}, \quad (34)$$

this makes any SD scheme to be introduced, and eventually a K-Best. At this moment, the critical point of neighbours generation in the reduced constellation has to be introduced. As previously presented, the set of possible values in the original constellation is affected by the matrix $\mathbf{T}^{-1}$. In particular, due to $\mathbf{T}$ properties introduced in the LR step, the scaling, rotating, and reflection operations may induce some missing (non-adjacent) or unbounded points in the reduced lattice, despite the regularity and bounds of the original constellation. In presence of noise, some candidates may not map to any legitimate constellation point in the original constellation. Therefore, it is necessary to take into account this effect by discarding vectors with one (or more) entries exceeding constellation boundaries. However, the vicinity of a lattice point in the reduced constellation would be mapped onto the same signal point. Consequently, a large number of solutions might be discarded, leading to inefficiency of any additional neighborhood study. Also note that it is not possible to prevent this aspect without exhaustive search complexity since $\mathbf{T}^{-1}$ applies on the whole $\hat{\mathbf{z}}$ vector while it is treated layer by layer.

Zhao *et al.* W. Zhao, and G.B. Giannakis (2006) propose a radius expression in the reduced lattice from the radius expression in the original constellation through the Cauchy-Schwarz inequality. This idea leads to an upper bound of the explored neighborhood and accordingly a reduction in the number of tested candidates. However, this proposition is not enough to correctly generate a neighborhood because of the classical - and previously introduced - problematic of any fixed radius.

A zig-zag strategy inside of the radius constraint works better S. Roger, A. Gonzalez, V. Almenar, and A.M. Vidal (2009); W. Zhao, and G.B. Giannakis (2006). Qi *et al.* X.-F. Qi, and K. Holt (2007) propose a predetermined set of displacement $[\delta_1, \cdots, \delta_N]$ $(N > K)$ generating a neighborhood around the constrained DFD solution $[\mathcal{Q}_{\mathbb{Z}_{\mathbb{C}}}\{\tilde{\mathbf{z}}_{n_T}\} + \delta_1, \cdots, \mathcal{Q}_{\mathbb{Z}_{\mathbb{C}}}\{\tilde{\mathbf{z}}_{n_T}\} + \delta_N]$. The N neighbors are ordered according to their norms, by considering the current layer similarly to the SE technique, and the K candidates with the least metrics are stored. The problem of this technique lies in the number of candidates that has to be unbounded, and consequently set to a very large number of candidates N for the sake of feasibility. Roger *et al.* S. Roger, A. Gonzalez, V. Almenar, and A.M. Vidal (2009) proposed to replace the neighborhood generation by a zig-zag strategy around the constrained DFD solution with boundaries control constraints. By denoting boundaries in the original DSS constellation $\mathbf{x}_{\mathbb{Z},\,\min}$ and $\mathbf{x}_{\mathbb{Z},\,\max}$, the reduced constellation boundaries can be obtained through the relation $\mathbf{z}_{\mathbb{Z}} = \mathbf{T}^{-1}\mathbf{x}_{\mathbb{Z}}$ that implies $\mathbf{z}_{\max,\,l} = \max\{\mathbf{T}^{-1}_{l,\,:}\mathbf{x}_{\mathbb{Z}}\}$ for a given layer l. The exact solution is given in S. Roger, A. Gonzalez, V. Almenar, and A.M. Vidal (2009) for the real case and can be extended to the complex case:

$$\mathbf{z}_{\max,\, l} = \mathbf{x}_{\mathbb{Z},\, \max} \sum_{j \in P^l} \mathbf{T}^{-1}_{l,\, j} + \mathbf{x}_{\mathbb{Z},\, \min} \sum_{j \in N^l} \mathbf{T}^{-1}_{l,\, j},\ \mathbf{z}_{\min,\, l} = \mathbf{x}_{\mathbb{Z},\, \min} \sum_{j \in P^l} \mathbf{T}^{-1}_{l,\, j} + \mathbf{x}_{\mathbb{Z},\, \max} \sum_{j \in N^l} \mathbf{T}^{-1}_{l,\, j}, \tag{35}$$

where P^l and N^l stands for the set of indices j corresponding to positive and negative entries $(l,\ j)$ of $\mathbf{T}^{-1}$, respectively. By denoting the latter algorithm as the LRA-KBest-Candidate Limitation (LRA-KBest-CL), note that this solution is exact and does not induce any performance degradation.
The main advantages in the LRA-KBest are highlighted. While it has been shown that the LRA-KBest achieves the ML performance for a reasonable K, even for 16QAM and 64QAM constellations, as depicted in Figure 17, the main favorable aspect lies in the neighborhood study size that is independent of the constellation order. So the SD complexity has been reduced though the LR-Aid and would be feasible, in particular for 16QAM and 64QAM constellations that are required in the 3GPP LTE-A norm 3GPP (2009). Also, such a detector is less sensitive to ill-conditioned channel matrices due to the LR step. However, the detector offers limited benefits with the widely used QPSK modulations, due to nearby lattice points elimination during the quantization step, and the infinite lattice problematic in the reduced domain constellation search has not been solved convincingly and is up to now an active field of search.
Let us introduce the particular case of Zhang *et al.* W. Zhang, and X. Ma (2007a;b) that proposes to combine both LR and a neighborhood study in the original constellation.

6.4 A particular case

In order to reduce the SNR offset by avoiding the problematic neighborhood study in the reduced constellation, a by-solution has been provided W. Zhang, and X. Ma (2007a) based on the unconstrained LRA-ZF result. The idea here was to provide a soft-decision LRA-ZF detector by generating a list of solutions. This way, Log-Likelihood Ratios (LLR) can be obtained through the classical max-log approximation, if both hypothesis and counter-hypothesis have been caught, or through-among others-a LLR clipping B.M. Hochwald, and S. ten Brink (2003); D.L. Milliner, E. Zimmermann, J.R. Barry and G. Fettweis (2008).
The idea introduced by Zhang *et al.* corresponds in reality to a SD-like technique, allowing to provide a neighborhood study around the unconstrained LRA-ZF solution: $\mathbf{r}_{LRA-ZF} = \tilde{\mathbf{H}}^{\dagger}\mathbf{r}$. The list of candidates, that corresponds to the neighborhood in the reduced constellation, can be defined using the following relation:

$$\mathcal{L}_z = \{\tilde{\mathbf{z}} :\ \|\tilde{\mathbf{z}} - \mathbf{r}_{LRA-ZF}\|^2 < d_z\}, \tag{36}$$

where $\tilde{\mathbf{z}}$ is a hypothetical value for $\mathbf{z}$ and $\sqrt{d_z}$ is the sphere constraint. However, a direct estimation of $\hat{\mathbf{x}}$ may be obtained by left-multiplying by correct lines of $\mathbf{T}^{-1}$ at each detected symbol:

$$\mathcal{L}_x = \{\tilde{\mathbf{x}} :\ \|\mathbf{T}^{-1}\tilde{\mathbf{x}} - \mathbf{r}_{LRA-ZF}\|^2 < d_z\}, \tag{37}$$

where $\tilde{\mathbf{x}}$ is a hypothetical value for $\mathbf{x}$ and by noting that the sphere constraint remains unchanged.
The problem introduced by such a technique is how to obtain $\tilde{\mathbf{x}}$ layer by layer, since it would lead to non-existing symbols. A possible solution is the introduction of the QRD of $\mathbf{T}^{-1}$ in

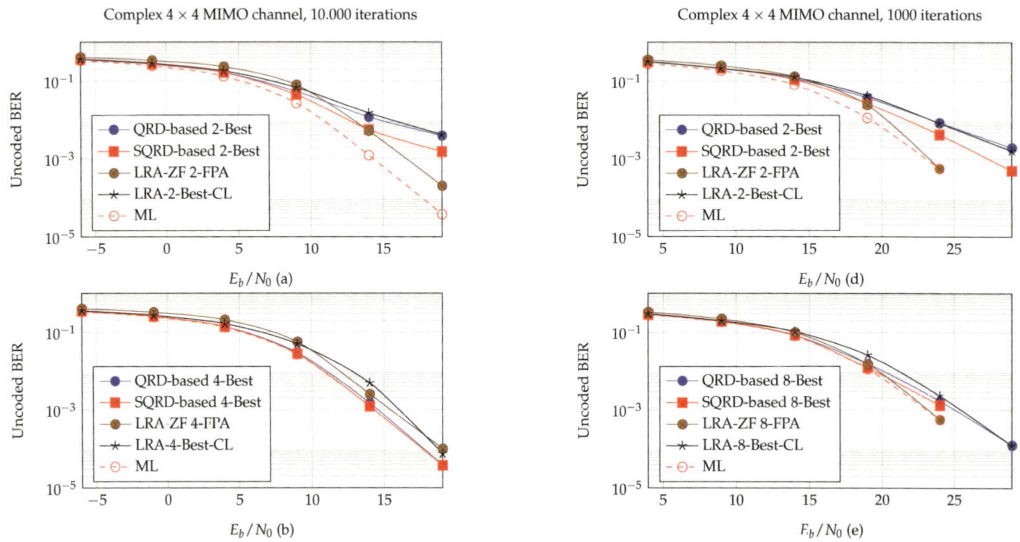

Fig. 17. Uncoded BER as a function of E_b/N_0, Complex Rayleigh 4×4 MIMO channel, QRD-based 2/4(8)-Best, SQRD-based 2/4(8)-Best, LRA-ZF 2/4(8)-FPA, LRA-2/4(8)-Best-CL and ML detectors, QPSK modulations at each layer (a-c) and 16QAM modulations at each layer (d-f).

order to make the current detected symbol within the symbols vector independent of the remaining to-detect symbols. This idea leads to the following expression:

$$\hat{\mathbf{x}} = \underset{\mathbf{x} \in \Omega_{\mathbb{C}}^{n_T}}{\operatorname{argmin}} \left\| \mathbf{Q}_{\mathbf{T}^{-1}}^{H} \mathbf{r}_{LRA-ZF} - \mathbf{R}_{\mathbf{T}^{-1}} \mathbf{x} \right\|^2 < d_z, \tag{38}$$

where $[\mathbf{Q}_{\mathbf{T}^{-1}}, \mathbf{R}_{\mathbf{T}^{-1}}] = QRD\{\mathbf{T}^{-1}\}$. Due to the upper triangular form of $\mathbf{R}_{\mathbf{T}^{-1}}$, $\hat{\mathbf{x}}$ can be detected layer by layer through the K-Best scheme such as the radius constraint can be eluded. Consequently, the problematic aspects of the reduced domain constellation study are avoided, and the neighborhood study is provided at the *cheap* price of an additional QRD. By denoting the latter algorithm as the LRA-ZF Fixed Point Algorithm (LRA-ZF-FPA), note that the problem of this technique lies in the Euclidean Distance expression which is not equivalent to the ML equation. The technique only aims at generating a neighborhood study for the Soft-Decision extension. There will be no significant additional performance improvement for larger K, as depicted in Figure 17.

6.5 Simulation results

In Figure 17 and in the case of a neighborhood study in the reduced domain, near-ML performance is reached for small K values, in both QPSK and 16QAM cases.

It is obvious to the reader that K is independent of the constellation order, which can be demonstrated. This aspect is essential for the OFDM extension since any SD-like detector has to be fully processed for each to-be-estimated symbols vector. Also, the solution offered by LRA-ZF-FPA is interesting in the sense that it allows to make profit of the LRA benefit with an additional neighborhood study in the original constellation. However, it does not reach the ML performance because of the non-equivalence of the metrics computation even in the case of a large K.

7. Conclusion

In this chapter, we have presented an up-to-date review, as well as several prominent contributions, of the detection problematic in MIMO-SM systems. It has been shown that, theoretically, such schemes linearly increase the channel capacity. However, in practice, achieving such increase in the system capacity depends, among other factors, on the employed receiver design and particularly on the de-multiplexing algorithms, a.k.a. detection techniques. In the literature, several detection techniques that differ in their employed strategies have been proposed. This chapter has been devoted to analyze the structures of those algorithms. In addition to the achieved performance, we pay a great attention in our analysis to the computational complexities since these algorithms are candidates for implementation in both latency and power-limited communication systems. The linear detectors have been introduced and their low performances have been outlined despite of their attracting low computational complexities. DFD techniques improve the performance compared to the linear detectors. However, they might require remarkably higher computations, while still being far from achieving the optimal performance, even with ordering. Tree-search detection techniques, including SD, QRD-M, and FSD, achieve the optimum performance. However, FSD and QRD-M are more favorable due to their fixed and realizable computational complexities. An attractive pre-detection process, referred to as lattice-basis reduction, can be considered in order to apply any detector through a close-to-orthogonal channel matrix. As a result, a low complexity detection technique, such as linear detectors, can achieve the optimum diversity order. In this chapter, we followed the lattice reduction technique with the K-best algorithm with low K values, where the optimum performance is achieved. In conclusion, in this chapter, we surveyed the up-to-date advancements in the signal detection field, and we set the criteria over which detection algorithms can be evaluated. Moreover, we set a clear path for future research via introducing several recently proposed detection methodologies that require further studies to be ready for real-time applications.

8. References

3GPP (2009). Evolved Universal Terrestrial Radio Access (E-UTRA); User Equipment (UE) radio transmission and reception v8.8.0, *TS 36.101*, 3rd Generation Partnership Project (3GPP).

A. Paulraj, R. Nabar, and D. Gore (2003). Introduction to Space-Time Wireless Communications, *UK, Cambridge: Cambridge University Press* .

A. Telatar (1999). Capacity of multi-antenna Gaussian channels, *Telecommunications, European Transactions on* 10(6): 585–595.

A.K. Lenstra, H.W. Lenstra, and L. Lovász (1982). Factoring polynomials with rational coefficients, *Mathematische Annalen* 261(4): 515–534.

B. Gestner, W. Zhang, X. Ma, and D.V. Anderson (2008). VLSI Implementation of a Lattice Reduction Algorithm for Low-Complexity Equalization, *Circuits and Systems for Communications, IEEE International Conference on* pp. 643–647.

B. Hassibi (2000). An efficient square-root algorithm for BLAST, *Acoustics, Speech, and Signal Processing, IEEE International Conference on* pp. 737–740.

B. Hassibi, and H. Vikalo (2001). On the expected complexity of sphere decoding, *Signals, Systems and Computers, Asilomar Conference on* pp. 1051–1055.

B. Schubert (2006). Analysis of sphere decoding in linear cooperative wireless relay networks., *Master Thesis, Berlin University of Technology* .

B.A. Lamacchia (1991). Basis Reduction Algorithms and Subset Sum Problems, *Technical report*, MSc Thesis, Massachusetts Institute of Technology.

B.M. Hochwald, and S. ten Brink (2003). Achieving Near-Capacity on a Multiple-Antenna Channel, *Communications, IEEE Transactions on* 51(3): 389–399.

C. Ling (2006). Approximate lattice decoding: Primal versus dual lattice reduction, *Information Theory, International Symposium on* .

C. Schnorr, and M. Euchner (1994). Lattice basis reduction: improved practical algorithms and solving subset sum problems, *Mathematical Programming* 66: 181–199.

C. Windpassenger (2004). Detection and precoding for multiple input multiple output channels., *PhD Dissertation, Erlangen-Nurnberg University* .

C. Windpassinger, and R.F.H. Fischer (2003). Low-complexity near-maximum-likelihood detection and precoding for MIMO systems using lattice reduction, *Information Theory Workshop, IEEE* pp. 345–348.

D. Seethaler, G. Matz, and F. Hlawatsch (2007). Low-complexity MIMO data detection using Seyseni£¡s lattice reduction algorithm, *Acoustics, Speech, and Signal Processing, IEEE International Conference on* 3: 53–56.

D. Shiu, and J. Kahn (1999). Layered space-time codes for wireless communications using multiple transmit antennas, *Communications, International Conference on* pp. 436–440.

D. Wübben, and D. Seethaler (2007). On the Performance of Lattice Reduction Schemes for MIMO Data Detection, *Signals, Systems and Computers, Asilomar Conference on* pp. 1534–1538.

D. Wübben, R. Böhnke, J. Rinas, V. Kühn, and K.-D. Kammeyer (2001). Efficient algorithm for decoding layered space-time codes, *Electronics Letters, IEEE* 37(22): 1348–1350.

D. Wübben, R. Böhnke, V. Kühn, and K.-D. Kammeyer (2003). MMSE extension of V-BLAST based on sorted QR decomposition, *Vehicular Technology Conference, IEEE* pp. 508–512.

D. Wübben, R. Böhnke, V. Kühn, and K.-D. Kammeyer (2004). MMSE-based lattice-reduction for near-ML detection of MIMO systems, *Smart Antennas, ITG Workshop on* pp. 106–113.

D.L. Milliner, E. Zimmermann, J.R. Barry and G. Fettweis (2008). Channel State Information Based LLR Clipping in List MIMO Detection, *Personal, Indoor and Mobile Radio Communications International, IEEE International Symposium on* pp. 1–5.

E. Agrell , T. Eriksson, A. Vardy, and K. Zeger (2002). Closest point search in lattices, *Information Theory, IEEE Transactions on* 48(8): 2201–2214.

H. Kawai, K. Higuchi, N. Maeda, and M. Sawahashi (2006). Adaptive control of surviving symbol replica candidates in QRM-MLD for OFDM-MIMO multiplexing, *IEEE Journal of Selected Areas in Communications* 24(6): 1130–1140.

H. Yao, and G.W. Wornell (2002). Lattice-reduction-aided detectors for MIMO communication systems, *Global Telecommunications Conference, IEEE* 1: 424–428.

H. Zhu, Z. Lei, and F. Chin (2004). An improved square-root algorithm for BLAST, *Signal Processing Letters, IEEE* 11(9): 772–775.

J. Anderson, and S. Mohan (1984). Sequential coding algorithms: a survey and cost analysis, *Communications, IEEE Transactions on* 32(2): 169–176.

J. Benesty, Y. Huang, and J. Chen (2003). A fast recursive algorithm for optimum sequential signal detection in a BLAST system, *Signal Processing, IEEE Transactions on* 51(7): 1722–1730.

J. Boutros, N. Gresset, L. Brunel, and M. Fossorier (2003). Soft-input soft-output lattice sphere decoder for linear channels, *Global Telecommunications Conference, IEEE* pp. 1583–1587.

J. Jaldén, and B. Ottersten (2005). On the complexity of sphere decoding in digital communications, *Signal Processing, IEEE Transactions on* 53(4): 1474–1484.

J. Jaldén, D. Seethaler, and G. Matz (2008). Worst-and average-case complexity of LLL lattice reduction in MIMO wireless systems, *Acoustics, Speech and Signal Processing, IEEE International Conference on* pp. 2685–2688.

J. Wang, and B. Daneshrad (2005). A comparative study of MIMO detection algorithms for wideband spatial multiplexing systems, *Wireless Communications and Networking, IEEE International Conference on* 1: 408–413.

K. Jeon, H. Kim, and H. Park (2006). An efficient QRD-M algorithm using partial decision feedback detection, *Signal, Systems, and Computers, Asilomar Conference on* pp. 1658–1661.

K.-W. Wong, C.-Y. Tsui, S.-K. Cheng, and W.-H. Mow (2002). A VLSI Architecture of a K-Best Lattice Decoding Algorithm For MIMO Channels, 3: 272–276.

K.J. Kim, J. Yue, R.A. Iltis, and J.D. Gibson (2005). A QRD-M/Kalman filter-based detection and channel estimation algorithm for MIMO-OFDM systems, *Wireless Communications, IEEE Transactions on* 4(2): 710–721.

L. Barbero, and J. Thompson (2008a). Extending a fixed-complexity sphere decoder to obtain likelihood information for turbo-MIMO systems, *Vehicular Technology, IEEE Transactions on* 57(5): 2804–2814.

L. Barbero, and J. Thompson (2008b). Fixing the complexity of the sphere decoder for MIMO detection, *Wireless Communications, IEEE Transactions on* 7(6): 2131–2142.

L. Lovász (1986). An Algorithmic Theory of Numbers, Graphs and Convexity, *PA: Society for Industrial and Applied Mathematics* .

L. Wang, L. Xu, S. Chen, and L. Hanzo (2008). MMSE Soft-Interference-Cancellation Aided Iterative Center-Shifting K-Best Sphere Detection for MIMO Channels, *Communications, IEEE International Conference on* pp. 3819–3823.

L.G. Barbero, T. Ratnarajah, and C. Cowan (2008). A comparison of complex lattice reduction algorithms for MIMO detection, *Acoustics, Speech and Signal Processing, IEEE International Conference on* pp. 2705–2708.

M. Mohaisen, and K.H. Chang (2009a). On improving the efficiency of the fixed-complexity sphere decoder, *Vehicular Technology Conference, IEEE* pp. 1–5.

M. Mohaisen, and K.H. Chang (2009b). On the achievable improvement by the linear minimum mean square error detector, *Communication and Information Technology, International Symposium on* pp. 770–774.

M. Mohaisen, and K.H. Chang (2010). Upper-lower bounded-complexity QRD-M for spatial multiplexing MIMO-OFDM systems, *Wireless Personal Communications* .

M. Mohaisen, H.S. An, and K.H.Chang (2009). Detection techniques for MIMO multiplexing: a comparative review, *Internet and Information Systems, KSII Transactions on* .

M. Mohaisen, K.H. Chang, and B.T. Koo (2009). Adaptive parallel and iterative QRDM algorithms for spatial multiplexing MIMO systems , *Vehicular Technology Conference* pp. 1–5.

M. Seysen (1993). Simultaneous reduction of a lattice basis and its reciprocal basis, *Combinatorica* 13(3): 363–376.

M. Shabany, and P.G. Gulak (2008). The Application of Lattice-Reduction to the K-Best Algorithm for Near-Optimal MIMO Detection, *Circuits and Systems, IEEE International Symposium on* pp. 316–319.

M. Taherzadeh, A. Mobasher, and A.K. Khandani (2005). LLL lattice-basis reduction achieves the maximum diversity in MIMO systems, *Information Theory, International Symposium on* .

P. Wolniansky, G. Foschini, G. Golden, and R. Valenzuela (1998). V-BLAST: an architecture for realizing very high data rates over the rich-scattering wireless channel, *Signals, Systems, and Electronics, URSI International Symposium on* pp. 295–300.

S. Aubert, M. Mohaisen, F. Nouvel, and K.H. Chang (2010). Parallel QR decomposition in LTE-A systems, *IEEE International Workshop on Signal Processing Advances in Wireless Communications* .

S. Haykin, and M. Moher (2005). Modern Wireless Communications, *USA, NJ, Pearson Prentice Hall* .

S. Roger, A. Gonzalez, V. Almenar, and A.M. Vidal (2009). On Decreasing the Complexity of Lattice-Reduction-Aided K-Best MIMO Detectors, *European Signal Processing Conference* pp. 2411–2415.

T. Cui, and C. Tellambura (2005). An efficient generalized sphere decoder for rank-deficient MIMO systems, *Communications Letters, IEEE* 9(5): 423–425.

U. Fincke, and M. Pohst (1985). Improved methods for calculating vectors of short length in a lattice, including complexity analysis, *Mathematics of Computation* 44(170): 463–471.

W. Van Etten (1976). Maximum likelihood receiver for multiple channel transmission systems, *Communications, IEEE Transactions on* pp. 276–283.

W. Zhang, and X. Ma (2007a). Approaching Optimal Performance By Lattice-Reduction Aided Soft Detectors, *Information Sciences and Systems, Conference on* pp. 818–822.

W. Zhang, and X. Ma (2007b). Designing Soft Detectors Based On Seysen's Algorithm, *Military Communications Conference, IEEE* 14(5): 1–7.

W. Zhao, and G.B. Giannakis (2006). Reduced Complexity Closest Point Decoding Algorithms for Random Lattices, *Wireless Communications, IEEE Transactions on* 5(1): 101–111.

X.-F. Qi, and K. Holt (2007). A Lattice-Reduction-Aided Soft Demapper for High-Rate Coded MIMO-OFDM Systems, *Signal Processing Letters, IEEE* 14(5): 305 –308.

Y.H. Gan, C. Ling, and W.H. Mow (2009). Complex Lattice Reduction Algorithm for Low-Complexity Full-Diversity MIMO Detection, *Signal Processing, IEEE Transactions on* 57(7): 2701–2710.

6

DFT Based Channel Estimation Methods for MIMO-OFDM Systems

Moussa Diallo[1], Maryline Hélard[2], Laurent Cariou[3] and Rodrigue Rabineau[4]
[1,3,4]*Orange Labs, 4 rue du Clos Courtel, 35512 Cesson-Sévigné Cedex*
[2]*INSA IETR, UEB, 20 Avenue des Buttes de Coesmes, CS 70839, 35708 Rennes Cedex 7*
France

1. Introduction

The great increase in the demand for high-speed data services requires the rapid growth of mobile communications capacity. Orthogonal frequency division multiplexing (OFDM) provides high spectral efficiency, robustness to intersymbol interference (ISI), as well as feasibility of low cost transceivers (Weinstein & Ebert, 1971). Multiple input multiple output (MIMO) systems offer the potential to obtain a diversity gain and to improve system capacity (Telatar, 1995), (Alamouti, 1998), (Tarokh et al., 1999). Hence the combination of MIMO and OFDM techniques (MIMO-OFDM) is logically widely considered in the new generation of standards for wireless transmission (Boubaker et al., 2001). In these MIMO-OFDM systems, considering coherent reception, the channel state information (CSI) is required for recovering transmitted data and thus channel estimation becomes necessary.

Channel estimation methods can be classified into three distinct categories: blind channel estimation, semi-blind channel estimation and pilot-aided channel estimation. In the pilot-aided methods, pilot symbols known from the receiver are transmitted as a preamble at the beginning of the frame or scattered throughout each frame in a regular manner. On the contrary, in blind methods, no pilot symbols are inserted and the CSI is obtained by relying on the received signal statistics (Winters, 1987). Semi-blind methods combine both the training and blind criteria (Foschini, 1996). In this paper, we focus our analysis on the time domain (TD) channel estimation technique using known reference signals. This technique is attractive owing to its capacity to reduce the noise component on the estimated channel coefficients (Zhao & Huang, 1997).

The vast majority of modern multicarrier systems contain null subcarriers at the spectrum extremities in order to ensure isolation from/to signals in neighboring frequency bands (Morelli & Mengali, 2001) as well as to respect the sampling theorem (3GPP, 2008). It was shown that, in the presence of these null subcarriers, the TD channel estimation methods suffer from the "border effect" phenomenon that leads to a degradation in their performance (Morelli & Mengali, 2001). A TD approach based on pseudo inverse computation is proposed in (Doukopoulos & Legouable, 2007) in order to mitigate this "border effect". However the degradation of the channel estimation accuracy persists when the number of the null subcarriers is large.

In this document, we look at various time domain channel estimation methods with this constraint of null carriers at spectrum borders. We show in detail how to gauge the importance

of the "border effect" depending on the number of null carriers, which may vary from one system to another. Thereby we assess the limit of the technique discussed in (Doukopoulos & Legouable, 2007) when the number of null carriers is large. Finally the DFT with the truncated singular value decomposition (SVD) technique is proposed to completely eliminate the impact of the null subcarriers whatever their number. A technique for the determination of the truncation threshold for any MIMO-OFDM system is also proposed.
The paper is organized as follows. Section 2 describes the studied MIMO-OFDM system, including the construction of the training sequences in the frequency domain and the least square (LS) channel estimation component. Then section 3.1 presents the main objectives (*noise reduction* and *interpolation*) of the classical DFT based channel estimation and its weakness regarding the "border effect". The pseudo inverse concept is then studied in section 4. Next, the DFT with truncated SVD is detailed in section 5. Finally, the efficiency of these channel estimators is demonstrated in section 6 for two distinct application environments: *indoor* and *outdoor* respectively applying 802.11n and 3GPP system parameters.
Notations: Superscript † stands for pseudo-inversion. Operator e represents an exponential function. $(.^{H})$ stands for conjugation and transpose. C denotes a complex number set and $j^2 = -1$. $\| \ \|$ denotes the Euclidean norm.

2. MIMO-OFDM system model

The studied MIMO-OFDM system is composed of N_t transmit and N_r receiving antennas. Training sequences are inserted in the frequency domain before OFDM modulation which is carried out for each antenna.
The OFDM signal transmitted from the i-th antenna after performing IFFT (OFDM modulation) on the frequency domain signal $X_i \in C^{N\times 1}$ at time index n can be given by:

$$x_i(n) = \sqrt{\frac{1}{N}} \sum_{k=0}^{N-1} X_i(k) e^{j\frac{2\pi kn}{N}}, \qquad 0 \leq (n,k) \leq N \tag{1}$$

where N is the number of IFFT points and k the subcarrier index.
The baseband time domain channel response between the transmitting antenna i and the receiving antenna j under the multipath fading environments can be expressed as (Van de Beek et al., 1995):

$$h_{ij}(n) = \sum_{l=0}^{L_{ij}-1} h_{ij,l}\delta(n - \tau_{ij,l}) \tag{2}$$

with L_{ij} the number of paths, $h_{ij,l}$ and $\tau_{ij,l}$ the complex time varying channel coefficient and delay of the l-th path.
The use of the cyclic prefix (CP) allows both the preservation of the orthogonality between the tones and the elimination of the ISI between consecutive OFDM symbols.
At the receiver side, after removing the CP and performing the OFDM demodulation, the received frequency domain signal can be expressed as follows by using (1) and (2):

$$R_j(k) = \sum_{i=0}^{N_t-1} X_i(k) H_{ij}(k) + \Xi(k) \tag{3}$$

where $H_{ij}(k)$ is the discrete response of the channel on subcarrier k between the i-th transmit antenna and the j-th receiving antenna and Ξ_k the zero-mean complex Gaussian noise after the

FFT process. Then Least Square (LS) channel estimation is performed by using the extracted pilots.
In SISO-OFDM, without exploiting any knowledge of the propagation channel statistics, the LS estimates regarding the pilot subcarrier k can be obtained by dividing the demodulated pilot signal $R_j(k)$ by the known pilot symbol $X(k)$ in the frequency domain (Zhao & Huang, 1997). The LS estimates regarding the pilot subcarrier k can be expressed as follows:

$$H_{LS}(k) = H(k) + \Xi(k)/X(k). \tag{4}$$

Nevertheless in the MIMO-OFDM system, from (3), an orthogonality between pilots is mandatory to obtain LS estimates for each receiver antenna without interference from the other antennas. In this paper, we consider the case where the pilots from different transmit antennas are orthogonal to each other in the frequency domain. It is important to note that this orthogonality can also be obtained in the time domain by using the cyclic shift delay (CSD) method (Auer, 2004).
The orthogonality between pilots in the frequency domain can be obtained by different ways:

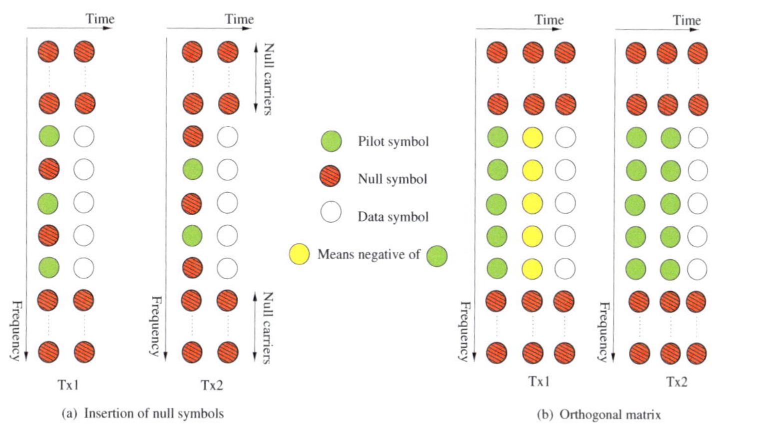

Fig. 1. Orthogonality between pilots in the frequency domain when $N_t = 2$.

- The orthogonality can be achieved with the use of transmission of pilot symbols on one antenna and of null symbols on the other antennas in the same instant (Fig. 1(a)). This solution is commonly and easily implemented in the presence of mobility as for instance in 3GPP/LTE (3GPP, 2008). Therefore LS estimates can only be calculated for M/N_t subcarriers, M representing the number of modulated subcarriers. Then interpolation must be performed in order to complete the estimation for all the subcarriers.
- The orthogonality can also be achieved using a specific transmit scheme represented by an orthogonal matrix. Fig. 1(b) represents the pilot insertion structure for a two transmit antenna system. The orthogonality between training sequences for antennas 1 and 2 is obtained by using the following orthogonal matrix.

$$\begin{bmatrix} 1 & -1 \\ 1 & 1 \end{bmatrix} \tag{5}$$

This technique is frequently used when the channel can be assumed constant at least over the duration of N_t OFDM symbols in the case of quite slow variations. For instance, this method is used in the wireless local area network (WLAN) IEEE802.11n system (802.11, 2007). LS estimates can be calculated for all the M modulated subcarriers with the use of full pilot OFDM symbols.

Assuming orthogonality between pilots, N_t LS estimation algorithms on pilot subcarriers can be applied per receive antenna:

$$H_{ij,LS}(k) = H_{ij}(k) + \Xi_{ij}(k)/X_{ij}(k). \tag{6}$$

Thus the LS estimation is computed for all the subchannels between the transmit and the receiver antennas. Nevertheless, it is important to note the two following points:

- From (6), it can be observed that the accuracy of LS estimated channel response is degraded by the noise component.
- To get an estimation for all the subcarriers, interpolation may be required depending on the pilots insertion scheme.

Time domain processing will then be used in order to improve the accuracy of the LS estimation of all the subchannels.

3. Classical DFT based channel estimation

In order to improve the LS channel estimation performance, the DFT-based method has been proposed first as it can advantageously target both noise reduction and interpolation purposes.

3.1 Main goals of DFT based channel estimation

3.1.1 Noise reduction

DFT-based channel estimation methods allow a reduction of the noise component owing to operations in the transform domain, and thus achieve higher estimation accuracy (Van de Beek et al., 1995) (Zhao & Huang, 1997). In fact, after removing the unused subcarriers, the LS estimates are first converted into the time domain by the IDFT (inverse discrete fourier transform) algorithm. A smoothing filter is then applied in the time domain assuming that the maximum multi-path delay is kept within the cyclic prefix (CP) of the OFDM symbol. As a consequence, the noise power is reduced in the time domain. The DFT is finally applied to return to the frequency domain. The smoothing process using DFT is illustrated in Fig.2.

3.1.2 Interpolation

DFT can be used simultaneously as an accurate interpolation method in the frequency domain when the orthogonality between training sequences is based on the transmission of scattered pilots (Zhao & Huang, 1997). The number ($N_p = M/N_t$) of pilot subcarriers, starting from the i-th subcarrier, is spaced every N_t subcarriers (Fig.1(a)). The LS estimates obtained for the pilot subcarriers given by (6) are first converted into the time domain by IDFT of length M/N_t. As the impulse response of the channel is concentrated on the CP first samples, it is possible to apply *zero-padding* (ZP) from M/N_t to $M-1$. The frequency channel response over the whole bandwidth is calculated by performing a M points DFT. It is obvious that N_t must satisfy the following condition:

$$N_t \leq \frac{M}{CP} \tag{7}$$

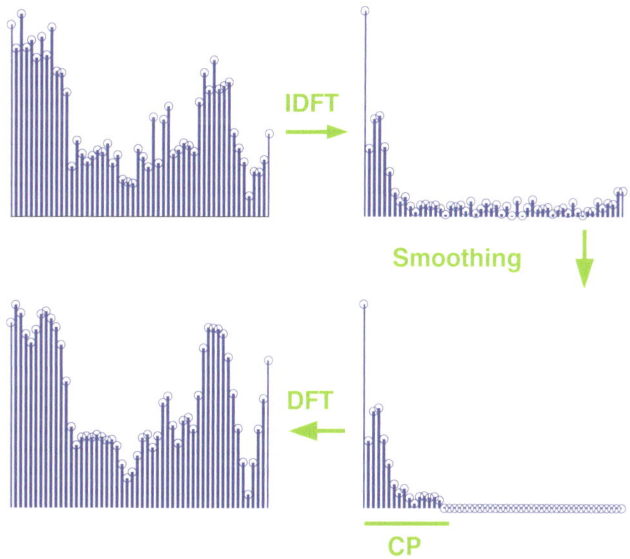

Fig. 2. Smoothing using DFT.

NB: In the rest of the paper, we will consider the case where $N_p = M$ for mathematical demonstrations in order to make reading easier. Otherwise if $N_p < M$ interpolation can be performed.

3.2 Drawback of DFT based channel estimation in realistic system

In a realistic context, only a subset of M subcarriers is modulated among the N due to the insertion of null subcarriers at the spectrum's extremities for RF mask requirements. The application of the smoothing filter in the time domain will lead to a loss of channel power when these non-modulated subcarriers are present at the border of the spectrum. That can be demonstrated by calculating the time domain channel response.

The time domain channel response of the LS estimated channel is given by (8). From (6) we can divide $h_{n,LS}^{IDFT}$ into two parts:

$$\begin{aligned} h_{ij,n,LS}^{IDFT} &= \sqrt{\frac{1}{N}} \sum_{k=\frac{N-M}{2}}^{\frac{N+M}{2}} H_{ij,k,LS} e^{j\frac{2\pi nk}{M}} \\ &= h_{ij,n}^{IDFT} + \xi_{ij,n}^{IDFT} \end{aligned} \tag{8}$$

where ξ_n^{IDFT} is the noise component in the time domain and h_n^{IDFT} is the IDFT of the LS estimated channel without noise. This last component can be further developed as follows:

$$\begin{aligned} h_{ij,n}^{IDFT} &= \sqrt{\frac{1}{N}} \sum_{k=N_b}^{N_e} \left(\sum_{l=0}^{L_{ij}-1} h_{ij,l} e^{-j\frac{2k\pi\tau_{ij,l}}{N}} \right) e^{-j\frac{2\pi kn}{N}} \\ &= \sqrt{\frac{1}{N}} \sum_{l=0}^{L_{ij}-1} h_{ij,l} \sum_{k=N_b}^{N_e} e^{-j\frac{2\pi k}{N}(\tau_{ij,l}-n)} \end{aligned} \tag{9}$$

where $N_b = (N - M)/2$ and $N_e = (N + M)/2 - 1$.

It can be seen from (9) that if all the subcarriers are modulated, i.e $M = N$, the last term of (9) $\sum_{k=\frac{N-M}{2}}^{\frac{N+M}{2}-1} e^{-j\frac{2\pi k}{N}(\tau_{ij,l}-n)}$ will verify:

$$\sum_{k=\frac{N-M}{2}}^{\frac{N+M}{2}-1} e^{-j\frac{2\pi k}{N}(\tau_{ij,l}-n)} = \begin{cases} N_p & n = \tau_{ij,l} \\ 0 & otherwise \end{cases} \tag{10}$$

where $\tau_{ij,l} = 0, 1, ..., L_{ij} - 1$ and $n = 0, ..., M - 1$.
From (10), we can safely conclude that:

$$h_{ij,n}^{IDFT} = 0 \quad n = L_{ij}, ..., M \tag{11}$$

Assuming $CP > L_{ij}$, we do not lose part of the channel power in the time domain by applying the smoothing filter of length CP.
Nevertheless, when some subcarriers are not modulated at the spectrum borders, i.e. $M < N$, the last term of (9) can be expressed as:

$$\sum_{k=\frac{N-M}{2}}^{\frac{N+M}{2}-1} e^{-j\frac{2\pi k}{N}(\tau_{ij,l}-n)} = \begin{cases} M & n = \tau_{ij,l} \\ \frac{1-e^{-j2\pi\frac{M}{N}(\tau_{ij,l}-n)}}{1-e^{-j\frac{2\pi}{N}(\tau_{ij,l}-n)}} & n \neq \tau_{ij,l} \end{cases} \tag{12}$$

where $\tau_{ij,l} = 0, 1, ..., L_{ij} - 1$ and $n = 0, ..., M - 1$.
The channel impulse response $h_{ij,n}^{IDFT}$ can therefore be rewritten in the following form:

$$h_{ij,n}^{IDFT} = \frac{1}{\sqrt{N}} \begin{cases} M.h_{ij,l=n} + \sum_{l=0,l\neq n}^{L_{ij}-1} h_{ij,l} \frac{1-e^{-j2\pi\frac{M}{N}(\tau_{ij,l}-n)}}{1-e^{-j\frac{2\pi}{N}(\tau_{ij,l}-n)}}. & n < L_{ij} \\ \sum_{l=0}^{L_{ij}-1} h_{ij,l} \frac{1-e^{-j2\pi\frac{M}{N}(\tau_{ij,l}-n)}}{1-e^{-j\frac{2\pi}{N}(\tau_{ij,l}-n)}}. & L_{ij} - 1 < n < M \end{cases} \tag{13}$$

We can observe that $h_{ij,n}^{IDFT}$ is not null for all the values of n due to the phenomenon called here "Inter-Taps Interference (ITI)". Consequently, by using the smoothing filter of length CP in the time domain, the part of the channel power contained in samples $n = CP, ..., M - 1$ is lost. This loss of power leads to an important degradation on the estimation of the channel response. In OFDM systems, Morelli shows that when null carriers are inserted at the spectrum extremities, the performance of the DFT based channel estimation is degraded especially at the borders of the modulated subcarriers (Morelli & Mengali, 2001). This phenomenon is called the "border effect". This "border effect" phenomenon is also observed in MIMO context (Le Saux et al., 2007).
In order to evaluate the DFT based channel estimation, the mean square error (MSE) performance for the different modulated subcarriers is considered in the following subsection.

3.3 MSE performance of DFT based channel estimation

In MIMO-OFDM context with N_t transmit antennas and N_r receive antennas, the (MSE) on the $k-th$ subcarrier is equal to:

$$MSE(k) = \frac{\sum_{i=0}^{N_t-1} \sum_{i=0}^{N_r-1} E\left[\left\|\widehat{H}(k) - H(k)\right\|^2\right]}{N_t N_r} \tag{14}$$

where $\widehat{H}(k)$ and $H(k)$ represent the estimated frequency channel response and the ideal one respectively.

MSE performance are provided here over frequency and time selective MIMO SCME (spatial channel model extension) channel model typical of macro urban propagation (Baum et al., 2005). The DFT based channel estimation is applied to a 2×2 MIMO system with the number of FFT points set equal to 1024. The orthogonality between pilots is obtained using null symbol insertion described in 2 and interpolation is performed to obtain channel estimates for all modulated subcarriers.

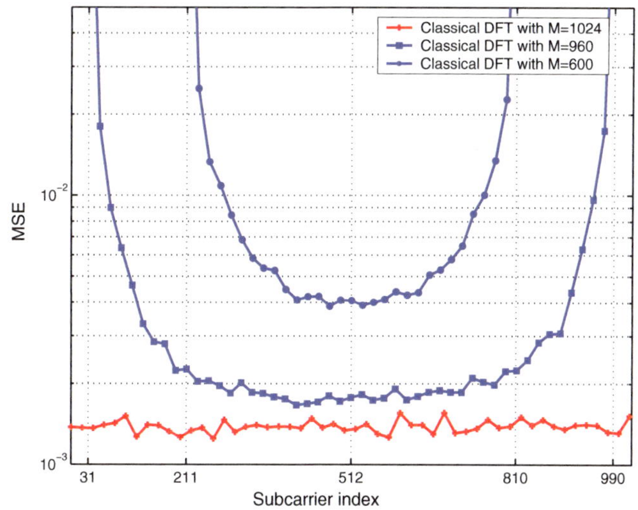

Fig. 3. Average mean square error versus subcarrier index for classical DFT based channel estimation. The number of modulated subcarrier are $M = 960$ and $M = 600$. $N_t = 2$, $N_r = 2$, $N = 1024$ and $SNR = 10dB$

Fig.3 shows the MSE performance of the different subcarriers when applying the classical DFT based channel estimation method depending on the number of modulated subcarriers. First, when all the subcarriers are modulated ($N = M = 1024$), there is no "border effect" and the MSE is almost the same for all the subcarriers. This is due to the fact that all the channel power is retrieved in the first CP samples of the impulse channel response (3.1). However when null subcarriers are inserted on the edge of the spectrum ($N \neq M$), the MSE performance is degraded by the loss of a part of the channel power in the time domain and then the "border effect" occurs. It is already noticeable that the impact of the "border effect" phenomenon increases greatly with the number of null subcarriers.
To mitigate this "'border effect" phenomenon, the pseudo inverse technique proposed in (Doukopoulos & Legouable, 2007) is studied in the next section.

4. DFT with pseudo inverse channel estimation

Classical DFT based channel estimation described in the previous section can be also expressed in a matrix form.

The unitary DFT matrix F of size $N \times N$ is defined with the following expression:

$$\mathrm{F} = \begin{bmatrix} 1 & 1 & 1 & \dots & 1 \\ 1 & W_N & W_N^2 & \dots & W_N^{N-1} \\ \vdots & \vdots & \vdots & \dots & \vdots \\ 1 & W_N^{N-1} & W_N^{2(N-1)} & \dots & W_N^{(N-1)(N-1)} \end{bmatrix} \tag{15}$$

where $W_N^i = e^{-j\frac{2\pi i}{N}}$.
To accommodate the non-modulated subcarriers, it is necessary to remove the rows of the matrix F corresponding to the position of those null subcarriers. Furthermore, in order to reduce the noise component in the time domain by applying smoothing filtering, only the first CP columns of F are used (Fig.2). Hence the *transfer matrix* becomes $F' \in C^{M \times CP}$:
$F' = F(\frac{N-M}{2} : \frac{N+M}{2} - 1, 1 : CP)$.
We can then express the impulse channel response, after the smoothing filter, in a matrix form:

$$h_{ij,LS}^{IDFT} = F'^{H} H_{ij,LS} \tag{16}$$

where $h_{ij,LS}^{IDFT} \in C^{CP \times 1}$, $H_{ij,LS} \in C^{M \times 1}$.
To reduce the "border effect" Doukopoulos propose the use of the following minimization problem (Doukopoulos & Legouable, 2007).

$$h_{ij,LS}^{pseudoinverse} = \arg\min_{h_{ij}} \left\| F' - H_{ij,LS} \right\|^2 \tag{17}$$

The idea that lies behind the above minimization problem is to reduce the "border effect", as illustrated in the figure 4, by minimizing the Euclidean norm between $H_{ij,LS}$ and $H_{ij,LS}^{IDFT}$.
The pseudo inverse of the matrix F' which is noted $F'^{\dagger} \in C^{CP \times M}$, provides a solution to equation 17. It can be used to transform the LS estimates in the time domain as proposed by equation 18 instead of F'^{H} as previously proposed in equation 16. The use of the pseudo inverse allows the minimization of the power loss in the time domain which was at the origin of the "border effect".

$$h_{ij,LS}^{pseudoinverse} = F'^{\dagger} H_{ij,LS} \tag{18}$$

The pseudo inverse which is sometimes named *generalized inverse* was described by Moore in 1920 in *linear algebra* (Moore, 1920). This technique is often used for the resolution of linear equations system due to its capacity to minimize the *Euclidean norm* and then to tend towards the exact solution. The pseudo inverse $F'^{\dagger}$ of F' is defined as unique matrix satisfying all four following criteria (Penrose, 1955).

$$\begin{cases} 1 : F' F'^{\dagger} F' = F' \\ 2 : F'^{\dagger} F' F'^{\dagger} = F'^{\dagger} \\ 3 : (F' F'^{\dagger})^H = F' F'^{\dagger} \\ 4 : (F'^{\dagger} F')^H = F^{\dagger'} F' \end{cases} \tag{19}$$

4.1 Pseudo inverse computation using SVD

The pseudo inverse can be computed simply and accurately by using the singular value decomposition (Moore, 1920). Applying SVD to the matrix $F^{'}$ consists in decomposing $F^{'}$ in the following form:

$$F^{'} = USV^{H} \tag{20}$$

where $U \in C^{M \times M}$ and $V \in C^{CP \times CP}$ are unitary matrices and $S \in C^{M \times CP}$ is a diagonal matrix with non-negative real numbers on the diagonal, called singular values.
The pseudo inverse of the matrix $F^{'}$ with singular value decomposition is:

$$F^{'\dagger} = VS^{\dagger}U^{H} \tag{21}$$

It is important to note that $S^{\dagger}$ is formed by replacing every singular value by its inverse.

4.2 Impact of pseudo inverse conditional number on channel estimation accuracy

The accuracy of the estimated channel response depends on the calculation of the pseudo inverse $F^{'\dagger}$. The conditional number (CN) can give an indication of the accuracy of this operation (Yimin et al., 1991). The higher the CN is, the more the estimated channel response is degraded.

4.2.1 Definition of the conditional number

It is defined as the ratio between the greatest and the smallest singular values of the transfer matrix $F^{'}$. By noting $s \in C^{CP \times 1}$ the vector which contains the elements (the singular values) on the diagonal of the matrix S, CN is expressed as follows:

$$CN = \frac{max(s)}{min(s)} \tag{22}$$

where $max(s)$ and $min(s)$ give the greatest and the smallest singular values respectively.

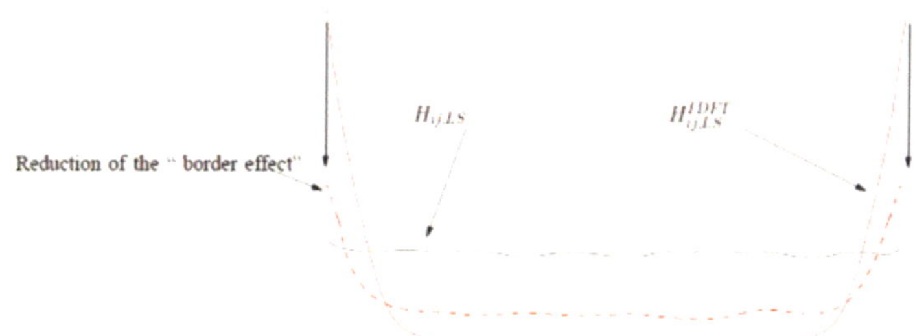

Fig. 4. Illustration of the "border effect" reduction

4.2.2 Behavior of the conditional number

It only evolves according to the number of modulated subcarriers. Fig.5 shows this behavior for different values of M when $N = 1024$ and $CP = 72$. When all the subcarriers are modulated (i.e when $M = N$), the CN is equal to 1. However when null carriers are inserted at the edge of the spectrum ($M < N$), the CN increases according to the number of non-modulated subcarriers ($N - N$) and can become very high.
We can note that if $M = 600$ as in 3GPP standard (where $N = 1024$, $CP = 72$ and $M = 600$), the CN is equal to $2.17\,10^{15}$.

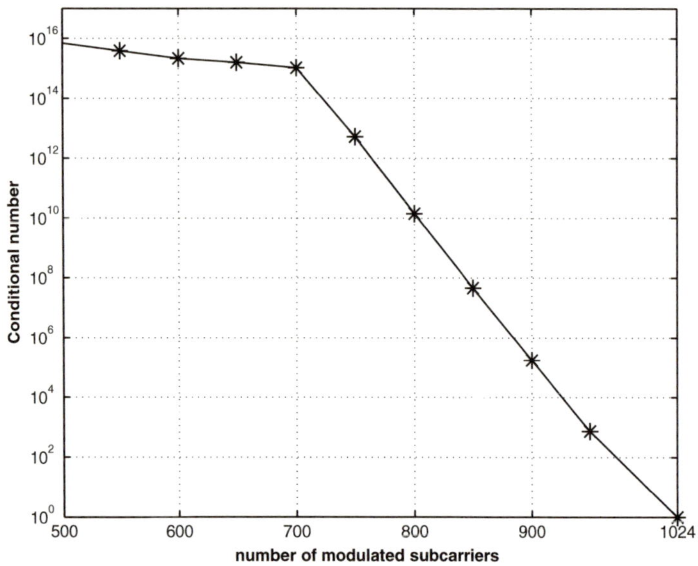

Fig. 5. Conditional number of F' versus the number of modulated subcarriers (M) where $CP = 72$ and $N = 1024$.

4.3 MSE performance of DFT with pseudo inverse channel estimation

System parameters for MSE performance evaluation are identical to those in section 3.3.

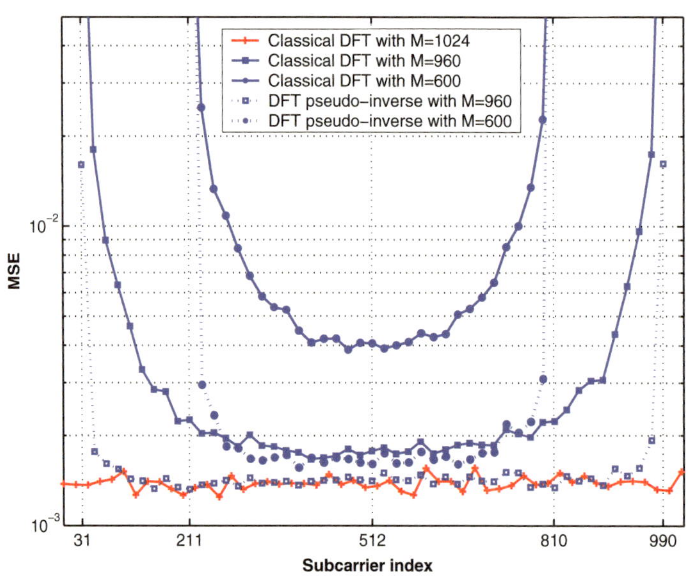

Fig. 6. Average mean square error versus subcarrier index for classical DFT and DFT pseudo inverse based channel estimation. The number of modulated subcarrier are $M = 960$ and $M = 600$. $N_t = 2$, $N_r = 2$, $N = 1024$ and $SNR = 10dB$

Fig.6 shows the MSE performance for different subcarriers when applying either classical DFT or DFT with pseudo inverse channel estimation methods. When null subcarriers are inserted on the edges of the spectrum ($N \neq M$), the MSE performance of the classical DFT based channel estimation is degraded by the "border effect". As for CN, this "border effect" increases with the increasing number of null subcarriers. The DFT with pseudo inverse technique significantly reduces the "border effect" when the CN is low ($M = 960$ and $CN \simeq 100$). However, when the number of null subcarriers is large ($M = 600$) and the CN is largely increased ($CN = 2.17\ 10^{15}$), the MSE performance remains degraded.

5. DFT with truncated SVD channel estimation

The DFT with pseudo inverse technique previously described allows the reduction of the "border effect". But it remains insufficient when the number of null subcarriers is large. To further reduce this "border effect", it is necessary to attain a small CN in this operation. The aim of the proposed approach is to reduce both the "border effect" and the noise component by considering only the most significant singular values of matrix S.

5.1 Principle of DFT with truncated SVD channel estimation

To reduce the CN, the lowest singular values have to be eliminated. Hence, any singular value smaller than an optimized threshold (detailed in 5.2) is replaced by zero. The principle is depicted in Fig.7. The SVD calculation of matrix F' provides the matrices U, S and V (20). The matrix S becomes S_{T_h} where T_h is the number of considered singular values.

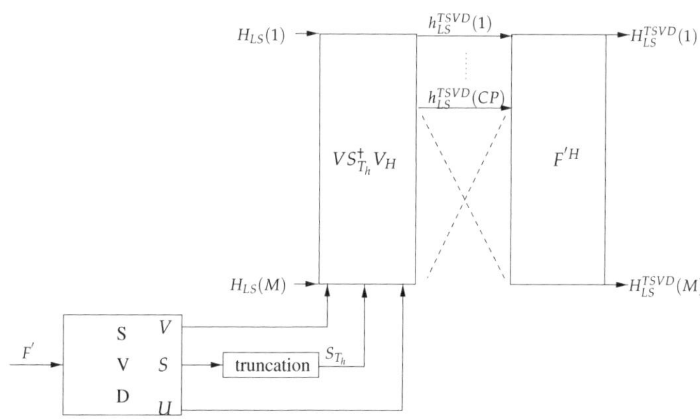

Fig. 7. Block diagram of DFT with Truncated SVD.

The channel impulse response after the smoothing process in the time domain can thus be expressed as follows:

$$h_{ij,LS}^{TSVD} = F_{T_h}^{'\dagger} H_{ij,LS} = VS_{T_h}^{\dagger} U^H H_{ij,LS} \tag{23}$$

$$h_{ij,LS,n}^{TSVD} = \begin{cases} F_{T_h}^{'\dagger} H_{ij,LS} = VS_{T_h}^{\dagger} U^H H_{ij,LS} & n < CP \\ 0 & otherwise \end{cases} \tag{24}$$

where $n = 0, ..., M-1$.
Finally F' of size $(CP \times M)$ is used to return to the frequency domain.

$$H^{ij,TSVD} = F' h_{ij,LS}^{TSVD} \tag{25}$$

It is important to note the two following points:

- On the one hand, the "border effect" is obviously further reduced due to the reduction of the CN.
- On the other hand, the suppression of the lowest singular values allows the noise component in the estimated channel response to be reduced. The rank of the matrix F' is CP, which means that the useful power of the channel is distributed into CP virtual paths with singular values as weightings. The paths corresponding to the weakest singular values are predominated by noise and their elimination benefits the noise component reduction.

5.2 Threshold determination for DFT with truncated SVD

5.2.1 Discussion

The choice T_h ($\in 1, 2, ..., CP$) can be viewed as a compromise between the accuracy of the pseudo-inverse calculation and the CN magnitude. Its value will depend only on the system parameters: the number and position of the modulated subcarriers (M), the smoothing window size (CP) and the number of FFT points (N). All these parameters are predefined and are known prior to any channel estimation implementation. It is thus feasible to optimize the T_h value prior to any MIMO-OFDM system implementation.
To determine a good value of T_h, it is important to master its effect on the channel estimation quality i.e. on the matrix $F'^{\dagger}_{T_h}$ (24):

- When all the singular values of F' are considered, the CN of the matrix $F'^{\dagger}_{T_h}$ is very high and the accuracy of the estimated channel response is then degraded by the "border effect".
- However if insufficient singular values are used, the estimated channel response is also degraded due to a very large energy loss, even if the CN of $F'^{\dagger}_{T_h}$ is minimized.

The optimum value of T_h which provides a good compromise between these two phenomenons exists, as illustrated in the next subsection, for any MIMO-OFDM system.

5.2.2 Illustration

As previously seen, the adjustment of T_h enables the improvement of the channel estimation quality and its value depends only on the system parameters. To assess the value of T_h, we study here the behavior of the singular values of $F'^{\dagger}_{T_h}$ according to T_h. The same system parameters are considered: a 2×2 MIMO system, number of FFT points equal to 1024, $CP = 72$ and two different values of M (600 and 960). The orthogonality between the pilots from the different transmit antennas is again obtained by using null symbol insertion. The number of singular values of the matrix $F'^{\dagger}_{T_h}$ is then M/N_t.
Fig.8 illustrates the behavior of the $M/N_t = 300$ singular values of the matrix $F'^{\dagger}_{T_h}$ for different values of T_h when $M = 600$. For $T_h = 46, 45, 44$ the singular values of $F'^{\dagger}_{T_h}$ are the same on the first T_h indexes and almost null for the other ones. We can consider that the rank of the matrix $F'^{\dagger}_{T_h}$ becomes T_h instead of CP and that its CN is equal to 1. Therefore the "border effect" will surely be mitigated and the noise component be minimized. When $T_h \leq 43$, all the singular values become null due to a too large loss of energy.
The behavior of the singular values of $F'^{\dagger}_{T_h}$ can not be considered as constant for $T_h > 46$ and then the CN is high.
Differently, a moderated "border effect" is obtained when the number of non-modulated subcarriers is not too large ($M = 960$). Fig.9 illustrates the behavior of the singular values

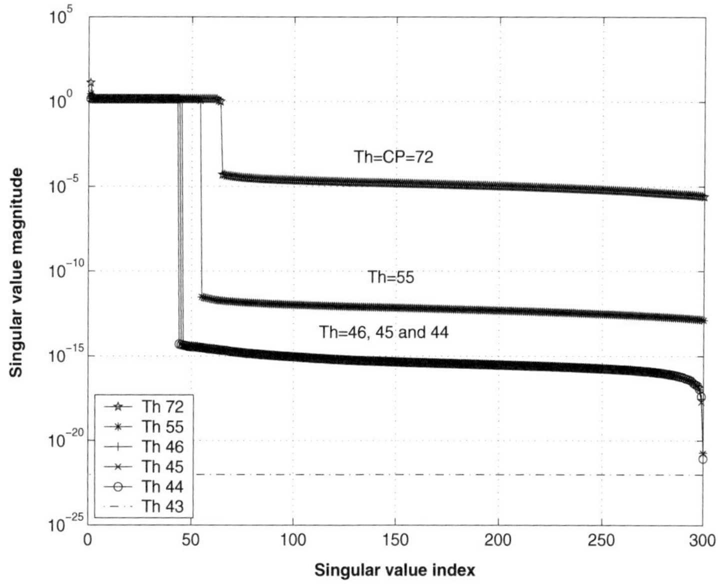

Fig. 8. Behavior of the $M/N_t = 300$ singular value of the matrix $F'^{\dagger}_{T_h}$ for $T_h = CP, 55, 46, 45, 44$ and 43. $N_t = 2$, $N_t = 2$, $CP = 72$, $N = 1024$ and $M = 600$.

of the matrix $F'^{\dagger}_{T_h}$ when $M = 960$. For $T_h = CP = 72$, the singular values of $F'^{\dagger}_{T_h}$ are the same on the first T_h indexes and almost zero for the other ones.

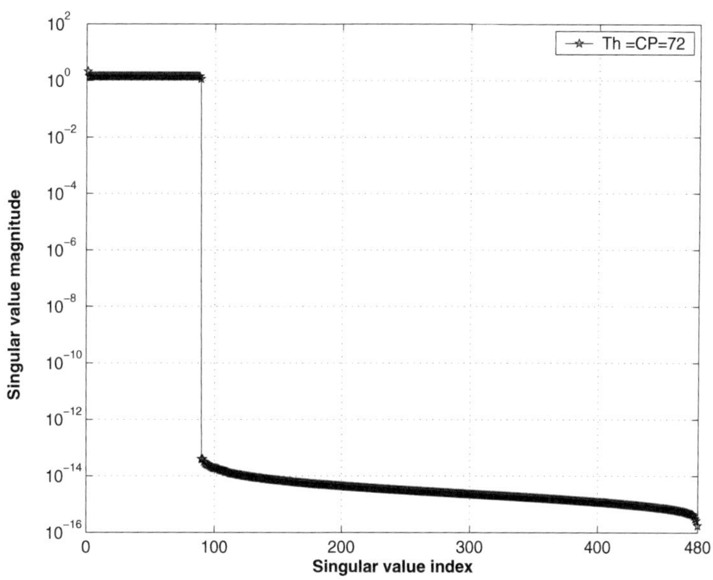

Fig. 9. Behavior of the $M/N_t = 480$ singular value of the matrix $F'^{\dagger}_{T_h}$ for $T_h = CP = 72$. $N_t = 2$, $N_t = 2$, $CP = 72$, $N = 1024$ and $M = 960$.

5.3 MSE performance of DFT with truncated SVD channel estimation

System parameters for MSE performance evaluation are identical to those in section 3.3.

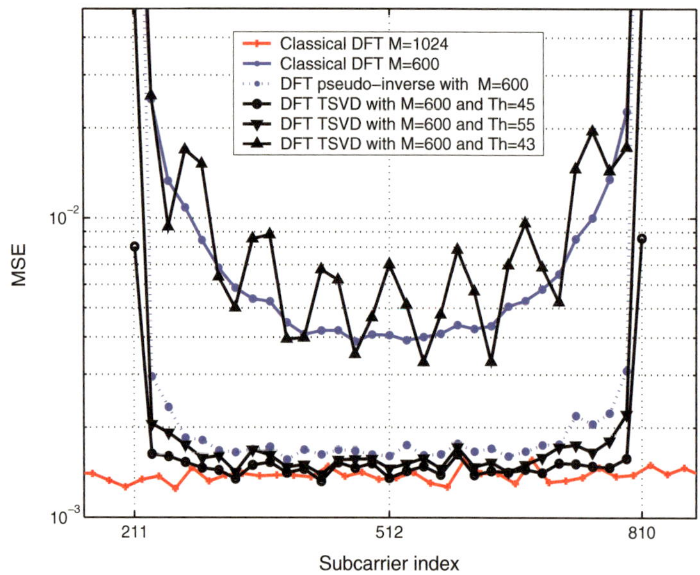

Fig. 10. Average mean square error versus subcarrier index for classical DFT, DFT pseudo inverse and DFT TSVD based channel estimation. The number of modulated subcarrier is $M = 600$. $N_t = 2$, $N_r = 2$, $N = 1024$ and $SNR = 10dB$

Fig.10 shows the MSE performance on the different subcarriers when applying either classical DFT, DFT with pseudo inverse channel or DFT with truncated SVD ($T_h = 43, 45, 55$) channel estimation methods. The DFT TSVD method with optimized T_h ($T_h = 45$) allows smaller MSE on all subcarriers to be obtained even at the edges of the spectrum. This is due to the minimization of the noise effect and the reduction of the CN provided by the TSVD calculation. Nevertheless this threshold has to be carefully assessed since a residual "border effect" is present when $T_h = 55$ and a large loss of channel power in the time domain brings poorer results when $T_h = 43$.

6. Applications

The performance of the three different DFT based channel estimation methods detailed in this paper are evaluated in the IEEE802.11n (typical *indoor*) and 3GPP/LTE (typical *outdoor*) system environments (downlink transmission).

6.1 Systems parameters

The simulation parameters are listed in Table 1 for both IEEE802.11n (802.11, 2007) and 3GPP/LTE (3GPP, 2008) configurations.

In IEEE802.11n, the channel (TGn channel models (Erceg et al., 2004)) can be assumed to be unchanged over the duration of 2 OFDM symbols due to quite slow variations. The orthogonality between training sequences on antennas 1 and 2 is obtained by using an orthogonal matrix. The estimated channel $H_{ij,LS}$ can be calculated for all the M modulated subcarriers due to the use of full pilot OFDM symbols.

Unlike in the IEEE802.11n system, in the 3GPP/LTE system, the time channel (SCME typical to urban macro channel model (Baum et al., 2005)) varies too much due to higher mobility. The orthogonality between training sequences in the 3GPP/LTE standard is thus based on the transmission on each subcarrier of pilot symbols on one antenna while null symbols are simultaneously sent on the other antennas. Therefore the LS channel estimates are calculated only for $\frac{M}{N_t} = 150$ subcarriers and interpolation is performed to obtain an estimation for all the modulated subcarriers.

system	802.11n	3GPP/LTE
Channel Model	TGn (Erceg et al., 2004)	SCME (Baum et al., 2005)
Sampling frequency (MHz)	20	15.36
Carrier frequency (GHz)	2.4	2
FFT size (N)	64	1024
OFDM symbol duration (μs)	4	71.35
Useful carrier (M)	52	600
Cyclic prefix (CP)	16	72
CP duration (μs)	0.80	4.69
MIMO scheme	SDM	double-Alamouti
MIMO rate (R_M)	1	1/2
$N_t \times N_r$	2×2	4×2
Modulation	QPSK	16QAM
Number of bit (m)	2	4
FEC	conv code (7,133,171)	turbo code (UMTS)
Coding Rate (R_c)	1/2	1/3

Table 1. Simulation Parameters

6.2 Simulation results

Perfect time and frequency synchronizations are assumed. Monte Carlo simulation results in terms of bit error rate (BER) versus $\frac{E_b}{N_0}$ are presented here for the different DFT based channel estimation methods: classical DFT, DFT with pseudo inverse and DFT with truncated SVD. The $\frac{E_b}{N_0}$ value can be inferred from the signal to noise ratio (SNR):

$$\frac{E_b}{N_0} = \frac{N_t}{mR_cR_M}.\frac{\sigma_s^2}{\sigma_n^2} = \frac{N_t}{mR_cR_M}.SNR \qquad (26)$$

where σ_{noise}^2 and σ_x^2 represent the noise and signal variances respectively. R_c, R_M and m represent the coding rate, the MIMO scheme rate and the modulation order respectively.

Fig.11 and Fig.12 show the performance results in terms of BER versus $\frac{E_b}{N_0}$ for perfect, least square (LS), classical DFT, DFT with pseudo inverse ($DFT - T_h = CP$) and DFT with truncated SVD for (several T_h) channel estimation methods in 3GPP/LTE and 802.11n system environments respectively.

In the context of 3GPP/LTE, the classical DFT based method presents poorer results due to the large number of null carriers at the border of the spectrum (424 among 1024). The conditional number is as a consequence very high ($CN = 2.17\ 10^{15}$) and the impact of the "border effect" is very important. For this reason, the DFT with pseudo inverse ($DFT - T_h = CP =$

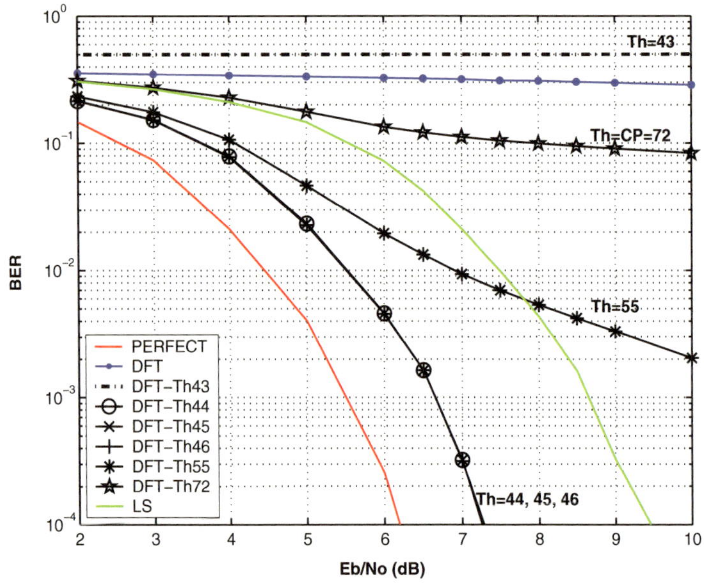

Fig. 11. BER versus $\frac{E_b}{N_0}$ for classical DFT, DFT pseudo inverse ($T_h = CP = 72$) and DFT with truncated SVD ($T_h = 55$, $T_h = 46$, $T_h = 45$, $T_h = 44$ and $T_h = 43$) based channel estimation methods in 3GPP context. $N_t = 4$, $N_r = 2$, $N = 1024$, $CP = 72$ and $M = 600$

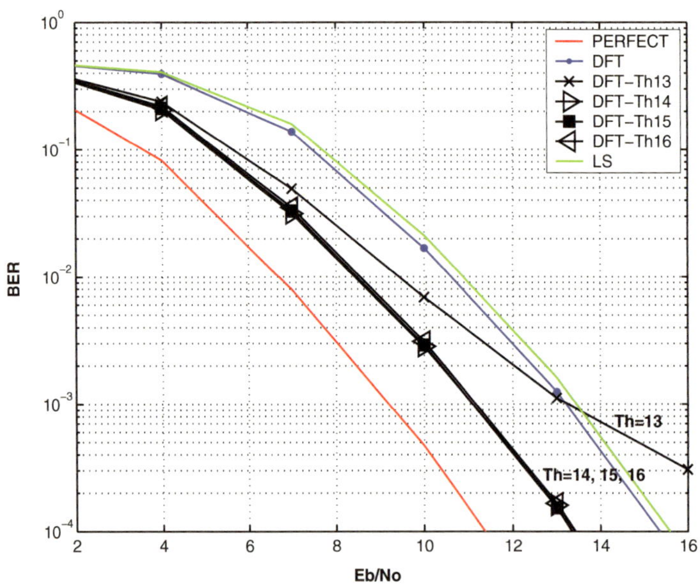

Fig. 12. BER versus $\frac{E_b}{N_0}$ for classical DFT, DFT pseudo inverse ($T_h = CP = 16$) and DFT with truncated SVD ($T_h = 15$, $T_h = 14$ and $T_h = 13$) based channel estimation methods in 802.11n context. $N_t = 2$, $N_r = 2$, $N = 1024$, $CP = 72$ and $M = 600$

72) can not greatly improve the accuracy of the estimated channel response. The classical DFT and the DFT with pseudo inverse estimated channel responses are thus considerably degraded compared to the *LS* one. The DFT with a truncated SVD technique and optimized T_h (T_h=46,45,44) greatly enhances the accuracy of the estimated channel response by both reducing the noise component and eliminating the impact of the "border effect" (up to 2dB gain compared to LS). This last method presents an error floor when $T_h = 55$ due to the fact that the "border effect" is still present and very bad results are obtained when T_h is small ($T_h = 43$) due to the large loss of energy.
Comparatively, in the context of 802.11n, the number of null carriers is less important and the classical DFT estimated channel response is not degraded even if it does not bring about any improvement compared to the LS. The pseudo inverse technique completely eliminates the "border effect" and thus its estimation ($DFT - T_h = CP = 16$) is already very reliable. DFT with a truncated SVD channel estimation method does not provide any further performance enhancement as the "border effect" is quite limited in this system configuration.

7. Conclusion

Several channel estimation methods have been investigated in this paper regarding the MIMO-OFDM system environment. All these techniques are based on DFT and are so processed through the time transform domain. The key system parameter, taken into account here, is the number of null carriers at the spectrum extremities which are used on the vast majority of multicarrier systems. Conditional number magnitude of the transform matrix has been shown as a relevant metric to gauge the degradation on the estimation of the channel response. The limit of the classical DFT and the DFT with pseudo inverse techniques has been demonstrated by increasing the number of null subcarriers which directly generates a high conditional number. The DFT with a truncated SVD technique has been finally proposed to completely eliminate the impact of the null subcarriers whatever their number. A technique which allows the determination of the truncation threshold for any MIMO-OFDM system is also proposed. The truncated SVD calculation is constant and depends only on the system parameters: the number and position of the modulated subcarriers, the cyclic prefix size and the number of FFT points. All these parameters are predefined and are known at the receiver side and it is thus possible to calculate the truncated SVD matrix in advance. Simulation results in 802.11n and 3GPP/LTE contexts have illustrated that DFT with a truncated SVD technique and optimized T_h is very efficient and can be employed for any MIMO-OFDM system.

8. References

Weinstein, S. B., and Ebert, P.M. (1971). Data transmission by frequency-division multiplexing using Discret Fourier Transform. *IEEE Trans. Commun.*, Vol. 19, Oct. 1971, pp. 628-634.

Telatar, I. E.. (1995). Capacity of Multi-antenna Gaussien Channel. *ATT Bell Labs tech. memo*, Jun. 1995.

Alamouti, S. (1998). A simple transmit diversity technique for wireless communications, *IEEE J. Select. Areas Communication*, Vol. 16, Oct. 1998, pp. 1451-1458.

Tarokh, V., Japharkhani, H., and Calderbank, A. R.. (1999). Space-time block codes from orthogonal designs. *IEEE Trans. Inform. Theory*, Vol. 45, Jul. 1999, pp. 1456-1467.

Boubaker, N., Letaief, K.B., and Murch, R.D. (2001). A low complexity multi-carrier BLAST architecture for realizing high data rates over dispersive fading channels. Proceedings of VTC 2001 Spring, 10.1109/VETECS.2001.944489, Taipei, Taiwan, May 2001.

Winters, J. H. (1987). On the capacity of radio communication systems with diversity in a Rayleigh fading environment. *I*EEE J. Select. Areas Commun., Vol.5, June 1987, pp. 871-878.

Foschini, G. J (1996). Layered space-time architecture for wireless communication in a fading environment when using multi-element antennas. *B*ell Labs Tech. J., Vol. 5, 1996, pp. 41-59.

Zhao, Y., and Huang, A. (1997). A Novel Channel Estimation Method for OFDM Mobile Communication Systems Based on Pilot Signals and Transform-Domain. Proceedings of IEEE 47th VTC, 10.1109/VETEC.1997.605966, Vol. 47, pp. 2089-2093, May 1997.

Morelli, M., and Mengali., U. (2001). A comparison of pilot-aided channel estimation methods for OFDM systems. *I*EEE Transactions on Signal Processing, Vol. 49, Jan. 2001, pp. 3065-3073.

3GPP (2008), 3GPP TS 36.300 V8.4.0: E-UTRA and E-UTRAN overall description, Mar. 2008.

Doukopoulos, X.G., and Legouable, R. (2007). Robust Channel Estimation via FFT Interpolation for Multicarrier Systems. *P*roceedings of IEEE 65th VTC-Spring, 10.1109/VETECS.2007.386, pages 1861-1865, 2007.

Van de Beek, J., Edfors, O., Sandell, M., Wilson, S. K. and Borjesson, P.O. (1995). On Channel Estimation in OFDM Systems. *P*roceedings of IEEE VTC 1995, 10.1109/VETEC.1995.504981, pp. 815-819, Chicago, USA, Sept. 1995.

Auer, G. (2004). Channel Estimation for OFDM with Cyclic delay Diversity. *P*roceedings of PIMRC 2004, 15th IEEE International Symposium on IEEE, 10.1109/PIMRC.2004.1368308, Vol. 3, pp. 1792-1796.

*D*raft-P802.11n-D2.0. IEEE P802.11nTM, Feb. 2007.

Le Saux, B., Helard, M., and Legouable, R. (2007). Robust Time Domaine Channel Estimation for MIMO-OFDM Downlink System. *P*roceedings of MC-SS, Herrsching, Allemagne, Vol. 1, pp 357-366, May 2007.

Baum, D.S., Hansen, J., and Salo J. (2005). An interim channel model for beyond-3g systems: extending the 3gpp spatial channel model (smc). *P*roceedings of VTC, 10.1109/VETECS.2005.1543924, Vol. 5, pp 3132-3136, May 2005.

Moore, E. H. (1920). On the reciprocal of the general algebraic matrix. *B*ulletin of the American Mathematical Society, Vol. 26, pp 394-395, 1920.

Penrose, R. (1955). A generalized inverse for matrices. *P*roceedings of the Cambridge Philosophical Society 51, pp 406-413, 1955.

Yimin W. and al. (1991). Componentwise Condition Numbers for Generalized Matix Inversion and Linear least sqares. *A*MS subject classification, 1991.

Erceg V. and al. (2004). TGn channel models. IEEE 802.11-03/940r4, May 2004.

7

Channels and Parameters Acquisition in Cooperative OFDM Systems

D. Neves[1], C. Ribeiro[1,2], A. Silva[1] and A. Gameiro[1]
[1]*University of Aveiro, Instituto de Telecomunicações,*
[2]*Instituto Politécnico de Leiria*
Portugal

1. Introduction

Cooperative techniques are promising solutions for cellular wireless systems to improve system fairness, extend the coverage and increase the capacity. Antenna array schemes, also referred as MIMO systems, exploit the benefits from the spatial diversity to enhance the link reliability and achieve high throughput (Foschini & Gans, 1998). On the other hand, orthogonal frequency division multiplexing (OFDM) is a simple technique to mitigate the effects of inter-symbol interference in frequency selective channels (Laroia et al., 2004). The integration of multiple antenna elements is in some situation unpractical especially in the mobile terminals because of the size constraints, and the reduced spacing does not guarantee decorrelation between the channels. An effective way to overcome these limitations is generate a virtual antenna-array (VAA) in a multi-user and single antenna devices environment, this is referred as cooperative diversity. The use of dedicated terminals with relaying capabilities has been emerging as a promising key to expanded coverage, system wide power savings and better immunity against signal fading (Liu, K. et al., 2009).
A large number of cooperative techniques have been reported in the literature the potential of cooperation in scenarios with single antennas. In what concerns channel estimation, some works have discussed how the channel estimator designed to point-to-point systems impacts on the performance of the relay-assisted (RA) systems and many cooperative schemes consider that perfect channel state information (CSI) is available (Muhaidat & Uysal, 2008), (Moco et al., 2009), (Teodoro et al., 2009), (Fouillot et al., 2010). Nevertheless, to exploit the full potential of cooperative communication accurate estimates for the different links are required. Although some work has evaluated the impact of the imperfect channel estimation in cooperative schemes (Chen et al., 2009), (Fouillot et al., 2010), (Gedik & Uysal, 2009), (Hadizadeh & Muhaidat, 2010), (Han et al., 2009), (Ikki et al. 2010), (Muhaidat et al., 2009), new techniques have been derived to address the specificities of such systems.
Channel estimation for cooperative communication depends on the employed relaying protocol, e.g., decode and forward (DF) (Laneman et al., 2004) when the relay has the capability to regenerate and re-encode the whole frame; amplify and forward (AF) (Laneman et al., 2004) where only amplification takes place; and what we term equalize and forward (EF) (Moco et al., 2010), (Teodoro et al., 2009), where more sophisticated filtering operations are used.

In the case of DF, the effects of the B→R (base station-relay node) channel are reflected in the error rate of the decided frame and therefore the samples received at the destination only depend on the R→U (relay node–user terminal) channel. In this protocol the relaying node are able to perform all the receiver's processes including channel estimation and the point-to-point estimators can be adopted in these cooperative systems. However the situation is different with AF and Equalize-and-Forward (EF) which are protocols less complex than the DF. In the former case (AF), B→R→U (base station-relay node-user terminal) channel is the cascaded of the B→R and R→U channels, which has a larger delay spread than the individual channels and additional noise introduced at the relay, this model has been addressed in (Liu M. et al., 2009), (Ma et al., 2009), (Neves et al., 2009), (Wu & Patzold, 2009), (Zhang et al., 2009), (Zhou et al., 2009).
Channel estimation process is an issue that impacts in the overall system complexity reason why it is desirable use a low complex and optimal estimator as well. This tradeoff has been achieved in (Ribeiro & Gameiro, 2008) where the MMSE in time domain (TD-MMSE) can decrease the estimator complexity comparatively to the frequency domain implementation. In (Neves et al., 2009) it is showed that under some considerations the TD-MMSE can provide the cascaded channel estimate in a cooperative system. Also regarding the receiver complexity (Wu & Patzold, 2009) proposed a criterion for the choice of the Wiener filter length, pilot spacing and power. (Zhang et al., 2009) proposed a permutation pilot matrix to eliminate inter-relay signals interference and such approach allows the use of the least square estimator in the presence of frequency off-sets. Based on the non-Gaussian dual-hop relay link nature (Zhou et al., 2009) proposed a first-order autoregressive channel model and derived an estimator based on Kalman filter. In (Liu, M. et al., 2009) the authors propose an estimator scheme to disintegrate the compound channel which implies insertion of pilots at the relay, in the same way (Ma et al., 2009) developed an approach based on a known pilot amplifying matrix sequence to improve the compound channel estimate taking into account the interim channels estimate. To separately estimate B→R and R→U channels (Sheu & Sheen. 2010) proposed an iterative channel estimator based on the expectation maximization. Regarding that the B→R and R→U links are independent and point-to-point links (Xing et al. 2010) investigated a transceiver scheme that jointly design the relay forward matrix and the destination equalizer which minimize the MSE. Concerning the two-way relay (Wang et al. 2010) proposed an estimator based on new training strategy to jointly estimate the channels and frequency offset. For MIMO relay channels (Pang et al. 2010) derived the linear mean square error estimator and optimal training sequences to minimize the MSE. However to the best of our knowledge channel estimation for EF protocol that use Alamouti coding from the base station (BS) to relay node (RN), equalizes, amplifies the signals and then forward it to the UT has not been considered from the channel estimation point of view in the literature. Such a scenario is of practical importance in the downlink of cellular systems since the BS has less constraints than user terminals (or terminals acting as relays) in what concerns antenna integration, and therefore it is appealing to consider the use of multiple antennas at the BS improving through the diversity achieved the performance in the B→R link.
However due to the Alamouti coding–decoding operations, the channel B→R→U is not just the cascade of the B→R and R→U channels, but a more complex channel. The channel estimator at the UT needs therefore to estimate this equivalent channel in order to perform the equalization. The derivation of proper channel estimator for this scenario is the objective of this chapter. We analyze the requirements in terms of channels and parameters estimation

to obtain optimal equalization. We evaluate the sensitivity of required parameters in the performance of the system and devise scheme to make these parameters available at the destination. We consider a scenario with a multiple antenna BS employing the EF protocol, and propose a time domain pilot-based scheme (Neves, et al. 2010) to estimate the channel impulse response. The B→R channels are estimated at the RN and the information about the equivalent channel inserted in the pilot positions. At the user terminal (UT) the TD-MMSE estimator, estimates the equivalent channel from the source to destination, taking into account the Alamouti equalization performed at the RN. The estimator scheme we consider operates in time domain because of the reduced complexity when compared against its implementation in frequency domain, e.g. (Ribeiro & Gameiro, 2008).
The remainder of this chapter is organized as follows. In Section 2, we present the scenario description, the relaying protocol used in this work and the corresponding block diagram of the proposed scheme. The mathematical description involving the transmission in our scheme is presented in Section 3. In Section 4, we present the channel estimation issues such as the estimator method used in this work and the channels and parameters estimates to be assess at the RN or UT. The results in terms of BER and MSE are presented in Section 5. Finally, the conclusion is pointed out in Section 6.

2. System model

2.1 Definition

Throughout the text index n and k denote time and frequency domain variables, respectively. Complex conjugate and the Hermitian transposition are denoted by $(.)^*$ and $(\cdot)^H$, respectively. $\mathrm{E}\{\cdot\}$ and $(*)$ correspondently denote the statistical expectation and the convolution operator. $\mathcal{N}(m,\sigma^2)$ refers to a complex Gaussian random variable with mean m and variance σ^2. $\mathrm{diag}(\cdot)$ stands for diagonal matrix, $|\cdot|$ denotes absolute value and $\mathbf{I}_Q$ denotes the identity matrix of size Q. Regular small letters denote variables in frequency domain while boldface small and capital letters denote matrices and vectors, respectively in frequency domain as well. Variables, vectors or matrixes in time domain are denoted by $(\sim)$. All estimates are denoted by $(\wedge)$.

2.2 Channel model

The OFDM symbol $\mathbf{x}=\mathbf{d}+\mathbf{p}$ where $\mathbf{p}$ corresponds to the pilots which are multiplexed with data $\mathbf{d}$ subcarriers. The element $x_{(k)}$ of the OFDM symbol vector is transmitted over a channel which the discrete impulse response is given by:

$$\tilde{h}_{(n)}=\sum_{g=0}^{G-1}\beta_l\delta\left(n-\tau_g\right), \tag{1.1}$$

where G is total number of paths, β_g and τ_g are the complex amplitude and delay of the gth path, respectively. β_g is modelled as wide-sense stationary uncorrelated scattering (WSSUS) process. The gth path has a variance σ_g^2 which is determined by the power delay profile and satisfies $\sum_{g=0}^{G-1}\sigma_g^2=1$. Although the channel is time-variant we assume it is constant during one OFDM symbol interval and its time dependence is not present in notation for simplicity.

2.3 Scenario description

The studied scenario, depicted in Fig. 1, corresponds to the proposed RA schemes for downlink OFDM-based system. The BS and the RN are equipped with M and L antennas, respectively. The BS is a double antenna array and the UT is equipped with a single antenna. Throughout this chapter we analyze two RA schemes: the RN as a single antenna or an array terminal. These scenarios are referred as $M \times L \times 1$ schemes.

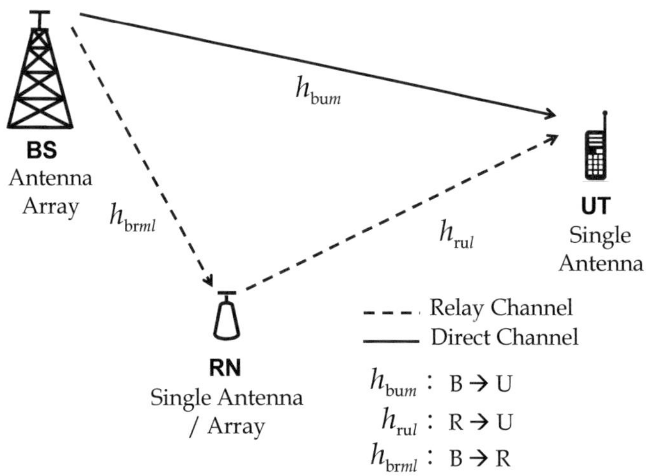

Fig. 1. Proposed RA scenario

The following channels per k subcarrier are involved in this scheme:

- $M \times 1$ MISO channel between the BS and UT (B→U): $h_{\mathrm{bu}m,(k)}$, $m = 1,2$
- $L \times 1$ MISO channel between the RN and UT (R→U): $h_{\mathrm{ru}l,(k)}$, $l = 1,2$
- $M \times L$ channel between the BS and RN (B→R): $h_{\mathrm{br}ml,(k)}$, $m = 1,2$ and $l = 1,2$

All the channels are assumed to exhibit Rayleigh fading, and since the RN and UT are mobile the Doppler's effect is considered in all channels and the power transmitted by the BS is equally allocated between the two antennas.

2.4 The Equalize-and-Forward (EF) relaying protocol

For the single antenna relay scenario, the amplify-and-forward protocol studied in (Moco et al., 2010) is equivalent to the RA EF protocol considered here. However, if the signal at the relay is collected by two antennas, doing just a simple amplify-and-forward it is not the best strategy. We need to perform some kind of equalization at the RN to combine the received signals before re-transmission. Since we assume the relay is half-duplex, the communication cycle for the aforementioned cooperative scheme requires two phases:

Phase I: the BS broadcasts its own data to the UT and RN, which does not transmit data during this stage.

Phase II: while the BS is idle, the RN retransmits to the UT the equalized signal which was received from the BS in phase I. The UT terminal receives the signal from the RN and after reception is complete, combines it with the signal received in phase I from the BS, and provides estimates of the information symbols.

2.5 The cooperative system

Fig. 2 shows the corresponding block diagram of the scenario depicted in Fig. 1, with indication of the signals at the different points. The superscripts (1) and (2) denote the first and the second phase of the EF protocol, respectively. In the different variables used, the subscripts u, r and b mean that these variables are related to the UT, RN and BS, respectively.

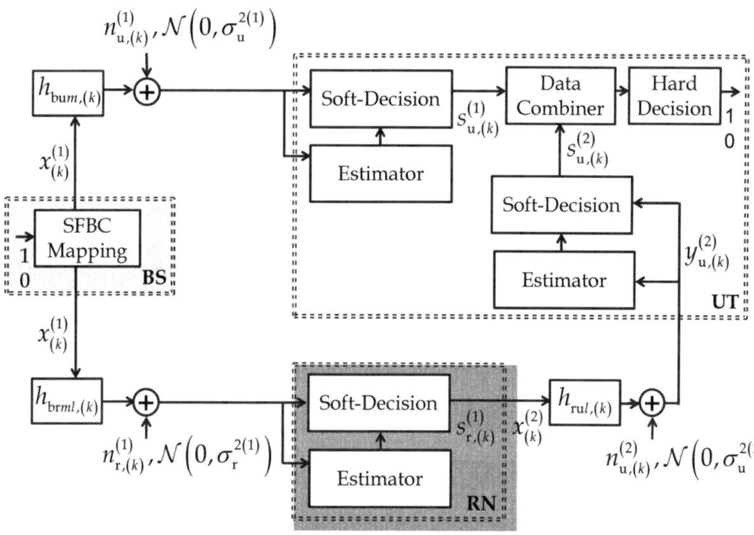

Fig. 2. The corresponding block diagram of the $M \times L \times 1$ RA scenario

Let $\mathbf{d} = (d_0\ d_1\ \cdots\ d_{N_d} - 1)^T$ be the symbol sequence to be transmitted where N_d is the number of data symbols, then for k even the SFBC (Teodoro el al., 2009) mapping rule is defined in Table 1. The symbols $d_{(k)}$ are assumed to have unit average energy, *i.e.* $|d_{(k)}|^2 = 1$, $\forall k$, and therefore the factor $1/\sqrt{2}$ used in the mapping, is to ensure that the total energy transmitted by the two antennas per subcarrier is normalized to 1.

Subcarrier	Antenna 1	Antenna 2
k	$d_{(k)}/\sqrt{2}$	$-d_{(k+1)}^*/\sqrt{2}$
$k+1$	$d_{(k+1)}/\sqrt{2}$	$d_{(k)}^*/\sqrt{2}$

Table 1. Two transmit antenna SFBC mapping

The pilot symbols are multiplexed with data and the BS broadcasts the information $x_{(k)}^{(1)}$ (data and pilot) to the RN and UT. This processing corresponds to the phase I of the EF relay protocol. At the UT, the direct channels are estimated and the data are SFBC de-mapped and equalized. These two operations are referred as soft-decision which the result is the soft-decision variable, in this case, $s_{u,(k)}^{(1)}$. At the RN, pilots and data are separated; based on pilots, the channels B→R are estimated and the soft-decision is performed. The result is the soft-decision variable $s_{r,(k)}^{(1)}$. Then, the new pilot symbols are multiplexed in $s_{r,(k)}^{(1)}$ and the information $x_{(k)}^{(2)}$ is transmitted / forwarded to the UT via R→U channel. This second transmission corresponds to the phase II of the EF protocol. At the final destination, the required channel is estimated and the soft-decision is performed in order to obtain the soft-decision variable $s_{\mathrm{u},(k)}^{(2)}$. After the phase II the UT has the soft-variable provided by both the BS and RN. These variables are combined and hard-decoded.

3. Mathematical description of the proposed cooperative scheme

The mathematical description for transmit and receive processing is described in this section. As this work is focused on channel estimation, this scheme is designed in order to be capable to provide all the channels and parameters that the equalization requires in both phases of the relaying protocol.

3.1 Phase I

During the first phase the information is broadcasted by the BS. The frequency domain (FD) signals received at the UT in data-subcarriers k and $k+1$ are given by

$$\begin{cases} y_{\mathrm{u},(k)}^{(1)} = \frac{1}{\sqrt{2}}\left(h_{\mathrm{bu}1,(k)}d_{(k)} - h_{\mathrm{bu}2,(k+1)}d_{(k+1)}^{*}\right) + n_{\mathrm{u},(k)}^{(1)} \\ y_{\mathrm{u},(k+1)}^{(1)} = \frac{1}{\sqrt{2}}\left(h_{\mathrm{bu}2,(k)}d_{(k)}^{*} + h_{\mathrm{bu}1,(k+1)}d_{(k+1)}\right) + n_{\mathrm{u},(k+1)}^{(1)} \end{cases}, \tag{3.1}$$

where $n_{\mathrm{u},(k)}^{(1)}$ is the additive white Gaussian noise with zero mean unit variance $\sigma_{\mathrm{u}}^{2(1)}$ and for $m=1,2$, $h_{\mathrm{bu}m,(k)}$ represent the channels between the BS and the UT terminal.
The FD signals received at the RN in data-subcarriers k and $k+1$ are expressed by:

$$\begin{cases} y_{\mathrm{r}l,(k)}^{(1)} = \frac{1}{\sqrt{2}}\left(h_{\mathrm{br}1l,(k)}d_{(k)} - h_{\mathrm{br}2l,(k+1)}d_{(k+1)}^{*}\right) + n_{\mathrm{r},(k)}^{(1)} \\ y_{\mathrm{r}l,(k+1)}^{(1)} = \frac{1}{\sqrt{2}}\left(h_{\mathrm{br}2l,(k)}d_{(k)}^{*} + h_{\mathrm{br}1l,(k+1)}d_{(k+1)}\right) + n_{\mathrm{r},(k+1)}^{(1)} \end{cases}, l=1,2 \tag{3.2}$$

where $h_{\mathrm{br}ml,(k)}$ represent the channels between the antenna m of the BS and antenna l of the RN terminal and $n_{\mathrm{r},(k)}^{(1)}$ is the additive white Gaussian noise with zero mean unit variance $\sigma_{\mathrm{r}}^{2(1)}$.
Since the data are SFBC mapped at the BS the SFBC de-mapping at the terminals RN and UT also includes the MRC (maximum ration combining) equalization which coefficients are functions dependent on the channels estimates. It is widely known that in the OFDM systems the subcarrier separation is significantly lower than the coherence bandwidth of the channel. Accordingly, the fading in two adjacent subcarriers can be considered flat and without loss of generality we can assume for generic channel $h_{(k)} = h_{(k+1)}$. Thus, in phase I the soft-decision variables at the UT follow the expression.

$$\begin{cases} s_{\mathrm{u},(k)}^{(1)} = g_{\mathrm{bu}1,(k)}^{*}y_{\mathrm{u},(k)}^{(1)} + g_{\mathrm{bu}2,(k)}y_{\mathrm{u},(k+1)}^{*(1)} \\ s_{\mathrm{u},(k+1)}^{(1)} = -g_{\mathrm{bu}2,(k)}y_{\mathrm{u},(k)}^{*(1)} + g_{\mathrm{bu}1,(k)}^{*}y_{\mathrm{u},(k+1)}^{(1)} \end{cases}, \tag{3.3}$$

where the equalization coefficients for $m = 1,2$ are given by $g_{\mathrm{bu}m,(k)} = \hat{h}_{\mathrm{bu}m,(k)} / \sqrt{2}\sigma_{\mathrm{u}}^2$. After some mathematical manipulation, these soft-decision variables may be expressed as:

$$\begin{cases} s_{\mathrm{u},(k)}^{(1)} = \Gamma_{\mathrm{bu},(k)} d_{(k)} + \dfrac{\hat{h}_{\mathrm{bu}1,(k)}^{*}}{\sqrt{2}} n_{\mathrm{u},(k)}^{(1)} + \dfrac{\hat{h}_{\mathrm{bu}2,(k)}}{\sqrt{2}} n_{\mathrm{u},(k)}^{(1)*} \\ s_{\mathrm{u},(k+1)}^{(1)} = \Gamma_{\mathrm{bu},(k+1)} d_{(k+1)} + \dfrac{\hat{h}_{\mathrm{bu}1,(k+1)}^{*}}{\sqrt{2}} n_{\mathrm{u},(k+1)}^{(1)} - \dfrac{\hat{h}_{\mathrm{bu}2,(k+1)}}{\sqrt{2}} n_{\mathrm{u},(k+1)}^{(1)*} \end{cases}, \tag{3.4}$$

where $\Gamma_{\mathrm{bu}} = \dfrac{1}{2}\sum_{m=1}^{2} \left| \hat{h}_{\mathrm{bu}m} \right|^2$.

The soft-decision variables should be kept in a buffer, waiting for the information to be provided by the RN in the second phase of the protocol. The mathematical formulation of the next phase varies according to the number of antennas at the RN, *i.e.* L, and these cases are separately presented in the next sub-sections.

3.2 Phase II

3.2.1 RN equipped with a single antenna

The FD soft-decision variables at the RN in data-subcarriers k and $k+1$ are expressed by:

$$\begin{cases} s_{\mathrm{r},(k)}^{(1)} = g_{\mathrm{br}11}^{*} y_{\mathrm{r}1,(k)}^{(1)} + g_{\mathrm{br}21} y_{\mathrm{r}1,(k+1)}^{*(1)} \\ s_{\mathrm{r},(k+1)}^{(1)} = -g_{\mathrm{br}21} y_{\mathrm{r}1,(k)}^{*(1)} + g_{\mathrm{br}11}^{*} y_{\mathrm{r}1,(k+1)}^{(1)} \end{cases}, \tag{3.5}$$

where the equalization coefficient are given by $g_{\mathrm{br}ml,(k)} = \left(\hat{h}_{\mathrm{br}ml,(k)} / \sqrt{2} \right)$, for $m = 1,2$ and $l = 1$. After some mathematical manipulation, these soft-decision variables are given by:

$$\begin{cases} s_{\mathrm{r},(k)}^{(1)} = \Gamma_{\mathrm{br},(k)} d_{(k)} + \dfrac{\hat{h}_{\mathrm{br}11,(k)}^{*}}{\sqrt{2}} n_{\mathrm{r},(k)}^{(1)} + \dfrac{\hat{h}_{\mathrm{br}21,(k)}}{\sqrt{2}} n_{\mathrm{r},(k)}^{(1)*} \\ s_{\mathrm{r},(k+1)}^{(1)} = \Gamma_{\mathrm{br},(k+1)} d_{(k+1)} + \dfrac{\hat{h}_{\mathrm{br}11,(k+1)}^{*}}{\sqrt{2}} n_{\mathrm{r},(k+1)}^{(1)} - \dfrac{\hat{h}_{\mathrm{br}21,(k+1)}}{\sqrt{2}} n_{\mathrm{r},(k+1)}^{(1)*} \end{cases}, \tag{3.6}$$

where $\Gamma_{\mathrm{br},(k)} = \dfrac{1}{2}\sum_{m=1}^{2} \left| \hat{h}_{\mathrm{br}ml} \right|^2$.

In order to transmit a unit power signal the RN normalizes the expression in (3.6) by considering the normalization factor $\alpha_{(k)}$ which is given by:

$$\alpha_{(k)} = \frac{1}{\sqrt{\Gamma_{\mathrm{br},(k)}^{2} + \Gamma_{\mathrm{br},(k)} \sigma_{\mathrm{r}}^{2(1)}}} \tag{3.7}$$

During the phase II, the normalized soft-decision variable is sent via the second hop of the relay channel R→U. The FD signal received at the UT per k subcarrier is expressed according to:

$$y_{u,(k)}^{(2)} = \alpha_{(k)} s_{r,(k)}^{(1)} h_{ru,(k)} + n_{u,(k)}^{(2)}, \tag{3.8}$$

where $n_{u,(k)}^{(2)}$ is the additive white Gaussian noise which is zero mean and has unit variance $\sigma_u^{2(2)}$.

The signal in (3.8) is equalized using the coefficients $g_{rul,(k)} = \left(\alpha_{(k)} \Gamma_{br,(k)} \hat{h}_{ru,(k)} / \sigma_t^2\right)$ which after some mathematical manipulation is given by:

$$\begin{aligned} s_{u,(k)}^{(2)} &= \alpha_{(k)}^2 \Gamma_{br,(k)}^2 \frac{\left|\hat{h}_{ru1,(k)}\right|^2}{\sigma_t^2} d_{(k)} + \alpha_{(k)} \Gamma_{(k)} \frac{\hat{h}_{ru1,(k)}^*}{\sigma_t^2} n_{u,(k)}^{(2)} \\ &+ \alpha_{(k)}^2 \Gamma_{br,(k)} \left|\hat{h}_{rul,(k)}\right|^2 \frac{1}{\sqrt{2\sigma_t^2}} \left(\hat{h}_{br11,(k)}^* n_{r,(k)}^{(1)} + \hat{h}_{br21,(k)} n_{r,(k)}^{*(1)}\right). \end{aligned} \tag{3.9}$$

The equalization coefficient $g_{ru1,(k)}$ is a function dependent on the channel estimate $\hat{h}_{ru1,(k)}$ and the variance of the total noise. Moreover, the statistics of the total noise is conditioned to the channel realization $\hat{h}_{brml,(k)}$. Therefore the variance of the total noise can be computed as conditioned to these channel realizations or averaged over all the channel realizations. We denote by $\sigma_{t(h_{ru1},(k))}^2$ the noise variance conditioned to the specific channel realization per k subcarrier and by σ_t^2 the unconditioned noise variance. The noise variance of the total noise conditioned to channel realizations is found to be:

$$\sigma_{t,(h_{ru,(k)})}^2 = \alpha_{(k)}^2 \Gamma_{br,(k)} \left|h_{ru1,(k)}\right|^2 \sigma_u^{2(2)} + \sigma_u^{2(1)}. \tag{3.10}$$

3.2.2 RN equipped with a double antenna array

When the array is equipped with two antennas the soft-decision variables are expressed by:

$$\begin{cases} s_{r,(k)}^{(1)} = \sum_{l=1}^{2} \left(g_{br1l,(k)}^* y_{rl,(k)}^{(1)} + g_{br2l,(k)} y_{rl,(k+1)}^{(1)*}\right) \\ s_{r,(k+1)}^{(1)} = \sum_{l=1}^{2} \left(-g_{br2l,(k)} y_{rl,(k)}^{(1)*} + g_{br1l,(k)} y_{rl,(k+1)}^{(1)}\right) \end{cases}, \; l = 1,2 \tag{3.11}$$

where the equalization coefficients are expressed by $g_{brml,(k)} = \left(\hat{h}_{brml,(k)} / \sqrt{2}\right)$.

The RN transmits to the UT a unit power signal following the normalization factor in (3.7). However, in this scenario $\Gamma_{br,(k)}$ is expressed by $\Gamma_{br,(k)} = \frac{1}{2} \sum_{m=1}^{2} \sum_{l=1}^{2} \left|\hat{h}_{brml,(k)}\right|^2$. The FD signals received at the UT are given by:

$$\begin{cases} y_{u,(k)}^{(2)} = \frac{1}{\sqrt{2}} \left(h_{ru1,(k)} \alpha_{(k)} s_{r,(k)}^{(1)} - h_{ru2,(k+1)} \alpha_{(k+1)}^* s_{r,(k+1)}^{(1)*}\right) + n_{u,(k)}^{(2)} \\ y_{u,(k+1)}^{(2)} = \frac{1}{\sqrt{2}} \left(h_{ru2,(k)} \alpha_{(k)}^* s_{r,(k)}^{(1)*} + h_{ru1,(k+1)} \alpha_{(k+1)} s_{r,(k+1)}^{(1)}\right) + n_{u,(k+1)}^{(2)} \end{cases}, \tag{3.12}$$

where $\hat{h}_{\mathrm{ru}l,(k)}$ represent the channels between the RN and the UT terminal. The soft-decision variables found at the UT in phase II are expressed by:

$$\begin{cases} s_{\mathrm{u},(k)}^{(2)} = g_{\mathrm{ru}1}^{*} y_{\mathrm{u},(k)}^{(2)} + g_{\mathrm{ru}2} y_{\mathrm{u},(k+1)}^{*(2)} \\ s_{\mathrm{u},(k+1)}^{(2)} = -g_{\mathrm{ru}2} y_{\mathrm{u},(k)}^{*(2)} + g_{\mathrm{ru}1}^{*} y_{\mathrm{u},(k+1)}^{(2)} \end{cases}, \tag{3.13}$$

where the equalization coefficients are given by $g_{\mathrm{ru}l,(k)} = \left(\alpha_{(k)} \Gamma_{\mathrm{br},(k)} \hat{h}_{\mathrm{ru}l,(k)} \big/ \sqrt{2}\sigma_{\mathrm{t}}^{2}\right)$, for $l = 1,2$. In this scenario the soft-decision variables also depend on the variance of the total noise σ_{t}^{2} which conditioned to the channel specific realization can be calculated from (3.12) and is expressed by,

$$\sigma_{\mathrm{t},h_{\mathrm{ru}l,(k)}}^{2} = \alpha_{(k)}^{2} \Gamma_{\mathrm{br},(k)} \Gamma_{\mathrm{ru},(k)} \sigma_{\mathrm{r}}^{2(1)} + \sigma_{\mathrm{u}}^{2(2)} \tag{3.14}$$

where $\Gamma_{\mathrm{ru},(k)} = \frac{1}{2}\sum_{l=1}^{2} \left|\hat{h}_{\mathrm{ru}l,(k)}\right|^{2}$.

The UT combines the signals received from the RN and the BS. By performing this combining the diversity of the relay path is exploited. This processing is conducted by taking into account $s_{\mathrm{u},(k)}^{(1)} + s_{\mathrm{u},(k)}^{(2)}$ with ensure and MRC combining. The result corresponds to the variable to be hard-decoded.

4. Channel estimation

4.1 Time Domain Minimum Mean Square Error (TD-MMSE) estimator

The TD-MMSE (Ribeiro & Gameiro, 2008) corresponds to the version of the MMSE estimator which was originally implemented in FD. This estimator comprises the least square (LS) estimation and the MMSE filtering, both processed in time domain (TD). The TD-MMSE is a pilot-aided estimator, *i.e.* the channel estimation is not performed blindly. It is based on pilot symbols which are transmitted by the source and are known at destination.

The pilot subcarriers convey these symbols that are multiplexed with data subcarriers according to a pattern, Fig. 3, where N_f and N_t correspond to distance between two consecutives pilots in frequency and in time, respectively. N is the number of OFDM symbols and N_c is the number of subcarriers. The pattern presented in Fig. 3 is adopted during the transmission stages of the envisioned cooperative scheme.

It is usual the pilot symbols assume a unitary value and be constant during an OFDM symbol transmission. Thus, at k subcarrier the element p of the vector $\mathbf{p}$ may be expressed by a pulse train equispaced by N_f with unitary amplitude. The corresponding expression in TD is also given by a pulse train with elements in the instants $\left(n - mN_c/N_f\right)$, for $m \in \left\{0; N_f - 1\right\}$, according to the following expression.

$$p_{(k)} = \sum_{m=0}^{N_f} \delta_{\left(k - mN_f\right)} \xleftrightarrow{F^{-1}} \tilde{p}_{(n)} = \frac{1}{N_f} \sum_{m=0}^{N_f - 1} \delta_{\left(n - mN_c/N_f\right)}. \tag{4.1}$$

The transmitted signal is made-up of data and pilot components. Consequently, at the receiver side the component of the received signal in TD is given by the expression in (4.2).

$$\tilde{y}_{(n)} = \sum_{k=0}^{N_c-1} \tilde{h}_{(k)} d_{(n-k)} + \frac{1}{N_f} \sum_{m=0}^{N_f-1} \tilde{h}_{\left(n-mN_c/N_f\right)} + \tilde{n}_{(n)}, \tag{4.1}$$

where $n_{(n)}$ is the complex white Gaussian noise.

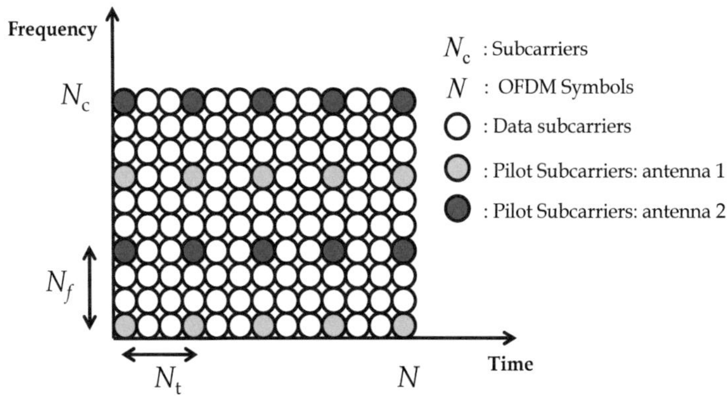

Fig. 3. Pilot pattern

In order to perform the LS estimation in TD, *i.e.* TD-LS, the received signal is convolved with the TD pilot symbols $\tilde{p}_{(n)}$. This convolution corresponds to multiply by 1 the subcarriers at frequencies N_f as by design these are the positions reserved to the pilots thus the data component in the received signal vanishes. The resulting CIR estimate $\hat{\tilde{h}}_{LS}$ is made-up of N_f replicas of the CIR separated by N_c/N_f.

$$\hat{\tilde{h}}_{\text{LS}} = \underbrace{\sum_{m=0}^{N_f-1} \tilde{h}_{\left(n-mN_c/N_f\right)}}_{\text{CIR}} + \underbrace{\sum_{m=0}^{N_f-1} \tilde{n}_{\left(n-mN_c/N_f\right)}}_{\text{Noise}}, \; n = 0,1,\cdots,\frac{N_c}{N_f}-1. \tag{4.3}$$

Besides to estimate the CIR the TD-MMSE in (Ribeiro & Gameiro, 2008) can estimate the noise variance as well. It corresponds to an essential requirement when the UT has no knowledge of this parameter. Since we know that the CIR energy is limited to the number of taps, or the set of the taps $\{G\}$, the noise variance estimate $\hat{\sigma}_n^2$ can be calculated by take into account the samples out of the number of taps, *i.e.* $n \notin \{G\}$ and by averaging the number of OFDM symbols N. Thus $\hat{\sigma}_n^2$ is given by:

$$\hat{\sigma}_n^2 = \frac{N_c/N_f}{\left[\left(N_c/N_f\right)-G\right]N} \sum_{i=1}^{N} \sum_{n \notin \{G\}} \left|\hat{\tilde{h}}_{(n)}\right|^2. \tag{4.4}$$

The MMSE filter can improve the LS estimates by reducing its noise variance. The TD-MMSE filter corresponds to a diagonal matrix with non-zero elements according to the number of taps, *i.e.* G, thus it can be implemented simultaneously with the TD-LS estimator and this operation simplifies the estimator implementation. The MMSE filter implemented by the $\left(N_c/N_f\right)\times\left(N_c/N_f\right)$ matrix and for a generic channel h it is expressed by:

$$\mathbf{W}_{MMSE,\tilde{h}} = \mathbf{R}_{\tilde{h}\hat{\tilde{h}}}\mathbf{R}^{-1}_{\hat{\tilde{h}}\hat{\tilde{h}}}, \tag{4.5}$$

where $\mathbf{R}_{\hat{\tilde{h}}\hat{\tilde{h}}}$ is the $(N_c/N_f)\times(N_c/N_f)$ filter input correlation matrix, $\mathrm{E}\left\{\hat{\tilde{\mathbf{h}}}\hat{\tilde{\mathbf{h}}}^H\right\}$, which is given by $\mathbf{R}_{\tilde{h}\tilde{h}} + \sigma_n^2\mathbf{I}_{N_c/N_f}$, and $\mathbf{R}_{\tilde{h}\hat{\tilde{h}}}$ is the $(N_c/N_f)\times(N_c/N_f)$ filter input-output cross-correlation matrix, $\mathrm{E}\left\{\tilde{\mathbf{h}}\hat{\tilde{\mathbf{h}}}^H\right\}$, which is given by $\mathbf{R}_{\tilde{h}\tilde{h}} = \mathrm{diag}\left(\left[\sigma_0^2,\ \sigma_1^2,\ \cdots,\ \sigma_{G-1}^2,\ 0,\ \cdots,\ 0\right]\right)$.

4.2 Channels and parameters estimates

According to the scenario previously presented, there are channels which correspond to point-to-point links: $h_{\mathrm{bu}m,(k)}$ and $h_{\mathrm{br}ml,(k)}$. Therefore, these channels can be estimated by using conventional estimators. However, for the RN→UT links, and since the EF protocol is used, it is necessary to estimate a version of $h_{\mathrm{ru}l,(k)}$, which depends on $\alpha_{(k)}$ and $\Gamma_{(k)}$, the equivalent channel $h_{\mathrm{eq}} = \alpha_{(k)}\Gamma_{(k)}h_{\mathrm{ru}l,(k)}$. Note that the UT has no knowledge of $\alpha_{(k)}$ and $\Gamma_{(k)}$. These factors are dependent on $h_{\mathrm{br}ml,(k)}$ which the UT has no knowledge as well. However, the channels $h_{\mathrm{br}ml,(k)}$ are estimated at the RN, and based on that, $\alpha_{(k)}\Gamma_{(k)}$ is calculated. Therefore, we propose to transmit the factor $\alpha_{(k)}\Gamma_{(k)}$ at the pilot subcarriers as pilots. Consequently, the new pilots are no longer constant and that may compromise the conventional TD-MMSE performance, since this estimator was designed in time domain assuming the pilots are unitary with constant values at the destination. Although our approach enables the destination had knowledge of the non-constant pilot $\left(\alpha_{(k)}\Gamma_{(k)}\right)$, the result of the convolution between the received signal and these pilots, would result in the overlapped replicas of CIR, according to the Fig. 4.

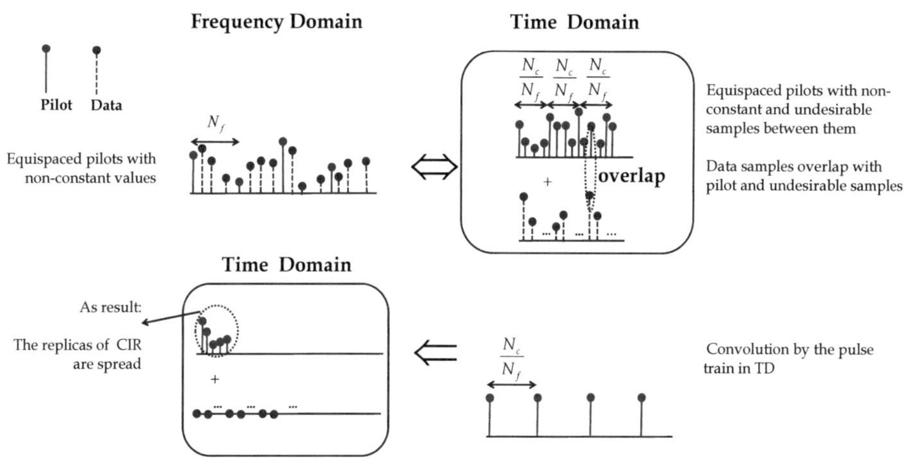

Fig. 4. Pilots with non-constant values result in the overlapped replicas of the CIR

However, the $\alpha_{(k)}$ expressions depend on the noise variance $\sigma_r^{2(1)}$ and the product $\alpha_{(k)}\Gamma_{(k)}$ tends to one for a high SNR value, according to (4.6). The same equation also suggests that the factor $\alpha_{(k)}\Gamma_{(k)}$ varies exponentially according to the SNR, as depicted in Fig. 5 for $L=1$.

$$\alpha_{(k)}\Gamma_{(k)} = \left(\frac{1}{\sqrt{\Gamma_{\mathrm{br},(k)}^2 + \Gamma_{\mathrm{br},(k)}\sigma_r^{2(1)}}}\right)\Gamma_{\mathrm{br},(k)} = \left(\frac{1}{\sqrt{\Gamma_{\mathrm{br},(k)}^2 + 0}}\right)\Gamma_{\mathrm{br},(k)} \cong 1. \tag{4.6}$$

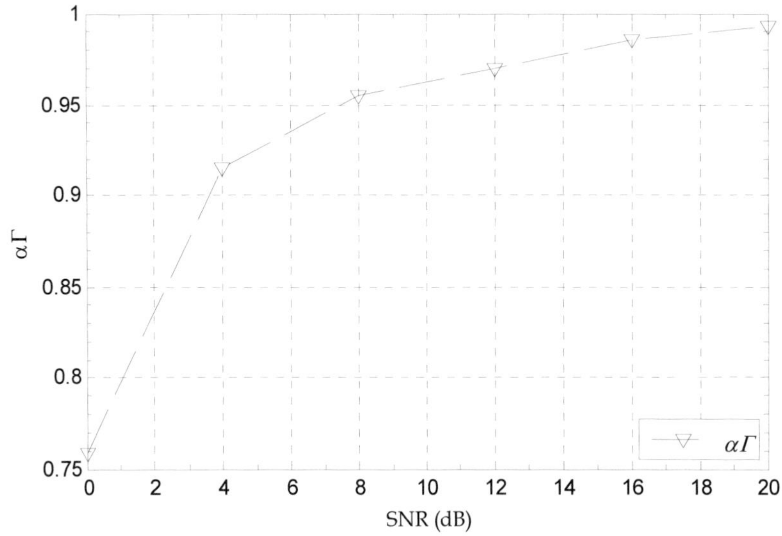

Fig. 5. $\alpha_{(k)}\Gamma_{(k)}$ vs. SNR

Other behaviour of the factor $\alpha_{(k)}\Gamma_{(k)}$ can be demonstrated in terms of SNR and subcarriers, as presented in Fig. 6. These plots show that in the first case, *i.e.* $\mathrm{SNR} = 20\mathrm{dB}$, the $\alpha_{(k)}\Gamma_{(k)}$ factor presents the amplitude close to 1 with some negligible fluctuation. However, in the second case, $\mathrm{SNR} = 2\mathrm{dB}$, the result is likely different to the previous one: the $\alpha_{(k)}\Gamma_{(k)}$ factor presents an amplitude also close to 1 but the fluctuation is not negligible.

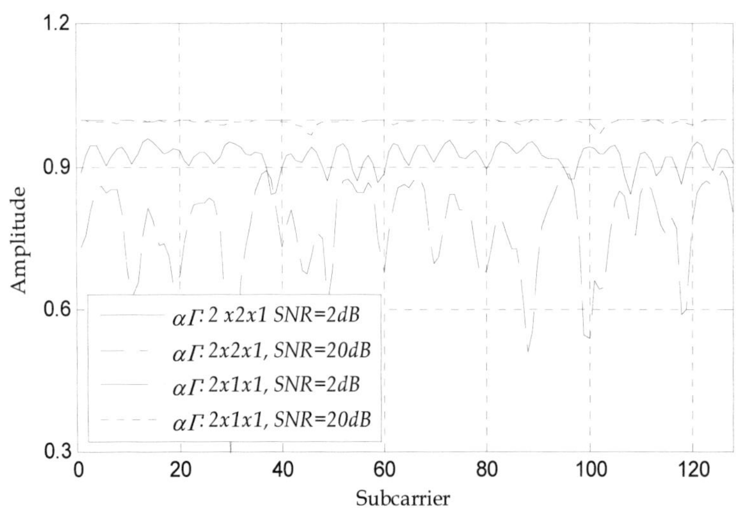

Fig. 6. $\alpha_{(k)}\Gamma_{(k)}$ per subcarriers

The results in Fig. 6 emphasize that transmit the factor $\alpha_{(k)}\Gamma_{(k)}$ in the pilot subcarriers may degrade the estimator performance and the causes are:

1. *Pilots with some fluctuation in amplitude:*

- As the amplitude of the pilots at the destination are constant and equal to one, the result of the estimation is a spread of the replicas of the CIR.

2. *Decreasing the amplitude of the pilots:*
   - The SNR of the pilots is decreased as well.
3. *The MMSE filter depends on the statistics of the channel B→R*

Despite we are considering the TD-MMSE estimator in our analysis, the causes presented previously degrade the performance of any other estimator scheme as well. In order to quantify how these effects can degrade the estimator performance we evaluated the impact of both of them, separately, in a SISO system, since the B→R and R→U channels correspond to point-to-point links.

First, we evaluate the case when the pilots have some fluctuation in amplitude. We consider that the pilots (originally with unit amplitude) had their amplitude disturbed by a noise with zero mean and variance equal to $\sigma_{\alpha\Gamma}^2 = \mathrm{E}\left\{\left|1-\alpha_{(k)}\Gamma_{(k)}\right|^2\right\}$, $\sigma_{\alpha\Gamma}^2$ quantifies how far the factor $\alpha_{(k)}\Gamma_{(k)}$ would be from the pilots with unitary amplitude. We can express $\sigma_{\alpha\Gamma}^2$ as:

$$\sigma_{\alpha\Gamma}^2 = 1 + \mathrm{E}\left\{\left(\alpha_{(k)}\Gamma_{(k)}\right)^2\right\} - 2\mathrm{E}\left\{\alpha_{(k)}\Gamma_{(k)}\right\}. \tag{4.7}$$

Therefore, the pilots are no longer constant and unitary. They have some fluctuation in amplitude which depend on $\alpha_{(k)}\Gamma_{(k)}$ and they are equal to $p_{\sigma_{\alpha\Gamma}^2} = 1 + z$, where z has a normal distribution with zero mean and power $\sigma_{\alpha\Gamma}^2$. The performance of a SISO system which the pilots correspond to $p_{\sigma_{\alpha\Gamma}^2}$ is shown in Fig. 7 (dash line). For reference, we also include the SISO performance for unit pilots, p_1. Since we are focus on the degradation of the estimator performance, the results are presented in terms of the normalized mean square error (MSE) and E_b/N_0, where E_b corresponds to the energy per bit received at UT and N_0 is the power spectrum density of the total noise which affects the information conveying signals. The normalized MSE for a generic channel h is given by:

$$\mathrm{MSE}_h = \frac{\mathrm{E}\left\{\left|\hat{h} - h\right|^2\right\}}{\mathrm{E}\left\{\left|h\right|^2\right\}}. \tag{4.8}$$

For low values of SNR, $[0-4]$, the major difference in performance between the two results is approximately $0.5\mathrm{dB}$, which is not a noticeable degradation and the estimator performance is not compromised.

The second effect to be evaluated is the decreasing of the amplitude of the transmitted pilot. In order to evaluate this effect we also consider a SISO system, for which the transmitted pilots assume different constant values, and we consider different values of the factor $\alpha_{(k)}\Gamma_{(k)}$ as pilot. The normalized MSE is given by (4.9). The result is shown in Fig. 8 and for reference we include the SISO performance for unit pilots, p_1, as well.

$$\mathrm{MSE}_{\mathrm{eq}} = \frac{\mathrm{E}\left\{\left|\hat{h}_{\mathrm{eq}} - h_{\mathrm{eq}}\right|^2\right\}}{\mathrm{E}\left\{\left|h_{\mathrm{eq}}\right|^2\right\}}. \tag{4.9}$$

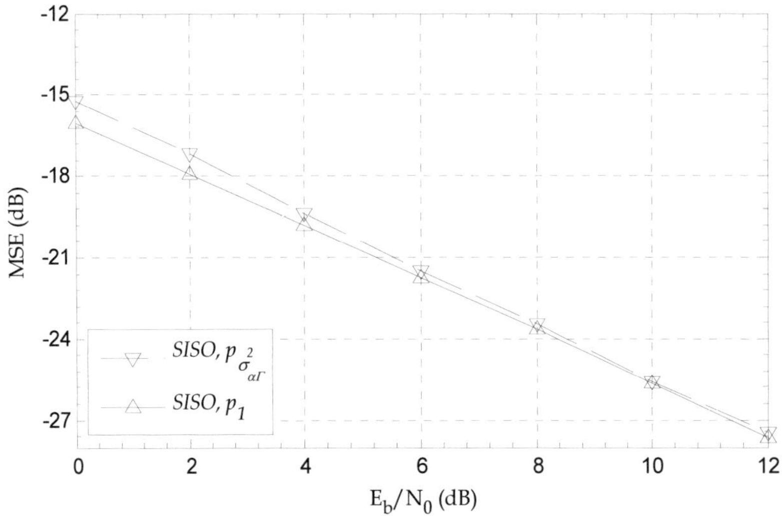

Fig. 7. Channel estimation MSE performance

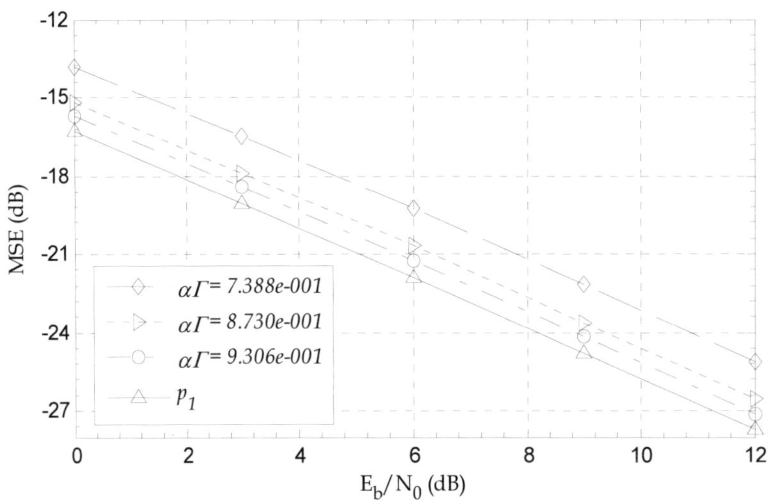

Fig. 8. Channel estimation MSE performance

The results show a constant shift in the MSE when the amplitude of the pilots is not unitary. The shift presents in all results is not a real degradation. It is caused by the normalization present in the MSE in (4.9). In fact, assuming a MSE without normalization the results are all the same.

Transmit the factor $\alpha_{(k)}\Gamma_{(k)}$ as pilot does not bring any noticeable degradation in the TD-MMSE performance comparing to transmitting unitary pilots. The major degradation occurs only when the pilots have some fluctuation in amplitude, as shown in Fig. 7, and solely for low SNR values.

The conventional MMSE filter in (4.5) is implemented to improve a channel estimate $\hat{h}$ when the required channel corresponds to h. However, according to our cooperative scheme, we need estimate an equivalent channel $h_{\text{eq}} = \alpha_{(k)}\Gamma_{(k)}h_{\text{rul},(k)}$. Since the factor $\widetilde{\alpha\Gamma}$ does not depend on $\tilde{h}_{\text{ru}}$, the MMSE filter input correlation matrix for the channel $\tilde{h}_{\text{eq}}$, referred $\mathbf{R}_{\hat{\tilde{h}}_{\text{eq}}\hat{\tilde{h}}_{\text{eq}}}$, is expressed by $E\left\{\hat{\tilde{\mathbf{h}}}_{\text{eq}}\hat{\tilde{\mathbf{h}}}_{\text{eq}}^{H}\right\} = \mathbf{R}_{\{\widehat{\alpha\Gamma}\widehat{\alpha\Gamma}\}}\mathbf{R}_{\tilde{h}_{\text{ru}}\tilde{h}_{\text{ru}}} + \sigma_n^2\mathbf{I}_{N_c/N_f}$ while the filter input-output cross-correlation, referred as $\mathbf{R}_{\tilde{h}_{\text{eq}}\hat{\tilde{h}}_{\text{eq}}}$, is given by $E\left\{\tilde{\mathbf{h}}_{\text{eq}}\hat{\tilde{\mathbf{h}}}_{\text{eq}}^{H}\right\} = \mathbf{R}_{\{\widehat{\alpha\Gamma}\widehat{\alpha\Gamma}\}}\mathbf{R}_{\tilde{h}_{\text{ru}}\tilde{h}_{\text{ru}}}$, both $\mathbf{R}_{\hat{\tilde{h}}_{\text{eq}}\hat{\tilde{h}}_{\text{eq}}}$ and $\mathbf{R}_{\tilde{h}_{\text{eq}}\hat{\tilde{h}}_{\text{eq}}}$ are $(N_c/N_f)\times(N_c/N_f)$ matrices. Thus the MMSE filter, when $\tilde{h}_{\text{eq}}$ is required, may be express as:

$$\mathbf{W}_{\text{MMSE},\tilde{h}_{\text{eq}}} = \mathbf{R}_{\{\widehat{\alpha\Gamma}\widehat{\alpha\Gamma}\}}\mathbf{R}_{\tilde{h}_{\text{ru}}\tilde{h}_{\text{ru}}}\left(\mathbf{R}_{\{\widehat{\alpha\Gamma}\widehat{\alpha\Gamma}\}}\mathbf{R}_{\tilde{h}_{\text{ru}}\tilde{h}_{\text{ru}}} + \sigma_n^2\mathbf{I}_{N_c/N_f}\right)^{-1}. \tag{4.10}$$

As we shown previously in (4.6) the factor $\alpha_{(k)}\Gamma_{(k)}$ tends to one for high values of SNR and examining (4.10), which depends on $\widehat{\alpha\Gamma}$, it is clear that (4.10) tends to (4.5) for high values of SNR as well. In order to show that several simulation were performed by taking into account $\mathbf{R}_{\hat{\tilde{h}}_{\text{eq}}\hat{\tilde{h}}_{\text{eq}}}$ and the noise variance $\sigma_r^{2(1)}$. According to Fig. 9 the results show that the maximum value in the $\mathbf{R}_{\hat{\tilde{h}}_{\text{eq}}\hat{\tilde{h}}_{\text{eq}}}$ matrix is close to -40dB for high values of the noise variance.

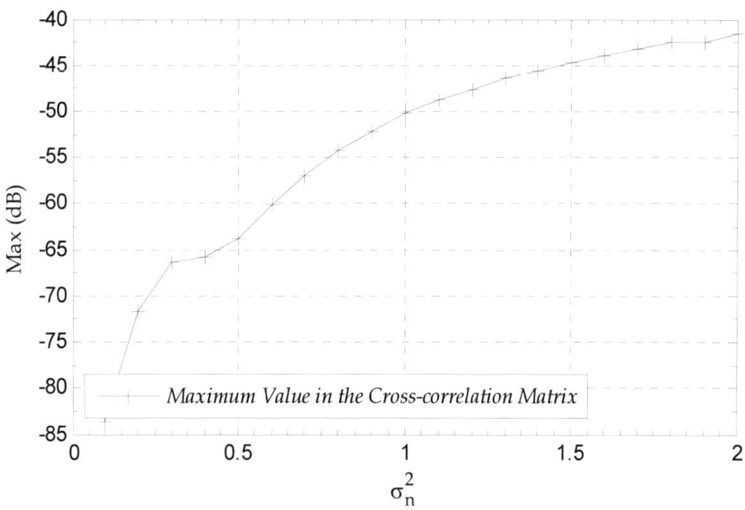

Fig. 9. Maximum value in the correlation matrix vs. noise variance

According to the results in Fig. 7 and Fig. 8 in terms of MSE, transmitting the factor $\widehat{\alpha\Gamma}$ brings, in the worst case, 0.5dB of degradation reason why there is no need to increase the system complexity by implementing the filter in (4.10). We have shown that our cooperative scheme allows the use of the TD-MMSE estimator without compromising its estimate. According to (4.6) the behaviour of the factor $\alpha_{(k)}\Gamma_{(k)}$ does not depend on number of taps of the channel, it depends only on the noise variance $\sigma_r^{2(1)}$. Therefore, this analysis is applied to any other channel without loss of generality. Besides to estimate the equivalent channel it is necessary estimate others factors $\alpha_{(k)}^2\Gamma_{(k)}\left|h_{\mathrm{ru},(k)}\right|^2$ and $\alpha_{(k)}^2\Gamma_{\mathrm{br},(k)}\Gamma_{\mathrm{ru},(k)}$ for $L=1$ and $L=2$, respectively. These factors are required parameters in the variance of the total noise $\sigma^2_{\mathrm{t},\left(h_{\mathrm{rul},(k)}\right)}$, previously presented in (3.10) and (3.14).

Although the UT does not have individual knowledge of $h_{\mathrm{brml},(k)}$, $\alpha_{(k)}$ and $\Gamma_{(k)}$ it has knowledge of the second moment of the expected value of the channels, *i.e.* for all channels $\mathrm{E}\left\{|h|^2\right\}=1$. Thus, we propose the use of the noise variance unconditioned to the channel realization, σ_{t}^2, instead of its instantaneous value $\sigma^2_{\mathrm{t}\left(h_{\mathrm{rul},(k)}\right)}$. Therefore, σ_{t}^2 is referred as the expectation value of the variance of the total noise. Also we consider that the channels have identical statistics, *i.e.* $\sigma_{\mathrm{u}}^{2(1)}=\sigma_{\mathrm{r}}^{2(1)}=\sigma_{\mathrm{u}}^{2(2)}$, thus σ_{t}^2 can be expressed numerically by (4.11) and (4.12) for $L=1$ and $L=2$, respectively.

$$\sigma_{\mathrm{t}}^2 \cong \frac{1}{1.5+\sigma_{\mathrm{u}}^{2(2)}}\sigma_{\mathrm{u}}^{2(2)}+\sigma_{\mathrm{u}}^{2(2)}. \tag{4.11}$$

$$\sigma_{\mathrm{t}}^2 \cong \frac{1}{5+2\sigma_{\mathrm{u}}^{2(2)}}2\sigma_{\mathrm{u}}^{2(2)}+\sigma_{\mathrm{u}}^{2(2)}. \tag{4.12}$$

If we consider the premise of the cooperative transmission, high SNR compared to the SNR of the direct link, we have $\sigma_{\mathrm{u}}^{2(2)} << 1.5$ for $L=1$ and consequently (4.11) may be express as $\sigma_{\mathrm{t}}^2 \cong \frac{5}{3}\sigma_{\mathrm{u}}^{2(2)}$.

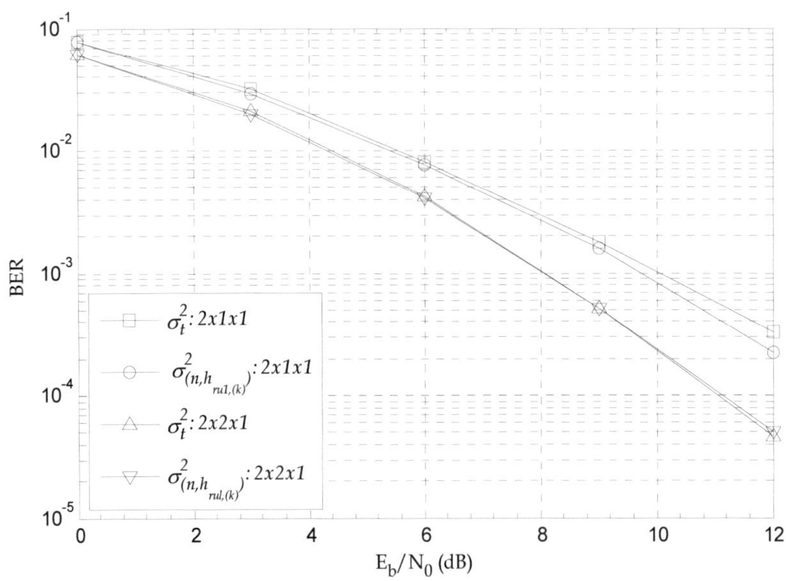

Fig. 10. System performance

To assess the validity of using the averaged noise variance instead of the conditioned one we plot in Fig. 10, the BER versus E_b/N_0 performance assuming perfect channel estimation is available at the receiver but considering the cases where the noise variance used is the conditioned one and the averaged ones. The results refer to a channel as referred in Section 5 but similar results were obtained with other models.

The performance penalty by using the averaged noise variance is less than 0.8dB which is a tolerable penalty to pay in order to obtain the variance of the total noise regarding the low complexity implementation. Therefore we consider the use of σ_t^2 in our schemes.

5. Results

5.1 Simulation parameters

In order to evaluate the performance of the presented RA schemes we considered a typical scenario, based on LTE specifications (3GPP TS, 2007). In the simulations we used the ITU pedestrian channel model B at speed $v = 10\text{km/h}$. The transmitted OFDM symbol carried pilot and data with a pilot separation $N_f = 4$ and $N_t = 1$.

We focus our analysis on the 2×1×1 and 2×2×1 scenarios and the simulations were performed assuming that the channels are uncorrelated, the receiver is perfectly synchronized and the insertion of a long enough cyclic prefix in the transmitter ensures that the orthogonality of the subcarriers in maintained after transmission. We use the TD-MMSE to estimate all the noise variances as well.

The results are presented in terms of BER and MSE, both as function of E_b/N_0. The normalized MSE is defined according to (4.9). The MSE performance of the cooperative channel is evaluated by averaging the MSE's of the direct and the relaying channel (Kim et al., 2007). Since the direct channel corresponds to a MISO its MSE is obtained also by averaging the MSE of the B→U channels, both normalized. The MSE of the relaying channel corresponds to the MSE of the equivalent channel $h_{\text{eq}} = \alpha_{(k)}\Gamma_{(k)}h_{\text{rul},(k)}$ which is calculated according to (5.1). Thus the resulting MSE, i.e. the MSE of the cooperative channel, is given by:

$$\text{MSE} = \frac{1}{2}\left(\frac{1}{2}(\text{MSE}_h) + \text{MSE}_{h_{\text{eq}}}\right). \tag{5.1}$$

5.2 Performance evaluation

In order to validate the use of the proposed scheme, some channel estimation simulations were performed using the TD-MMSE estimator. Fig. 11 depicts the BER attained with perfect CSI and the TD-MMSE estimator when the RN was employing the proposed pilots. The difference of performance is minimal in most of the cases and in the 2×1×1 scheme which is in the worst case this difference is 0.5dB.

Fig. 12 depicts the normalized MSE's performance of the 2×1×1 scheme. These results show that the proposed pilot allocation, at the RN, according to Section I.B, allows the TD-MMSE satisfactory estimate the required channel. When comparing the channel estimator for the link with relay against the one of the direct link, there is some penalty which accounts for the additional noise added at the relay. The relative penalty decreases as E_b/N_0 increases and can be verified to converge to 2.2dB which is the factor of $5/3$ that relates the total and individual noises in the asymptotic case of high SNR. According to Fig. 13, this penalty is smaller in the 2×2×1 scheme, since the factor $\alpha_{(k)}\Gamma_{(k)}$ presents a flatter behavior.

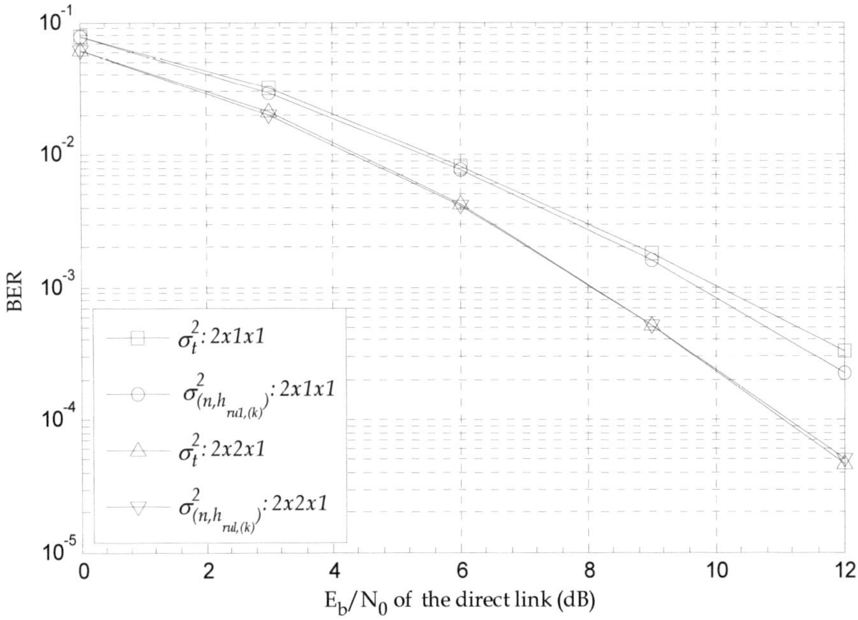

Fig. 11. System performance: RA 2×1×1 and RA 2×2×1 schemes

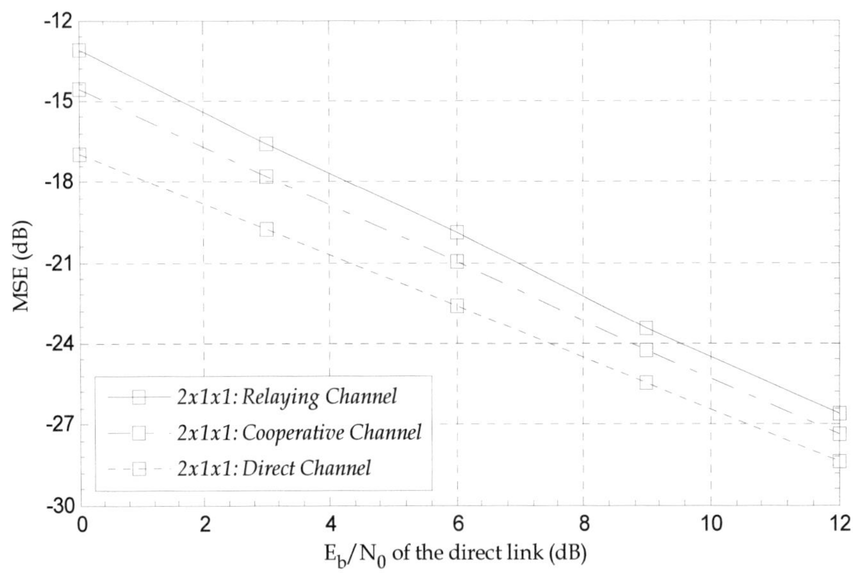

Fig. 12. Channel estimation MSE performance: RA 2×1×1 scheme

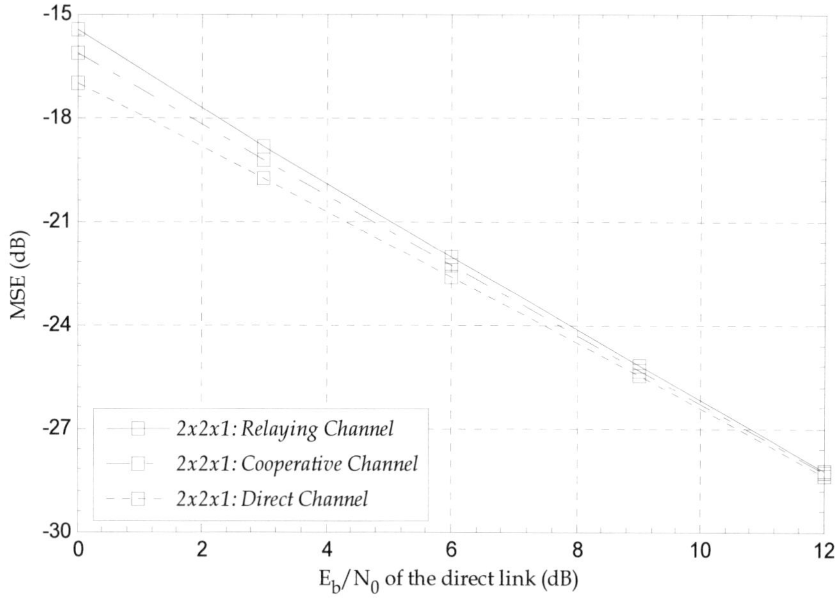

Fig. 13. Channel estimation MSE performance: RA 2×2×1 scheme

6. Conclusion

In this chapter we considered two problems of channel estimation in a scenario where spatial diversity provided by SFBC is complemented with the use of a half-duplex relay node employing the EF protocol. The channel estimation scheme was based on the TD-MMSE which led to a significant complexity reduction when compared to its frequency domain counterpart. We proposed a scheme where the estimates of the B→R link are inserted in the pilot positions in the R→U transmission. For the estimation of the equivalent channel, *i.e.* B→R→U, at the destination we analyzed several simplifying options enabling the operation of channel estimation namely the use of averaged statistics for the overall noise and the impact of the fluctuations in the amplitude of the equivalent channel. In the RA 2×1×1 scheme is shown that in the asymptotic case of high SNR, and equal noise statistics at the relay and destination the penalty in the estimation equivalent channel is 2.2dB relatively to the case of a direct link using the same pilot density. This difference in performance is smaller in the RA 2×2×1 scheme since the equivalent channel presents a flatter behaviour. The resulting estimation was assessed in terms of the BER of the overall link through simulation with channel representative of a real scenario and the results have shown its effectiveness despite a moderate complexity.

7. Acknowledgments

The authors wish to acknowledge the support of the European project Enhanced Wireless Communication Systems Employing Cooperative Diversity - CODIV, FP7/ICT/2007/215477, Portuguese Cooperative and Antenna Diversity for Broadband Wireless Networks - CADWIN, PTDC/EEA - TEL/099241/2008 and Portuguese Foundation for Science and Technology (FCT) grant for the first author.

8. References

Chen, Z.; Peng, M.; Wang, W. & Chen, H. H. (2009). Cooperative base station beamforming in WiMAX systems, *IET Communications*, Vol. 4, No 9, June, 2009, 1049-1058, ISSN: 1751-8628.

Foschini, G. J. & Gans, M. J. (1998). On limits of wireless communications in a fading environment when using multiple antenna. *Wireless Personal Communications Magazine*, Vol. 6, No. 3, 1998, 311-335, DOI: 10.1023/A: 1008889222784.

Fouillot, P.; Icart, I. & Martret, C. J. (2010). Performance analysis of a two-relay assisted transmission scheme, in *Proceedings on European Wireless Conference*, 116-122, Print ISBN: 978-1-4244-5999-5, France, April, 2010, IEEE, Lucca.

Gedik, B. & Uysal, M. (2009). Impact of imperfect channel estimation on the performance of amplify-and-forward relaying, *IEEE Trans. on Wireless Communication*, Vol. 8, No 3, March, 2009, 1468-1479, ISSN: 1536-1276.

Hadizadeh, H. & Muhaidat, S. (2010). Impact of imperfect channel estimation on the performance of inter-vehicular cooperative networks, in *Proceedings on Biennial Symposium on Communication*, 373-376, Print ISBN: 978-1-4244-5709-0, Canada, May, 2010, IEEE, Kingston.

Han, S.; Ahn, S.; Oh, E. & Hong, D. (2009). Effect of channel-estimation error on BER performance in cooperative transmission, *IEEE Trans. on Vehicular Technology*, Vol. 58, No. 4, May, 2009, 2083-2088, ISSN: 0018-9545.

Ikki, S. S.; Feteiha, M. & Uysal, M. (2010). Performance analysis of cooperative diversity networks with imperfect channel estimation over rician fading channels, in *Proceedings on International Conference on Communications*, 1–5, ISSN: 1550-3607, South Africa, May, 2010, IEEE, Cape Town.

Kim, K.; Kim, H. & Park, H. (2007). OFDM channel estimation for the amply–and–forward cooperative channel, in *Proceedings on Vehicular Technology Conference-Spring*, 1642–1646, ISSN: 1550-2252, Ireland, April, 2007, IEEE, Dublin.

Laneman, J. N.; Tse, D. N. C. & Wornell, G. W. (2004). Cooperative diversity in wireless networks: efficient protocols and outage behaviour, *IEEE Trans. Inform. Theory*, Vol. 50, No. 12, December, 2004, 3062–3080, ISSN: 0018-9448.

Laroia, R. & Uppala, Li, S. (2004). Designing a mobile broadband wireless access network. *IEEE Signal Processing Magazine*, Vol. 21, No. 5, September, 2004, 20-28, ISSN: 1053-5888.

Liu, K. J. R.; Sadek, A. K.; Su, W. & Kwasinski, A. (2009). *Cooperative Communications and Networking*, Cambridge University Press, ISBN: 978-0-521-89513-2, New York.

Liu, M.; Zhang, J.; Zhang, Y. & Liu, Y. (2009). A channel estimation scheme for amplify-and-forward OFDM relay networks, in *Proceedings on Vehicular Technology Conference-Fall*, 1-5, ISSN: 1090-3038, United States, September, 2009, IEEE, Anchorage.

Ma, J.; Orlik, P.; Zhang, J. & Li, G. Y. (2009). Pilot matrix design for interim channel estimation in two-hop MIMO AF relay systems, in *Proceedings on International Conference on Communications*, 1-5, ISSN: 1938-1883, Germany, June, 2009, IEEE, Dresden.

Moço, A.; Lima, H.; Silva, A. & Gameiro, A. (2009). Multiple antenna relay-assisted schemes for the uplink OFDM based systems, in *Proceedings on International*

Symposium on Wireless Communication Systems, 633–637, Print ISBN: 978-1-4244-3584-5, Italy, September, 2009, IEEE, Tuscany.

Moço, A.; Teodoro, S.; Silva, A.; Lima, H. and Gameiro, A. (2010), Single and multiple antenna, relay-assisted techniques for uplink and downlink OFDM systems, *IARIA International Journal on Advances in Systems and Measurements,* Vol. 3, No. 1&2, 2010, pp. 22-34.

Muhaidat, H. & Uysal, M. (2008). Cooperative diversity with multiple-antenna nodes in fading relay channels, *IEEE Trans. on Wireless Communications,* Vol. 7, No. 8, August, 2008, 3036–3046, ISSN: 1536-1276.

Muhaidat, H.; Uysal, M. & Adve, R. (2009), Pilot-symbol-assisted detection scheme for distributed orthogonal space-time block coding, *IEEE Trans. on Wireless Communication,* Vol. 3, No. 3, March, 2009, 1057–1061, ISSN: 1536-1276.

Neves, D.; Ribeiro, C.; Silva, A. & Gameiro A. (2009). Channel estimation schemes for OFDM relay-assisted systems, in *Proceedings on Vehicular Technology Conference-Spring,* 1-5, ISSN: 1550-2252, Spain, April, 2009, IEEE, Barcelona.

Neves, D.; Ribeiro, C.; Silva, A. & Gameiro A. (2010). A Time Domain Channel Estimation Scheme for Equalize-and-Forward Relay-Assisted Systems, in *Proceedings on Vehicular Technology Conference-Fall,* 1-5, ISSN: 1090-3038, Canada, September, 2010, IEEE, Ottawa.

Pang, J.; Shen, G.; Wang, D.; Jiang, L. & Wang, W. (2010). Channel Estimation and Optimal Training Design for Amplify and Forward MIMO Relay Channel under Spatial Fading Correlation, in *Proceedings on Vehicular Technology Conference-Fall,* 1-5, ISSN: 1090-3038, Canada, September, 2010, IEEE, Ottawa.

Ribeiro, C. & Gameiro, A. (2008). An OFDM symbol design for reduced complexity mmse channel estimation, *Academic Publisher on Journal of Communication,* Vol. 3, No. 4, September, 2008, 26-33, ISSN: 796-2021.

Sheu, J. S. & Sheen, W. H. (2010). An EM algorithm-based channel estimation for OFDM amplify-and-forward relaying systems, in *Proceedings on International conference on Communications,* 1-5, ISSN: 1550-3607, South Africa, May, 2010, IEEE, Cape Town.

Teodoro, S.; Silva, A.; Gil, J. M. & Gameiro, A. (2009). Virtual MIMO schemes for downlink space–frequency coding OFDM systems, in *Proceedings on Personal Indoor and Mobile Radio Communications,* 1322-1326, E-ISBN: 978-1-4244-5123-4, Japan, September, 2009, IEEE, Tokyo.

Wang, G.; Gao, F. & Tellambura, C. (2010). Superimposed pilots aided joint CFO and channel estimation for ZP-OFDM modulated two-way, in *Proceedings on Vehicular Technology Conference-Fall,* 1-5, ISSN: 1090-3038, Canada, September, 2010, IEEE, Ottawa.

Wu, Y. & Patzold, M. (2009). Parameter optimization for amplify-and-forward relaying with imperfect channel estimation, in *Proceedings on Vehicular Technology Conference-Spring,* 1-5, ISSN: 1550-2252, Spain, April, 2009, IEEE, Barcelona.

Xing, C.; Ma, S.; Wu, Y. C. & Ng, T. S. (2010). Transceiver design for dual-hop non-regenerative MIMO-OFDM relay systems under channel uncertainties, *IEEE Transc. on Signal Processing,* Vol. PP, No. 99, August, 2010, 1-1, ISSN: 1053-587X.

Zhang, Z.; Zhang, W. & Tellambura, C. (2009). Cooperative OFDM channel estimation in the presence of frequency offsets, *IEEE Trans. on Vehicular Technology,* Vol. 58, No 7, September, 2009, 3447–3459, ISSN: 0018-9545.

Zhou, X.; Lamahewa, T. A. & Sadeghi, P. (2009). Kalman filter-based channel estimation for amplify and forward relay communications, in *Conference Records of Asilomar Conference Signals Systems and Computers*, 1498-1502, ISSN: 1058-6393, United States, November, 2009, IEEE, Pacific Grove.

3GPP TS 36.201 V8.1.0 (2007). *3rd Generation Partnership Project*, Technical Specification Group Radio Access Network Evolved Universal Terrestrial Radio Access (E-UTRA), LTE Physical Layer - General Description, November, 2007.

8

Fast Power and Channel Adaptation for Mobile Users in OFDMA Multi-Cell Scenarios

L. Reggiani[1], L. Galati Giordano[2] and L. Dossi[3]
[1]*Politecnico di Milano*
[2,3]*IEIIT CNR - Milano*
Italy

1. Introduction

Methods for interference mitigation and adaptive multi-user resource allocation are among the most promising technological breakthroughs that should improve capacity of the last generation broadband wireless systems. In multi-cell scenarios characterized by radio technologies based on OFDMA (Orthogonal Frequency Division Multiple Access), co-channel interference is the performance limiting factor and techniques that reduce transmission power, act indirectly on the co-channel interference levels. In this work we investigate multi-cell scenarios with mobile users in which power and channel adaptation act as positive factors on the interference reduction. We are interested particularly on the role of closed loop power control and of simple techniques for fast channel assignment in multi-cell configurations with low reuse factors (possibly between 1 and 3). From the numerical findings presented in this chapter, it turns out that spectral efficiency can be improved by algorithms that are simple and fast enough to be exploited for mobile users, at least at pedestrian velocities.

The ambitious capacity requirements for future broadband wireless networks expose system designers to the challenging compromise between the scarcity of spectral resources and the impairments introduced by radio propagation randomness. At the heart of this challenge there is the ability to exploit radio resources as efficiently as possible in all available dimensions (space, time, frequency or channel, power, modulation and coding). Since mitigation of interference and resource allocation strategies are usually studied separately, in this work we aim to investigate more deeply the relations between the application of simple techniques for allocating power and channels to mobile users and their impact on interference and, consequently, on multi-cell overall performance.

Adaptive allocation techniques have already demonstrated their ability to mitigate the effect of deep fading by encouraging channel access to users temporarily experiencing better propagation conditions, taking advantage by the so-called *'multiuser diversity gain'* (Knop & Humblet (1995)). In our previous works (Galati et al. (2010); Reggiani et al. (2007); Galati et al. (2008)) we have shown how resource allocation algorithms in multi-cell environments have a non-trivial impact on downlink mutual interference among cells when the simple mechanism of power adaptation is applied to the system sub-channels: when the power, after channel assignment, is reduced to the minimum level required for the selected modulation and coding profile, all the network acquires an advantage from the interference reduction, giving rise to a *'virtuous loop'* in which performance parameters as spectral efficiency or coverage turn out to be improved significantly. So the decrease of transmitted power levels, possibly due to the efficient use of resources, reveals an interference mitigation impact even if the algorithms or the allocation strategies are not directly designed or applied for this final effect. More precisely, in (Reggiani et al. (2007)) different specific adaptive allocation algorithms were applied, on a cell-by-cell basis, to a multi-cell system with fixed terminals and air interface parameters compliant to IEEE 802.16 (WiMax) AMC standard (IEEE Std 802.16 (2004); IEEE Std 802.16e (2005)).

In this chapter, we overcome the fixed users assumption of our previous works and we examine a downlink/uplink scenario with mobile users focused on OFDMA physical layer architectures as in the recent LTE (Long Term Evolution) standard and in the previous IEEE 802.16. In particular, we discuss the performance outcomes considering, at the same time, the updating rate of the algorithms, which is clearly one of the main constraints for their application to mobile users. As already mentioned, we examine two relevant issues: (i) the role of fast, closed loop, power adaptation in the multi-cell interference mitigation and (ii) the positive impact of very fast and simple allocation techniques, well suited to the application within a mobile scenario. The key point under the analysis of these techniques is the power reduction that is made possible either as a response to channel variations (power control) or as effect of advantageous channel assignments. The final, positive impact is a reduction of the overall network interference. So the novel contributions reported in this work are (i) the evaluation of allocation techniques and fast loop power adaptation on the multi-cell interference scenario with mobile users and (ii) the principle of set partitioning applied to allocation strategies for enhancing the response rate of these techniques (Sect. 3).

The chapter is organized as follows: Sect. 2 describes the system model and the channel assumptions, Sect. 3 reports the proposed fast adaptive strategies and Sect. 4 resumes the assumptions and procedures adopted in the analysis and in the simulations. Finally Sect. 5 illustrates the numerical results in terms of spectral efficiency, allocation updating rate and complexity reduction.

2. System model

The simulation scenario is a network of cell sites, each one divided into three hexagonal sectors and with a frequency reuse factor equal to 3 (Fig. 1). Performance is simulated only for users belonging to the center sector; the surrounding base stations (BS) generate two tiers of interferers. Base stations are equipped with directional antennas, whose patterns are taken from the 3*GPP* standard document (3GPP TR 25.996 (2003)) while mobile terminals are equipped with omnidirectional antennas. The frequency selective time variant channel model is derived from (3GPP ETSI TR 101 112 (1998)), which is dedicated to test environments and deployment models. We select one vehicular model with a power delay profile of type $Veh - A$, corresponding to a delay spread of 370 ns and two pedestrian models of type $Ped - A$ and $Ped - B$, characterized by delay spreads of 45 ns and 750 ns respectively. All the models are described by a power delay profile with 6 or 4 taps at fixed delays and a classic Doppler spectrum whose bandwidth depends on the user velocity v. Antenna and channel parameters are summarized in Table 1, the tapped-delay line parameters of vehicular and pedestrian test environments in Table 2.

Our case study is a scalable-OFDMA (S-OFDMA) system, in which the total available bandwidth is divided in N_{sc} sub-carriers, each one associated to a bandwidth that is much smaller than the coherent channel bandwidth but greater than the inverse of the channel coherence time. Then sub-carriers are grouped into N_S sub-channels and a user is assigned

Base station equipment	
Antenna Gain	17 dBi
3-dB beamwidth	70°
Maximum BS power	35 dBm
User station equipment	
Antenna Gain	0 dBi
Radiation Pattern	Omnidirectional
Noise figure	5.2 dB
Channel model	
Path loss	$PL(d) = 136.7 + 39.7log_{10}(d)$ [PL in dB; d in km]
Center frequency	3.5 GHz
Fast fading	Veh-A, Ped-A or Ped-B model

Table 1. Antenna, channel model and equipment parameters.

one or more sub-channels for its traffic. The number N_U of users admitted to the service will be smaller or equal to N_S. A sub-channel can be constituted by adjacent or distributed sub-carriers. In our simulations, we will consider adjacent sub-carrier allocation modes (named as 'AMC' in IEEE 802.16 standard). For example, in case of AMC permutation zone, each sub-channel consists of a group of 18 contiguous sub-carriers (16 data and 2 pilots) where fading is assumed correlated. With a bandwidth of 10 MHz per sector and an FFT (Fast Fourier Transform) size equal to 1024, the sub-carrier spacing Δf is 11.16 kHz and the number of sub-channels per sector is $N_S = 48$.
As the fading is assumed flat in each sub-channel, the instantaneous channel state indicator (CSI) of the i-th sub-channel, seen by the k-th user, is defined as

$$\beta_{k,i} = \frac{P_N + I_k}{H_{k,i}}, \tag{1}$$

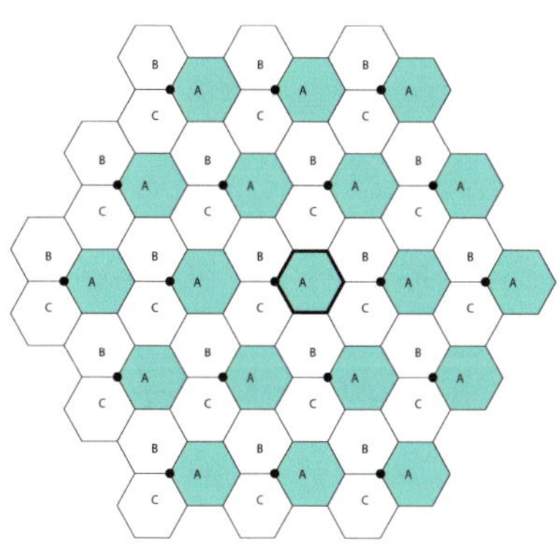

Fig. 1. Network topology: the sectors identified by the same letter (A, B or C) employ the same frequency band. Circles denote the base stations and the center sector is delimited by a bold line.

Veh-A Model		
Tap	Rel. Delay (ns)	Avg. Power (dB)
1	0	0.0
2	310	−1.0
3	710	−9.0
4	1090	−10.0
5	1730	−15.0
6	2510	−20.0

Ped-A Model		
Tap	Rel.Delay (ns)	Avg. Power (dB)
1	0	0.0
2	110	−9.7
3	190	−19.2
4	410	−22.8

Ped-B Model		
Tap	Rel.Delay (ns)	Avg. Power (dB)
1	0	0.0
2	200	−0.9
3	800	−4.9
4	1200	−8.0
5	2300	−7.8
6	3700	−23.9

Table 2. Test Environment Tapped-Delay-Line Parameters

where $H_{k,i}$ is the channel gain, including path loss, fast fading and antenna gains, P_N is the noise power and I_k is the instantaneous interference power generated by the surrounding co-channel interferers. We do not generally assume the ideal instantaneous estimation of I_k in Equation 1 but only of an average interference power $\overline{I}_k$ since random fluctuations of I_k appear unpredictable without a distributed control among the interfering BSs. Moreover, as we are considering mobile terminals, imperfect estimation of channel gains $H_{k,i}$ can derive from the fact that the algorithm updating time (i.e. the interval between two consecutive applications of the allocation algorithm) is not negligible w.r.t. the channel coherence time. Assuming that only an average interference power $\overline{I}_k$ can be known at the algorithm processing unit (typically at the BS), the interference variations, not perfectly estimated, are compensated by an interference (and noise) margin FM_I, while the variations experienced by the channel gain, due to Doppler effect, are compensated by a Doppler margin ΔFF, resulting in the new CSI definition

$$\gamma_{k,i} = \Delta FF \cdot FM_I \frac{P_N + \overline{I}_k}{H_{k,i}}, \tag{2}$$

which is the general input of the allocation algorithms. The two functions 1 and 2 are directly related to the signal-to-interference noise ratio (SINR) at the i-th sub-channel for the k-th user by means of the transmitted power $P_{k,i}$.

If we consider, for each modulation and coding profile $l = 1..L$, the number c_l of information bits provided per carrier and the lower SINR bound α_l for its correct functioning, each sub-channel i is supposed to transmit successfully $c_{k,i} = c_l$ information bits per carrier for the k-th user if the algorithm is able to assign a power $P_{k,i} : SINR_{k,i} = P_{k,i}/\gamma_{k,i} = \alpha_l$. In our simulations, the thresholds α_l ($1 \leq l \leq L$) are assumed equally spaced between the minimum and the maximum values specified in modulation and coding profiles for IEEE 802.16 S-OFDMA standard ($\alpha_1 = 2.88dB$ and $\alpha_L = 17.50dB$). For the sake of generality, instead of considering a number c_l of information bits provided per carrier that depends on the chosen modulation and coding, we assume that an user, whose SINR achieves the threshold α_l, transmits with the theoretical Shannon efficiency

$$\eta_l[bit/s/Hz] = log_2(1 + 10^{\alpha_l/10}). \tag{3}$$

If none of the thresholds is exceeded, sub-channel i is switched off for the k-th user.

3. Fast adaptive techniques

In OFDMA systems, smart allocation of radio resources is a crucial aspect for achieving excellent performance levels. In Fig. 2 we can observe a typical structure of an OFDMA time-frequency layer: the set of sub-carriers and symbol times is divided into resource blocks (RB), which constitute the minimum amount of resources that can be assigned to an user connection. In fact each user is assigned a set of RBs, generally but not necessarily contiguous (Fig. 2(a)). The sub-carriers of the same RB are interested by the same modulation, coding profile and power. For the sake of simplicity but without loss of generality w.r.t. the scope of this study, we assume a resource division based on a one-dimensional approach, where a user is assigned the same sets of sub-carriers for the entire allocation time T_{UPD} (Fig. 2(b)). So a resource block is equivalent to a sub-channel and an user can share separate sub-channels in the same T_{UPD}.
The allocation techniques require knowledge of the function reported in Equation 1 with an updating time T_{UPD} that should be shorter than the coherence time of the channel; this constitutes the main limiting factor in mobile applications. When T_{UPD} is comparable or greater than the coherence time of the channel, the algorithm performance degrades rapidly as the channel gain in that updating interval can experience heavy fluctuations due to the Doppler effect. In order to compensate this degradation, our simulations pre-compute the 'Doppler margin' ΔFF in Equation 2 as the channel gain variation that each sub-channel exceeds, during an entire period T_{UPD}, with a probability equal to 0.10. So ΔFF is used in the new CSI estimate as in Equation 2. As expected, this Doppler margin depends both on the updating time and on the channel coherence time, which is function of the mobile user velocity v.
In the sequel, we present a study on the impact of fast closed loop power control on interference (Sect. 3.1) and of techniques for exploiting multi-user diversity in channel assignments also in a mobile context (Sect. 3.2).

3.1 Closed loop power adaptation

In OFDMA systems, power adaptation is performed either in open or closed loop modalities. It is well known that a fast closed loop mechanism plays a crucial role in CDMA cellular systems for limiting the *intra-cell* interference, especially on the uplink. Here we are interested to the impact of fast power adaptation on OFDMA systems with low reuse factors as a positive contribution to limiting *extra-cell* interference. This is true for the uplink direction, as already observed in (Schoenen & Qin (2009); Li et al. (2008); Tee et al. (2007)), but also for the downlink side, on which we have focused our analysis. The analytical procedure presented in Sect. 5 will take into account the multi-cell interference and it will assume ideal channel knowledge and

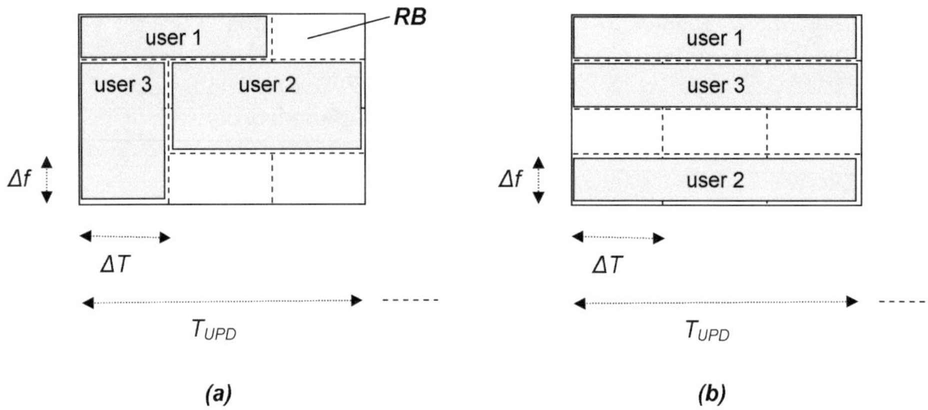

Fig. 2. Time-frequency resource organization in an OFDMA system: two-dimensional (a) and one-dimensional scenarios.

adaptation. On the other hand, the simulations will reveal the impact of these techniques for several levels of the updating time (w.r.t. the channel coherence) and with other impairments. The closed loop power adaptation (CLPA) is able to adjust the power at the base station (downlink) or at the terminal (uplink) at the level that is exactly necessary to achieve the maximum profile threshold α_l ($1 \leq l \leq L$) compatible with the power assigned to each sub-channel $P_S = P_{BS}/N_S$, being P_{BS} the total maximum BS transmission power. In CLPA there is no allocation based on the channel state indicator but users are assigned to available sub-channels randomly. In practice, in our simulations, each user is assigned to N_S/N_U sub-channels: the selected modulation profile $0 \leq l \leq L$ is given by

$$l : \begin{cases} \frac{P_S}{\gamma_{k,i}} \geq \alpha_l \\ \frac{P_S}{\gamma_{k,i}} < \alpha_{l+1} \end{cases} \tag{4}$$

where $\alpha_0 = -\infty$, being $l = 0$ associated to the absence of transmission, and $\alpha_{L+1} = +\infty$. Then power is adapted to the value

$$P_{k,i} = \alpha_l \cdot \gamma_{k,i} \leq P_S. \tag{5}$$

Disequalities 4 and 5 are checked every T_{UPD} seconds and either powers or modulation profiles can be changed according to the channel variations reported by the updated coefficients $\gamma_{k,i}$.

3.2 Fast channel assignment

As mentioned in Sect. 1, smart resource allocation in OFDMA systems is usually performed for fixed users since the assignment procedure is hardly compatible with challenging mobility constraints. The algorithm updating time is affected by the necessity of transmitting CSI at the base station (downlink) and by the computing time of the algorithm itself. Here we propose a very simple approach in which the base station does not operate on the complete group of sub-channels and users but only on small subsets. In other words, the base station does not assign a channel to a user but subsets of N_S/P channels to N_U/P users with $1 \leq P \leq N_U$ (also with small values, e.g. $N_S/P = 3$ or 4) reserving one or two bits to the fast assignment communication to the users inside each subset. This solution is a partitioning procedure into P sets of sub-channels and P sets of users, applied to the general problem of channel assignment and necessary for speeding up the process (Fig. 3). Set partitioning is fixed and, by means of this procedure, the initial problem is reduced to a computational complexity that can be expressed as

$$C_L = P \cdot f_C\left(\frac{N_S}{P}, \frac{N_U}{P}\right) << f_C\left(N_S, N_U\right), \tag{6}$$

with f_C denoting a measure of the computational load of the allocation algorithm adopted in the system.
In fact, this kind of algorithm reduction to a minimum assignment problem is the way for allowing fast and light adaptation to the channel variations for mobile users. The numerical results will highlight the performance trade-offs that can be obtained at several degrees of updating time and partitioning factors. The assignment should follow the same updating time T_{UPD} of the power control in Sect. 3.1; this means that, each T_{UPD} seconds, the system reallocates the radio resources to the available users (not necessarily the same of the previous period T_{UPD}).

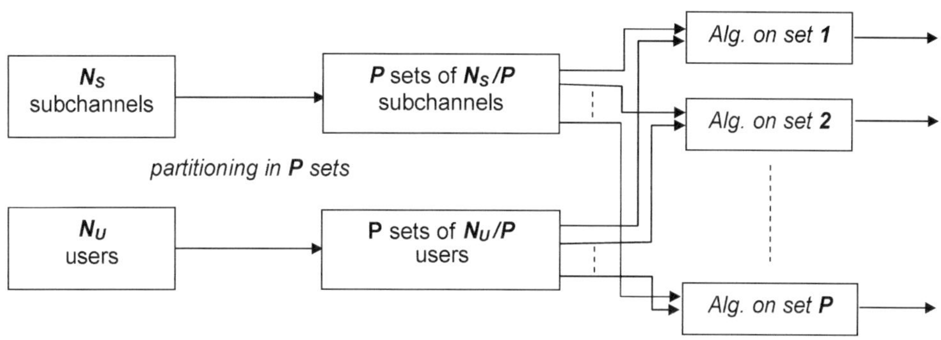

Fig. 3. Partitioned channel assignment for reduced complexity and increased speed.

3.3 Test-bed allocation strategy

In (Galati et al. (2008)) it is shown that, in a fading environment, the impact of a generic allocation strategy on SINR distribution can be described, in its dominant aspects, by a single parameter I_D. The use of the parameter I_D allows us to not consider, in this study, a particular allocation algorithm, which is not our objective here (a high number of examples are present in the literature), but to focus our attention on the impact produced on the network. So we investigate the overall system by using this simple parameter and by avoiding long and useless discussions about the details of numerous solutions. In practice, when we are interested on a particular allocation solution, we can estimate its I_D in order to have immediately a measure of its impact on network performance. As allocation algorithms usually operate with many different parameters and constraints, this I_D assignment could require an a-posteriori estimation. In the simulations of this work and in our comparative study, we simulate a simple allocation algorithm where we can modulate a-priori the value of parameter I_D from 1 to its maximum value, which is, for independent fading among the users, equal to the number N_U of active users in each cell (maximum order of multi-user diversity). We remark that this choice allows to separate the numerical results of this work from a specific choice of resource allocation algorithms for both analysis and simulations. In

this test-bed algorithm, multi-user diversity of order I_D is provided by a strategy that assigns a sub-group of I_D users to each sub-channel and selects, in each sub-group, the user with the best SINR.
Another rule is introduced in the assignment of users to sub-groups, in order to give each user the same number of chances to transmit. Given N_S sub-channels, N_U users and a diversity order I_D, with $N_S = k * N_U$ a fair rule assigns each user to $(N_S * I_D)/N_U = k * I_D$ sub-groups. The assignment slots can be structured as a matrix with N_S rows and I_D columns and the assignment of slots to the users is actuated filling the matrix by rows with k repetitions of an ordered list of the users $U_1 ... U_{N_U}$, as can be seen in Fig. 4(a). In the sequel this algorithm will be denoted as TAB_{ID}. After this operation, the power per sub-channel is adjusted according to Equation 5.
We observe that, when $I_D = 1$, the algorithm corresponds to the absence of any allocation strategy since the users are allocated to the sub-channels without SINR selections. So $I_D = 1$ can be considered as the realization of the CLPA mechanism described in Sect. 3.1. Moreover, the application of the set partitioning principle on TAB_{ID} (Fig. 4(b)) highlights the main impact that a complexity reduction procedure has on the algorithm effectiveness, i.e. a reduction of the multi-user diversity. As can be observed in Fig. 4(b), after set partitioning, the effective I_D of the reduced complexity algorithm becomes

$$I_{D,eff} = min \left\{ I_D, \frac{N_U}{P} \right\}. \tag{7}$$

Our overall framework, with the algorithm options, is sketched in Fig. 5.

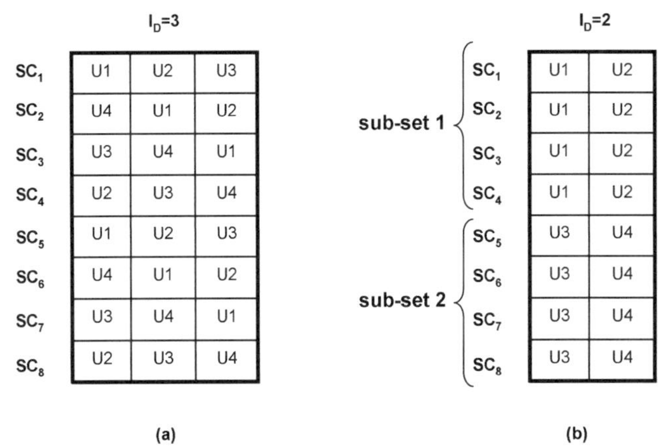

Fig. 4. Allocation of groups of users to sub-channels sc_i with TAB_{ID} strategy for $N_U = 4$, $N_S = 8$ and $I_D = 3$ with $P = 1$ (a) and $P = 2$ (b). For each sub-channel, TAB_{ID} will select the user with the best SINR in the corresponding row.

4. Multi-cell analysis

In multi-cell networks, SINR levels are affected by the power reduction caused by the adaptive techniques. In order to understand the behavior of the overall multi-cell system two different approaches have been implemented: (i) a completely simulative one (denoted as $F1$), in which the TAB_{ID} algorithm runs in the simulation environment as detailed in (Reggiani et al. (2007)) and (ii) an iterative analytical approach denoted as $F2$ and already presented in (Galati et al.

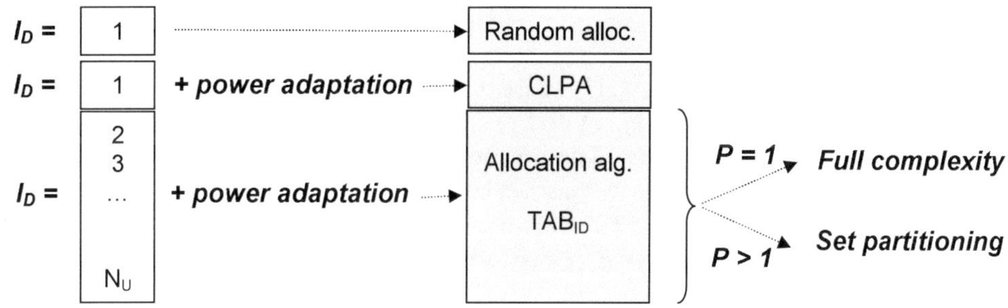

Fig. 5. Overview of the allocation options used for the numerical results.

(2008)), which reproduces the algorithm effect on SINR distribution by means of I_D, computes the power reduction and recursively applies it to the power of interfering BSs. In other words, $F2$ reproduces successive applications (over consecutive T_{UPD}) of a generic allocation algorithm until the system has achieved its stationary interference and SINR levels. In this work, numerical results will be focused on the final spectral efficiency for different algorithm parameters (P in Sect. 3.2, updating time), channels and user velocity (Sec. 2). The analysis is characterized by the following assumptions:

- All the active users are at distance d from the six reference BS and at distance D from six interfering BSs and no shadowing is present.
- Identically independent distributed (i.i.d.) fading A_n is applied on the generic $n-th$ link ($n = 0$ for the reference link, $n = 1, ..., 6$ for interfering links) with a probability density function $f_{A_n}(x) = f_A(x)$.
- At the first iteration ($i = 0$), the transmitted power per sub-channel in all BSs is fixed to a nominal value $P_{TX}(0) = P_S = P_{BS}/N_S$.
- At $i-$th iteration, the power reduction $\rho(i)$, which results from Equation 5 and is described by its probability density function $f_{\rho(i)}(x)$, is computed and applied to the nominal value P_S in all the co-channel BSs, modifying their transmission power $P_{TX}(i)$.
- Channel fading is assumed non-ergodic, constant in each user transmission block.

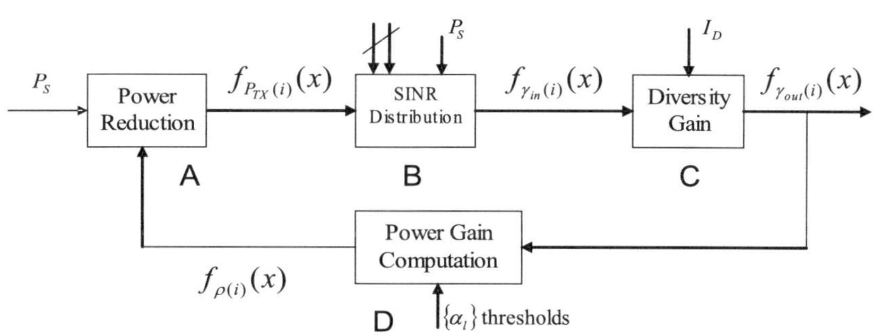

Fig. 6. Block diagram of the recursive loop for the analytical procedure $F2$.

So, in this scenario, at $i-$th iteration, the SINR value $\gamma_{in}(i)$ is computed in model $F2$ as

$$\gamma_{in}(i) = \frac{S}{I(i) + N} = \frac{P_S \cdot PL_0 \cdot A_0}{\sum_{n=1}^{6} P_{TX,n}(i) \cdot PL_n \cdot A_n + N} \tag{8}$$

where N, i.e. the additive white Gaussian noise power, PL_0 and PL_n, i.e. the path loss of the reference and interfering links, are deterministic parameters, while the fading A_0, A_n and the transmission power $P_{TX,n}(i)$ in the $n-$th BS co-channel are statistical variables with probability density functions $f_A(x)$ and $f_{P_{TX}}(x)$ respectively. The term $I(i)$ denotes the interference at the $i-$th iteration step. The functional block diagram of the recursive system is shown in Fig. 6: the distribution of $\gamma_{in}(i)$ is computed in block B and it is processed in block C through the parameter I_D, producing the cumulative distribution function of $\gamma_{out}(i)$ as $F_{\gamma_{out}}(x) = [F_{\gamma_{in}}(x)]^{I_D}$, with $F_\gamma(z) = \int_{-\infty}^{z} f_\gamma(x)dx$. The distribution of $\gamma_{out}(i)$ goes into block D that computes the distribution of power gain $\rho(i)$ (i.e. the power adaptation). Finally block A closes the loop, receiving the power gain distribution $f_{\rho(i)}(x)$ and applying it to nominal transmission power P_S of the interfering BSs. The distribution of the updated power $P_{TX}(i)$ is used for the new distribution $\gamma_{in}(i+1)$ in the next iteration. If the initial distribution $f_{\gamma_{in}}(x)$ cannot be derived analytically, it is obtained by simulation ($F1$) at the first iteration and then it is processed by $F2$ to produce final distributions $f_{\gamma_{out}}(x)$.

5. Numerical results

Simulations have been performed in different configuration scenarios, mobile users at a fixed distance d_{FIX} from the BS, at different distances d from the BS, in the downlink or in the uplink. However some common parameters are adopted in the simulations: each BS is set to a nominal power equal to $P_{BS} = 35$ dBm, the number of users is fixed to $N_U = 12$ and the number of available sub-channels is equal to $N_S = 48$. Moreover a set of 6 SINR thresholds α_l is defined among a minimum value $\alpha_1 = 2.88$ dB and a maximum $\alpha_L = 17.50$ dB. Channel fading is modelled by $Veh - A$ power delay profile for users' velocities from $v = 0$ km/h to $v = 60$ km/h, while two different pedestrian models ($Ped - A$ and $Ped - B$) are used from 0 to 20 km/h. The system performance is computed and analyzed in terms of achievable spectral efficiency η_{out} at different mobile terminals velocities, different updating times ($T_{UPD} = [5, 10, 20, 40]$ ms), in presence or not of smart radio allocation techniques and Closed Loop Power Adaptation (CLPA). The maximum spectral efficiency in the analyzed system is equal to $max(\eta_{out}) = 10log_2(1 + 10^{(\alpha_L/10)}) = 5.839$ [bit/s/Hz].
Figs. 7-11 have been obtained in the downlink configuration. In Figs. 7-8, we show the validity of the the analytical model $F2$ introduced in Sect. 4 w.r.t. the results obtained with intensive simulation ($F1$). Performance is shown in terms of the spectral efficiency η_{out} that can be achieved using different fading models, $Ped - A$ and $Ped - B$ respectively, at different velocities of the mobile terminals (from $v = 0$ to $v = 15$ km/h) and when the updating time of the allocation strategies is progressively increased, i.e. producing a new allocation configuration each $T_{UPD} = [5, 10, 20, 40]$ ms. In fact, $T_{UPD} = 5$ ms means that the allocation algorithm is able to take decisions in each OFDMA frame (in IEEE 802.16 standard each frame has a duration equal to 5 ms) and to distribute the available resources according to the channel state conditions (sub-channels, modulation profile and power) among the active users. Here we apply $I_D = N_U = 12$, that corresponds to the configuration that is able to exploit the highest multi-user diversity order, in order to produce the maximum performance. As we can see in Fig. 7 spectral efficiency estimated by means of the $F2$ procedure, whose results are shown with continuous lines, fits very well the values computed by means of intensive simulations $F1$, whose results are reported with filled markers. We can notice that η_{out} progressively decreases when T_{UPD} increases as the allocation strategy loses its ability to react to the time-varying channel conditions, especially when the updating time is higher.

Similar considerations can be done for Fig. 8, which has been derived using pedestrian channel model $Ped - B$.

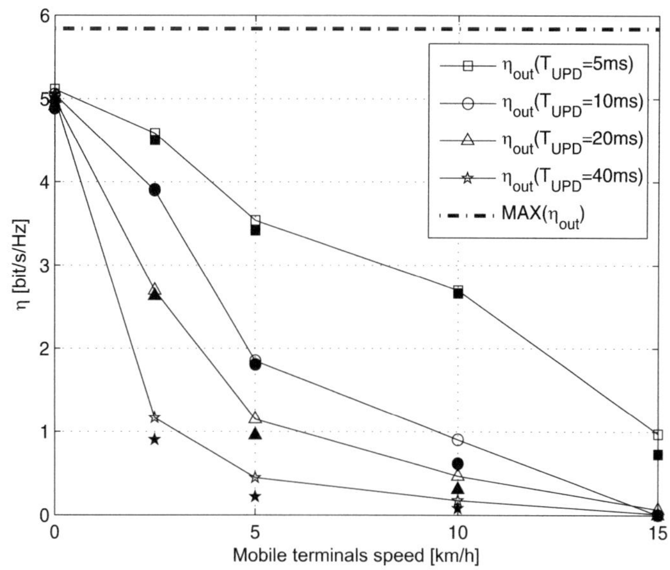

Fig. 7. Spectral efficiency (η_{out}) as a function of user velocity v [km/h] with fast fading defined by pedestrian channel model $Ped - A$. Results are obtained with the analytical approach $F2$ (continuous lines) and compared to performance computed with intensive simulations $F1$ (filled markers).

Figs. 9-11 show results that are similar to those reported in Figs. 7-8 since we highlight the achievable spectral efficiency as a function of terminals velocity and algorithm updating time. However we want to stress the advantages of smart dynamic resource allocation algorithms ($I_D = N_U = 12$) over a simple mechanism of power adaptation, referred as CLPA in Sect. 3.1, which corresponds to the absence of any allocation strategies (diversity order parameter $I_D = 1$). In order to have a complete comparison, we draw also the achievable spectral efficiency values when sub-channels are assigned randomly and power adaptation mechanism is not applied. This represents the worst case with $I_D = 1$ (so absence of allocation strategy) and transmission power applied to each sub-channel always fixed to the maximum available value $P_S = P_{BS}/N_S$; it is clear the advantage provided by CLPA and particularly by even simple allocation strategies. In Fig. 9, we can observe the performance obtained with a $Veh - A$ fading model and at several velocities, from $v = 0$ to $v = 60$ km/h. It is clear how with $v > 20 - 30$ km/h, forming sub-channels from contiguous sub-carriers, as in the AMC configuration, is not able to react effectively to the severe channel conditions; in these cases, interference averaging strategies like mechanisms of channel permutation are more advantageous solutions (e.g. the PUSC or FUSC configurations in IEEE 802.16 standard). In fact, at high speeds, even the adoption of advanced smart allocation solutions is not effective. However, if we consider slow mobile terminals movements with average velocity within $v = 15$ km/h, we can notice that we achieve a considerable gain over the simple CLPA strategy when we apply radio resource allocation algorithms. For pedestrian users, Fig. 10 and Fig. 11 highlight the value of η_{out} in the presence of $Ped - A$ and $Ped - B$ power delay profiles respectively. We can notice that, with

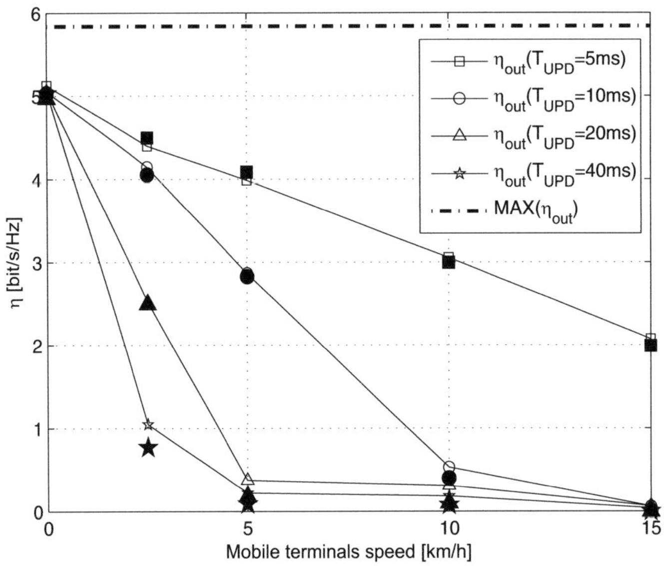

Fig. 8. Spectral efficiency (η_{out}) as a function of variable mobile terminals speed v [km/h] with fast fading defined by pedestrian channel model $Ped - B$. Results are obtained with the analytical approach $F2$ (continuous lines) and compared to performance computed with intensive simulations $F1$ (filled markers).

$T_{UPD} = 40$ ms and speed $v > 5$ km/h, the additional complexity introduced by the smart allocation strategy makes no sense as we can obtain the same performance with the simple CLPA or even random allocation with no power adaptation at all. In fact, the updating rate $1/T_{UPD}$ has to be faster for making the algorithm react to the rapidly changing conditions of the wireless channel. Nevertheless, with updating time $T_{UPD} < 10$ ms (corresponding to a new resource allocation each two OFDM frames in IEEE 802.16 standard), we can see that smart algorithms are strongly recommended for achieving a satisfactory transmission rate up to velocities around 5 km/h.

Although radio resource allocation solutions have demonstrated their ability to increase the spectral efficiency of mobile users, it has still to be considered their impact on the computational complexity. In other words, it should be evaluated the level of complexity that can be supported by the processing units, giving rise to the trade-off between performance and sustainable computational complexity. In Figs. 12 and 13 we point out the relation between computational complexity and performance by using performance evaluations expressed as a function of the partitioning factor P. Curves with constant spectral efficiency ($\eta_{out} = 1, 2, 3, 4, 5$ [bit/s/Hz]) are depicted as a function of the partitioning factor of sub-channels and users (P), velocity and updating time. We can notice that the best value $\eta_{out} = 5$ can be achieved only if we adopt the algorithm at the maximum complexity ($P = 1$) and with fixed users ($v = 0$ km/h). In general, a high level of complexity corresponds to higher levels of η_{out} even with mobile users. However, we observe also that a complexity reduction might allow a faster updating time, which always guarantees higher performance. So, in these figures, we can appreciate the overall trade-off among computational complexity, expressed by the partitioning factor P, updating time and achievable η_{out}. This kind of simulation or analysis

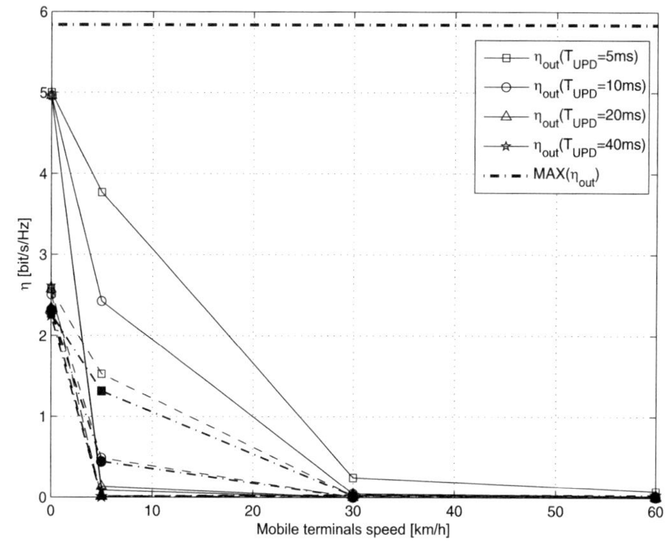

Fig. 9. Spectral efficiency (η_{out}) as a function of users velocity (v) and updating time (T_{UPD}) with vehicular channel model $Veh - A$. Performance of radio resource allocation algorithm with $I_D = N_U$ (continuous lines '–'), simple CLPA with $I_D = 1$ (dashed lines '- -'), and random allocation without any power adaptation (dotted lines '-.-') are compared.

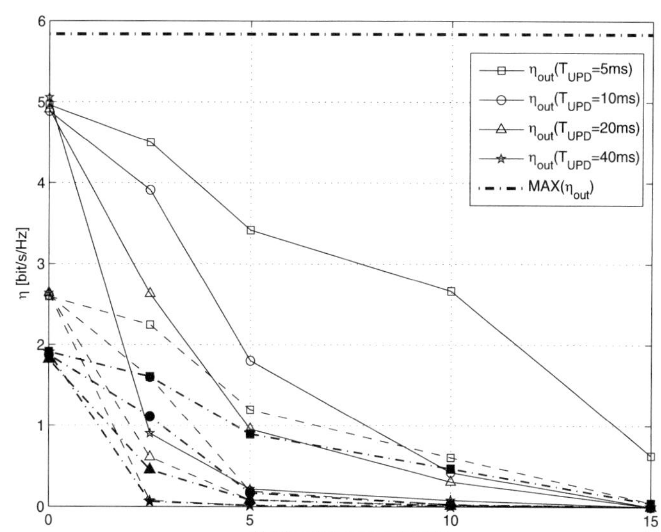

Fig. 10. Spectral efficiency (η_{out}) as a function of users velocity (v) and updating time (T_{UPD}) with pedestrian channel model $Ped - A$. Performance of radio resource allocation algorithm with $I_D = N_U$ (continuous lines '–'), simple CLPA with $I_D = 1$ (dashed lines '- -'), and random allocation without any power adaptation (dotted lines '-.-') are compared.

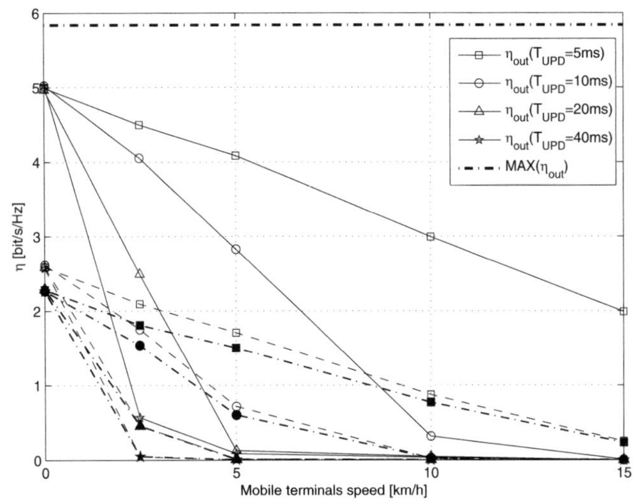

Fig. 11. Spectral efficiency (η_{out}) as a function of users velocity (v) and updating time (T_{UPD}) with pedestrian channel model $Ped - B$. Performance of radio resource allocation algorithm with $I_D = N_U$ (continuous lines '–'), simple CLPA with $I_D = 1$ (dashed lines '- -'), and random allocation without any power adaptation (dotted lines '-.-') are compared.

reveals the possible design choices that can be adopted in a multi-cellular system, according to the system updating or response time and to the BS processing power.

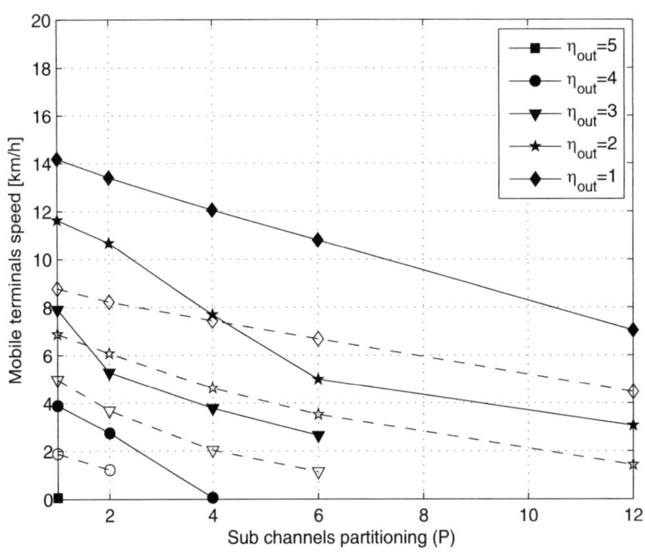

Fig. 12. Curves at fixed η_{out} as a function of user velocity (v), partitioning factor (P) and updating time $T_{UPD} = 5$ ms (continuous line '–') and $T_{UPD} = 10$ ms (dashed line '- -') in the presence of pedestrian channel model $Ped - A$.

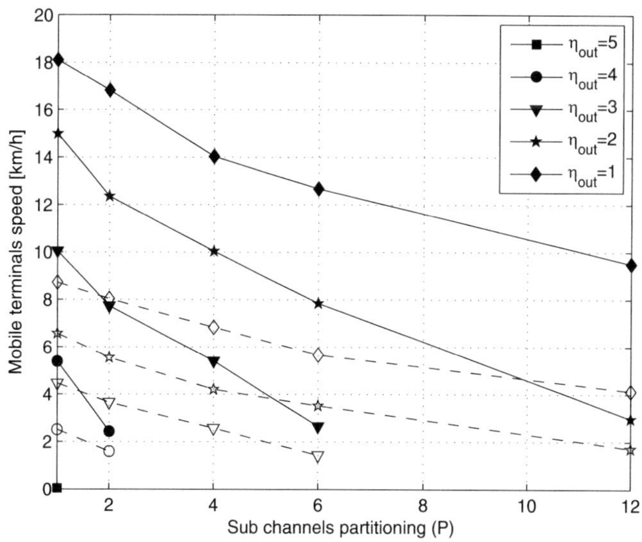

Fig. 13. Curves at fixed η_{out} as a function of user velocity (v), partitioning factor (P) and updating time $T_{UPD} = 5$ ms (continuous line '–') and $T_{UPD} = 10$ ms (dashed line '- -') in the presence of pedestrian channel model$Ped - B$.

When mobile users are at different and varying distances d from the BSs, the numerical findings confirm the same performance behavior described below. Also uplink simulations show similar results even if with lower values of η_{out}.

6. Conclusions

In the chapter, we have investigated the impact of allocation strategies on multi-cell networks with mobile users. The power reduction that can be achieved by means of the multi-user diversity exploitation has a beneficial impact on the overall network interference with a successive improvement of spectral efficiency. This positive effect is present, even if clearly reduced, also when only an efficient power adaptation loop, without smart allocation, is implemented in the network. On the other hand, it is shown how the users velocity has a strong impact on the updating time that is necessary for maintaining a satisfactory performance. This trade-off is completed by the algorithm complexity, which is another fundamental parameter that affects the updating ability of the system. In this context, a set partitioning technique is presented as a step for reducing the complexity order of smart allocation towards very fast algorithms that are compatible with low updating times.

7. References

Knop, R. & Humblet, P. (1995). Information Capacity and Power Control in Single-cell Multiuser Communications. *Proceedings of IEEE ICC 1995*, June 1995, Seattle (WA, USA)

Galati Giordano, L. & Reggiani, L. & Dossi, L. (2010). Radio Resource Management for New Generation Wireless Networks: Smart Allocation Techniques and Interference Evaluation. *VDM Verlag Dr. Müller, ISBN 978-3-639-26548-4, 112 pages,* June 2010

Reggiani, L. & Galati Giordano, L. & Dossi, L. (2007). Multi-User Sub-Channel, Bit and Power Allocation In IEEE 802.16 systems. *Proceedings of IEEE VTC-2007 Spring*, April 2007

Galati Giordano, L. & Reggiani, L. & Dossi, L. (2008). Interferente Evaluation in Multi-Cell Environment with Resource Allocation Algorithms. *Proceedings of IEEE VTC-2008 Spring*, May 2008

IEEE Std 802.16-2004. *IEEE Standard for Local and Metropolitan Area Networks: Air Interface for Fixed Broadband Wireless Access Systems*, October 2004

IEEE Std 802.16e-2005 & IEEE Std 802.16-2004/Cor1-2005. *IEEE Standard for Local and Metropolitan Area Networks IEEE: Air Interface for Fixed and Mobile Broadband Wireless Access Systems and Corrigendum 1*, February 2006

3GPP TR 25.996. *Universal Mobile Telecommunication System (UMTS): Spatial Channel Model for Multiple Input Multiple Output (MIMO) Simulations*, Release 6, September 2003

ETSI TR 101 112 (1998-04). *Selection Procedures for the Choice of Radio Transmission Technologies of the UMTS Ű Annex B: Test Environments and deployment models*, UMTS 30.03, version 3.2.0, 1998

Schoenen, R. & Qin, F. (2009). Adaptive Power Control for 4G OFDMA systems on Frequency Selective Fading Channels. *Proceedings of WiCOM 2009*, March 2009

Li, Z. & Wang, Y. & Yang, D. (2008). A Novel Power Control Scheme in OFDMA Uplink. *Proceedings of ICSP 2008*, May 2008, Leipzig (Germany)

Tee, L.K. & van Rensburg, C. & Tsai, J.-A. (2007). Uplink Power Control for an OFDMA Mobile Cellular System. *Proceedings of IEEE VTC-2007 Fall*, September 2007, Baltimore (MD, USA)

9

Statistical Properties of the Capacity of Double Nakagami-*m* Channels for Applications in V2V Dualhop Communication Systems

Gulzaib Rafiq[1], Bjørn Olav Hogstad[2] and Matthias Pätzold[3]

[1,3]*University of Agder*

[2]*University of Navarra*

[1,3]*Norway*

[2]*Spain*

1. Introduction

The design and analysis of cascaded fading models has been an active area of research in recent years due to its applications in numerous real world scenarios such as keyhole channels (Salo et al., 2006; Zlatanov et al., 2008), and multihop communication systems (Andersen, 2002; Karagiannidis et al., 2007; Talha & Pätzold, 2007; Velkov et al., 2009). It is shown in (Chizhik et al., 2002; Ercerg et al., 1997) that in the presence of a keyhole, the fading between each transmit and receive antenna pair in a multi-input multi-output (MIMO) system can be characterized using a double[1] Rayleigh process. Afterwards, this model has been extended to the double Nakagami-*m* fading model in (Shin & Lee, 2004). In (Salo et al., 2006), the authors have listed a few real world scenarios which give rise to the keyhole effect. Two such scenarios include diffraction through the street edges in urban microcellular environments (Ercerg et al., 1997) and traversal of the propagation paths through a narrow space for the case when the distance between the rings of scatterers around the transmitter and receiver is large (Gesbert et al., 2002).
Multihop communication systems on the other hand fall under the category of cooperative diversity techniques (Laneman et al., 2004; Sendonaris et al., 2003). In such systems, the wireless nodes (in a cooperative network) assist each other by relaying the information from the source mobile station (SMS) to the destination mobile station (DMS), hence improving the network coverage quite significantly. If however, the wireless nodes are assumed to be moving with relatively high speed, the concept of multihop communication can be applied to vehicle-to-vehicle (V2V) communication systems, where the source vehicle (SV) or the traffic control center (TCS) communicates with a destination vehicle (DV) via relay vehicles (RV). Figure 1 depicts these two scenarios for the case of dualhop communication, where the information from the TCS (or the SV) is received at the DV via an RV. V2V communication has received a lot of attention in recent years due to its applications in traffic safety and road traffic flow (Gradinescu et al., 2007). The statistical analysis of the received signal envelope under non-line-of-sight (NLOS) propagation conditions in an amplify-and-forward based

[1] Throughout this chapter, we will refer to a double process as the product of two independent but not necessarily identical processes.

dualhop communication system can be found in (Patel et al., 2006), where the overall channel between the transmitter and the receiver is modeled using a double Rayleigh process. This model is then extended to the double Rice channel model in (Talha & Pätzold, 2007), by taking the line-of-sight propagation conditions into account. The statistical properties of the capacity of double Rice channels have been analyzed in (Rafiq & Pätzold, 2008). However, the Nakagami-m process is considered to be a more general channel model as compared to the Rice and Rayleigh channel models. Hence, to generalize all the aforementioned works in the regime of multihop communication, the authors of (Karagiannidis et al., 2007) have presented the statistical analysis of the $N{*}$Nakagami-m model (i.e., a product of N Nakagami-m processes). Moreover, second order statistics for the double Nakagami-m process can be found in (Zlatanov et al., 2008). Though a lot of papers have been published in the literature employing the cascaded fading channel model, the statistical properties of the capacity of double Nakagami-m channels have not been investigated so far. Such channels find applications both in V2V communication systems employing dualhop communication, and keyhole channels (Zlatanov et al., 2008).

In this chapter[2], we have studied the statistical properties of the capacity of double Nakagami-m channels. Specifically, the influence of the severity of fading on the statistical properties of the capacity of double Nakagami-m channels is analyzed. We have derived exact analytical expressions for the probability density function (PDF), the cumulative distribution function (CDF), the level-crossing rate (LCR), and the average duration of fades (ADF) of the channel capacity. Here, the LCR and ADF of the channel capacity are important characteristic quantities which provide insight into the temporal behavior of the channel capacity (Giorgetti et al., 2003), (Hogstad & Pätzold, 2004). Our analysis has revealed that if the fading severity in one or both links of double Nakagami-m channels decreases (i.e., increasing the value of the severity parameter m in one or both of the cascaded Nakagami-m processes), the mean channel capacity increases, while the ADF of the channel capacity decreases. Moreover, this effect results in an increase in the LCR of the channel capacity at lower signal levels.

2. The double Nakagami-m channel model

In this chapter, we have considered Scenario 2 in Fig. 1, where the channel between the SV and the DV via an RV is represented as a concatenation of the SV-RV and RV-DV channels (Patel et al., 2006; Talha & Pätzold, 2007). Moreover, we have assumed that the fading in the SV-RV link and the RV-DV link is characterized by Nakagami-m processes denoted by $\chi_1(t)$ and $\chi_2(t)$, respectively. Hence, the overall fading channel describing the SV-DV link is modelled by a double Nakagami-m process given by (Kovacs et al., 2002; Zlatanov et al., 2008)

$$\Xi(t) = A_{\mathrm{RV}}\,\chi_1(t)\chi_2(t) \tag{1}$$

where A_{RV} is a real positive constant representing the relay gain. The PDF $p_{\chi_i}(z)$ of the Nakagami-m process $\chi_i(t)$ $(i = 1, 2)$ is given by (Nakagami, 1960)

$$p_{\chi_i}(z) = \frac{2m_i^{m_i} z^{2m_i-1}}{\Gamma(m_i)\Omega_i^{m_i}} e^{-\frac{m_i z^2}{\Omega_i}}, \quad z \geq 0 \tag{2}$$

[2] The material in this chapter is based on "Statistical Properties of the Capacity of Double Nakagami-m Channels", by Gulzaib Rafiq, Bjørn Olav Hogstad and Matthias Pätzold which appeared in the proceedings of 5th IEEE International Symposium on Wireless Pervasive Computing, ISWPC 2010, Modena, Italy, May 2010.

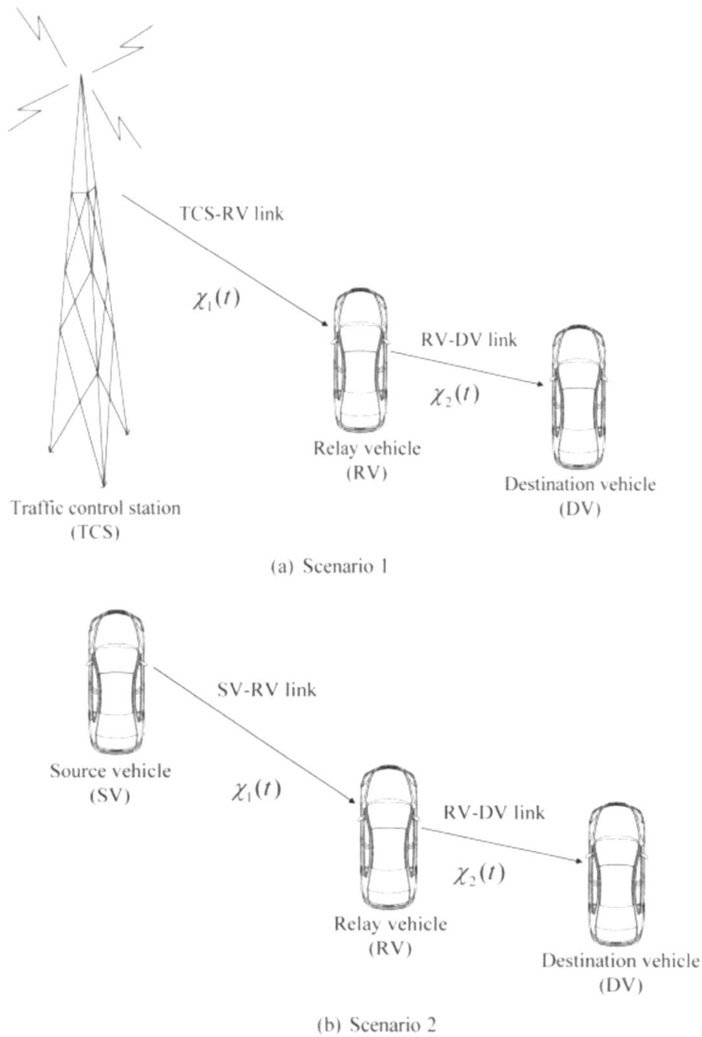

Fig. 1. The propagation scenarios describing double Nakagami-m fading channels.

where $\Omega_i = E\left\{\chi_i^2(t)\right\}$, $m_i = \Omega_i^2 / \mathrm{Var}\left\{\chi_i^2(t)\right\}$, and $\Gamma(\cdot)$ represents the gamma function (Gradshteyn & Ryzhik, 2000). The parameter m_i controls the severity of the fading. Increasing the value of m_i, decreases the severity of fading and vice versa.
The PDF of double Nakagami-m processes $\Xi(t)$ is given by (Karagiannidis et al., 2007)

$$p_{\Xi}(z) = \frac{4z^{m_1+m_2-1}}{\prod_{i=1}^{2} \Gamma(m_i) \left(\acute{\Omega}_i / m_i\right)^{(m_1+m_2)/2}} K_{m_1-m_2}\left(2z \prod_{i=1}^{2} \sqrt{\frac{m_i}{\acute{\Omega}_i}}\right), \quad z \geq 0 \tag{3}$$

where $\acute{\Omega}_1 = A_{\mathrm{RV}}^2 \Omega_1$, $\acute{\Omega}_2 = \Omega_2$, and $K_n(\cdot)$ denotes the modified Bessel function of the second kind of order n (Gradshteyn & Ryzhik, 2000, Eq. (8.432/1)). In order to derive the expressions for the PDF, CDF, LCR, and ADF of the capacity of double Nakagami-m channels, we need the joint PDF $p_{\Xi^2\dot{\Xi}^2}(z, \dot{z})$ of the squared process $\Xi^2(t)$ and its time derivative $\dot{\Xi}^2(t)$, as well as the PDF $p_{\Xi^2}(z)$ of $\Xi^2(t)$. The joint PDF $p_{\Xi^2\dot{\Xi}^2}(z,\dot{z})$ can be found by following the procedure presented in (Zlatanov et al., 2008) for the joint PDF $p_{\Xi\dot{\Xi}}(z,\dot{z})$ and then by using the concept of transformation of random variables (Papoulis & Pillai, 2002, Eq. (7-8)), which results in

$$p_{\Xi^2\dot{\Xi}^2}(z,\dot{z}) = \frac{1}{4z}\, p_{\Xi\dot{\Xi}}(\sqrt{z}, \frac{\dot{z}}{2\sqrt{z}})$$

$$= \frac{z^{m_2-3/2}}{\sqrt{2\pi}} \left[\prod_{i=1}^{2} \frac{m_i^{m_i}}{\Omega_i^{m_i}\Gamma(m_i)}\right] \int_0^\infty \frac{x^{2m_1-2m_2-1}}{\sqrt{\frac{z\beta_1}{x^2}+x^2\beta_2}}\, e^{-\frac{zm_2}{x^2\Omega_2}}\, e^{-\left(\frac{\dot{z}^2}{8z\left(\frac{z\beta_1}{x^2}+x^2\beta_2\right)}+\frac{x^2 m_1}{\Omega_1}\right)} dx,$$

$$z \geq 0, |\dot{z}| < \infty \tag{4}$$

where

$$\beta_1 = \frac{\Omega_1 \pi^2}{m_1}\left(f_{\max_1}^2 + f_{\max_2}^2\right) \tag{5a}$$

and

$$\beta_2 = \frac{\Omega_2 \pi^2}{m_2}\left(f_{\max_2}^2 + f_{\max_3}^2\right). \tag{5b}$$

Here, $f_{\max_1}$, $f_{\max_2}$, and $f_{\max_3}$ represent the maximum Doppler frequencies of the SV, RV, and DV, respectively. The expression for the PDF $p_{\Xi^2}(z)$ can be obtained by integrating the joint PDF $p_{\Xi^2\dot{\Xi}^2}(z,\dot{z})$ over $\dot{z}$. Alternatively, the PDF $p_{\Xi^2}(z)$ can also be found from the PDF $p_{\Xi}(z)$ in (3) as follows

$$p_{\Xi^2}(z) = \frac{1}{2\sqrt{z}}\, p_{\Xi}(\sqrt{z})$$

$$= \frac{2z^{\frac{m_1+m_2-2}{2}}}{\prod_{i=1}^{2}\Gamma(m_i)\left(\Omega_i/m_i\right)^{(m_1+m_2)/2}} K_{m_1-m_2}\left(2\sqrt{z}\prod_{i=1}^{2}\sqrt{\frac{m_i}{\Omega_i}}\right), \quad z \geq 0. \tag{6}$$

The expressions presented in (4) and (6) will be used in the next section to calculate the PDF and LCR of the channel capacity.

3. Statistical properties of the capacity of double Nakagami-m channels

The instantaneous capacity $C(t)$ of double Nakagami-m channels is defined as (Nabar et al., 2004)

$$C(t) = \frac{1}{2}\log_2\left(1+\gamma_s\,|\Xi(t)|^2\right) = \frac{1}{2}\log_2\left(1+\gamma_s\,\Xi^2(t)\right) \text{ (bits/s/Hz)} \tag{7}$$

where γ_s denotes the average received signal-to-noise ratio (SNR) at the DV. The factor 1 /2 in (7) is due to the fact that the RV in Fig. 1 operates in a half-duplex mode, and hence the signal transmitted from the SV is received at the DV in two time slots. Equation (7) can be considered as a mapping of a random process $\Xi(t)$ to another random process $C(t)$. Hence, the expressions for the statistical properties of the channel capacity $C(t)$ can be found by using the results for the statistical properties of the process $\Xi(t)$ obtained in the previous section. The PDF $p_C(r)$ of the channel capacity $C(t)$ can be found in closed form with the help of the PDF

$p_{\Xi^2}(z)$ and by applying the concept of transformation of random variables (Papoulis & Pillai, 2002, Eq. (7-8)) as

$$p_C(r) = \left(\frac{2^{2r+1}\ln(2)}{\gamma_s}\right) p_{\Xi^2}\left(\frac{2^{2r}-1}{\gamma_s}\right)$$
$$= \frac{2^{2r+2}\ln(2)\left(\left(2^{2r}-1\right)/\gamma_s\right)^{(m_1+m_2)/2}}{\left(2^r-1\right)\prod_{i=1}^{2}\Gamma(m_i)\left(\acute{\Omega}_i/m_i\right)^{(m_1+m_2)/2}} K_{m_1-m_2}\left(2\sqrt{\frac{2^{2r}-1}{\gamma_s}}\prod_{i=1}^{2}\sqrt{\frac{m_i}{\acute{\Omega}_i}}\right), \quad r \geq 0. \quad (8)$$

The CDF $F_C(r)$ of the channel capacity $C(t)$ can now be derived by integrating the PDF $p_C(r)$ and by making the use of relationships in (Gradshteyn & Ryzhik, 2000, Eq. (9.34/3)) and (Adamchik & Marichev, 1990, Eq. (26)) as

$$F_C(r) = \int_0^r p_C(x)dx$$
$$= \frac{1}{\prod_{i=1}^{2}\Gamma(m_i)} G_{1,3}^{2,1}\left[\frac{2^{2r}-1}{\gamma_s}\prod_{i=1}^{2}\left(\frac{m_i}{\acute{\Omega}_i}\right)\middle|\begin{matrix}1\\ m_1, m_2, 0\end{matrix}\right], \quad r \geq 0 \quad (9)$$

where $G[\cdot]$ denotes the Meijer's G-function (Gradshteyn & Ryzhik, 2000, Eq. (9.301)). The LCR $N_C(r)$ of the channel capacity describes the average rate of up-crossings (or down-crossings) of the capacity through a certain threshold level r. In order to find the LCR $N_C(r)$, we first need to find the joint PDF $p_{C\dot{C}}(z,\dot{z})$ of $C(t)$ and its time derivative $\dot{C}(t)$. The joint PDF $p_{C\dot{C}}(z,\dot{z})$ can be obtained by using the joint PDF $p_{\Xi^2\dot{\Xi}^2}(z,\dot{z})$ given in (4) as

$$p_{C\dot{C}}(z,\dot{z}) = \left(\frac{2^{2z+1}\ln(2)}{\gamma_s}\right)^2 p_{\Xi^2\dot{\Xi}^2}\left(\frac{2^{2z}-1}{\gamma_s}, \frac{2\dot{z}\ln(2)}{\gamma_s/2^{2z}}\right)$$
$$= \frac{\left(2^{2z+1}\ln(2)\right)^2\left(2^{2z}-1\right)^{m_2-\frac{3}{2}}}{\sqrt{2\pi\gamma_s}\,\gamma_s^{m_2}\left[\prod_{i=1}^{2}\left(\frac{\acute{\Omega}_i}{m_i}\right)^{m_i}\Gamma(m_i)\right]}\int_0^\infty \frac{x^{2m_1-2m_2-1}}{\sqrt{\frac{(2^{2z}-1)\beta_1}{\gamma_s x^2}+x^2\beta_2}} e^{-\frac{\left(2^{2z+1}\ln(2)\dot{z}\right)^2}{8\gamma_s\left(2^{2z}-1\right)\left(\frac{(2^{2z}-1)\beta_1}{\gamma_s x^2}+x^2\beta_2\right)}}$$
$$\times e^{-\left(\frac{x^2 m_1}{\acute{\Omega}_1}+\frac{(2^{2z}-1)m_2}{\gamma_s x^2\acute{\Omega}_2}\right)}dx \quad (10)$$

for $z \geq 0$ and $|\dot{z}| < \infty$. Finally, the LCR $N_C(r)$ can be found as follows

$$N_C(r) = \int_0^\infty \dot{z}\, p_{C\dot{C}}(r,\dot{z})d\dot{z}$$
$$= \sqrt{\frac{8}{\pi}}\left(\frac{2^{2r}-1}{\gamma_s}\right)^{m_2-\frac{1}{2}}\left[\prod_{i=1}^{2}\frac{m_i^{m_i}}{\acute{\Omega}_i^{m_i}\Gamma(m_i)}\right]\int_0^\infty e^{-\frac{(2^{2r}-1)m_2}{\gamma_s x^2\acute{\Omega}_2}}\frac{\sqrt{\frac{(2^{2r}-1)\beta_1}{\gamma_s x^2}+x^2\beta_2}}{x^{1+2m_2-2m_1}}e^{-\frac{x^2 m_1}{\acute{\Omega}_1}}dx \quad (11)$$

for $r \geq 0$. The ADF $T_C(r)$ of the channel capacity $C(t)$ denotes the average duration of time over which the capacity is below a given level r (Hogstad & Pätzold, 2004; 2007). The ADF

$T_C(r)$ of the channel capacity can be expressed as (Hogstad & Pätzold, 2007)

$$T_C(r) = \frac{F_C(r)}{N_C(r)} \tag{12}$$

where $F_C(r)$ and $N_C(r)$ are given by (9) and (11), respectively.

4. Statistical properties of the capacity of double rayleigh channels

The double Rayleigh channel follows as a special case of the double Nakagami-m channel when $m_i = 1$ $(i = 1, 2)$. Hence, by letting $m_i = 1$ $(i = 1, 2)$ in (8), (9), and (11), the PDF, CDF, and LCR of the capacity of double Rayleigh channels can be expressed as

$$p_C(r)\,|_{m_i=1} = \frac{2^{2r}\ln(2)}{\gamma_s\sigma_2^2\sigma_{\mathrm{RV}}^2} K_0\left(\sqrt{\frac{2^{2r}-1}{\gamma_s\sigma_2^2\sigma_{\mathrm{RV}}^2}}\right), \quad r \geq 0 \tag{13}$$

$$F_C(r)\,|_{m_i=1} = 1 - \sqrt{\frac{2^{2r}-1}{\gamma_s\sigma_2^2\sigma_{\mathrm{RV}}^2}} K_1\left(\sqrt{\frac{2^{2r}-1}{\gamma_s\sigma_2^2\sigma_{\mathrm{RV}}^2}}\right), \quad r \geq 0 \tag{14}$$

and

$$N_C(r)\,|_{m_i=1} = \frac{\sqrt{2^{2r}-1}}{\sqrt{2\pi\gamma_s}\sigma_2^2\sigma_{\mathrm{RV}}^2} \int_0^\infty \sqrt{\beta_2 + \beta_1\left(\frac{2^r-1}{x^4\gamma_s}\right)}\, e^{-\frac{2^{2r}-1}{2\sigma_2^2\gamma_s x^2}} e^{-\frac{x^2}{2\sigma_{\mathrm{RV}}^2}} dy, \quad r \geq 0 \tag{15}$$

respectively. The ADF of the capacity $C(t)$ of double Rayleigh channels can be found using (12), (14), and (15). In (13)–(15), $\sigma_{\mathrm{RV}}^2 = A_{\mathrm{RV}}^2\sigma_1^2$ and σ_i^2 $(i = 1, 2)$ represent the variances of the underlying Gaussian processes in the corresponding Rayleigh processes $\chi_i(t)\,|_{m_i=1}$ $(i = 1, 2)$.

5. Numerical results

In this section, we will discuss the analytical results obtained in the previous section. The validity of the theoretical results is confirmed with the help of simulations. For comparison purposes, we have also shown the results for double Rayleigh channels, which represent a special case of double Nakagami-m channels. In order to generate Nakagami-m processes $\chi_i(t)$, we have used the following relationship (Yacoub et al., 1999)

$$\chi_i(t) = \sqrt{\sum_{l=1}^{2\times m_i} \mu_{i,l}^2(t)} \tag{16}$$

where $\mu_{i,l}(t)$ $(l = 1, 2, ..., 2m_i;\, i = 1, 2)$ are the underlying independent and identically distributed (i.i.d.) Gaussian processes, and m_i is the parameter of the Nakagami-m distribution associated with the ith link of the dualhop communication systems. The Gaussian processes $\mu_{i,l}(t)$, each with zero mean and variances σ_0^2, were simulated using the sum-of-sinusoids model (Pätzold, 2002). The model parameters were computed using the generalized method of exact Doppler spread (GMEDS$_1$) (Pätzold et al., 2009). The number of sinusoids for the generation of Gaussian processes $\mu_{i,l}(t)$ was chosen to be $N = 29$. The parameter Ω_i was chosen to be equal to $2m_i\sigma_0^2$. Unless stated otherwise, the values of the maximum Doppler frequencies $f_{\max_1}$, $f_{\max_2}$, and $f_{\max_3}$ were taken to be 0, 91, and 125 Hz, respectively. The SNR

γ_s was set to 15 dB. The parameters A_{RV} and σ_0 were chosen to be unity. Finally, using (16), (1), and (7), the simulation results for the statistical properties of the channel capacity were found. The PDF and CDF of the channel capacity of double Nakagami-m channels are presented in Figs. 2 and 3, respectively. Both figures illustrate the fact that increasing the value of the severity parameter m_i (i.e., a decrease in the level of the severity of fading) in one or both links of the double Nakagami-m channels results in an increase in the mean channel capacity. This result is specifically presented in Fig. 4, where the mean channel capacity is studied for different values of the severity parameter m_i $(i = 1, 2)$. It can also be seen that double Rayleigh channels ($m_i = 1$; $i = 1, 2$) have a lower mean channel capacity as compared to the mean channel capacity of double Nakagami-m channels ($m_i = 2$; $i = 1, 2$). Moreover, it can also be observed from Figs. 2 and 3 that increasing the value of the severity parameter m_i decreases the variance of the channel capacity.

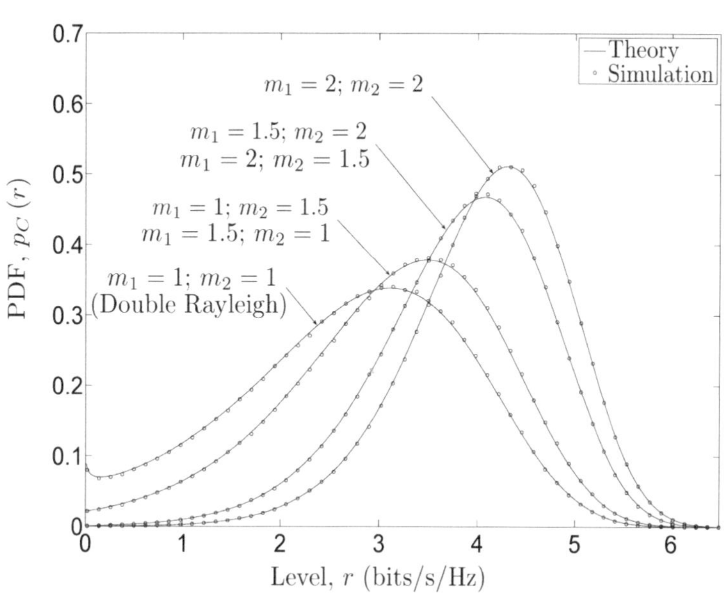

Fig. 2. The PDF $p_C(r)$ of the capacity of double Nakagami-m channels.

Figure 5 presents the LCR $N_C(r)$ of the capacity $C(t)$ of double Nakagami-m channels. It is observed that an increase in the level of severity of fading in one or both links of double Nakagami-m channels increases the LCR $N_C(r)$ of the channel capacity at low levels r. Hence, at low levels r, the LCR $N_C(r)$ of the capacity of double Rayleigh channels ($m_i = 1$; $i = 1, 2$) is higher as compared to that of double Nakagami-m channels ($m_i = 2$; $i = 1, 2$). However, the converse statement is true for higher levels r. The ADF of the capacity of double Nakagami-m channels is shown in Fig. 6. It is evident from this figure that the ADF of the capacity decreases with an increase in the value of the severity parameter m_i $(i = 1, 2)$.

Figures 7 and 8 study the influence of the maximum Doppler frequencies of the RV and the DV on the LCR and ADF of the channel capacity. It can clearly be observed in Figs. 7 and 8 that the LCR and ADF are strongly dependent on the Doppler frequencies of the RV and the DV. This means that the mobility of the RV and the DV has a significant influence on the LCR and ADF of the channel capacity. It is observed that increasing the maximum Doppler frequencies

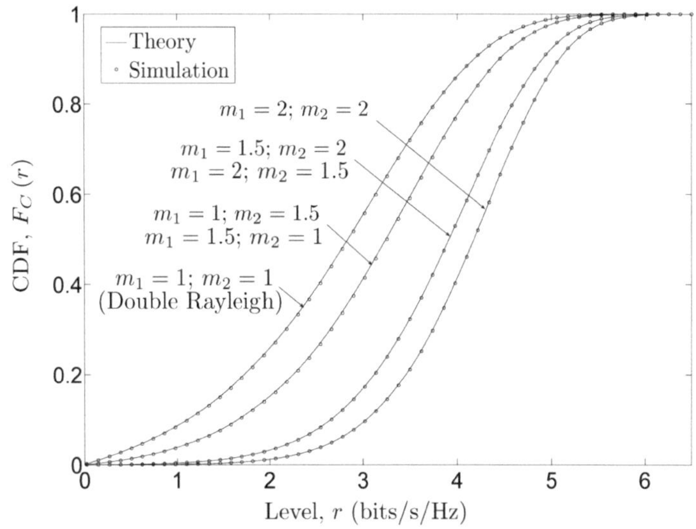

Fig. 3. The CDF $F_C(r)$ of the capacity of double Nakagami-m channels.

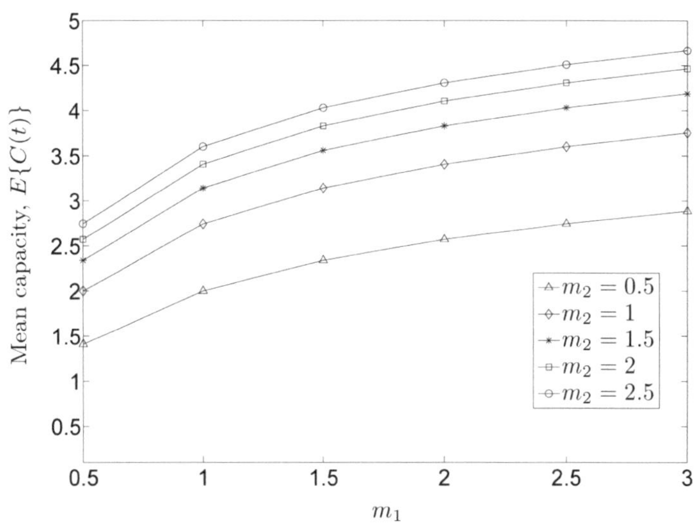

Fig. 4. The mean channel capacity of double Nakagami-m channels for different levels of fading severity.

$f_{\max_2}$ and $f_{\max_3}$ results in a significant increase in the LCR. However, the ADF decreases by increasing the maximum Doppler frequencies of the RV and the DV.

6. Conclusion

This Chapter presents the derivation of exact analytical expressions for the statistical properties of the capacity of double Nakagami-m channels, which finds applications in V2V communication systems employing dualhop communication, and keyhole channels. We have studied the influence of the severity of fading on the PDF, CDF, LCR, and ADF of the channel

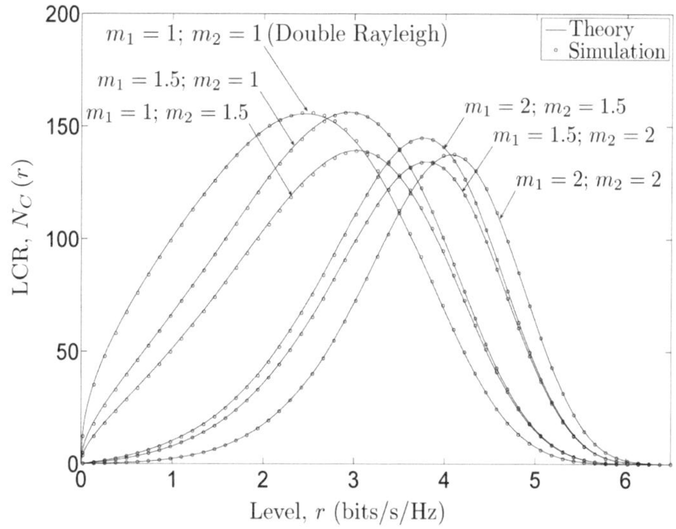

Fig. 5. The LCR $N_C(r)$ of the capacity of double Nakagami-m channels.

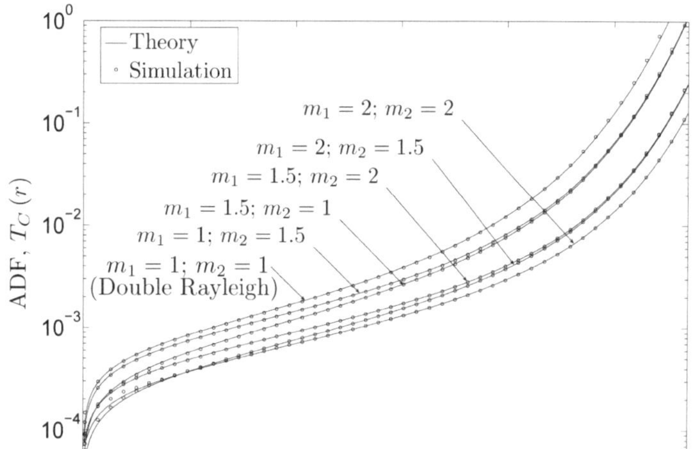

Fig. 6. The ADF $T_C(r)$ of the capacity of double Nakagami-m channels.

capacity. It is observed that an increase in the severity of fading in one or both links of double Nakagami-m channels decreases the mean channel capacity, while it results in an increase in the ADF of the channel capacity. Moreover, at lower signal levels, this effects increases the LCR of the channel capacity. Results also show that the mobility of the RV and DV has a significant influence on the LCR and ADF of the channel capacity. Specifically, an increase in the maximum Doppler frequencies of the RV and DV increases the LCR, while it has an opposite influence on the ADF of the channel capacity. The presented exact results are validated with the help of simulations, whereby a very good fitting is observed.

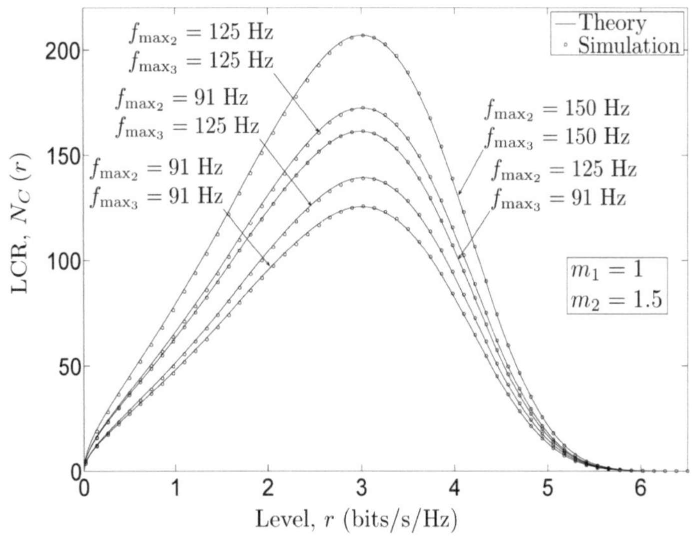

Fig. 7. The LCR $N_C(r)$ of the capacity of double Nakagami-m channels.

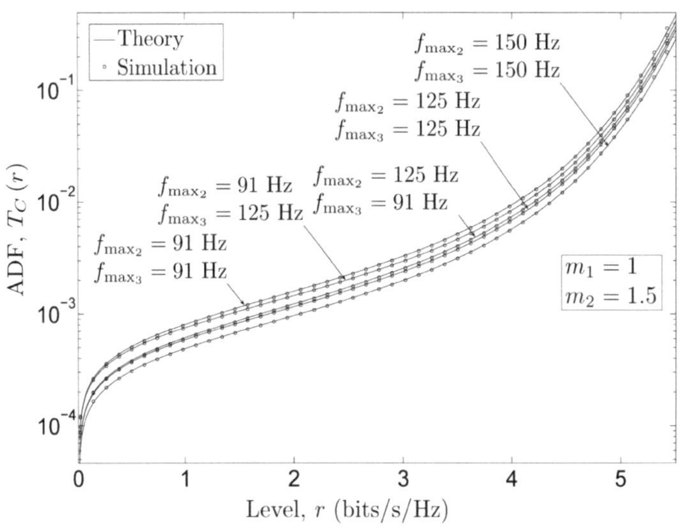

Fig. 8. The ADF $T_C(r)$ of the capacity of double Nakagami-m channels.

7. Acknowledgment

The contribution of G. Rafiq and Prof. M. Pätzold in this chapter was partially supported by the Research Council of Norway (NFR) through the project 176773/S10 entitled "Optimized Heterogeneous Multiuser MIMO Networks – OptiMO".

The contribution of Dr. B. O. Hogstad was supported in part by the Basque Government through the MIMONET project (PC2009-27B), and by the Spanish Ministry of Science and Innovation through the projects COSIMA (TEC2010-19545-C04-02) and COMONSENS (CSD2008-00010).

8. References

Adamchik, V. S. & Marichev, O. I. (1990). The algorithm for calculating integrals of hypergeometric type functions and its realization in REDUCE system, *Proc. Int. Symp. Symbolic and Algebraic Computation, ISSAC '90*, Tokyo, Japan, pp. 212–224.

Andersen, J. B. (2002). Statistical distributions in mobile communications using multiple scattering, *Proc. 27th URSI General Assembly*, Maastricht, Netherlands.

Chizhik, D., Foschini, G. J., Gans, M. J. & Valenzuela, R. A. (2002). Keyholes, correlations, and capacities of multielement transmit and receive antennas, *IEEE Trans. Wireless Commun.* 1(2): 361–368.

Ercerg, V., Fortune, S. J., Ling, J., Rustako Jr., A. J. & Valenzuela, R. A. (1997). Comparison of a computer-based propagation prediction tool with experimental data collected in urban microcellular environment, *IEEE J. Select. Areas Commun.* 15(4): 677–684.

Gesbert, D., Bölcskei, H., Gore, D. A. & Paulraj, A. J. (2002). Outdoor MIMO wireless channels: Models and performance prediction, *IEEE Trans. Commun.* 50(12): 1926–1934.

Giorgetti, A., Smith, P. J., Shafi, M. & Chiani, M. (2003). MIMO capacity, level crossing rates and fades: The impact of spatial/temporal channel correlation, *J. Commun. Net.* 5(2): 104–115.

Gradinescu, V., Gorgorin, C., Diaconescu, R. & Cristea, V. (2007). Adaptive traffic lights using car-to-car communication, *Proc. 65th IEEE Semiannual Vehicular Technology Conference, IEEE VTC 2007-Spring*, Dublin, Ireland, pp. 21–25.

Gradshteyn, I. S. & Ryzhik, I. M. (2000). *Table of Integrals, Series, and Products*, 6th edn, Academic Press.

Hogstad, B. O. & Pätzold, M. (2004). Capacity studies of MIMO models based on the geometrical one-ring scattering model, *Proc. 15th IEEE Int. Symp. on Personal, Indoor and Mobile Radio Communications, PIMRC 2004*, Vol. 3, Barcelona, Spain, pp. 1613–1617.

Hogstad, B. O. & Pätzold, M. (2007). Exact closed-form expressions for the distribution, level-crossing rate, and average duration of fades of the capacity of MIMO channels, *Proc. 65th Semiannual Vehicular Technology Conference, IEEE VTC 2007-Spring*, Dublin, Ireland, pp. 455–460.

Karagiannidis, G. K., Sagias, N. C. & Mathiopoulos, P. T. (2007). N*Nakagami: A novel stochastic model for cascaded fading channels, *IEEE Trans. Commun.* 55(8): 1453–1458.

Kovacs, I. Z., Eggers, P. C. F., Olesen, K. & Petersen, L. G. (2002). Investigations of outdoor-to-indoor mobile-to-mobile radio communication channels, *Proc. IEEE 56th Veh. Technol. Conf., VTC'02-Fall*, Vancouver BC, Canada, pp. 430–434.

Laneman, J. N., Tse, D. N. C. & Wornell, G. W. (2004). Cooperative diversity in wireless networks: Efficient protocols and outage behavior, *IEEE Trans. Inform. Theory* 50(12): 3062–3080.

Nabar, R. U., Bölcskei, H. & Kneubühler, F. W. (2004). Fading relay channels: Performance limits and space-time signal design, *IEEE J. Select. Areas Commun.* 22: 1099–1109.

Nakagami, M. (1960). The m-distribution: A general formula of intensity distribution of rapid fading, *in* W. G. Hoffman (ed.), *Statistical Methods in Radio Wave Propagation*, Oxford, UK: Pergamon Press.

Papoulis, A. & Pillai, S. U. (2002). *Probability, Random Variables and Stochastic Processes*, 4th edn, New York: McGraw-Hill.

Patel, C. S., Stüber, G. L. & Pratt, T. G. (2006). Statistical properties of amplify and forward relay fading channels, *IEEE Trans. Veh. Technol.* 55(1): 1–9.

Pätzold, M. (2002). *Mobile Fading Channels*, Chichester: John Wiley & Sons.

Pätzold, M., Wang, C.-X. & Hogstad, B. (2009). Two new methods for the efficient generation of multiple uncorrelated Rayleigh fading waveforms, *IEEE Trans. Wireless Commun.* 8(6): 3122Ű–3131.

Rafiq, G. & Pätzold, M. (2008). On the statistical properties of the capacity of amplify-and-forward channels under LOS conditions, *Proc. IEEE 11th Int. Conf. Communication Systems, IEEE ICCS 2008*, Guangzhou, China, pp. 1614–1619.

Salo, J., El-Sallabi, H. M. & Vainikainen, P. (2006). Impact of double-Rayleigh fading on system performance, *Proc. 1st IEEE Int. Symp. on Wireless Pervasive Computing, ISWPC 2006*, Phuket, Thailand.

Sendonaris, A., Erkip, E. & Aazhang, B. (2003). User cooperation diversity—Part I: System description, *IEEE Trans. Commun.* 51(11): 1927–1938.

Shin, H. & Lee, J. H. (2004). Performance analysis of space-time block codes over keyhole Nakagami-m fading channels, *IEEE Trans. Veh. Technol.* 53(2): 351–362.

Talha, B. & Pätzold, M. (2007). On the statistical properties of double Rice channels, *Proc. 10th International Symposium on Wireless Personal Multimedia Communications, WPMC 2007*, Jaipur, India, pp. 517Ű–522.

Velkov, Z. H., Zlatanov, N. & Karagiannidis, G. K. (2009). On the second order statistics of the multihop Rayleigh fading channel, *IEEE Trans. Commun.* 57(6): 1815–1823.

Yacoub, M. D., Bautista, J. E. V. & de Rezende Guedes, L. G. (1999). On higher order statistics of the Nakagami-*m* distribution, *IEEE Trans. Veh. Technol.* 48(3): 790–794.

Zlatanov, N., Velkov, Z. H. & Karagiannidis, G. K. (2008). Level crossing rate and average fade duration of the double Nakagami-m random process and application in MIMO keyhole fading channels, *IEEE Communications Letters* 12(11): 822–824.

10

Resource Allocation and User Scheduling in Coordinated Multicell MIMO Systems

Edgar Souza[1], Robson Vieira[2], Mari Kobayashi[3] and Mérouane Debbah[4]
[1,2]*Instituto Nokia de Tecnologia*
[3,4]*Ecole Supérieure d'Electricité*
[1,2]*Brazil*
[3,4]*France*

1. Introduction

Multiple-input multiple-output (MIMO) systems have become one of the most studied topics in the field of wireless communications because of the well-known potential for increasing spectral efficiency when compared to single-antenna systems. However, in the high interference regime, like in cellular systems, spatial multiplexing MIMO techniques can lose much of their effectiveness. Recently, some techniques have been developed in order to reduce intercell interference in MIMO systems and interference reduction based on base station cooperation seems to be a promising one.

When the multiple base stations can fully cooperate, the multicell downlink system reduces a classical MIMO broadcast channel with per-antenna power constraints. In this case, the optimal strategy to maximize the multicell throughput is the joint dirty-paper coding (DPC) (Caire & Shamai (2003)). Since practical implementation of DPC is still a problem, some sub-optimal solutions have been proposed and some works can be found in the literature (see Andrews et al. (2007) and references therein).

One of the basic requirements for most of the proposed base station cooperation schemes is the need of perfect channel state information at both receivers and network backbone for the joint processing at the central station. Therefore, there is a need of a two-step feedback/training from user terminals to each base station and then from base stations to the backbone. In the end, channel knowledge at the backbone is typically imperfect due to delays in feedback link, imperfect training sequence and etc. Besides, the channel learning at the central station requires a great amount of overhead, which is not desired nor recommended. In the work of Marsch & Fettweis (2008) the authors provide a comprehensive study of the multicell downlink optimization with limited backbone capacity and they analyze the problem of finding the optimal power allocation and beamforming matrices for different scenarios. Since the statistics are expected to change only in long-term basis, Kobayashi et al. (2009) proposed a power allocation strategy based on partial channel state information at the central station, which requires a very small additional backbone capacity.

The power allocation minimizing outage probability strategy for single-carrier with partial channel state information at transmitter for a cooperative downlink transmission system was proposed by Kobayashi et al. (2009). However, it was observed that the solution applied for a single-carrier condition cannot be directly applied to the multicarrier case because the frequency diversity gains were inferior to the loss due to the division of power among the carriers. Hence, Souza et al. (2009b) proposed a power allocation strategy that minimizes the outage probability based on the knowledge of channel statistics for a multicarrier system. The multicarrier power allocation strategy that was proposed exploits the multiplexing gain of cooperative MIMO and the frequency diversity gain provided by the multicarrier transmission scheme.

User scheduling in coordinated multicell MIMO systems is also a problem of a paramount importance. We consider the relevant case of a large number of user terminals and propose a simple scheduling scheme called Distributed Diversity Scheduling (DDS) which efficiently chooses a subset user terminals while limiting the amount of the backhaul communication. More precisely, each base station with local channel state information chooses its best set of user terminals over a predefined partition and reports the corresponding index and value to the central station. The central station then decides and informs the selected set to all base stations. Finally, the selected user terminals are served exactly in the same manner as the previous case of small number of terminals. It is proved that the scheduling algorithm scales optimally in the number of base stations, user terminals and transmit antennas per base station.

Within this context, this chapter aims at presenting some recent advances on adaptive resource allocation strategies for cooperative MIMO-OFDM networks. Such strategies allocate resources as a function of the time-varying channel state information and QoS parameters. In this chapter, we will present our main contributions that were obtained during the last years, indicating some research trends and future directions. Power allocation strategies are presented for single-carrier and multiple carrier networks according to the knowledge of the central station with respect to the wireless channels between the base stations and the terminals (perfect, partial and no channel state information). Additionally, a user scheduling scheme is presented for the case where all user terminals cannot be served at the same time. The so-called Distributed Diversity Scheduling (DDS) scheme selects a subset of user terminals and achieves a diversity gain that scales optimally with the number of cooperative base stations as well as user terminals, even under limited backhaul capacity.

2. Model of multicell MIMO systems

In this section, the multicell MIMO downlink model, where base stations are connected to a central station and communicate simultaneously with K single-antenna user terminals, is presented. It is assumed that the base stations are connected to a common backbone via a possibly error-free wired line, which enables some cooperation between base stations. Furthermore, the base stations do not communicate directly to each other and each one is equipped with M antennas. It is also assumed that each base station knows perfectly the channels while the central station may have perfect, partial or even no channel state information (this will be discussed with more details later on). Figure 1 illustrates an example of the multicell downlink system.

The model assumes multi-carrier transmission with Orthogonal Frequency Division Multiplexing (OFDM) in the wireless channels between the base stations and the user terminals. For each carrier n, these frequency flat fading channels are mutually independent and distributed as $\mathbf{h}_{bk}[n] \sim \mathcal{N}_{\mathbb{C}}(\mathbf{0}, \sigma_{bk}^2[n]\mathbf{I}_M)$. For an OFDM system model with N carriers, let $\{\mathbf{g}_{bk}[n]\}$ be the precoding vectors, $\{s_{bk}[n]\}$ be the transmit symbols and $\{p_{bk}[n]\}$ be the transmit powers from base station b to user terminal k at the n-th carrier. Base station b forms

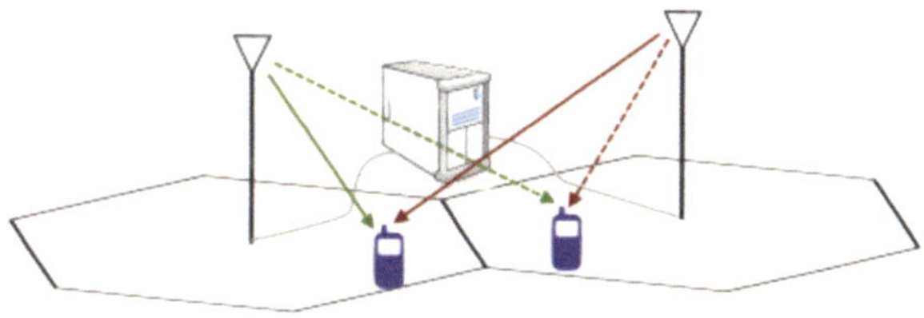

Fig. 1. System model.

its transmit vector at the n-th carrier as follows:

$$\mathbf{x}_b[n] = \sum_{k=1}^{K} \sqrt{p_{bk}[n]}\mathbf{g}_{bk}[n]s_{bk}[n], \tag{1}$$

and is subject to the power constraint $\sum_k \sum_n p_{bk}[n] \leq P_b$, where P_b is maximum transmit power.

If a distributed zero-forcing beamforming scheme is applied for a small number of terminals ($K \leq M$), then it can be shown that the model is equivalent to a system with K parallel MISO channels. In this case, the k-th user terminal achieves a diversity gain of $B(M-K+1)$. Let $n_k[n]$ be the equivalent zero-mean white Gaussian noise, then the received signal at the n-th carrier is given by:

$$y_k[n] = \sum_{b=1}^{B} \sqrt{p_{bk}[n]}a_{bk}[n]s_{bk}[n] + n_k[n], \tag{2}$$

where $a_{bk}[n] = \mathbf{h}_{bk}^H[n]\mathbf{g}_{bk}[n]$ represents the channel gain between base station b and terminal k at carrier n. The unitary precoding vector $\mathbf{g}_{bk}[n]$ is orthogonal to $\mathbf{h}_{bj}[n]$ for $j \neq k$ and the random variable $|a_{bk}[n]|^2$ is chi-squared distributed with $2(M-K+1)$ degrees of freedom. Assuming that each user terminal k perfectly knows the channel state $\mathbf{a}_n^k = (a_{1k}[n], \ldots, a_{Bk}[n])$, it decodes the space-time code and achieves the following rate:

$$R_k = \frac{1}{N} \sum_{n=1}^{N} \log\left(1 + \sum_{b=1}^{B} |a_{bk}[n]|^2 p_{bk}[n]\right). \tag{3}$$

The capacity region of the K parallel MISO channels in Equation (2) for a fixed set of power $\mathbf{p}_n^k = (p_{1k}[n], \ldots, p_{Bk}[n])$ and channel state $\mathbf{a}_n^k = (a_{1k}[n], \ldots, a_{Bk}[n])$ for all k and n is given by:

$$\mathcal{R}(\mathbf{a};\mathbf{p}) = \left\{\mathbf{R} \in \mathbb{R}_+^K : R_k \leq \frac{1}{N} \sum_{n=1}^{N} \log\left(1 + \sum_{b=1}^{B} |a_{bk}[n]|^2 p_{bk}[n]\right) \forall k\right\}, \tag{4}$$

where $\mathbf{a} = \{\mathbf{a}_n^k\}$ and $\mathbf{p} = \{\mathbf{p}_n^k\}$ for notation simplicity. The above region is convex (rectangular for $K = 2$ and $N = 1$). Let $\mathcal{P}$ denote a power allocation policy $\mathbf{a} \mapsto \mathbf{p}$ that maps the channel state $\mathbf{a}$ into the power vector $\mathbf{p}$ with component $\mathcal{P}_{bk}(\mathbf{a}) = p_{bk}$ and $\mathcal{F}$ denote the feasibility set satisfying $\sum_k \sum_n \mathcal{P}_{bk}(\mathbf{a}) \leq P_b, \forall b$ for any channel realization $\mathbf{a}$. Then, the capacity region of the K parallel MISO channels (2) under the individual base station power constraints $\mathbf{P} = (P_1, \ldots, P_B)$ for a fixed channel state $\mathbf{a}$ is given by:

$$\mathcal{C}(\mathbf{a};\mathbf{p}) = \bigcup_{\mathcal{P} \in \mathcal{F}} \mathcal{R}(\mathbf{a};\mathbf{p}), \tag{5}$$

The capacity region (5) is convex and its boundary can be explicitly characterized by solving the weighted sum rate maximization as specified in the next section.

3. Resource allocation strategies for single carrier systems

The purpose of the resource allocation problems is to optimize the power distribution over the carriers of all user terminals for a given target rate tuple γ. According to the system model that was presented in the previous section, equations (3) and (4) provide generic expressions for the rate achieved by user terminal k and capacity region of the parallel MISO channels rate, respectively. For the special case of single carrier networks, these equations reduce to:

$$R_k = \log\left(1 + \sum_{b=1}^{B} |a_{bk}|^2 \, p_{bk}\right), \tag{6}$$

$$\mathcal{R}(\mathbf{a};\mathbf{p}) = \left\{ \mathbf{R} \in \mathbb{R}_+^K : R_k \leq \log\left(1 + \sum_{b=1}^{B} |a_{bk}|^2 \, p_{bk}\right) \forall k \right\}. \tag{7}$$

The next subsections present resource allocation strategies for single carrier systems. The strategies depend on the assumptions regarding the channel state information that is available at the transmitters. Strategies for three different assumptions will be presented in this section. First we consider the case of perfect channel state information (CSI) at the central stations. Then, we consider the case of partial channel state information at the central station. Finally, we review the case when the central station has absolutely no channel state information. Each assumption requires a different approach and, at the end, we compare the performance of the resource allocation strategies with respect to the outage probability of the system.

3.1 Perfect channel state information

When perfect (or full) channel state information, i.e. knowledge of all channel realizations is available at the central station, the optimal power allocation is found by an iterative algorithm which is a generalization of the classical waterfilling algorithm (see Lee & Jindal (2007) and Yu et al. (2004)). Ideally, the transmitter could adjust its powers such that the outage of the system is reduced to zero. In this case, the problem is focused on the search for power allocation policies that provide the rate tuple proportional to the target rate tuple (rate-balancing). It is remarked that the zero-outage performance is not achieved when the transmitter is under limited power constraints. This policy equalizes the individual outage probability of all user terminals and thus provides the strict fairness among them. It is worthy to note that fairness is one of the most desired properties.
The objective is to find the set of powers $\{p_{bk}\}$ that satisfy the following condition:

$$\frac{R_k(\mathbf{p})}{R_1(\mathbf{p})} = \frac{\gamma_k}{\gamma_1} \stackrel{\Delta}{=} \alpha_k, \quad k = 2, \ldots, K \tag{8}$$

where $\frac{\gamma_k}{\gamma_1} = \alpha_k$ and $\alpha_1 = 1$. More precisely, according to Lee & Jindal (2007) the optimal power allocation is a solution of:

$$\min_{\sum_k \theta_k = 1} \; \max_{\mathbf{R} \in \mathcal{C}(\mathbf{a},\mathbf{P})} \sum_{k=1}^{K} \theta_k \frac{R_k}{\alpha_k}. \tag{9}$$

Notice that the inner problem for a fixed set $\{\theta_k\}$ is convex since the objective function (weighted sum rate) is concave and the constraints are linear. As discussed by Boyd & Vandenberghe (2004), it is necessary and sufficient to solve the Karush-Kuhn-Tucker (KKT) conditions given by:

$$\frac{\theta_k}{\alpha_k} \frac{|a_{bk}|^2}{1 + \sum_{b=1}^{B} |a_{bk}|^2 p_{bk}} = \frac{1}{\mu_b}, \quad k = 1, \ldots, K, \quad b = 1, \ldots, B \tag{10}$$

where μ_b denotes the Lagrangian variable to be determined such that $\sum_k p_{bk} \leq P_b$. Although a closed-form solution does not exist, the multiuser waterfilling algorithm for the MIMO multiple access channel proposed by Yu et al. (2004) can be easily modified to solve the KKT conditions (10) iteratively. Then, at each iteration it is found a new set of K powers $(p_{b1}, \ldots, p_{bK})$ related to base station b by treating the powers of the other base stations constant and computing:

$$p_{bk} = \left[\frac{\theta_k \mu_b}{\alpha_k} - \frac{1 + \sum_{j \neq b} |a_{jk}|^2 p_{jk}}{|a_{bk}|^2} \right]_+, \quad \forall k. \tag{11}$$

The outer problem consists of minimizing the solution of the inner problem with respect to $\theta_2, \ldots, \theta_K$. Since the problem at hand is convex, a subgradient method can be suitably applied as suggested by Bertsekas (1999). The overall algorithm implements the rate-balancing by allocating the rates proportional to the target rate tuple and it is summarized as follows.

Algorithm 1 Resource allocation for single carrier systems with perfect CSI

1: Initialize $\theta_k \in [0, 1]$ for $k > 1$.
2: **repeat**
3: for the fixed set of weights θ_k calculate via waterfilling approach (11):

$$\arg \max_{\mathbf{R} \in \mathcal{C}(\mathbf{a}, \mathbf{P})} \sum_{k=1}^{K} \theta_k \frac{R_k}{\alpha_k} \tag{12}$$

4: Calculate new subgradients
5: Update the weights θ_k
6: **until** convergence of (9)

3.2 Partial channel state information

For the special case of partial (or statistical) channel state information at the central station, the optimal power allocation is proposed by Kobayashi et al. (2009) and is the solution of the following optimization problem:

$$\begin{aligned} \text{minimize} \quad & P_{\text{out}}(\gamma, \mathbf{p}) = 1 - \prod_{k=1}^{K} \Pr\left(R_k > \gamma_k\right) \\ \text{subject to} \quad & \sum_{k=1}^{K} \sum_{n=1}^{N} p_{bk}[n] \leq P_b. \end{aligned} \tag{13}$$

The expression for the outage probability of the system for a fixed power allocation $\mathbf{p}$ is given by:

$$P_{\text{out}}(\gamma, \mathbf{p}) = 1 - \prod_{k=1}^{K} \left[1 - \Pr\left(\Delta_k(\mathbf{p}^k) < c_k \right) \right], \tag{14}$$

where $\Delta_k(\mathbf{p}^k) = \sum_b |a_{bk}|^2 p_{bk}$ and $c_k = 2^{\gamma_k} - 1$. It can be shown that, for a fixed set of powers $\mathbf{p}^k = (p_{1k}, \ldots, p_{Bk})$ of user terminal k, that $\Delta_k(\mathbf{p}^k)$ is a Hermitian quadratic form of a Gaussian random variable and its characteristic function is:

$$\Phi_{\Delta_k(\mathbf{p}^k)}(s) = \prod_{b=1}^{B} \frac{1}{(1 + s\alpha_{bk} p_{bk})^{M-K+1}}, \tag{15}$$

where $\alpha_{bk} = \sigma_{bk}/(M-K+1)$. The widely used upper bound is the Chernoff bound and for fixed powers $\mathbf{p}^k$ it is defined as:

$$\Pr\left(\Delta_k(\mathbf{p}^k) < c_k\right) \le \min_{\lambda \ge 0} e^{\lambda c_k} \Phi_{\Delta_k(\mathbf{p}^k)}(\lambda) \triangleq \overline{F}_{\Delta_k(\mathbf{p}^k)}(c_k, \mathbf{p}^k). \tag{16}$$

Using the expression (16) of the Chernoff upper bound for each user terminal k, the outage probability of the system for a fixed $\mathbf{p}$ is upper-bounded by:

$$P_{\text{out}}(\gamma, \mathbf{p}) \le 1 - \prod_{k=1}^{K} \left(1 - \overline{F}(c_k, \mathbf{p}^k)\right). \tag{17}$$

The power optimization based on the exact outage probability does not seem to be feasible and the algorithm proposed by Kobayashi et al. (2009) searches for the power allocation that minimizes the Chernoff upperbound. In that work, the authors conjecture that the proposed algorithm converges to the optimal solution that minimizes the exact outage probability, although there is no formal proof.
Hence, the corresponding optimization problem is formed as follows:

$$\begin{aligned}
\text{maximize} \quad & f(\{\lambda_k\}, \{p_{bk}\}) \triangleq \prod_{k=1}^{K} (1 - h_k(\lambda_k, \mathbf{p}^k)) \\
\text{subject to} \quad & \sum_{k=1}^{K} p_{bk} \le P_b, \quad b = 1, \ldots, B \\
& \lambda_k \ge 0, \quad k = 1, \ldots, K \\
& p_{bk} \ge 0, \quad b = 1, \ldots, B, \; k = 1, \ldots, K
\end{aligned} \tag{18}$$

where the function $h_k(\lambda_k, \mathbf{p}_k)$ is defined as:

$$h_k(\lambda_k, \mathbf{p}^k) = \frac{e^{\lambda_k c_k}}{\prod_{b=1}^{B} (1 + \lambda_k \alpha_{bk} p_{bk})^{M-K+1}}. \tag{19}$$

The solution of (18) is found by a two-step approach that explores the structure of the objective function. It is observed that the maximization of f with respect to $\{\lambda_k\}$ can be decoupled into the minimization of h_k over λ_k for each k, where h_k is convex in λ_k. In addition, since f is concave in $\{p_{bk}\}$, the overall problem is convex.
The first step consists in minimizing the monotonically decreasing function h_k with respect to λ_k. When the transmission powers are fixed, the optimal value of λ_k is given by the solution of the following polynomial of degree B:

$$\frac{c_k}{M-K+1} = \sum_{b=1}^{B} \frac{\alpha_{bk} p_{bk}}{1 + \lambda_k \alpha_{bk} p_{bk}}. \tag{20}$$

The second step is the maximization of the concave function $f(\lambda, \mathbf{p})$ with respect to p_{bk}. The Lagrangian function associated with $f(\lambda, \mathbf{p})$ is formed with the introduction of the Lagrangian multipliers $\{\mu_b\}$ and the following Karush-Kuhn-Tucker (KKT) conditions conditions for $k = 1, \ldots, K$ are obtained:

$$\frac{h_k(\lambda_k, \mathbf{p}_k)}{1 - h_k(\lambda_k, \mathbf{p}_k)} \frac{\alpha_{bk} \lambda_k}{1 + \lambda_k \alpha_{bk} p_{bk}} = \mu_b. \tag{21}$$

Finally, the iterative algorithm which solves the optimization problem (18), i.e. minimizes the Chernoff upperbound, is listed below. Although a formal proof has not been provided yet, it is conjectured that Algorithm 2 converges to its optimal solution. At each iteration, λ_k is determined as a unique solution for all k and a fixed set of powers. Regarding the power iteration, since the objective function (21) is concave in p_{bk} when fixing all other powers, a sequential update of the powers $\mathbf{p}_1, \mathbf{p}_2, \ldots, \mathbf{p}_B, \mathbf{p}_1 \ldots$ shall converge under individual base station power constraints.

Algorithm 2 Resource allocation for single carrier systems with partial CSI

1: Initialize $\mathbf{p}$
2: **repeat**
3:    **for** each base station **do**
4:       Update λ with the solution of the polynomial (20)
5:       Update $\mathbf{p}$ by evaluating the KKT conditions (21)
6:    **end for**
7: **until** convergence of $f(\lambda, \mathbf{p})$

3.3 No channel state information

When there is no channel state information at the central station, the strategy is to equally divide the total available power of each base station among all user terminals. Thus, there is no optimization problem here. Assuming that the maximum power of each base station is equal to P and defining $p \triangleq p_{bk} = P/K$, then:

$$\Delta_k = \sum_{b=1}^{B} |a_{bk}|^2 p_{bk} = p \sum_{b=1}^{B} |a_{bk}|^2 \triangleq \Delta_k^e, \tag{22}$$

where Δ_k^e is a chi-squared random variable with $4(M-K+1)$ degrees of freedom. If it is assumed that $\sigma_{bk} = 1$ for all links, then the cumulative distribution function of Δ_k^e is given by the following expression:

$$F_{\Delta_k^e}(y) = 1 - \exp\left[-\left(\frac{M-K+1}{p}y\right)\right] \sum_{k=0}^{2(M-K+1)-1} \frac{1}{k!}\left(\frac{M-K+1}{p}y\right)^k. \tag{23}$$

Let $c_k = 2^{\gamma_k} - 1$. Under these assumptions, the outage probability of the system is given by:

$$\begin{aligned} P_{\text{out}}(\gamma, \mathbf{p}) &= 1 - \prod_{k=1}^{K} \Pr\left(\log\left(1 + \sum_{b=1}^{B} |a_{bk}|^2 p_{bk}\right) > \gamma_k\right) \\ &= 1 - \prod_{k=1}^{K} \Pr\left(p \sum_{b=1}^{B} |a_{bk}|^2 > 2^{\gamma_k} - 1\right) \\ &= 1 - \prod_{k=1}^{K} \Pr\left(\Delta_k^e > c_k\right) \\ &= 1 - \prod_{k=1}^{K} \left(1 - F_{\Delta_k^e}(c_k)\right). \end{aligned} \tag{24}$$

3.4 Performance of resource allocation strategies

Figure 2 shows the outage probability performance versus signal-to-noise ratio for $K = 2$ user terminals and $M \in \{2,4\}$ antennas. The target rate is fixed to $\gamma = [1,3]$ bpcu (bits per channel use). The three different power allocation strategies are compared and the baseline case without network MIMO, where each base station sends a message to its corresponding user terminal in a distributed fashion, is also shown. The base station cooperation schemes provides a diversity gain of $2(M - K + 1)$, i.e. 2 and 6 with 2 and 4 antennas, respectively. These gains are twice as large as the case without network MIMO. Moreover, the schemes provide a additional power gains compared to equal power allocation.

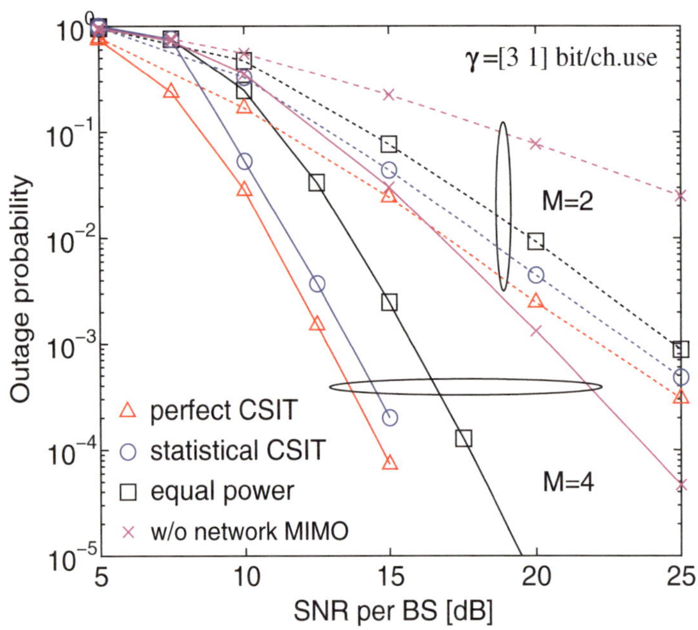

Fig. 2. Outage probability versus signal-to-noise ratio for $M \in \{2,4\}$ antennas per base station.

In Figure 3, it is plotted the individual outage probability under the same setting as Figure 2 only for $M = 2$. Assuming perfect channel state information at the central station, the proposed waterfilling allocation algorithm guarantees identical outage probability for both user terminals by offering the strict fairness. Under partial channel state information, the algorithm provides a better outage probability to user terminal 1 but keeps the gap between two user terminals smaller than the equal power allocation.

In real networks, there is a need to identify the best situations for the use of coordinated multicell MIMO. In order to identify the situations where coordinated transmission provides higher gains, a simulation campaign similar to the one done by Souza et al. (2009a) was configured. The basic simulation scenario consists of two cells, which contain a two-antenna base station each. Single-antenna user terminals are uniformly distributed in the cells. At each simulation step, the base stations transmit the signal to two randomly chosen user terminals, one terminal at each cell. The channel model that was adopted in these simulations is based on the sum-of-rays concept and it is described by IST-WINNER II (2007).

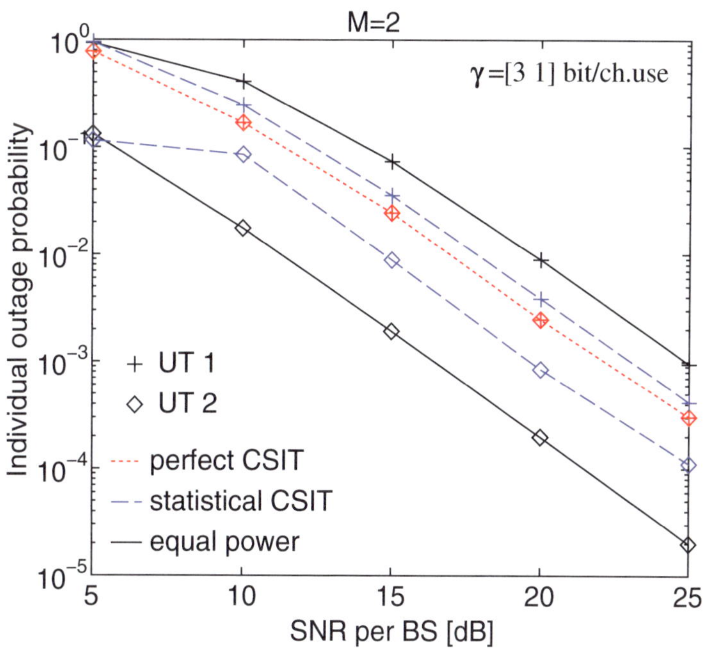

Fig. 3. Individual outage probabilities of each user terminal (UT) vs. SNR with $B = K = 2$ and $M = 2$.

Basically, the system has two transmit modes. In normal mode, each base station transmits to only one user terminal by performing spatial multiplexing. In coordinated mode, signal is transmitted according to the model that is described in Section 2. Let r_c be the cell radius. The transmit mode of the system is chosen by the function $r : [0, 1] \rightarrow \mathbb{R}^+$ which is given by the following expression:

$$r(\xi) = (1 - \xi) r_c. \tag{25}$$

The system operates in coordinated mode if and only if the chosen user terminals are inside the shadowed area of Figure 4; otherwise, the system operates in normal mode. The size of the shadowed area is controlled by the variable ξ in equation (25): if $\xi = 0$ the system operates in normal mode; if $\xi = 1$ the system operates in coordinated mode regardless the position of the user terminals; for other values of ξ it is possible to control the size of the shadowed area. Hence, the coordinated transmit mode may be enabled for the user terminals that are on the cell edges and, consequently, the normal transmit mode is enabled for the user terminals that are in the inner part of the cells.

Figure 5 shows the performance of the system when $\gamma = [1, 1]$ bpcu for given values of ξ. It is observed that the system performs best when $\xi = 1$, because under this configuration the coordinated mode provides more significant gains for all user terminals. In addition, it is seen that the gains of the coordinated mode are not significant when transmit powers are low. Under this power conditions, it is better for the system to operate in normal transmit mode because it would reduce the load of the feedback channels and signaling between the central and the base stations. The coordinated transmit mode outperforms the normal mode only when transmit powers are higher.

The distance between base stations and user terminals impacts the performance of the system and this is shown in Figure 6. The results in this figure refer to normal and coordinated

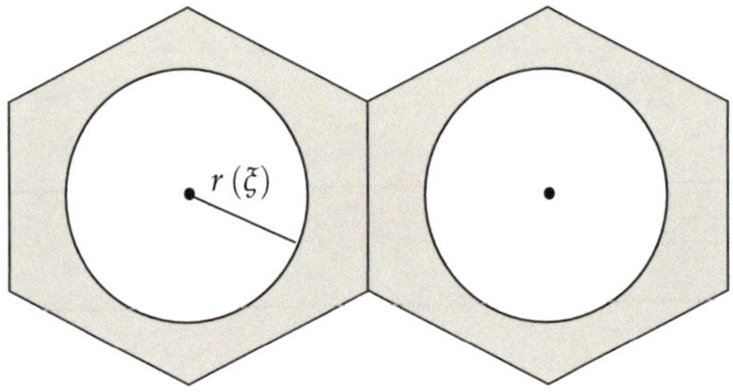

Fig. 4. Simulation scenario.

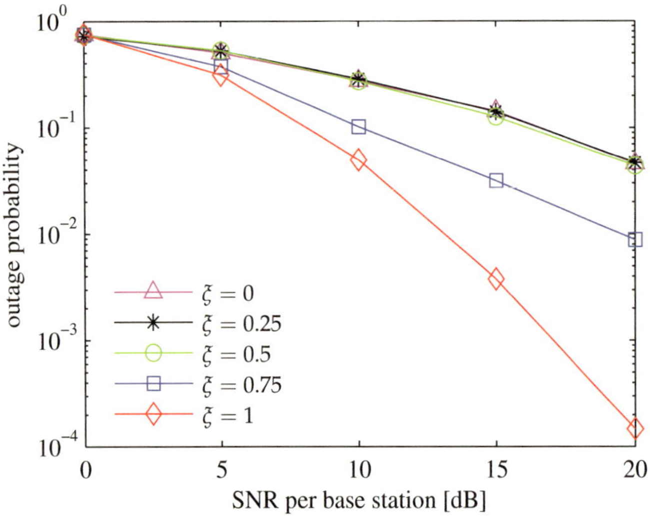

Fig. 5. Outage probability versus SNR per base station for $\gamma = [1, 1]$ bpcu

transmit modes for three given value of base stations' transmit powers. In all cases it is seen that gains of the coordinated mode decrease when distances increase. It is evident that the user terminals which are in the cell edges and experience bad propagation conditions cannot squeeze similar gains from the coordinated mode as the user terminals which are in the inner area of the cells.

For example, if the user terminals of the communication system are required to operate at a fixed outage probability of 10^{-3}, the results such as the ones in Figure 6 may provide systems' administrators with insights into the choice of the transmit mode and transmit powers of each base stations. In this example, if the system operates in normal transmit mode, signal-to-noise ratio would have to be equal or greater than 20 dB for the system to provide the performance which is required by the user terminals and this would be achieved only for distances less than 350 meters. However, the coordinated mode allows the system to serve the same set of user terminals in lower signal-to-noise ratio (around 10 dB in this case). On the other hand, if the base stations transmit with the same power and the system operates in coordinated mode, then it would be possible to serve all terminals with this required outage probability value.

Figure 7 shows the outage probability maps of the simulation scenario for the case $\xi = 1$. The base stations are positioned in $(x_1, y_1) = (750, 750)$m and in $(x_2, y_2) = (2250, 750)$m and transmit power of each base station is 10 dB. The blue squares indicate the areas where

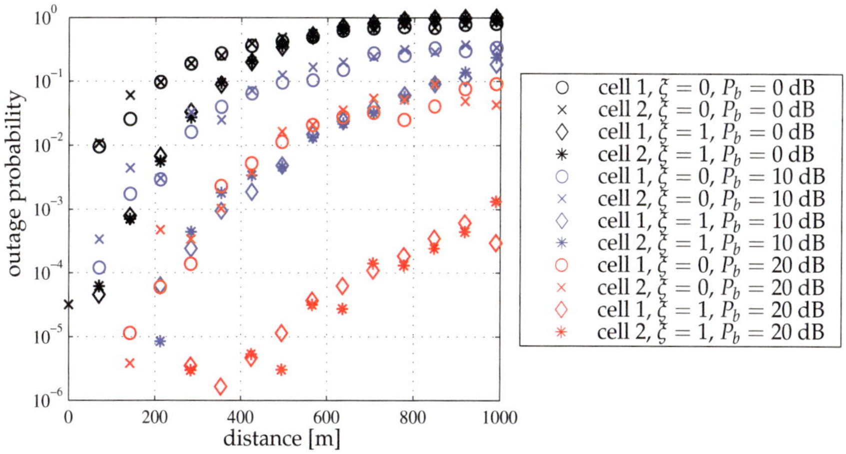

Fig. 6. Outage probability versus distance for given values of transmit power.

user terminals achieve the lowest outage probability values and the red squares indicate where user terminals have higher outage values. Figure 7a shows that the cells have similar performance when user terminals have the same target rate. On the other hand, Figure 7b shows the case when the user terminal in cell 2 (on the right side) requires three times the target rate of the one in cell 1. Cell 2 has worse performance the cell 1 because equal power allocation is performed.

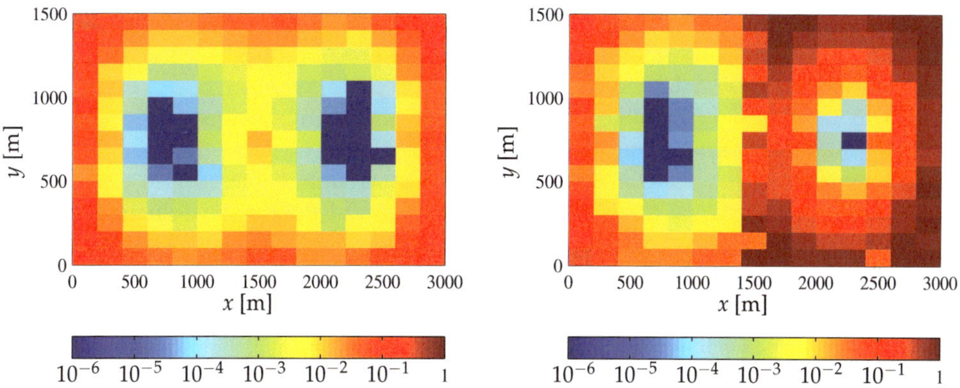

Fig. 7. Outage probability map for a) $\gamma = [1, 1]$ bits/s/Hz and b) $\gamma = [1, 1]$ bpcu

4. Resource allocation strategies for multiple carrier systems

This section is dedicated to the study of allocation strategies for multiple carrier systems. There are much more variables that impact the performance of these systems when compared to single carrier systems. That is why the challenge of allocating resources for such systems deserves special attention.

The difficulties encountered in this general case will be discussed in the next subsections, where we assume similar assumptions regarding channel state information (perfect, partial and no channel state information) similarly to the assumptions that were made for the case of single carrier systems.

4.1 Perfect channel state information

For the case where perfect channel state information is available at the central station, optimal power allocation is also found by a generalization of the classical waterfilling algorithm. The power allocation problem is modeled by a mathematical optimization problem that is solved using classical techniques. This is the case where the outage of the system is equal to the probability of γ being outside the capacity region $\mathcal{C}(\mathbf{a}, \mathbf{P})$:

$$P_{\text{out}} = 1 - \Pr\left(\gamma \in \mathcal{C}(\mathbf{a}, \mathbf{P})\right). \tag{26}$$

As well as for single carrier systems, the power allocation is performed by the algorithm that equalizes the individual outage probabilities. Hence, the following optimization problem has to be solved:

$$\min_{\sum w_k = 1} \max_{\mathbf{R} \in \mathcal{C}(\mathbf{a}, \mathbf{P})} \sum_{k=1}^{K} w_k \frac{R_k}{\alpha_k}, \tag{27}$$

where R_k is given by equation (3). Again, the inner optimization problem consists of maximizing the total system's capacity for a fixed $\mathbf{w} = (w_1, w_2, \ldots, w_K)$ and its solution can be found by applying the dual decomposition technique presented by Boyd & Vandenberghe (2004). The outer problem is identical to the one of the single carrier systems and it also consists of calculating subgradients and updating the weights. The overall algorithm is the same as Algorithm 1 and shall not be repeated in this subsection.

4.2 Partial channel state information

A feasible closed-form solution for the power allocation problem in multiple carrier systems with network MIMO has not been found yet. The proposal made by Souza et al. (2009b) consists of an iterative algorithm that finds the optimal number of allocated carriers as well as the optimal power allocation in multicell MIMO systems based on heuristics.

The solution to this problem was inspired by studies which demonstrated that, when $N \geq 2$ and considering the statistical channel knowledge, a closed-form for the outage probability can result in a complex and a numerical ill conditioned solution. The initial studies of Souza et al. (2009b) also included the analysis of Monte Carlo simulation results. of a very simple scenario with two base stations (equipped with two antennas each) and two user terminals. For this scenario the outage probability for different values of SNR and carriers, when the target rate tuple is $\gamma = [1, 1]$ bits per channel use and the both links have the same noise power, was evaluated. Results are presented in Figure 8. It is observed that the optimal strategy sometimes consists of allocating only a few carriers, even when more carriers are available. Hence, depending on SNR values, the distribution of power among carriers can result in rate reduction and increased outage of the system. Besides, frequency diversity gain only can be explored after a certain SNR value which is dependent of the number of carriers considered.

It was observed that the solution found in single-carrier case cannot be directed applied to the multicarrier case because the gains provided by frequency diversity were inferior to the loss due to the division of power between carriers. The proposed algorithm exploits this trade off and minimizes the outage probability of the system.

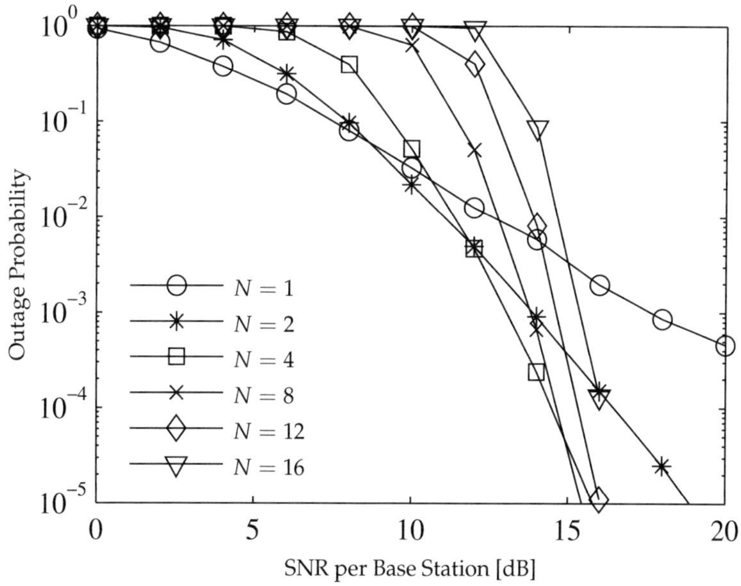

Fig. 8. Outage probability as a function of SNR

The heuristic solution is presented below. Let $\{p^*_{bk}[n]\}$ be the optimal power allocation for the multiple carrier case and $\{\theta^*_{bk}\}$ be the auxiliary variables that completely describe the power allocation so that the transmit power of each base station and each carrier can be defined as:

$$p^*_{bk}[n] = \theta^*_{bk}\frac{P_b}{N}, \tag{28}$$

with $\sum_k \theta^*_{bk} = 1$ for $b = 1, \ldots, B$. The solution is based on iterative calculations of the variables that represent the optimal power allocation and it is described by the Algorithm 3. Initially, equal power allocation is applied for each terminal and the optimal number number of carrier is defined as the total number of available carriers. In the next step, the optimal number of carriers is calculated based on the outage probability metric. Finally, for each carrier the optimal power allocation is obtained minimizing the Chernoff upperbound. The number of allocated carriers and the transmission powers are updated iteratively and minor optimization problems are solved until the convergence of the algorithm.

4.3 No channel state information

If there is no channel state information at the central station, the strategy is similar to the case of single-carrier networks and the total available power is divided among all carriers of the user terminals. Again, there is no optimization problem. The transmit power from base station b to user terminal k at carrier n is $p = P_b/KN$ and it means that $\Delta^e_{kn} = p\sum_b |a_{bk}[n]|^2$. Hence, the outage probability of the system is:

$$\begin{aligned} P_{\text{out}}(\gamma, \mathbf{p}) &= 1 - \prod_{k=1}^{K} \Pr\left(\frac{1}{N}\sum_{n=1}^{N} \log\left(1 + \Delta^e_{kn}\right) > \gamma_k\right) \\ &= 1 - \prod_{k=1}^{K} \Pr\left(\prod_{n=1}^{N} \left(1 + \Delta^e_{kn}\right) > 2^{N\gamma_k}\right), \end{aligned} \tag{29}$$

Algorithm 3 Resource allocation for multiple carrier systems with partial CSI

Initialize $\theta_{bk} = 1/K$ for $b = 1, \dots, B$ and $k = 1, \dots, K$
Initialize $N_{\text{opt}} = N$
repeat
  Calculate $p_{bk}[n] = \theta_{bk} P_b / N_{\text{opt}}$
  Find N_{opt} which minimizes the outage probability
  Update $p_{bk}[n] = \theta_{bk} P_b / N_{\text{opt}}$
  for each carrier **do**
    Solve the single carrier optimization problem (18)
  end for
  Update θ_k for $k = 1, \dots, K$
until convergence

where Δ^e_{kn} is a chi-squared random variable with $4(M - K + 1)$ degrees of freedom and its cumulative distribution function is given by:

$$F_{\Delta^e_{kn}}(y) = 1 - \exp\left[-\left(\frac{M-K+1}{p}y\right)\right] \sum_{k=0}^{2(M-K+1)-1} \frac{1}{k!}\left(\frac{M-K+1}{p}y\right)^k. \tag{30}$$

It is quite difficult to evaluate the analytical expression (29), but approximated values of the outage probability of the system may be easily found with Monte Carlo simulations.

4.4 Performance of resource allocation strategies

We considered a simulation scenario that consists of $B = 2$ base stations with $M = 2$ antennas each and $K = 2$ single-antenna terminals. Since $\theta_{b2} = 1 - \theta_{b1}$ in this case, it is sufficient to find the variables θ_{11} and θ_{21}. So, the results are presented in terms of the optimal values of θ_{11} and θ_{21} and the optimal number of allocated carriers N_{opt}.

The optimal values of θ_{bk} and N_{opt}, for the scenario where the target rate tuple is $\gamma = [1, 1]$ bits per channel use and when the both links have the same noise power, are presented in the Figure 9. As expected, θ_{11} and θ_{21} have the same values since the channel conditions and target rates are the same. Besides, as already observed, in order to minimize the outage probability, the optimal number of allocated carriers N_{opt} was found and it is greater than 1 only when SNR is above a certain value (around 9 dB in these simulations). Hence, in this scenario, both terminals are allocated with equal power and the system outage is minimized only for the optimal number of allocated carriers.

On the other hand, when the terminals have different rate requirements ($\gamma = [1, 3]$ bpcu), more power is allocated to the terminal with the highest target rate in order to minimize the outage probability (see Figure 10). However, this power difference only happens when SNR is greater than a certain value (9 dB in this case) because in the low SNR regime the single carrier optimization subproblem cannot be solved. In this scenario, the minimum system outage is only achieved with one allocated carrier, more carriers are allocated only when SNR values are greater than 19dB.

Figure 11 presents the results for the scenario where noise power of the links is different (asymmetric links). The noise power is modeled as follows: $\sigma_{ii} = \alpha \sigma_{ij}$ for $\alpha < 1$, $i, j = 1, 2$ and $i \neq j$. Considering $\alpha = 0.5$, it is possible to see that the algorithm allocates more power to the links which are in better conditions. This fact is observed specially for intermediate values of SNR; in the high SNR regime the allocation approximates to the equal power allocation

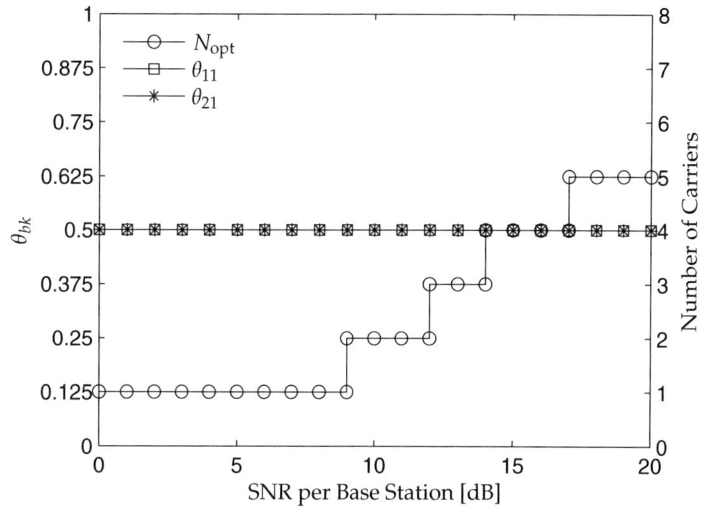

Fig. 9. Optimal values of θ_{11}, θ_{21} and N_{opt} for $\gamma = [1, 1]$ bpcu and symmetric links

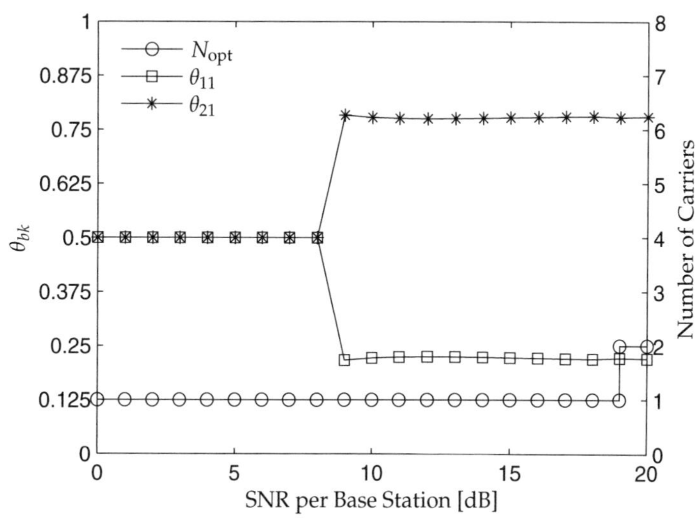

Fig. 10. Optimal values of θ_{11}, θ_{21} and N_{opt} for $\gamma = [1, 3]$ bpcu and symmetric links

because the difference of performance of the links decreases as the total available power increases.

Finally, Figure 12 shows the performance of the multicarrier system with perfect and partial channel state information. These curves represent the performance that may be achieved with the respective optimal allocation strategies together with the optimization of number of allocated carriers. Is has to be remarked that, for a given number of carriers, strategies for perfect and partial channel state information present similar trend as equal power allocation (see Figure 8).

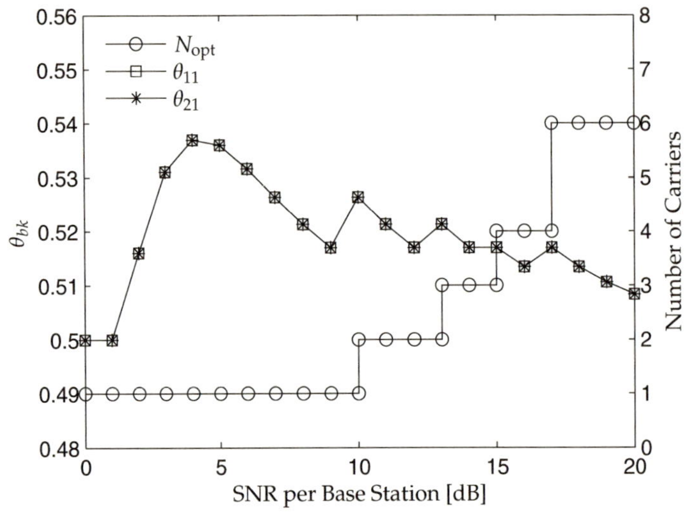

Fig. 11. Optimal values of θ_{11}, θ_{21} and N_{opt} for $\gamma = [1, 1]$ bpcu and asymmetric links

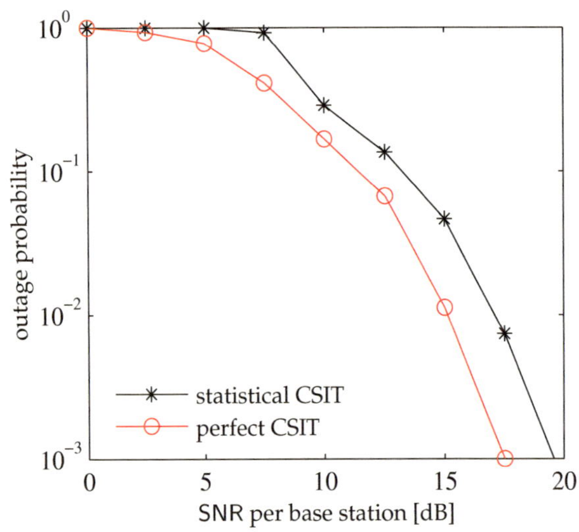

Fig. 12. Outage probability versus SNR with optimal number of allocated carriers.

5. Distributed diversity scheduling

In this section it is considered the importance of user scheduling when the number of user terminals is greater than the number of transmit antennas per base station. In order to apply the zero-forcing beamforming for each base station in a distributed manner, a set of $\tilde{K} < M$ user terminals shall be selected beforehand. It is assumed that the user scheduling is handled by the central station together with the power allocation for a system with B base stations with M antennas each. In this section, the Distributed Diversity Scheduling (DDS) scheme that was proposed by Kobayashi et al. (2010) is presented. This scheme achieves a diversity gain of $B\frac{K}{\tilde{K}}\left(M - \tilde{K} + 1\right)$ and scales optimally with the number of cooperative base stations as

well as user terminals while limiting the amount of side information necessary at the central station and at the base stations.
Basically, the proposed scheduling scheme can be described as follows. Assuming local channel state information, each base station chooses its best set of user terminals over the predefined partition and reports the corresponding index and value to the central station. For its part, the central station decides and informs the selected set to all base stations. Finally, the selected user terminals are served exactly in the same manner as the previous case of $K < M$. Let $\mathcal{S},\mathcal{U}$ denote the set of all K users, the $\tilde{K}$ selected users, with $|\mathcal{S}| = K$, $|\mathcal{U}| = \tilde{K}$, respectively. In addition, let $\mathcal{Q}(\tilde{K})$ be the set of all possible user selections, i.e., $\mathcal{Q}(\tilde{K}) = \{\mathcal{U}\,|\mathcal{U} \subseteq \mathcal{S},\, |\mathcal{U}| = \tilde{K}\}$ for $\tilde{K} \leq M$. Then, the equivalent channel from the base stations to the selected users is:

$$y_k = \mathbf{a}^k \mathbf{u}^k + z_k, \quad k \in \mathcal{U}, \tag{31}$$

which is a MISO channel with $\mathbf{a}^k = [a_{1k} \cdots a_{Bk}]$ and $\mathbf{u}^k = \left[\sqrt{p_{1k}}s_{1k} \cdots \sqrt{p_{Bk}}s_{Bk}\right]^T$. For convenience, we only consider the diversity order of the worst user and refer it as the diversity of the system hereafter. Since the diversity order of a given channel depends solely on the Euclidean norm of the channel matrix, the following user selection scheme maximizes the diversity of the system:

$$\mathcal{U}^* = \arg_{\mathcal{U}} \max_{\mathcal{U} \in \mathcal{Q}} \min_{k \in \mathcal{U}} \|\mathbf{a}^k\|^2. \tag{32}$$

Unfortunately, this scheduling scheme has two major drawbacks: 1) it requires perfect knowledge at the central station on $\{\mathbf{a}_k\}$, which is crucial for the scheduling, is hardly implementable as aforementioned, and 2) the maximization over all $|\mathcal{Q}(\tilde{K})| = \binom{K}{\tilde{K}}$ possible sets $\mathcal{U}$ grows in polynomial time with K.
To overcome the first drawback, the following selection scheme is used:

$$\mathcal{U}_{\mathrm{d}} = \arg_{\mathcal{U}} \max_{b=1\ldots B} \max_{\mathcal{U} \in \mathcal{Q}} \min_{k \in \mathcal{U}} |a_{bk}|^2. \tag{33}$$

This means that base station b selects the set $\mathcal{U}$ that maximizes $\min_{k \in \mathcal{U}} |a_{bk}|^2$ and sends both the index of the set and the corresponding maximum value to the central station. Upon the reception of B values and the corresponding sets from the B base stations, the central station makes a decision by selecting the largest one. Therefore, only a very small amount of information is sent through the links between the base station and the central station. To address the second drawback, the choices of $\mathcal{U}$ are narrowed down to $\kappa = K/\tilde{K}$ possibilities (It is assumed that κ is integer for simplicity of demonstration, but it can be shown that the same conclusion holds otherwise):

$$\mathcal{P}_{\mathcal{S}} = \{\mathcal{U}_1, \mathcal{U}_2, \ldots, \mathcal{U}_\kappa\}, \quad \bigcup_i \mathcal{U}_i = \mathcal{S},\, \mathcal{U}_i \cap \mathcal{U}_j = \emptyset,\, \forall\, i \neq j,\, |\mathcal{U}_i| = \tilde{K},\, \forall\, i. \tag{34}$$

In other words, $\mathcal{P}_{\mathcal{S}}$ is partition of the set of all users $\mathcal{S}$. Furthermore, it is assumed that the partition $\mathcal{P}_{\mathcal{S}}$ is fixed by the central station and known to all base stations. Hence, the proposed scheduling scheme selects the following set of users:

$$\mathcal{U}_{\mathrm{d}} = \arg_{\mathcal{U}} \max_{b=1\ldots B} \max_{\mathcal{U} \in \mathcal{P}_{\mathcal{S}}} \min_{k \in \mathcal{U}} |a_{bk}|^2. \tag{35}$$

To summarize, the scheduling scheme works as described in Algorithm 4.
An example of two base stations and six user terminals is shown in Figure 13. In this example, in order to serve two user terminals simultaneously, a partition of three sets is

Algorithm 4 Distributed Diversity Scheduling (DDS)

1: Central station fixes a partition $\mathcal{P}_{\mathcal{S}}$ and informs it to all base stations
2: Base station b finds $\max_{\mathcal{U}\in\mathcal{P}_{\mathcal{S}}} \min_{k\in\mathcal{U}} |a_{bk}|^2$ and sends this value and the index of the maximizing set $\mathcal{U}$ to the central station
3: Central station chooses the highest value and broadcasts the index of the selected set $\mathcal{U}_{\mathrm{d}}$ as defined in (35)
4: All base stations serve the user terminals in $\mathcal{U}_{\mathrm{d}}$ simultaneously

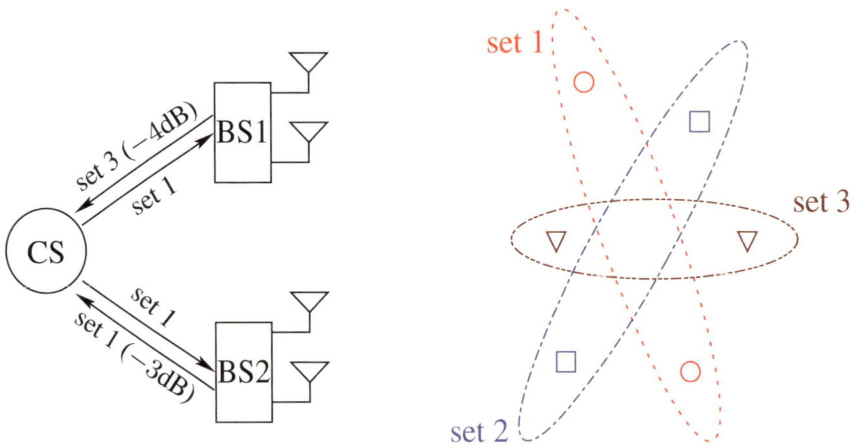

Fig. 13. An example scenario of user scheduling with two base stations and six user terminals.

fixed by the central station. With local channel state information, each base station compares the coefficients $\min_{k\in\mathcal{U}} |a_{bk}|^2$ for all three sets $\mathcal{U}$, finds out the largest one and sends the corresponding "index(value)" pair to the central station. The central station compares the values and broadcasts the index of the selected set (set 1 in this example).
Figure 14 shows the outage probability versus signal-to-noise ratio when there are more users than the number of served users, i.e. $K \geq \tilde{K} = 2$. Assuming the same setting as Figure 2 for $M = 4$, the distributed diversity scheme is applied to select a set of two users among $K \in \{2, 4, 6\}$. Once the user selection is done, any power allocation policy presented in Sections 3 and 4 can be applied. However, it is non-trivial (if not impossible) to characterize the statistics of the overall channel gains in the presence of any user scheduling. Hence, it is illustrated here only the performance with equal power allocation in the absence of channel state information. As a matter of fact, any smarter allocation shall perform between the waterfilling allocation and the equal power allocation. It is observed in the figure that diversity gain increases significantly as the number K of users in the system increases.

6. Conclusions

In this chapter, we reviewed the litterature on the power allocation problems for coordinated multicell MIMO systems. It was seen that the optimal resource allocation is given by the waterfilling algorithm when the central station knows all channel realizations. Assuming a more realistic scenario, we also reviewed the solutions for the case of partial channel state information, i.e. local channel knowledge at each base station and statistical channel

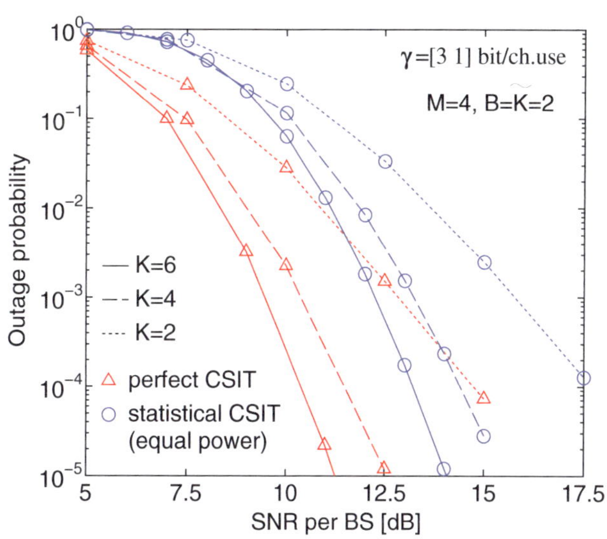

Fig. 14. Outage probability vs. SNR for many users with $B = \tilde{K} = 2$ and $M = 4$.

knowledge at the central station. Under this setting, it was presented an outage-efficient strategy which builds on distributed zero-forcing beamforming to be performed at each base station and efficient power allocation algorithms at the central station.

In addition, in the case of a small number of users $K \leq M$, it was proposed a scheme that enables each user terminal to achieve a diversity gain of $B(M - K + 1)$. On the other hand, when the number of users is larger than the number of antennas ($K \geq M$), the proposed distributed diversity scheduling (DDS) can be implemented in a distributed fashion at each base station and requires only limited amount of the backbone communications. The scheduling algorithm can offer a diversity gain of $B\frac{K}{\tilde{K}}(M - \tilde{K} + 1)$ and this gain scales optimally with the number of cooperative base stations as well as the number of user terminals. The main finding is that limited base station cooperation can still make network MIMO attractive in the sense that a well designed scheme can offer high data rates with sufficient reliability to individual user terminal. The proposed scheme can be suitably applied to any other interference networks where the transmitters can perfectly share the messages to all user terminals and a master transmitter can handle the resource allocation.

7. References

Andrews, J. G., Choi, W. & Jr., R. W. H. (2007). Overcoming interference in spatial multiplexing MIMO cellular networks, *IEEE Wireless Communications* 14(6): 95–104.

Bertsekas, D. P. (1999). *Nonlinear Programming*, Athena Scientific, Belmont, MA, USA.

Boyd, S. & Vandenberghe, L. (2004). *Convex optimization*, Cambridge University Press.

Caire, G. & Shamai, S. (2003). On the achievable throughput of a multiantenna Gaussian broadcast channel, *IEEE Transactions on Information Theory* 49(7): 1691–1706.

IST-WINNER II (2007). D1.1.2 WINNER II channel models, [online] Available: https://www.ist-winner.org/.

Kobayashi, M., Debbah, M. & Belfiore, J.-C. (2009). Outage efficient strategies for network MIMO with partial CSIT, *IEEE International Symposium on Information Theory*.

Kobayashi, M., Yang, S., Debbah, M. & Belfiore, J.-C. (2010). Outage efficient strategies for network MIMO with partial CSIT. submitted to IEEE Transactions on Signal Processing.

Lee, J. & Jindal, N. (2007). Symmetric capacity of MIMO downlink channels, *IEEE International Symposium on Information Theory*, IEEE, Seattle, WA, EUA.

Marsch, P. & Fettweis, G. (2008). On multicell cooperative transmission in backhaul-constrained cellular systems, *Annales des Télécommunications* 63(5-6): 253–269.

Souza, E. B., Vieira, R. D. & Carvalho, P. H. P. (2009a). Análise da probabilidade de interrupção para sistemas MIMO cooperativo, *Simpósio Brasileiro de Telecomunicações*, SBrT, Blumenau, Brazil. [in portuguese].

Souza, E. B., Vieira, R. D. & Carvalho, P. H. P. (2009b). Outage probability for multicarrier cooperative MIMO with statistical channel knowledge, *IEEE Vehicular Technology Conference*, IEEE, Anchorage, Alaska, USA.

Yu, W., Rhee, W., Boyd, S. & Cioffi, J. (2004). Iterative water-filling for Gaussian vector multiple-access channels, *IEEE Transactions on Information Theory* 50(1): 145 – 152.

11

Hybrid Evolutionary Algorithm-based Schemes for Subcarrier, Bit, and Power Allocation in Multiuser OFDM Systems

Wei-Cheng Pao, Yung-Fang Chen and Yun-Teng Lu
Department of Communication Engineering, National Central University
Taiwan, R.O.C.

1. Introduction

Multiuser orthogonal frequency division multiplexing (OFDM) is a very promising multiple access technique to efficiently utilize limited RF bandwidth and transmit power in wideband transmission over multipath fading channels. When a wideband spectrum is shared by multiple users in multiuser OFDM-based systems, different users may experience different fading conditions at all subcarriers. Each user is assigned a subset of all subcarriers by some allocation algorithm. Thus, multiuser diversity can be achieved by adaptively adjusting subcarrier, bit, and power allocation depending on channel status among users at different locations (Wong et al., 1999a). In (Wong et al., 1999a), Wong applies a Lagrangian optimization technique and an iterative algorithm to solve the subcarrier, bit and power allocation problem. The suboptimal scheme for the NP-hard joint optimization problem is decoupled into two steps while it has a high computational complexity. A sub-optimal algorithm has been proposed to solve a related problem (Wong et al., 1999b). In (Wong et al., 1999b), Wong presents a real-time subcarrier allocation (SA) algorithm. It is a two-phase algorithm, including the constructive initial assignment (CIA) and the subcarrier swapping steps. The initial subcarrier allocation algorithm needs to pre-determine the numbers of subcarriers for each user before the allocation process starts. The performance of the SA-based algorithm will be compared in the simulation. In (Kim et al., 2006), Kim shows that the allocation problem in (Wong et al., 1999a) can be transformed into an integer programming (IP) problem. The branch-and-bound algorithm (Wolsey, 1998) can be employed to find the optimal solution of the allocation problem which has exponential computational complexity in the worse cases. We utilize the approach to obtain the optimal solution as the performance bound for comparison.
Evolutionary algorithms (EA) are used to solve extremely complex search and optimization problems which are difficult to solve through simple methods. EAs are intended to provide a better solution, as it is based on the natural theory of evolution. Evolutionary algorithm (EA)-based schemes have been applied to solve subcarrier, bit, and power allocation problems (Wang et al., 2005) (Reddy et al., 2007) (Reddy & Phora, 2007) (Pao & Chen, 2008). In general, chromosomes can be designed with binary, integer, or real representation. The chromosome lengths are related to the number of subcarriers. Each element in the chromosome is a subcarrier allocated to a user. In this research, a subset of subcarriers can be assigned to one user depending upon the availability of subcarriers at a particular time

(Wong et al., 1999a) (Wong et al., 1999b) (Kim et al., 2006) (Wang et al., 2005) (Reddy et al., 2007) (Reddy & Phora, 2007) (Pao & Chen, 2008). In the methods (Wang et al., 2005) (Reddy et al., 2007) (Reddy & Phora, 2007), chromosomes with "good" genes are added in the initial population to improve the convergence rate. This concept is also adopted in the scheme (Pao & Chen, 2008) which generates a chromosome with good genes by employing the CIA method (Wong et al., 1999b) in the proposed ES-based schemes. It is believed that a better initial subcarrier assignment could achieve better performance (Wang et al., 2005) (Reddy et al., 2007) (Reddy & Phora, 2007) (Pao & Chen, 2008). The ES-based scheme (Pao & Chen, 2008) uses the integer representation for the solutions. In this paper, an encoding scheme is discussed about the mapping between the binary-string representation and the integer representation in the solution.

In this paper, a hybrid evolutionary algorithm (HEA) is proposed to solve the subcarrier, bit, and power allocation problem. The HEA is an EA-based approach coupled with a local search algorithm. The concept of a HEA can be found in (Miller et al., 1993) (Kassotakis et al., 2000) (Quintero & Pierre, 2008). A local search algorithm intends to perform optimization locally, while an EA tries to achieve optimum globally. There are two EA-based natural selection schemes (NSS), *NSS-I* and *NSS-II*, presented for the allocation problem and compared in this paper. *NSS-I* is a novel scheme proposed in the paper and *NSS-II* is adapted from (Wang et al., 2005). Every step of these two evolutionary algorithm-based natural selection schemes is addressed in details in this paper. The recombination operation in the proposed *NSS-I* considers the difference between two chromosomes. The recombination operation in *NSS-I* would re-assign a specific subcarrier to different users. A better solution could be found through the proposed natural selection scheme. The similarity and the difference between chromosomes are not considered in the recombination operation in *NSS-II* (Wang et al., 2005). It makes the recombination operation in *NSS-I* more efficient than that in *NSS-II* (Wang et al., 2005). One local refinement strategy is proposed to re-assign the "free" subcarriers to provide better performances, and aims to enhance the convergence rate. The simulation results show that the proposed hybrid evolutionary algorithm (HEA)-based scheme with the integer representation converges fast, and the performance is better and close to that of the optimum solution with the judicious designs of the recombination operation, the mutation operation, and the local refinement strategy. Besides, an adaptive scheme for time-varying channels is also proposed to provide competitive performance with the reduction of the population sizes and the number of generations.

2. System model

Assume the system has K users and N subcarriers. A subset of N subcarriers is assigned to a user, and the number of bits is also determined on downlink transmission. $\{h_{n,k}\}$ denotes the channel gains over all N subcarriers for the k-th user at subcarrier n. The number of bits of the n-th subcarrier assigned to user k is $r_{n,k}$. The required received power f_k at a particular data error rate is a function of bits per symbol $r_{n,k}$. It is not allowed a subcarrier is shared among different users. Therefore, we define

$$\rho_{n,k} = \begin{cases} 1, & if\ r_{n,k} \neq 0 \\ 0, & if\ r_{n,k} = 0 \end{cases} \tag{1}$$

The variable $\rho_{n,k}$ is either 1 or 0, and the sum of all $\rho_{n,k}$ is equal to 1 for any particular n. The required transmit power can be expressed as

$$P = \sum_{n=1}^{N} \sum_{k=1}^{K} \frac{f_k\left(r_{n,k}\right)}{h_{n,k}^2} \rho_{n,k} \tag{2}$$

Data rates $\{R_1, R_2, ..., R_K\}$ are predetermined parameters for each user. The bit error rate must be ensured at a certain level to meet the service quality. The subcarrier, bit and power allocation problem for the minimization of total transmit power is formulated as

$$\underset{r_{n,k}, \rho_{n,k}}{Min} \quad \sum_{n=1}^{N} \sum_{k=1}^{K} \frac{f_k\left(r_{n,k}\right)}{h_{n,k}^2} \rho_{n,k}$$

$$subject\ to\ \sum_{k=1}^{K} \rho_{n,k} = 1,\ for\ n = 1,2,...N \text{ and } \sum_{n=1}^{N} r_{n,k} = R_k,\ for\ k = 1,2,...,K \tag{3}$$

This nonlinear optimization problem could be solved by employing integer programming (IP) (Kim et al., 2006). Therefore, the formulation to achieve the optimal bound for the performance comparison is briefly summarized as follows (Kim et al., 2006). If $r_{n,k} \in \{0,1,...,M\}$, then

$$f_k\left(r_{n,k}\right) \in \{0, f_k(1), ..., f_k(M)\} \tag{4}$$

A new indicator variable $\gamma_{n,k,r}$ is defined as

$$\gamma_{n,k,r} = \begin{cases} 1, & if\ \rho_{n,k} = 1 \text{ and } r_{n,k} = r \\ 0, & \text{otherwise} \end{cases} \tag{5}$$

where $r \in \{0,1,...,M\}$. For a particular subcarrier n, the value of $\gamma_{n,k,r}$ must be 0 or 1. $\gamma_{n,k,r}$ is an indicator. $\gamma_{n,k,r} = 1$ means that the nth subcarrier is assigned to the kth user with the rth modulation. $\sum_{r=0}^{M} \gamma_{n,k,r} = 1$ means that only one type of modulation is employed for that user. For other users, $\sum_{r=0}^{M} \gamma_{n,k',r} = 0$ because the nth subcarrier is not assigned to the other users. $\sum_{r=0}^{M} \gamma_{n,k,r}$ is either 1 or 0. Therefore, when more than one subcarrier (by observing different n) is assigned to one user, $\sum_{r=0}^{M} \gamma_{n,k,r}$ is either 1 or 0. The problem formulation is presented for the use of the branch-and-bound to solve the problem. It follows that (4) is rewritten as

$$f_k\left(r_{n,k}\right) = \sum_{r=0}^{M} \gamma_{n,k,r} f_k(r) \tag{6}$$

The required transmit power (2) can be re-expressed as

$$P = \sum_{n=1}^{N} \sum_{k=1}^{K} \left\{ \sum_{r=0}^{M} \gamma_{n,k,r} \frac{f_k(r)}{h_{n,k}^2} \right\} \rho_{n,k} \tag{7}$$

The indicators $\gamma_{n,k,r}$ and $\rho_{n,k}$ are related as

$$\rho_{n,k} = \sum_{r=0}^{M} \gamma_{n,k,r} \tag{8}$$

and we have

$$\gamma_{n,k,r} \cdot \rho_{n,k} = \gamma_{n,k,r} \ . \tag{9}$$

(7) is rewritten as a linear cost function:

$$P = \sum_{n=1}^{N} \sum_{k=1}^{K} \sum_{r=0}^{M} \gamma_{n,k,r} \frac{f_k(r)}{h_{n,k}^2} \tag{10}$$

A linear integer programming-based branch-and-bound algorithm (Wolsey, 1998) can be employed to solve this problem. In general, integer programming is a full-search approach which needs exponential time. Therefore, we proposed a suboptimal scheme to solve the subcarrier, bit, and power allocation problem for OFDMA systems.

3. Encoding schemes

One encoding scheme is required in the implementation of the EA-based schemes. Here, an encoding scheme is presented for the mapping of the possible solutions between the binary-string representation and the integer representation. In the proposed scheme, the chromosome lengths are related to the number of subcarriers. Each element in the chromosome contains a user index which means a subcarrier allocated to the user. A subset of subcarriers can be assigned to a user.

First, we introduce the encoding scheme with integer representations. Referring to Table 1, each chromosome represents one solution. The chromosome with N elements denotes a subcarrier assignment solution to the optimization problem. Each element in the chromosome represents a subcarrier assignment, and its value is coded as an integer at the range of 1 to K that stands for an index of a user. For example, the first element in the first chromosome is 2. It means that the 1st subcarrier is assigned to the 2nd user. The union of these chromosomes is called the population as shown in Table 1. The chromosome lengths of N are equal to the number of total subcarriers, and are not varied for different numbers of users. The size of the population should make the possible solutions diverse enough for finding the optimal solution.

N subcarriers

	1st	2nd	3rd	4th	5th	6th	7th	Nth
Chromosomes	2	8	5	1	4	2	3	…
	1	3	6	2	2	4	1	…
	6	1	5	6	2	3	…	…
	4	…	…	…	…	…	…	…
	….	…	…	…	…	…	…	…

Table 1. The integer representation

In general, chromosomes can be designed with binary, integer, or real-valued representation. Encoding methods are discussed in details in (Coley, 2003). Referring to Table 1, we assume there are 8 users in the example. The number of bits for encoding the user index into the binary-string representation should be sufficient to fulfill the presentation needs associated with the number of users. For example, the maximum number of users which can be represented with 3 encoding bits is $2^3 = 8$.

User	1	2	3	4	5	6	7	8
Binary	000	001	010	011	100	101	110	111

Table 2. A typical example of the encoded binary-string representations for 8 users

Table 3 is an example of a three-bit encoding scheme for the mapping between the binary-string representation and the integer representation. By taking the first chromosome in Table 1 for example, the binary-string representation based on Table 2 is obtained as:

Subcarrier	1st	2nd	3rd	4th	5th	6th	7th	Nth
Real	2	8	5	1	4	2	3	...
Binary	001	111	100	000	011	001	010	...

Table 3. A typical example of a three-bit encoding scheme

The encoded binary-strings are equally distributed among users. One or more than one encoded binary-string representations may be mapped to a particular user. Table 4 shows an example when the number of user is less than the number of the combinations with 3 encoding bits. Referring to Table 4, "000" and "001" stand for user 1; "010" and "011 stand for user 2; "100" and "101" stand for user 3; "110" and "111" stand for user 4. If each user has the equal number of the encoded binary-string representations, then it means each user has equal subcarrier assignment probability during the iteration process. Under some circumstances, each user would not have the same assignment probability, such as 6 users with 3 bits to encode. If the length of the binary-string associated with the number of bits is not long enough, the unequal assignment probability would happen. Long binary-string encoding is preferred with this mapping method. Other mapping method can also be designed to avoid this requirement of the long binary-string encoding such as the modulus operation.

User	1		2		3		4	
Binary	000	001	010	011	100	101	110	111

Table 4. A typical example of the encoded binary-string representations for 4 users

Note that the number of encoding bits is predetermined and related to the number of users. Both representations for the solution are able to be utilized in the following proposed EA-based allocation schemes.

4. Proposed evolutionary algorithm-based allocation schemes

The evolutionary algorithm (EA) (Back, 1996) (Spears, 2000) is proposed to search optimum solutions, which can be traced to at least the 1950s. Two typical approaches in evolutionary computing methodologies are "evolutionary strategy (ES)" and "genetic algorithm (GA)".

GA emphasizes on recombination process; ES makes use of both mutation and recombination procedures. The basic structure of the evolutionary algorithm consists of four operations, including selection, recombination, mutation, and fitness evaluation.

Fig. 1 illustrates the processing block diagram of the proposed HEA-based scheme in this paper. The scheme provided in Table 1 is one subcarrier for one gene to construct a chromosome and procedures are followed in all papers (Wang et al., 2005) (Reddy et al., 2007) (Reddy & Phora, 2007) (Pao & Chen, 2008). The encoding design can be integer, binary, or real-value. The subcarrier allocation method of Wong et al. (Wong et al., 1999b) is utilized to create one chromosome with good genes in the initial population, where the numbers of subcarriers are predetermined for all users. In this paper, one subcarrier for one gene and random allocation creates the other chromosomes in the initial population. The predetermined numbers of subcarriers for users would be adjusted while performing the proposed algorithm.

There are two natural selection schemes for the allocation problem presented in this paper. Every component of these two schemes is discussed in details as the following.

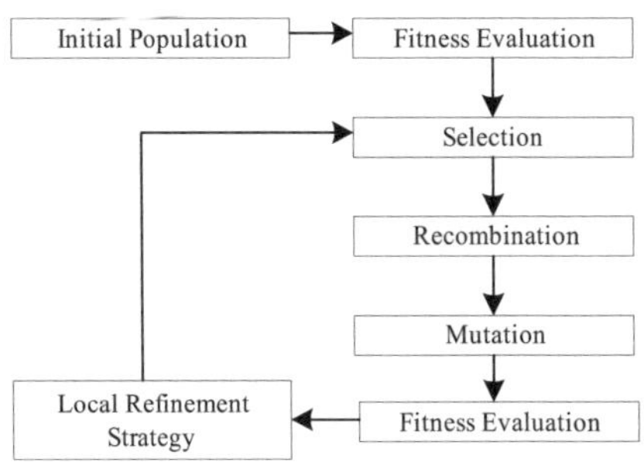

Fig. 1. Block diagram of the hybrid evolutionary algorithm-based scheme

A. Natural Selection Scheme I (NSS-I)

The natural selection scheme (*NSS-I*) is proposed for solving the allocation problem with the judicious designs of the recombination operation, the mutation operation, and the local refinement strategy.

A.1 Initialization of the Population

The initial population of size A_1 is composed of one chromosome generated by the CIA procedure (Wong et al., 1999b) and A_1-1 chromosomes created by randomization. The major merit of the CIA procedure is to select subcarriers with better channel conditions for the users according to the predetermined number of subcarriers per user. According to the CIA procedure (Wong et al., 1999b), the numbers of subcarriers assigned to users have to be determined in the beginning. However, in the problem under consideration, the optimum number of assigned subcarriers for each user is unknown. The numbers of subcarriers assigned to users are equal in the initialization phase. Note that the numbers of the subcarriers are temporarily preset in the initial population and they would be adjusted as the number of the generations increases for approaching those of the optimal solution. It does not mean the numbers of subcarriers for the final solution have to be determined in the beginning with the proposed scheme. Table 1 shows a typical example of the integer representation in the EA-based allocation scheme.

A.2 Selection

The fitness function is defined as the transmit power evaluated by a bit loading approach. After sorting, μ chromosomes associated with the μ largest values of fitness evaluation are selected from the population of size A_1, and saved in the mating pool, while the rest $A_1 - \mu$ chromosomes are deleted. The number of chromosomes μ that are kept in each generation is fixed. The new offsprings of size $A_2 = A_1 - \mu$ will be generated in the next step. The population size of $A_1 = \mu + A_1 - \mu$ to evaluate the fitness is not changed for each generation. Besides, the best one chromosome is kept independent at each generation in order to hold the convergence of the solutions. If the current best chromosome is superior to the one in the last generation, the best chromosome is replaced. The problem under consideration is to find the allocation solution which minimizes the transmit power. Due to the property of the minimization problem in communications with wireless channels under consideration, the scheme keeps one best chromosome to make the solution continue to improve in the sense of the minimization of power.

A.3 Recombination

Two chromosomes (x_f, x_m) are randomly and iteratively selected from the mating pool of μ chromosomes to produce two new offsprings (x'_f, x'_m) (Back, 1996) (Siu et al., 2005). The differences between genes are utilized in the recombination procedure. The rule can be represented as

$$x'_f \leftarrow x_f + g_r k(x_f - x_m) \text{ and } x'_m \leftarrow x_m + g_r k(x_f - x_m) \tag{11}$$

where the step size g_r has the inequality $0 < g_r \le 1$. The coefficient k is defined as

$$k = \begin{cases} 1, \ \mathit{if} \ U(0,1) < p_c \\ 0, \ \mathit{if} \ U(0,1) > p_c \end{cases} \tag{12}$$

where $U(0,1)$ is a value generated by a random number generator with the uniform distribution of [0,1]. From equation (11), some elements in chromosomes x_f and x_m would be changed if the genes are different. That is, a specific subcarrier is assigned to different users. The parameters, g_r and p_c, are used to change the value of the element in the chromosome, i.e. to re-assign a specific subcarrier to another user. A better solution could be found through the natural selection scheme. Note that elements in (x'_f, x'_m) are rounded off to integers. The μ chromosomes and the new offsprings of size $A_2 = A_1 - \mu$ are merged to be a set of A_1 chromosomes for mutation.

A.4 Mutation

The concept of mutation is to prevent the solution trapped into a local minimum. Mutation points are randomly selected from the population of size A_1. The mutation probability P_m is defined by the fuzzy logic technique that provides an effective concept for dealing with the problem of uncertainty and imprecision (Lee, 1990). The fuzzy logic controller (FLC) is thus utilized based on this concept to provide the adaptive mutation probability during the optimization process. Average variance alleles (AVA) (Herrera et al., 1994) are used to

determine the mutation probability. An AVA is employed to measure the difference among chromosomes by monitoring the population diversity and correlation. An AVA is a dynamic parameter in the process of the EA-based natural selection schemes. The mutation probability P_m is defined by the fuzzy logic controller (FLC) with the AVA parameter as following:
If the current AVA is greater than the averaged AVA, the mutation probability should increase and vise versa. AVA is calculated per each generation. AVA is low when the A_1 chromosomes are quite similar. The adaptive mutation probability adjustment scheme is given as the following:

$$P_m = \begin{cases} P_m + \delta \dfrac{AVA}{\overline{AVA}}, & if\ AVA \geq \overline{AVA} \\ P_m - \delta \dfrac{\overline{AVA}}{AVA}, & if\ AVA < \overline{AVA} \end{cases} \tag{13}$$

where $AVA = \sum_{j=1}^{N} \sum_{i=1}^{A_1} \left(\overline{S}_i - \overline{S}_j \right)^2 / NA_1$, $\overline{S}_i = \sum_{j=1}^{N} S_{ij} / N$, and $\overline{S}_j = \sum_{i=1}^{A_1} S_{ij} / A_1$; "$-$" denotes an average value. S, i.e. the population of size A_1, is an A_1-by-N sample matrix whose element has the sample value $S_{i,j}$. δ is a constant to control the amount of the change in the mutation probability for every generation. The mutation probability P_m is held in the range of $\left[P_m^{Lower}, P_m^{Upper} \right]$. If the instant probability P_m is larger than the upper mutation probability P_m^{Upper} or smaller than the lower mutation probability P_m^{Lower}, the current probability P_m is reset to the initial value to continue the process. The number of mutation points is $P_m \cdot A_1$, which is rounded off to an integer. The mutation points are randomly selected from the population of size A_1. The values of the elements for the selected mutation points are randomly changed to an integer value at the range of 1 to *K*. The mutated population of size A_1 together with the best one chromosome is merged to have the population of size $A_1 + 1$.

A.5 Fitness Evaluation

The fitness is defined as the required transmit power. In the evaluation of the fitness for each chromosome in the population of $A_1 + 1$, a bit loading approach by using the water-filling technique is employed. It is the optimal algorithm to load bits in a single-user environment (Hughes-Hartogs, 1989) (Lai et al., 1999). As each chromosome has the solution of the subcarrier assignment for each user, the water-filling technique can be applied to solve the bit loading problem. After the fitness evaluation, one chromosome with the worst fitness is deleted if the best chromosome is updated. If the best chromosome is not replaced, the one identical to the best chromosome in the population of size A_1 is deleted. The population size is still A_1 for the next generation.
Finally, we iteratively perform the previous steps until the algorithm reaches the convergence condition.

B. Natural Selection Scheme II (NSS-II)

The second natural selection scheme is adapted from (Wang et al., 2005) with further investigation and detailed discussions. The performance will be compared with that of the proposed *NSS-I* in the simulation results. The related recombination and the mutation operations are addressed below.

B.1 Initialization of the population

The scheme presented in (Wang et al., 2005) has suggested that good genes should be added in the initial population to improve the convergence rate. For fair comparisons, the initial populations of the size B_1 are also composed of one chromosome generated by the CIA procedure (Wong et al., 1999b) and $B_1 - 1$ chromosomes generated by the randomized assignment. The representation for the solution is the same as in Table 1 with integer-values.

B.2 Selection

After sorting, B_2 chromosomes associated with the B_2 largest values of the fitness evaluation are selected from the population of size B_1. B_2 is the population size of the selected chromosomes. The remaining $B_1 - B_2$ chromosomes are saved in the mating pool. Besides, the best one chromosome is also kept independently in order to hold the convergence of solutions.

B.3 Recombination

One half of the $B_1 - B_2$ chromosomes are selected as parents by the following procedure, and then the two-point crossover by the rank weighting (Spears, 2000) is performed in our implementation. The first step is to give a random value for each chromosome. After sorting

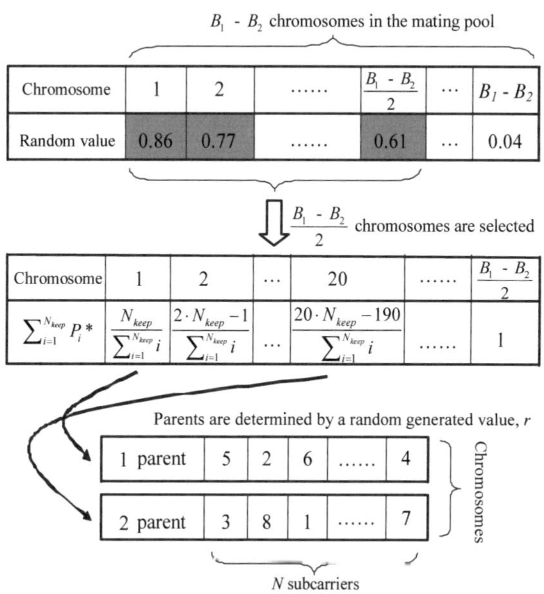

Fig. 2. Illustration of the recombination operation

the random values associated with the chromosomes, the first half of the $B_1 - B_2$ chromosomes are selected as parent chromosomes while the others are deleted. After choosing chromosomes, the rank weighting method is to produce a probability P_i^* for each chromosome and the cumulative probability is obtained by $\sum_{i=1}^{N_{keep}} P_i^*$. The chromosomes are selected from the mating pool to produce new offsprings. Here is the definition of the probability for each chromosome, P_i^*:

$$P_i^* = \left(N_{keep} - i + 1\right) \Big/ \sum_{i=1}^{N_{keep}} i \tag{14}$$

Where i stands for an index of a chromosome; N_{keep} is the value of $(B_1 - B_2)/2$. A randomly generated value, *r*, is to check if *r* is between P_i^* and P_{i+1}^*. If it is true, chromosome i associated with P_i^* is selected as one parent. The above process of generating a random number *r* is repeated in order to find a pair of chromosomes. Figure 2 illustrates the procedure of the parent chromosomes' selection. The two chromosomes are selected as parents to perform the two-point crossover. There are $(B_1 - B_2)/2$ offsprings generated. The recombination operation in *NSS-I* considers the similarity and the difference between two chromosomes. It makes *NSS-I* more efficient than the recombination operation in *NSS-II*. The effect can be observed in the simulation.

B.4 Mutation

The concept of the mutation is to prevent the solution trapped into a local minimum. The mutation points are randomly selected from the merged population of size B_1, which is composed of the B_2 chromosomes, the $(B_1 - B_2)/2$ parent chromosomes, and the $(B_1 - B_2)/2$ new offsprings. The population size in the mutation step is $B_1 = B_2 + \frac{B_1 - B_2}{2} + \frac{B_1 - B_2}{2}$.

The mutation probability P_m^* is adjusted adaptively. The mutation probability is increased if no better chromosomes are found in the consecutive generation. The parameters are set in the same manner as (Wang et al., 2005). The mutation probability P_m^* is held in the range of $\left[P_m^{Lower*}, P_m^{Upper*}\right]$. The mutated population of size B_1 together with the best one chromosome is the population of size $B_1 + 1$.

B.5 Fitness Evaluation

Before the evaluation of the fitness for each chromosome, the best one chromosome in the last generation is added. Again, in the evaluation of the fitness for each chromosome in the population of size $B_1 + 1$, the water-filling technique (Hughes-Hartogs, 1989) (Lai et al. 1999) is used to load bits to subcarriers. After the fitness evaluation, one chromosome with the worst fitness is deleted if the best chromosome is updated. If the best chromosome is not replaced, the one identical to the best chromosome in the population of size B_1 is deleted. The population size is B_1 for the next generation. Finally, we repeat the previous steps until the algorithm achieves the convergence condition.

C. Local Refinement Strategy

Referring to Fig. 1, the local refinement strategy (LRS) is performed after the fitness evaluation. For a given solution associated with a chromosome, some subcarrier assigned to a user may not be employed if the situation of no modulation (no transmission) on that subcarrier occurs. It means that the channel condition of this specific subcarrier is relatively bad for the user. The specific subcarrier is set to be free in our proposed approaches. The step of LRS is proposed to enhance the performance by re-allocating the free subcarrier to another user. It is to assign the specific subcarrier to a user who provides the maximum total transmit power reduction. After the operation of the bit loading for the chromosomes, then we check if there is a free subcarrier $\tilde{n}$ which has been assigned to user $\tilde{k}$. The free subcarrier is assigned to user k which results in the maximum required total transmit power reduction. After the subcarrier re-assignment operation, the process continues until there is no free subcarrier.

D. Adaptive Scheme

As wireless channels vary gradually but slowly for this type of allocation problems, the channel fading condition could be similar in adjacent allocation time slots. Based on this concept, the solutions obtained in adjacent allocation time slots could be similar. Owing to the property of wireless channels, the final solution of the previous processing time slot may be included in the initial population for the current time slot allocation process to speed up the convergence rate. If a better solution is applied in the initial population, the evolutionary algorithm-based schemes could converge faster and the population size could be reduced in the adaptive manner. The computational time would be further improved because of less possible candidate solutions and less generations to provide competitive performance. With the adaptive manner, the proposed algorithm may be performed with a large number of generations and a large population size in the first allocation time slot. In the followed consecutive allocation time slots, the final solutions in the last allocation time slot are utilized as a portion of the initial population in the current allocation time slot. As a smaller size of population and a smaller number of generations are employed to perform the algorithm to obtain the final solution, the computation cost can be reduced.

5. Simulation results

The frequency selective wireless channel model used in (Dong et al., 2001) is adopted. Perfect channel estimation is assumed. Each user has the same requested data rate. The channel power of the received signal for each user is varied because of the various path losses at the different locations. The set of the switching levels in (Torrance & Hanzo, 1996) for the modulation types is employed. All the following experiments are conducted under the channel conditions with the same statistics. The simulation parameters are listed in Table 5. In the initialization of the population, the numbers of subcarriers assigned to users are equal. The subcarriers are assigned to users by the CIA procedure (Wong et al., 1999b) and the randomized assignment. For fair comparisons, the sizes A_1 and B_1 of the initial populations are the same. The ratio $\mu / A_1 \approx 1/7$ is selected in our proposed scheme according to a selection mechanism (Siu et al., 2005). The mutation probability in the *NSS-I* scheme can be set small because the new offspring of size A_2 are formed by μ chromosomes with higher fitness. The parameters for *NSS-II* are set in the same manner as (Wang et al., 2005). In the simulation, the chromosome lengths are equal to the number of total subcarriers. Each element in the chromosome contains a user index which means a subcarrier allocated to the user. The chromosome length is fixed. Table 3 is an example of a 3-bit encoding scheme for the binary-string representation. By taking the first chromosome in Table 1 for example, the binary-string representation is obtained by using Table 2. This is the way we encode the chromosomes in the simulation.

The simulation results are displayed for the performance comparison among various schemes including the Wong's subcarrier allocation (SA) (Wong et al., 1999b) algorithm plus the optimum bit loading with the equal numbers of subcarriers for each user. The Wong's subcarrier allocation (SA) algorithm (Wong et al., 1999b) includes the constructive initial assignment (CIA) and the subcarrier swapping two steps to provide sub-optimal performance. CIA, which is adopted as an initial subcarrier allocation method for one chromosome, needs to pre-assign the numbers of subcarriers for each user. It is temporarily set for the proposed scheme and will be adjusted during the process. The optimal allocation solutions as the performance bound are obtained by using the linear integer programming-based branch-and-bound algorithm (BnB) (Kim et al., 2006) (Wolsey, 1998). In the following simulation, 8-bit encoding is employed for the schemes with the binary-string representation.

Number of users	3, 4, 6, 8
Modulation types	0, QPSK, 16QAM, 64QAM
Symbol rate per subcarrier	250k symbols/s
Total Allowed Data rate	36 Mbits/s
Number of subcarriers	48
Center frequency	5 GHz
Frame duration	10 ms
$\{A_1, A_2, \mu, B_1, B_2, g_r, p_c\}$	$\{36, 31, 5, 36, 6, 0.5, 0.95\}$
$\{P_m^{Upper*}, P_m^*, P_m^{Lower*}, P_m^{Upper}, P_m, P_m^{Lower}, \delta\}$	$\{0.5, 0.1, 0.1, 0.05, 0.025, 0.01, 0.001\}$

Table 5. Parameters of Simulations

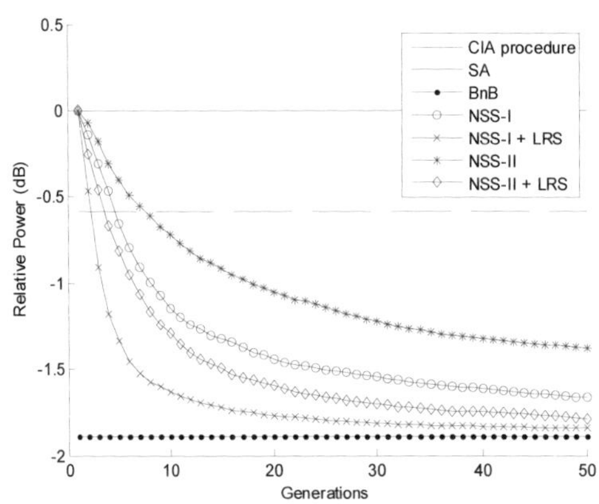

Fig. 3. Convergence comparison in terms of required transmit power for the proposed EA-based scheme with the real-valued representation for 6 users, the largest channel power difference among users = 30dB, and target BER=10^{-2}

The local refinement strategy, labeled *LRS*, is used in both EA-based schemes, *NSS-I* and *NSS-II* (Wang et al., 2005), with the integer representation. The local refinement strategy is proposed to enhance the convergence rate to the optimum solution. The convergence curves are plotted by averaging 100 trial runs with 50 generations to show the performance improvement of LRS in Fig. 3. The relative required transmit power for the best one chromosome is reduced as the number of generations increases because the best chromosome is kept independently. The convergence rates of *NSS-I* + *LRS* and *NSS-II* + *LRS* are faster than those of *NSS-I* and *NSS-II,* respectively. The simulation results in Fig. 3 demonstrate that the LRS truly improves the performance, and the performance of *NSS-I* is better than that of *NSS-II*. The result also shows that the proposed scheme, *NSS-I* + *LRS,* with the integer representation converges fast and is close to the optimum performance. After 50 generations, the performance of *NSS-I+LRS* is very close to that of *BnB*. By taking about 0.1 db point away from the BnB performance as an example, the number of the generations with the proposed scheme is about 25 generations. On the other hand, the other schemes require more than 50 generation to achieve that performance.

In Figs. 4 to 7, there are eight allocation schemes for performance comparison in the simulation, including five EA-based schemes, one suboptimal scheme (Wong et al., 1999b) (Wong et al., 1999b), the linear IP-based branch and bound scheme for the optimum performance bound (Kim et al., 2006), and an up-to-date genetic algorithm-based scheme (Reddy & Naraghi-Pour, 2007):**(1)** Scheme I: *NSS-II* + 8 bits encoding; **(2)** Scheme II: *NSS-II* + *LRS* + 8 bits encoding; **(3)** Scheme III: *NSS-II* + The integer encoding; **(4)** Scheme VI: *NSS-II* + *LRS* + The integer encoding; **(5)** Scheme V: *NSS-I* + *LRS* + The integer encoding; **(6)** SA (Wong et al., 1999b); **(7)** BnB (Kim et al., 2006); **(8)** GA (Reddy & Naraghi-Pour, 2007).

The simulation results are averaged over 100 trial runs with 200 generations and are shown in Figs. 4 to 7 for different channel power differences, different target BERs, and different numbers of users. Figs. 4 to 7 reveal the similar performance trend can be observed in the results. The detailed comments regarding the performance comparison are as follows:

1. The difference between Scheme I and Scheme III is the solution presentation. The performance of Scheme III with the integer representation is better than that of Scheme I with the binary-string representation at the same generation. It also reveals that the scheme with the integer representation outperforms the scheme with the binary-string representation. The EA-based schemes with the integer representation are more suitable for solving the resource allocation problems. Even if the local refinement strategy is employed, similar results with the effect of the representation can be observed (Scheme II and VI).
2. The effects of the design in the recombination and the mutation operations are also investigated. The performances of Scheme VI and Scheme V indicate that the recombination and the mutation operations affect the performance. It reveals that the proposed recombination operation and the mutation operation can achieve better performances. The performance of the proposed Scheme V is very close to that of the optimum solution.
3. The proposed local refinement strategy can improve the performance. In addition to the convergence rate improvement, Scheme II and Scheme VI outperform Scheme I and Scheme III in terms of transmit power. The performances of the proposed Scheme V are even closer to those of the optimum solutions.
4. Besides the fast convergence feature demonstrated in Fig. 3, the simulation results in Figs. 4 to 7 show that the performances of the proposed HEA-based scheme with the integer representation, Scheme V, are close to those of the optimum solutions with the judicious designs of the recombination operation, the mutation operation, and the local refinement strategy.
5. The GA-based scheme (Reddy & Naraghi-Pour, 2007) outperforms Scheme I and Scheme III in terms of transmit power when a fewer number of users is considered. When the proposed local refinement strategy is employed to Scheme I and Scheme III, i.e. Scheme II and Scheme IV, their performances are greatly improved and better than those of GA-based scheme (Reddy & Naraghi-Pour, 2007). The major proposed hybrid evolutionary algorithm-based scheme, Scheme V, performs best and the performances are the closest to those of the optimum solutions in these simulations.

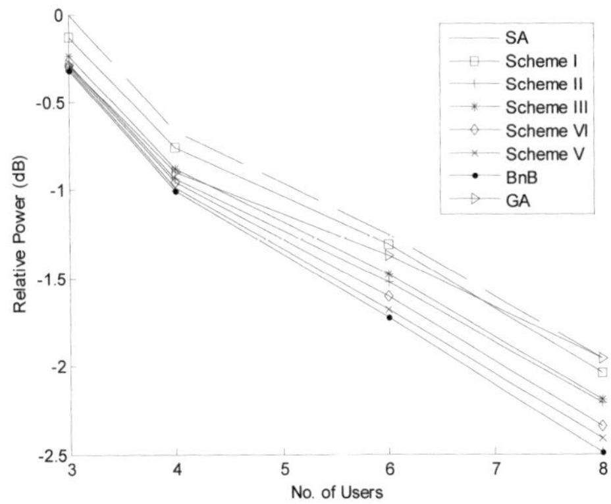

Fig. 4. Performance comparison in terms of required transmit power among eight allocation schemes, the largest channel power difference among users = 15dB, and target BER= 10^{-2}

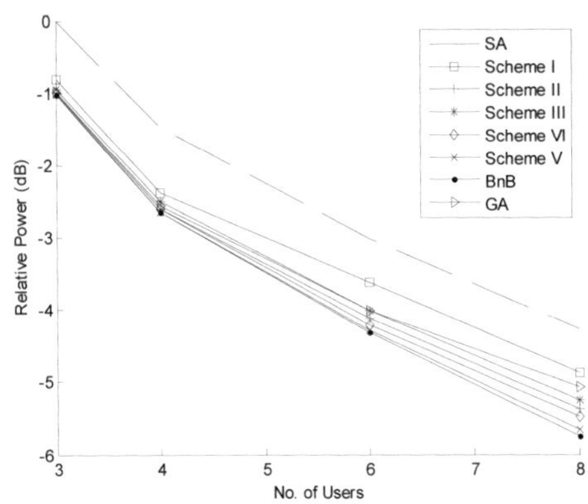

Fig. 5. Performance comparison in terms of required transmit power among eight allocation schemes, the largest channel power difference among users = 30dB, and target BER= 10^{-2}

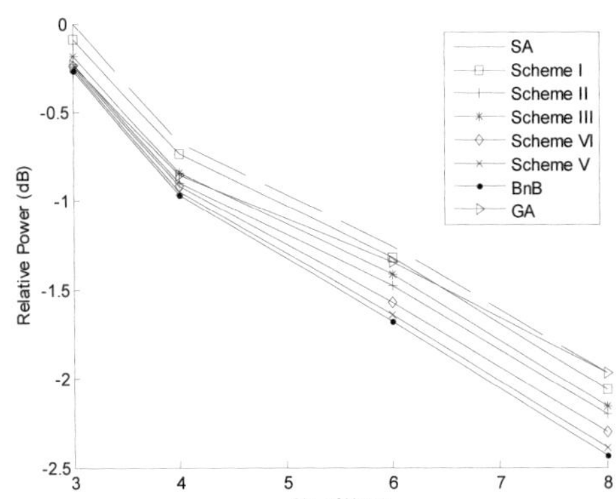

Fig. 6. Performance comparison in terms of required transmit power among eight allocation schemes, the largest channel power difference among users = 15dB, and target BER= 10^{-2}

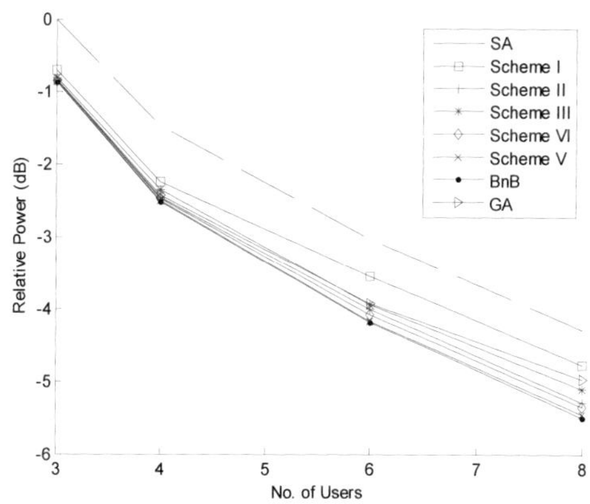

Fig. 7. Performance comparison in terms of required transmit power among eight allocation schemes, the largest channel power difference among users = 30dB, and target BER= 10^{-2}

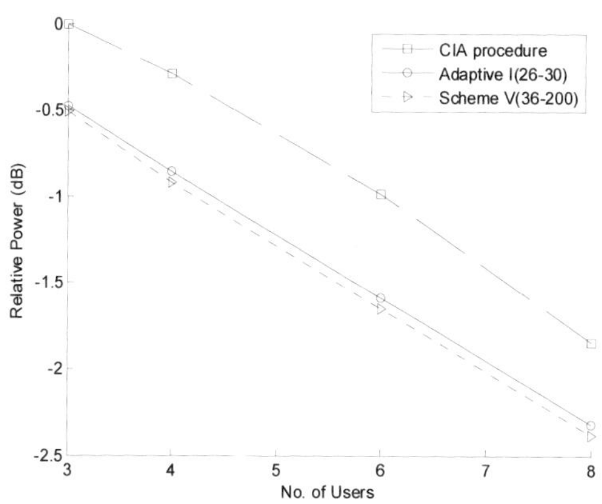

Fig. 8. Performance comparison with the adaptive scheme in terms of required transmit power at mobile speed=1km/h, the largest channel power difference among users = 15dB, and target BER= 10^{-2}

The simulations are conducted to demonstrate the efficacy of the adaptive scheme which can reduce the number of the generations and the population size to provide competitive performance in various adaptive situations. The mobile speed is 1km/h. The allocation period is on a per frame basis. The scheme labeled "Adaptive I" considers that the final solutions of the previous processing frame are included in the initial population for the current frame allocation process. The proposed scheme labeled "Scheme V" is compared to demonstrate the efficacy of the adaptive scheme in the reduced computational cost to provide competitive performance. In the figures, the two parameters (a-b) inside the parenthesis denote the population size and the number of generations. The population size A_1 =26; 30 generations are executed; 5 best solutions obtained after 200 generations in the previous frame are included in the initial population. As performed with fewer generations and a smaller population size, there is performance degradation between Adaptive I and Scheme V. However, the performance degradation is less than 0.1 dB in Fig.

8. Fig. 9 reveals the similar performance trend but even closer to that of the Scheme V, where the largest channel power difference among users is increased.

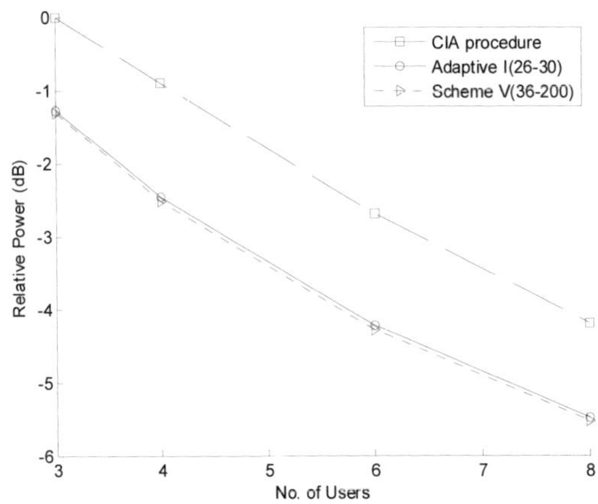

Fig. 9. Performance comparison with the adaptive scheme in terms of required transmit power at mobile speed=1km/h, the largest channel power difference among users = 30dB, and target BER= 10^{-2}

In Figs. 10-11, 100 ms is the duration to perform the algorithm. The scheme "Adaptive II" obtains the solution for the first frame with 200 generations. The algorithm is performed with the smaller number of generations and the smaller size of population for the followed consecutive 9 frames. As indicated in the simulation results, the adaptive scheme tracks the channel variations well and provides very competitive performance compared to the scheme with the full number of generations = 200 and the full size of population = 36 of scheme V.

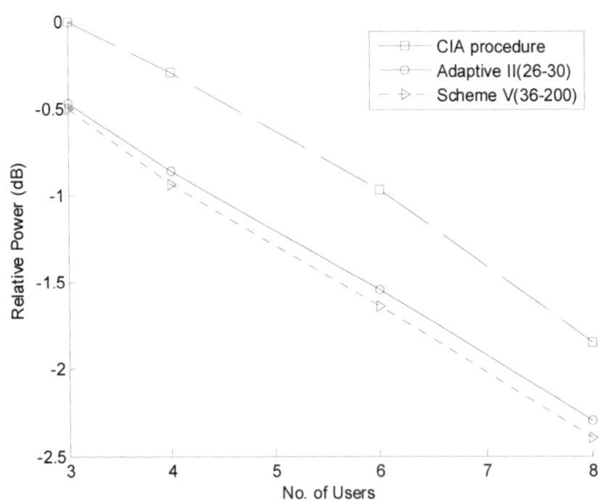

Fig. 10. Performance comparison to demonstrate the tracking capability in terms of required transmit power at mobile speed=1km/h, the largest channel power difference among users = 15dB, and target BER= 10^{-2}

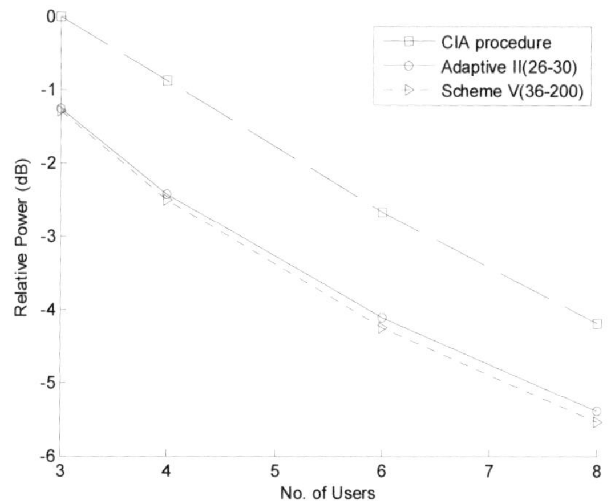

Fig. 11. Performance comparison to demonstrate the tracking capability in terms of required transmit power at mobile speed=1km/h, the largest channel power difference among users = 30dB, and target BER= 10^{-2}

In the next simulation to demonstrate the tracking capacity with the scheme, "Adaptive III", the frame duration is switched to 5 ms which is employed in WIMAX standard and a carrier frequency of 1.95 GHz in the PCS (Personal Communication Services) band is adopted. The channel duration for the simulation is 1 second. The algorithm is performed with 200 generations for the solution of the first frame while being performed with 30 generation for those of the rest of the frames to provide the solutions. As displayed in Figs. 12-13, the performances are very close to those of Scheme V with the full number of generations and the full size of population per each frame operation.

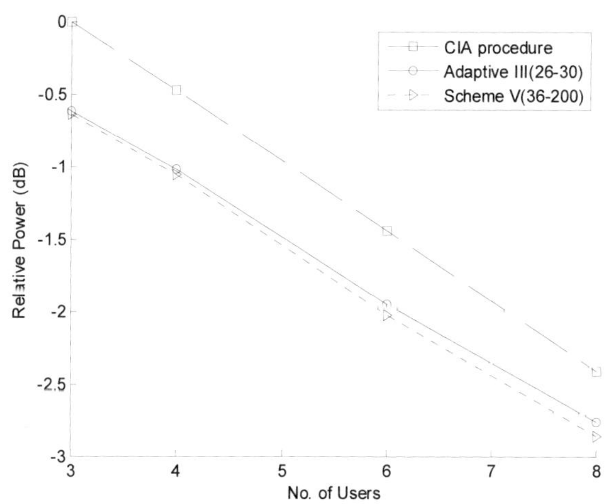

Fig. 12. Performance comparison to demonstrate the tracking capability in terms of required transmit power at mobile speed=1km/h, the largest channel power difference among users = 15dB, and target BER= 10^{-2}. Frame duration = 5 ms, carrier frequency = 1.95 GHz

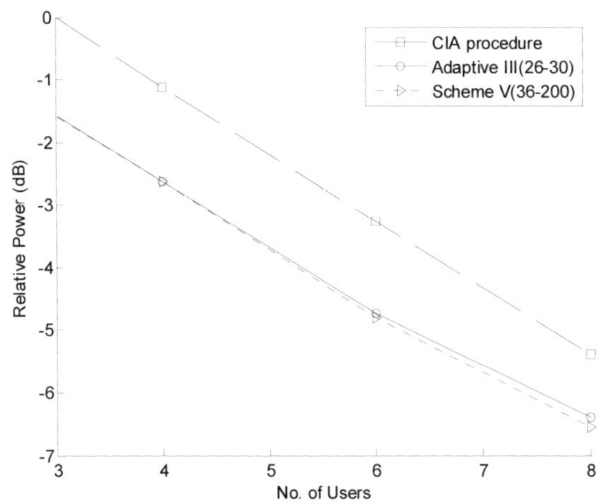

Fig. 13. Performance comparison to demonstrate the tracking capability in terms of required transmit power at mobile speed=1km/h, the largest channel power difference among users = 30dB, and target BER= 10^{-2}. Frame duration = 5 ms, carrier frequency = 1.95 GHz

6. Conclusion

This paper proposes a hybrid evolutionary algorithm-based scheme to solve the subcarrier, bit, and power allocation problem. The hybrid evolutionary algorithm is an evolutionary algorithm-based approach coupled with a local refinement strategy. It is presented to improve the performance and offers the faster convergence rate. Simulation results show that the proposed hybrid evolutionary algorithm-based scheme with the integer representation converges fast, and the performance is close to that of the optimum solution with the judicious designed of the recombination operation, the mutation operation, and the local refinement strategy. An adaptive scheme for time-varying channels is also proposed to obtain the solution having competitive performance with the reduction of the population sizes and the number of generations.

7. References

Back, T. (1996). Evolutionary Algorithm in Theory and Practice: Evolution Strategies, Evolutionary Programming, Genetic Algorithm, Oxford University Press, ISBN: 0-19-509971-0.

Coley, D. A (2003). *An Introduction to Genetic Algorithms for Scientists and Engineers*, World Scientific, ISBN: 981-02-3602-6.

Dong, L.; Xu, G. & Ling, H. (2001). Prediction of fast fading mobile radio channels in wideband communication systems, *Proceedings of IEEE Global Telecommunications Conf.*, pp. 3287-3291, ISBN: 0-7803-7206-9, Nov. 2001.

Herrera, F.; Herrera-Viedma, E.; Lozanzo, M. & Verdegay, J. L. (1994). Fuzzy tools to improve genetic algorithms, *Proceedings of Second European Congress on Intelligent Techniques and Soft Computing*, pp. 1532-1539, 1994.

Hughes-Hartogs, D. (1989). Ensemble modem structure for imperfect transmission media, U.S. Patents Nos. 4,679227 (July 1987), 4,731,816 (March 1988) and 4,833,796 (May 1989).

Kassotakis, I. E.; Markaki, M. E. & Vasilakos, A. V. (2000). A hybrid genetic approach for channel reuse in multiple access telecommunication networks, *IEEE J. Selected Areas in Communications*, Vol. 18, No. 2, pp. 234-243, ISSN: 0733-8716.

Kim, I.; Park, I.-S. & Lee, Y. H. (2006). Use of linear programming for dynamic subcarrier and bit allocation in Multiuser OFDM, *IEEE Trans. on Vehicular Technology*, Vol. 55, No. 4, pp. 1195-1207, ISSN: 0018-9545.

Lai, S. K.; Cheng, R. S.; Lataief, K. B. & Murch, R. D. (1999). Adaptive trellis coded MQAM and power optimization for OFDM transmission, *Proceedings of IEEE Vehicular Technology Conf.*, pp. 290-294., ISBN: 0-7803-5565-2, May 1999.

Lee, C. C. (1990). Fuzzy logic in control system: fuzzy logic controller. I, *IEEE Trans. on Systems, Man and Cybernetics*, Vol. 20, No. 2, pp. 404-418, ISSN: 0018-9472.

Miller, J. A.; Potter, W. D.; Ganham, R. V. & Lapena, C. N. (1993). An evaluation of local improvement operators for genetic algorithms, *IEEE Trans. on Systems, Man and Cybernetics*, Vol. 23, No. 5, pp. 1340-1351, ISSN: 0018-9472.

Pao, W. C. & Chen, Y. F. (2008). Evolutionary strategy-based approaches for subcarrier, bit, and power allocation for multiuser OFDM systems, *Proceedings of IEEE Vehicular Technology Conf.*, pp. 1702-1706, ISBN: 978-1-4244-1644-8, May 2008.

Quintero, A. & Pierre, S. (2008). On the design of large-scale UMTS mobile networks using hybrid genetic algorithms, *IEEE Trans. on Vehicular Technology*, Vol. 57, No. 4, pp. 2498-2508, ISSN: 0018-9545.

Reddy, Y. B. & Naraghi-Pour, M. (2007). Genetic algorithm approach for adaptive power and subcarrier allocation in multi-user OFDM systems, *Intelligent Computing: Theory and Applications V (SPIE Defense & Security Symposium)*, April 2007.

Reddy, Y. B. & Phoha, V. V. (2007). Genetic algorithm approach for resource allocation in multi-user OFDM systems, *Proceedings of IEEE Int. Conf. on Communication Systems Software and Middleware*, pp. 1-6, ISBN: 1-4244-0613-7, Jan. 2007.

Reddy, Y. B.; Gajendar, N.; Taylor, P. & Madden, D. (2007). Computationally efficient resource allocation in OFDM systems: genetic algorithm approach, *Proceedings of IEEE Int. Conf. on Information Technology*, pp. 36-41, ISBN: 0-7695-2776-0, April 2007.

Siu, S.; Ho, Chia-Lu & Lee, Chien-Min (2005). TSK-based decision feedback equalizer using an evolutionary algorithm applied to QAM communication systems, *IEEE Trans. on Circuits and Systems – II: Express Brief*, Vol. 52, No. 9, pp. 596-600, ISSN: 1549-7747.

Spears, William M. (2000). Evolutionary Algorithms: The Role of Mutation and Recombination, Springer, ISBN: 978-3-540-66950-0.

Torrance, J. M. & Hanzo, L. (1996). Optimization of switching levels for adaptive modulation in slow Rayleigh fading, *Electronic Letters*, Vol. 32, pp. 1167-1169, ISSN: 0013-5194.

Wang, Y.; Chen, F. & Wei, G. (2005). Adaptive subcarrier and bit allocation for multiuser OFDM system based on genetic algorithm, *Proceedings of IEEE Int. Conf. on Communications, Circuits and Systems*, pp. 242-246, ISBN: 0-7803-9015-6, May 2005.

Wolsey, Laurence A. (1998). *Integer Programming*, John Wiley & Sons, ISBN: 978-0-471-28366-9.

Wong, C. Y.; Cheng, R. S.; K. B. Lataief & Murch, R. D. (1999a). Multiuser OFDM with adaptive subcarrier, bit, and power allocation, *IEEE J. Selected Areas in Communications*, Vol. 17, No. 10, pp. 1747-1758, ISSN: 0733-8716.

Wong, C. Y.; Tsui, C. Y.; Cheng, R. S. & Letaief, K. B. (1999b). A real-time sub-carrier allocation scheme for multiple access downlink OFDM transmission, *Proceedings of IEEE Vehicular Technology Conf.*, pp. 1124-1128, ISBN: 0-7803-5435-4, Amsterdam, Sept. 1999.

12

Reduced-Complexity PAPR Minimization Schemes for MC-CDMA Systems

Mariano García Otero and Luis A. Paredes Hernández
Universidad Politécnica de Madrid
Spain

1. Introduction

Multicarrier Code-Division Multiple Access (MC-CDMA) (Hara & Prasad, 1997), which is based on a combination of an CDMA scheme and Orthogonal Frequency Division Multiplexing (OFDM) signaling (Fazel & Kaiser, 2008), has attracted much attention in forthcoming mobile communication systems, because of its intrinsic spectrum efficiency and interference suppression capabilities. In MC-CDMA, information symbols of many users are spread using orthogonal codes and combined in the frequency domain; this results in a relatively low symbol rate and thus non-selective fading in each subcarrier.

However, one main drawback of any kind of multicarrier modulation is the inherent high value of the Peak-to-Average Power Ratio (PAPR) of the transmitted signals, because they are generated as an addition of a large number of independent signals. If low power consumption at the transmitter is a strict requirement, one would like the RF High Power Amplifier (HPA) to operate with a low back-off level (i.e. with operation point near saturation state); as a consequence of this, signal peaks will frequently enter the nonlinear part of the input-output characteristic of the HPA, thus causing severe nonlinear artifacts on the transmitted signals such as intermodulation distortion and out-of-band radiation. Therefore, reducing the PAPR is crucial in multicarrier systems, especially when transceivers are fed by batteries (such as in mobile devices), because of the intrinsic limitations in power consumption.

There has been a lot of research work about PAPR reduction techniques in multicarrier systems. Among these, we have clipping and filtering schemes (Li & Cimini, 1997), block coding algorithms (Jones et al., 1994), the Partial Transmit Sequences (PTS) (Cimini & Sollenberger, 2000; Jayalath & Tellambura, 2000), and Selected Mapping (SLM) approaches (Bäuml et al., 1996; Breiling et al., 2001), and the Tone Reservation (TR) (Tellado & Cioffi, 1998), and the Tone Injection (TI) techniques (Han et al., 2006). In general, reducing the PAPR is always done either at the expense of distorting the transmitted signals, thus increasing the BER at the receiver, or by reducing the information data rate, usually because high PAPR signals are somehow discarded and replace by others with lower PAPR before been transmitted.

All the previously mentioned methods have been originally proposed for single-user multicarrier schemes such as OFDM. Although most of them are also applicable with minor modifications to MC-CDMA systems (Ruangsurat & Rajatheva, 2001; Ohkubo & Ohtsuki, 2002), other families of algorithms can be developed after carefully considering the different structure of the generated MC-CDMA signals. Between these, probably the most popular are those based on dynamically selecting an "optimal" set of codes (those that give the lowest possible PAPR), according to the number of active user in the system (Ochiai & Imai, 1998; Kang et al., 2002; Alsusa & Yang, 2006).

In this chapter, we further explore a PAPR reduction technique previously proposed by the authors, namely the User Reservation (UR) approach (Paredes Hernández & García Otero, 2009). The UR technique is based on the addition of peak-reducing signals to the signal to be transmitted; these new signals are selected so that they are orthogonal to the original signal and therefore can be removed at the receiver without the need of transmitting any side information and, ideally, without penalizing the bit error rate (BER). In the UR method, these peak-reducing signals are built by using spreading codes that are either dynamically selected from those users that are known to be idle, or deliberately reserved *a priori* for PAPR reduction purposes.

The concept of adding orthogonal signals for peak power mitigation has been previously proposed to reduce PAPR in Discrete MultiTone (DMT) and OFDM transmissions (Tellado & Cioffi, 1998; Gatherer & Polley, 1997), and also in CDMA downlink systems (Väänanen et al., 2002). However, the implementation of this idea in the context of MC-CDMA communications poses particular problems that are discussed in this chapter. Our aim is also to develop strategies to alleviate the inherent complexity of the underlying minimization problem.

2. PAPR properties of MC-CDMA signals

In an MC-CDMA system, a block of M information symbols from each active user are spread in the frequency domain into $N = LM$ subcarriers, where L represents the spreading factor. This is accomplished by multiplying every symbol of the block for user k (where $k \in \{0,1,\dots,L-1\}$) by a spreading code $\{c_l^{(k)}, l = 0,1,\dots,L-1\}$, selected from an set of L orthogonal sequences, thus allowing a maximum of L simultaneous users to share the same radio channel. The spreading codes are the usual Walsh-Hadamard (WH) sequences, which are the columns of the Hadamard matrix of order L, $\mathbf{C}_L$. For L a power of 2, the Hadamard matrix is constructed recursively as

$$\mathbf{C}_2 = \begin{bmatrix} 1 & 1 \\ 1 & -1 \end{bmatrix} \tag{1a}$$

$$\mathbf{C}_n = \mathbf{C}_{n/2} \otimes \mathbf{C}_2 \quad \text{for } n = 4,8,\dots,L/2,L \tag{1b}$$

where the symbol $\otimes$ denotes the Kronecker tensor product.

We will assume in the sequel that, of the L maximum users of MC-CDMA system, only $K_A < L$ are "active", i.e., are transmitting information symbols, while the other $K_I = L - K_A$ remain inactive or "idle". We will further assume that there is a "natural" indexing for all the users based on their WH codes, being the index associated to a given user the number of the column that its code sequence occupies in the order-L Hadamard matrix. For notational convenience, we will assume that column numbering begins at 0, so that

$$\mathrm{C}_L = \begin{bmatrix} \mathrm{c}_L^{(0)} & \mathrm{c}_L^{(1)} & \cdots & \mathrm{c}_L^{(L-1)} \end{bmatrix} \tag{2}$$

with $\mathbf{c}_L^{(k)} = \left[c_0^{(k)}, c_1^{(k)}, \ldots, c_{L-1}^{(k)}\right]^T$ and $(\cdot)^T$ denotes transpose. In this situation, the indices of the active users belong to a set Ω_A, while the indices of the inactive users constitute a set Ω_I. The cardinals of the sets Ω_A and Ω_I are, thus, K_A and K_I, respectively.

In the downlink transmitter, the data symbols of the K_A active users are spread by their specific WH sequences and added together. The complex envelopes are then interleaved in the frequency domain, so that the baseband transmitted signal is

$$s(t) = \sum_{k \in \Omega_A} \sum_{l=0}^{L-1} \sum_{m=0}^{M-1} a_m^{(k)} c_l^{(k)} e^{j2\pi(Ml+m)t/T}, \quad 0 \le t < T \tag{3}$$

where $\{a_m^{(k)}, m = 0,1,\ldots,M-1\}$ are the data symbols in the block for the kth active user and T is the duration of the block. Actually, the modulation of the subcarriers is performed in discrete-time by means of an Inverse Fast Fourier Transform (IFFT).

The PAPR of a signal can be defined as the ratio of peak envelope power to the average envelope power

$$PAPR = \frac{\max\limits_{0 \le t < T} |s(t)|^2}{E[|s(t)|^2]} \tag{4}$$

where $E(\cdot)$ represents the expectation operation, and $E[|s(t)^2]$ is the average power of $s(t)$. In practice, the computation of the peak power is performed on the discrete-time version of $s(t)$.

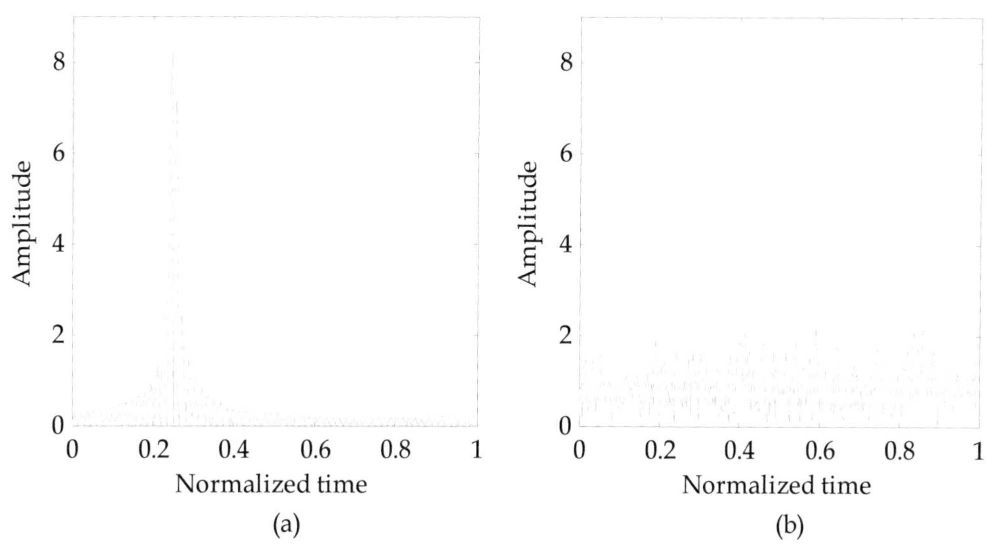

Fig. 1. Examples of amplitude envelopes in MC-CDMA. (a) Single user. (b) Full load

As the PAPR is a random variable, an adequate statistic is needed to characterize it. A common choice is to use the Complementary Cumulative Distribution Function (CCDF), which is defined as the probability of the PAPR exceeding a given threshold

$$CCDF(x) = \Pr(PAPR > x) \tag{5}$$

It should be noticed that the distribution of the PAPR of MC-CDMA signals substantially differs from other multicarrier modulations. For instance, in OFDM schemes, the subcarrier complex envelopes can be assumed to be independent random variables, so that, by applying the Central Limit Theorem, the baseband signal is usually assumed to be a complex Gaussian process. However, in MC-CDMA the subcarrier envelopes generally exhibit strong dependencies, because of the poor autocorrelation properties of WH codes. This fact, in turn, translates into a baseband signal that is no longer Gaussian-like, but instead has mostly low values with sharp peaks at regular intervals. This effect is particularly evident when the number of K_A active users is low.

Fig. 1 shows examples of amplitude envelopes for an MC-CDMA system, with L=32 and M=4 (N=128 subcarriers), and where the two extreme conditions are considered, single user (K_A=1) and full load (K_A=32).

We can see from Fig. 1 that we should expect higher PAPR values as the load of the system decreases.

3. PAPR reduction by user reservation

Our approach to PAPR reduction is based on "borrowing" some of the spreading codes of the inactive users set, so that an adequate linear combination of these codes is added to the active users before the IDFT operation. The coefficients of such linear combination ("pseudo-symbols") should be chosen so that the peaks of the signal are reduced in the time domain. As the added signals are orthogonal to the original ones, the whole process is transparent at the receiver side.

3.1 System model

Fig.2 shows a block diagram of the proposed MC-CDMA downlink transmitter. We can see that the binary information streams of the K_A active users are first converted into sequences of symbols belonging to a QAM constellation, and the symbol sequence of each user is subsequently spread by its unique code. Notice also from Fig.2 that, unlike a conventional MC-CDMA system, the codes belonging to the left K_I inactive users are also used to spread a set of pseudo-symbols computed from the current active users' symbols, and then the whole set of spread sequences are added together before the frequency-domain interleaving and OFDM modulation steps.

With the addition of K_I inactive users for PAPR reduction purposes, our MC-CDMA downlink complex envelope signal for $0 \le t < T$, can be expressed as

$$s(t) = \sum_{k\in\Omega_A} \sum_{l=0}^{L-1} \sum_{m=0}^{M-1} a_m^{(k)} c_l^{(k)} e^{j2\pi(Ml+m)t/T} + \sum_{k\in\Omega_I} \sum_{l=0}^{L-1} \sum_{m=0}^{M-1} a_m^{(k)} c_l^{(k)} e^{j2\pi(Ml+m)t/T} \tag{6}$$

If we sample $s(t)$ at multiples of T_s=T/NQ, where Q is the oversampling factor, we will obtain the discrete-time version of (6), which can be rewritten in vector notation as

$$\mathbf{s} = \mathbf{W}_{NQ}^{N}\left(\mathbf{C}_L^A \otimes \mathbf{I}_M\right)\mathbf{a}^A + \mathbf{W}_{NQ}^{N}\left(\mathbf{C}_L^I \otimes \mathbf{I}_M\right)\mathbf{a}^I \tag{7}$$

where the components of vector $\mathbf{s}$ are the NQ samples of the baseband signal $s(t)$ in the block, $\{s_n \equiv s(nT_s), n=0,1,\ldots,NQ-1\}$, $\mathbf{a}^A$ is the vector of K_AM symbols of the K_A active users to be transmitted, $\mathbf{a}^I$ is the vector of $K_I M$ pseudo-symbols of the K_I idle users to be determined, $\mathbf{W}_{NQ}^N$ is a $NQ \times N$ matrix formed by the first N columns of the Inverse Discrete Fourier Transform (IDFT) matrix of order NQ

$$\mathbf{W}_{NQ}^N = \begin{bmatrix} 1 & 1 & \cdots & 1 \\ 1 & e^{j2\pi\frac{1\times 1}{NQ}} & \cdots & e^{j2\pi\frac{1\times(N-1)}{NQ}} \\ \vdots & \vdots & \ddots & \vdots \\ 1 & e^{j2\pi\frac{(NQ-1)\times 1}{NQ}} & \cdots & e^{j2\pi\frac{(NQ-1)\times(N-1)}{NQ}} \end{bmatrix} \tag{8}$$

$\mathbf{C}_L^A$ is a $L \times K_A$ matrix whose columns are the WH codes of the active users, $\mathbf{C}_L^I$ is a $L \times K_I$ matrix whose columns are the WH codes of the idle users, and $\mathbf{I}_M$ is the identity matrix of order M.

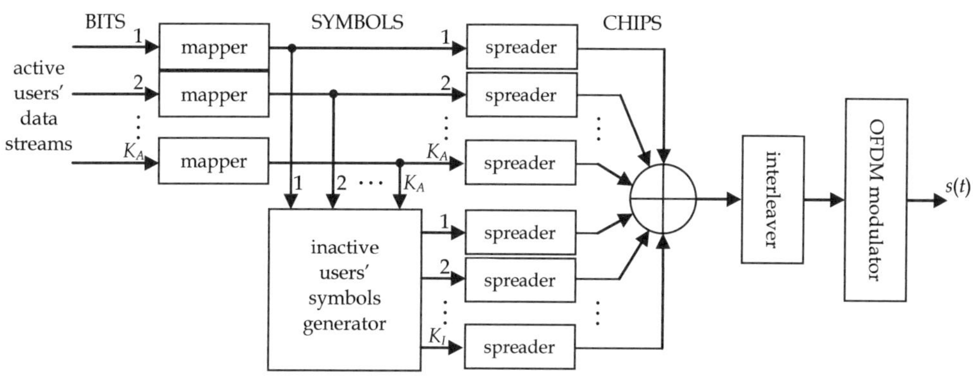

Fig. 2. MC-CDMA downlink transmitter with addition of idle users for PAPR reduction

Thus, our objective is to find the values of the pseudo-symbols $\mathbf{a}^I$ that minimize the peak value of the amplitudes of the components of vector $\mathbf{s}$ in (7).

3.2 Quadratic programming method

Our optimization problem can be formulated as

$$\min_{\mathbf{a}^I} \max_{0 \le n \le NQ-1} |s(nT_s)| = \min_{\mathbf{a}^I} \|\mathbf{s}\|_\infty = \min_{\mathbf{a}^I} \left\|\mathbf{H}^A \mathbf{a}^A + \mathbf{H}^I \mathbf{a}^I\right\|_\infty \tag{9}$$

where $\|\cdot\|_\infty$ denotes ℓ_∞ norm, and $\mathbf{H}^A$ and $\mathbf{H}^I$ are, respectively, the following $NQ \times K_A M$ and $NQ \times K_I M$ matrices:

$$\mathbf{H}^A = \mathbf{W}_{NQ}^N \left(\mathbf{C}_L^A \otimes \mathbf{I}_M\right) \tag{10a}$$

$$\mathbf{H}^I = \mathbf{W}_{NQ}^N \left(\mathbf{C}_L^I \otimes \mathbf{I}_M\right) \tag{10b}$$

The minimization involved in (9) may be formulated as a Second-Order Cone Programming (SOCP) convex optimization problem (Sousa et al., 1998)

$$\begin{array}{ll} \text{minimize} & z \\ \text{subject to} & |s_n| \le z,\ 0 \le n \le NQ-1, \\ & \mathbf{s}=\mathbf{H}^A\mathbf{a}^A+\mathbf{H}^I\mathbf{a}^I \\ \text{in variables} & z \in \mathbb{R}\ ,\ \mathbf{a}^I \in \mathbb{C}^{K_I M} \end{array} \tag{11}$$

Solving (11) in real-time can be a daunting task and we are, thus, interested in reducing the complexity of the optimization problem. Two approaches will be explored in the sequel:

a. Reducing the dimension of the optimization variable $\mathbf{a}^I$.
b. Using suboptimal iterative algorithms to approximately solve (11).

4. Dimension reduction

We will see in the next subsections that not all the inactive users are necessary to enter the system in (6) to reduce the PAPR. This is a consequence of the specific structure of the Hadamard matrices.

4.1 Periodic properties of WH sequences

The particular construction of Hadamard matrices imposes their columns to follow highly structured patterns, thus making WH codes to substantially depart from ideal pseudo-noise (PN) sequences. The most important characteristic of WH sequences that affects their Fourier properties is the existence of inner periodicities, i.e., groups of binary symbols (1 or –1) that are replicated along the whole length of the code. This periodic behavior of WH codes in the frequency domain leads to the appearance of characteristic patterns in the time domain, with many zero values that give the amplitude of the resulting signal a "peaky" aspect (see Fig 1a). This somewhat "sparse" nature of the IDFT of WH codes is, in turn, responsible of the high PAPR values we usually find in MC-CDMA signals.

For the applicability of our UR technique, it is important to characterize the distribution of the peaks in the IDFTs of WH codes. This is because PAPR reduction is possible only if we add in (7) those inactive users whose WH codes have time-domain peaks occupying exactly the same positions as those of the active users, so that, with a suitable choice of the pseudo-symbols, a reduction of the amplitudes of the peaks is possible. As we will see, this characterization of WH sequences will lead us to group them in sets of codes, where the elements of a given set share the property that any idle user with a code belonging to the set can be used to reduce the peaks produced by other active users with codes of the same set.

A careful inspection of the recursive algorithm (1) for generating the Hadamard matrix of order n, $\mathbf{C}_n$ (with n a power of two), shows that two columns of this matrix are generated using a single column of the matrix of order $n/2$, $\mathbf{C}_{n/2}$. If we denote as $\mathbf{c}_{n/2}^{(k)}$ the kth column of $\mathbf{C}_{n/2}$ ($k=0,1,\ldots,n/2-1$), it can be seen that the two columns of the matrix $\mathbf{C}_n$ generated by $\mathbf{c}_{n/2}^{(k)}$ are, respectively:

$$\mathbf{c}_n^{(k)} = \begin{bmatrix} \mathbf{c}_{n/2}^{(k)} \\ \mathbf{c}_{n/2}^{(k)} \end{bmatrix}, \qquad k=0,1,\ldots,n/2-1 \tag{12a}$$

$$\mathbf{c}_n^{(n/2+k)} = \begin{bmatrix} \mathbf{c}_{n/2}^{(k)} \\ -\mathbf{c}_{n/2}^{(k)} \end{bmatrix}, \qquad k=0,1,\ldots,n/2-1 \tag{12b}$$

We can see from (12a) that the columns of the Hadamard matrix of order $n/2$ are simply repeated twice to form the first $n/2$ columns of the Hadamard matrix of order n. This, in turn, has two implications:

Assertion 1. Any existing periodic structure in $\mathbf{c}_{n/2}^{(k)}$ is directly inherited by $\mathbf{c}_n^{(k)}$.

Assertion 2. In case $\mathbf{c}_{n/2}^{(k)}$ has no inner periodicity, a new repetition pattern of length $n/2$ is created in $\mathbf{c}_n^{(k)}$.

On the other hand, (12b) implies that the last $n/2$ columns of the order n Hadamard matrix are formed by appending to the columns of the order $n/2$ matrix these same columns but with the sign of their elements changed; therefore, the periodicities in the columns of the original matrix are now destroyed by the copy-and-negate operation in the last $n/2$ columns. Nevertheless, we can easily see that (12b), when iterated in alternation with (12a), introduces another significant effect:

Assertion 3. Any repetitive structure in $\mathbf{c}_n^{(k)}$ is always composed of two equal-length consecutive substructures with opposite signs.

If we denote as P the minimum length of a pattern of binary symbols that is repeated an integer number of times along any given column of the Hadamard matrix of order n (period length), we can see by inspection that the first column (formed by n 1s) has $P=1$ and the second column (formed by a repeated alternating pattern of 1s and –1s) has $P=2$, respectively; then, by recursively applying assertions 1 and 2, we can build the following table:

WH code index	Period
0	1
1	2
2, 3	4
4 to 7	8
⋮	⋮
$L/4$ to $L/2-1$	$L/2$
$L/2$ to $L-1$	L

Table 1. Periods of the WH codes of length L

Notice from Table 1 that, for an Lth order Hadamard matrix, we will have $\log_2 L+1$ different periods in its columns. Notice also that, for $P>1$, as the period is doubled the number of WH sequences with the same period length is also doubled.

4.2 Selection of inactive users

The periodic structure of the WH codes determines their behavior in the time domain, because the number and positions of the non-zero values of the IDFT of a sequence directly depends on the value of its period P. To illustrate this fact, Fig. 3 shows the sampled amplitude envelopes of the signals obtained in a single-user MC-CDMA transmitter with two different WH codes (corresponding to different columns of the Hadamard matrix $\mathbf{C}_L$)

Notice from Fig.3 that users whose codes have low indices tend to produce few scattered peaks in the time domain, while users using codes with higher values in their indices generate a high number of non-zero values in the time domain.

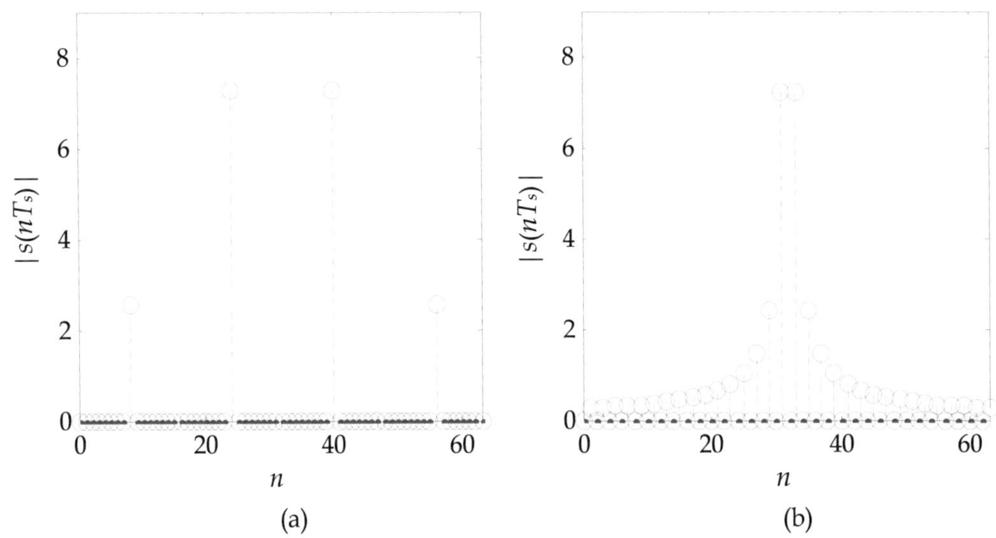

Fig. 3. Samples of the envelope of an MC-CDMA signal for a single user and different WH codes. (a) k=2. (b) k=16

It is clear from Fig. 3 that idle users can only mitigate the PAPR of signals generated by active users with the same periodic patterns in their codes. This is because only those users will be able to generate signals with their peaks located in the same time instants (and with opposite signs) as the peaks of the active users, so that these latter peaks can be reduced. Therefore we conclude that we will include in (11) only those idle users whose WH codes have the same period as any of the active users currently in the system; the choice of inactive users can be easily obtained with the help of Table 1, and the selection rule can be summarized as follows:

For every active user $k_A \in \Omega_A$ (with $k>1$), select for the optimization (11) only the inactive users $k_I \in \Omega_I$ such that $\lfloor \log_2 k_I \rfloor = \lfloor \log_2 k_A \rfloor$, where $\lfloor \cdot \rfloor$ denotes the "integer part".

5. Iterative clipping approaches

The SOCP optimization of (11) solved with interior-point methods requires $O((NQ)^{3/2})$ operations (Sousa et al, 1998). Although the structure of the matrices involved could be exploited to reduce the complexity, it is desirable to devise simpler suboptimal algorithms whose complexity only grows linearly with the number of subcarriers. This can be accomplished if we adopt a strategy of iterative clipping of the time-domain signal, so that, at the ith iteration, the signal vector is updated as

$$\mathbf{s}^{(i+1)} = \mathbf{s}^{(i)} + \mathbf{r}^{(i)} \tag{13}$$

where $\mathbf{r}^{(i)}$ is a "clipping vector" that is designed to reduce the magnitude of one or more of the samples of the signal vector. Notice that, as the clipping vector should cause no interference to the active users, it must be generated as

$$\mathbf{r}^{(i)} = \mathbf{H}^I \mathbf{b}^{(i)} \tag{14}$$

where $\mathbf{H}^I$ is defined in (10b) and $\mathbf{b}^{(i)} \in \mathbb{C}^{K_I M}$.
Now suppose that, at the ith iteration, we want to clip the set of samples of vector $\mathbf{s}^{(i)}$ $\{s_u^{(i)}, u \in U^{(i)}\}$, where $U^{(i)}$ is a subset of the indices $\{0, 1, \ldots, NQ-1\}$. Thus, in (13) we would like the clipping vector $\mathbf{r}^{(i)}$ to reduce the magnitudes of those samples without modifying other values in vector $\mathbf{s}^{(i)}$, so the *ideal* clipping vector should be of the form

$$\widehat{\mathbf{r}}^{(i)} = \sum_{u \in U^{(i)}} \alpha_u^{(i)} \boldsymbol{\delta}_u \tag{15}$$

where $\boldsymbol{\delta}_u$ is the length-NQ discrete-time impulse delayed by u samples

$$\boldsymbol{\delta}_u = [\underbrace{0,0,\ldots,0}_{u},1,\underbrace{0,0,\ldots,0}_{NQ-u-1}]^T \tag{16}$$

and $\{\alpha_u^{(i)}, u \in U^{(i)}\}$ is a set of suitably selected complex coefficients.
Notice, however, that, as we require vector $\mathbf{r}^{(i)}$ to be of the form (14) it is not possible, in general, to synthesize the set of required time-domain impulses using only symbols from the inactive users, so $\boldsymbol{\delta}_u$ must be replaced by another vector $\mathbf{d}_u$ generated as

$$\mathbf{d}_u = \mathbf{H}^I \mathbf{b}_u \tag{17}$$

so that the actual clipping vector would result in

$$\mathbf{r}^{(i)} = \sum_{u \in U^{(i)}} \alpha_u^{(i)} \mathbf{d}_u \tag{18}$$

which can be easily shown to be in agreement with restriction (14), with $\mathbf{b}^{(i)}$ obtained as

$$\mathbf{b}^{(i)} = \sum_{u \in U^{(i)}} \alpha_u^{(i)} \mathbf{b}_u \tag{19}$$

A straightforward way to approximate the impulse vector $\boldsymbol{\delta}_u$ is by minimizing a distance between vectors $\boldsymbol{\delta}_u$ and $\mathbf{d}_u$

$$\mathbf{b}_u = \arg\min_{\mathbf{b}} \left\| \mathbf{H}^I \mathbf{b} - \boldsymbol{\delta}_u \right\|_p \tag{20}$$

where $\|\cdot\|_p$ denotes the p-norm. When $p = 2$, we have the conventional least-squares (LS) solution

$$\mathbf{b}_u = (\mathbf{H}^{I*}\mathbf{H}^I)^{-1}\mathbf{H}^{I*}\boldsymbol{\delta}_u = \frac{1}{LNQ}\mathbf{H}^{I*}\boldsymbol{\delta}_u \tag{21}$$

where $(\cdot)^*$ denotes conjugate transpose and we used the fact: $\mathbf{H}^{I*}\mathbf{H}^I = LNQ\mathbf{I}_{K_I M}$. Replacing (21) in (17), we arrive at

$$\mathbf{d}_u = \mathbf{H}^I \mathbf{b}_u = \frac{1}{LNQ}\mathbf{H}^I \mathbf{H}^{I*}\boldsymbol{\delta}_u \tag{22}$$

so that $\mathbf{d}_u$ is the orthogonal projection of the impulse vector onto the subspace spanned by the columns of $\mathbf{H}^I$. Now, taking into account from (16) that $\boldsymbol{\delta}_u$ is just the uth column of $\mathbf{I}_{NQ}$, we conclude that the LS approximation to the unit impulse vector centered at position u is the uth column of the projection matrix

$$\mathbf{P}^I = \frac{1}{LNQ}\mathbf{H}^I\mathbf{H}^{I*} \tag{23}$$

Notice that, in general, matrix $\mathbf{P}^I$ of (23) is not a circulant matrix. This is in contrast with the POCS approach for PAPR mitigation in OFDM (Gatherer and Polley, 1997) and related methods, where the functions used for peak reduction are obtained by circularly shifting and scaling a single basic clipping vector. Of course, other norms can be chosen in the optimization (20). For instance, for $p=\infty$, the problem of finding the optimal vector $\mathbf{b}_u$ can be also cast as a SOCP. Notice that the set of vectors $\{\mathbf{d}_u, u \in U^{(i)}\}$ is pre-computed and stored off-line, and so the complexity of solving (20) is irrelevant.

Fig. 4 shows examples of the approximations to a discrete-time impulse we get using (20) and (17) for $p=2$ and $p=\infty$, respectively. We can see that the minimization of the ℓ_2 norm produces a signal that has spurious peaks with very high amplitudes; therefore, the addition of this approximate impulse to the original signal will induce the emergence of new peaks that need to be clipped, thus slowing the convergence of any iterative procedure.

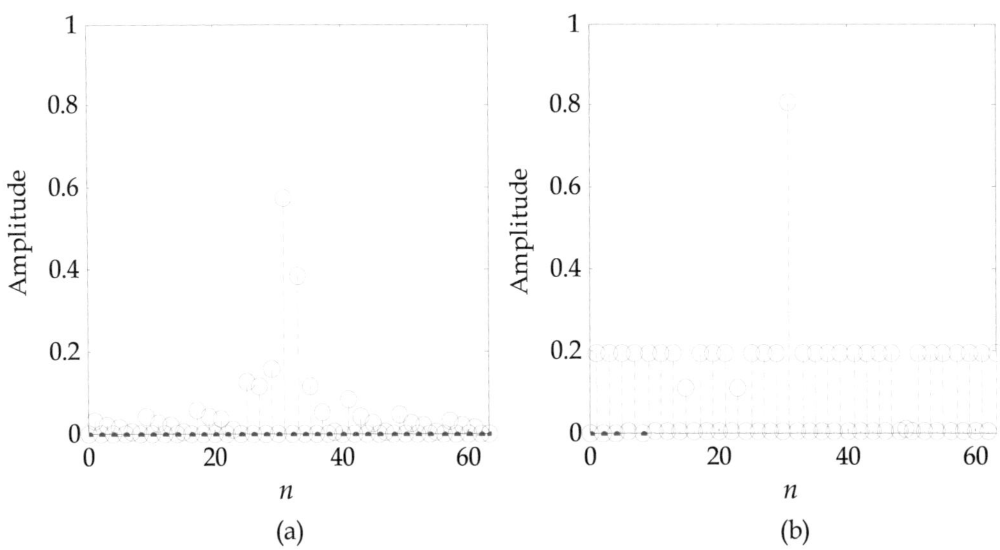

Fig. 4. Approximations to a discrete-time impulse using only inactive users. (a) Minimizing ℓ_2 norm. (b) Minimizing ℓ_∞ norm

Several approaches can be found in the literature for the iterative minimization of the PAPR in OFDM based on tone reservation. Among those, the two most popular are probably the SCR-gradient method (Tellado and Cioffi, 1998) and the active-set approach (Krongold, 2004). Both can be readily adapted to simplify the UR method for PAPR reduction in MC-CDMA as we will see in the sequel.

5.1 SCR-gradient

Following (Tellado & Cioffi, 1998), we define the clipping or soft limiter (SL) operator

$$clip(s_n) = \begin{cases} s_n, & |s_n| \leq A \\ A\dfrac{s_n}{|s_n|}, & |s_n| > A \end{cases} \tag{24}$$

with $A > 0$. Now, if all the components $\{s_n, n=0,1,\ldots,NQ-1\}$ of a given signal vector $\mathbf{s}$ are transformed by an SL, we can define the clipping noise as

$$\mathbf{z} = \mathbf{s} - clip(\mathbf{s}) \tag{25}$$

If inactive users are added to the original signal $\mathbf{s}^A$ (generated using only the symbols of the active users), so that the model for the clipped signal $\mathbf{s}$ is

$$\mathbf{s} = \mathbf{s}^A + \mathbf{s}^I = \mathbf{H}^A\mathbf{a}^A + \mathbf{H}^I\mathbf{a}^I \tag{26}$$

now the vector of pseudo-symbols of the inactive users $\mathbf{a}^I$ can be designed to reduce the clipping noise. To accomplish this, we define the signal to clipping noise power ratio (SCR) as

$$SCR = \frac{\mathbf{s}^{A*}\mathbf{s}^A}{\mathbf{z}^*\mathbf{z}} \tag{27}$$

The maximization of the SCR in (27) leads to a minimization of its denominator (the clipping noise power). This latter can be written, using (25), (26) and (24), and after some manipulations as:

$$\mathbf{z}^*\mathbf{z} = \sum_{|s_u|>A} \left(|s_u| - A\right)^2 \tag{28}$$

with

$$s_u = s_u^A + \boldsymbol{\delta}_u^T\mathbf{H}^I\mathbf{a}^I \tag{29}$$

where $\boldsymbol{\delta}_u$ was defined in (16). Now, instead of a direct minimization of (28) with respect to $\mathbf{a}^I$, we will try an iterative algorithm based on the gradient of the clipping noise power of the form:

$$\mathbf{a}^{I(i+1)} = \mathbf{a}^{I(i)} - \mu\nabla_{\mathbf{a}^{I(i)}}\left(\mathbf{z}^*\mathbf{z}\right) \tag{30}$$

where $\mu > 0$ and ∇ is the complex gradient operator (Haykin, 1996). Using (28) and (29), the gradient vector can be shown to be

$$\nabla_{\mathbf{a}^I}\left(\mathbf{z}^*\mathbf{z}\right) = 2\sum_{|s_u|>A}\left(1 - A/|s_u|\right)s_u\mathbf{H}^{I*}\boldsymbol{\delta}_u \tag{31}$$

According to (26), the recursion for the pseudo-symbols (30) can be equivalently translated to the signal vector

$$\mathbf{s}^{(i+1)} = \mathbf{s}^{(i)} - \mu \mathbf{H}^I \nabla_{\mathbf{a}^{I(i)}} \left(\mathbf{z}^* \mathbf{z} \right) \tag{32}$$

So that, substituting (31) in (32), and taking into account (22), we finally arrive at

$$\mathbf{s}^{(i+1)} = \mathbf{s}^{(i)} - \mu' \sum_{\left|s_u^{(i)}\right| > A} \left(1 - A / \left|s_u^{(i)}\right|\right) s_u^{(i)} \mathbf{d}_u \tag{33}$$

where $\mu' = 2\mu LNQ$, which is in agreement with (13) and (18) with

$$\begin{aligned} U^{(i)} &= \left\{ u : \left| s_u^{(i)} \right| > A \right\} \\ \alpha_u^{(i)} &= -\mu' \left(1 - A / \left|s_u^{(i)}\right|\right) s_u^{(i)}, \quad u \in U^{(i)} \end{aligned} \tag{34}$$

Notice that, for $\mu' = 1$ and "ideal" impulses ($\mathbf{d}_u \approx \boldsymbol{\delta}_u$), the signal would be soft-limited in one iteration to the maximum amplitude A.
The main drawbacks of this algorithm are the difficulty of *a priori* fixing a convenient clipping level A, and the probable slow convergence of the algorithm.

5.2 Active-set

The slow convergence of the SCR-gradient method is due to the use of non-ideal impulses $\mathbf{d}_u$ in the clipping process, because they must satisfy restriction (17). As they have non-zero values outside the position of their maximum (see Fig. 4), any attempt to clip a peak of the signal at a given discrete time u using $\mathbf{d}_u$ can potentially give rise to unexpected new peaks at another positions of the signal vector.
On the contrary, the active-set approach (Krongold & Jones, 2004) keeps the maximum value of the signal amplitude controlled, so that it is always reduced at every iteration of the algorithm. An outline of the active-set method follows below (Wang et al, 2008):

1. Find the component of $\mathbf{s}$ with the highest magnitude (peak value).
2. Clip the signal by adding inactive users so that the peak value is balanced with another secondary peak. Now we have two peaks with the same magnitude, which is lower than the original maximum.
3. Add again inactive users to simultaneously reduce the magnitudes of the two balanced peaks until we get three balanced peaks.
4. Repeat this process until either the magnitudes of the peaks cannot be further reduced significantly or a maximum number of iterations is reached.

Notice that, at the ith stage of the algorithm we have an *active set* $\{s_u^{(i)}, u \in U^{(i)}\}$ of signal peaks that have the same maximum magnitude, i.e.:

$$\begin{aligned} |s_u^{(i)}| &= A^{(i)}, \quad \text{if } u \in U^{(i)} \\ |s_u^{(i)}| &< A^{(i)}, \quad \text{if } u \in \bar{U}^{(i)} \end{aligned} \tag{35}$$

where $A^{(i)}$ is the peak magnitude and $\bar{U}^{(i)}$ is the complement of the set $U^{(i)}$. The problem at this point is, thus, to find a clipping vector $\mathbf{r}^{(i)}$ generated as (18) that, when added to the signal $\mathbf{s}^{(i)}$ as in (13), will satisfy two conditions:

a. The addition of the clipping vector must keep the magnitudes of the components of the current active set balanced.

b. The addition of the clipping vector should reduce the value of the peak magnitude until it reaches the magnitude of a signal sample that was previously outside the active set.

Both conditions can be easily met if we design the vector $\mathbf{r}^{(i)}$ in two stages: first, we obtain a vector $\mathbf{q}^{(i)}$ as a suitable combination of non-ideal impulses of the form (17) that satisfies condition a)

$$\mathbf{q}^{(i)} = \sum_{u \in U^{(i)}} \beta_u^{(i)} \mathbf{d}_u \tag{36}$$

and then, we compute a real number to scale vector $\mathbf{q}^{(i)}$ until condition b) is met. Therefore, the final update equation for the signal vector is

$$\mathbf{s}^{(i+1)} = \mathbf{s}^{(i)} + \mu^{(i)} \mathbf{q}^{(i)} \tag{37}$$

where $\mu^{(i)}$ is a convenient step-size.

A simple way to ensure that $\mathbf{q}^{(i)}$ satisfies condition a) is to force its components at the locations of the peaks to be of unit magnitude and to have the opposite signs to the signal peaks in the current active set

$$q_u^{(i)} = -\frac{s_u^{(i)}}{|s_u^{(i)}|} = -\frac{s_u^{(i)}}{A^{(i)}}, \quad u \in U^{(i)} \tag{38}$$

because then, according to (35), (37) and (38):

$$A^{(i+1)} = \left| s_u^{(i+1)} \right| = \left| s_u^{(i)} - \mu^{(i)} \frac{s_u^{(i)}}{A^{(i)}} \right| = \left| A^{(i)} - \mu^{(i)} \right|, \quad u \in U^{(i)} \tag{39}$$

So, taking into account (36) and (38), the set of coefficients $\{\beta_u^{(i)}, u \in U^{(i)}\}$ is obtained as the solution of a system of linear equations

$$\sum_{v \in U^{(i)}} d_{u,v} \beta_v^{(i)} = -\frac{s_u^{(i)}}{A^{(i)}}, \quad u \in U^{(i)} \tag{40}$$

where $d_{u,v}$ is the vth component of vector $\mathbf{d}_u$.

Once the vector $\mathbf{q}^{(i)}$ is computed, the step-size $\mu^{(i)}$ is determined by forcing the new peak magnitude $A^{(i+1)}$ to be equal to the highest magnitude of the components of $\mathbf{s}^{(i+1)}$ not in the current active set

$$A^{(i+1)} = \max_{n \in \bar{U}^{(i)}} \left| s_n^{(i+1)} \right| \tag{41}$$

So, we can consider the possible samples to be included in the next active set $\{s_n^{(i)}, n \in \bar{U}^{(i)}\}$ and associate a candidate positive step-size $\{\mu_n^{(i)} > 0, n \in \bar{U}^{(i)}\}$ to each of them. According to (39) and (37), the candidates verify the conditions

$$\left| A^{(i)} - \mu_n^{(i)} \right| = \left| s_n^{(i)} + \mu_n^{(i)} q_n^{(i)} \right|, \quad n \in \bar{U}^{(i)} \tag{42}$$

so that we select as step-size the minimum of all the candidates

$$\mu^{(i)} = \min\left\{\mu_n^{(i)}, n \in \bar{U}^{(i)}\right\} \tag{43}$$

and its associated signal sample enters the new active set. This choice ensures that no other sample exceeds the magnitude of the samples in the current active set because we have the smallest possible reduction in the peak magnitude.
Squaring (42) and rearranging terms, we find that $\mu_n^{(i)}$ satisfies a quadratic equation with two real roots, so we choose for $\mu_n^{(i)}$ the smallest positive root, given by (Erdogan, 2006)

$$\mu_n^{(i)} = \frac{\psi_n^{(i)} - \sqrt{\psi_n^{(i)2} - \xi_n^{(i)}\zeta_n^{(i)}}}{\xi_n^{(i)}} \tag{44}$$

with

$$\begin{aligned} \xi_n^{(i)} &= 1 - |q_n^{(i)}|^2 \\ \psi_n^{(i)} &= A^{(i)} + \Re\{s_n^{(i)} q_n^{(i)*}\} \\ \zeta_n^{(i)} &= A^{(i)2} - |s_n^{(i)}|^2 \end{aligned} \tag{45}$$

where $\Re(\cdot)$ denotes real part. The overall complexity of the active-set method can be alleviated if we reduce the number of possible samples to enter the active set, so that we need to compute only a small number of candidate step-sizes. For instance, in (Wang et al, 2008) the authors propose a technique based on the prediction at the ith stage of a tentative step-size $\hat{\mu}^{(i)}$, and so the candidate samples are only those that verify the condition

$$\left|s_n^{(i)} + \hat{\mu}^{(i)} q_n^{(i)}\right| > A^{(i)} - \hat{\mu}^{(i)}, \quad n \in \bar{U}^{(i)} \tag{46}$$

6. Experimental results

The performance of the UR algorithm was tested by simulating the system of Fig. 2 under the conditions listed in Table 2

Modulation	QPSK
Spreading codes	Walsh-Hadamard
Spreading factor (L)	32
Data symbols per user in a frame (M)	4
Number of subcarriers (N)	128
Oversampling factor (Q)	4

Table 2. Simulation parameters

The transformation of the signal in the time-domain induced by the UR method is illustrated in Fig.5, which depicts the amplitude of the signal to be transmitted for an MC-CDMA system with only one active user. It can be seen that the value of the peak is substantially reduced with respect to the original signal, and it is also evident from Fig. 3 that the ℓ_∞ norm minimization performed by algorithm (12) forces, in this case, the resulting signal vector to have a characteristic pattern of "equalized" maximum amplitude values.

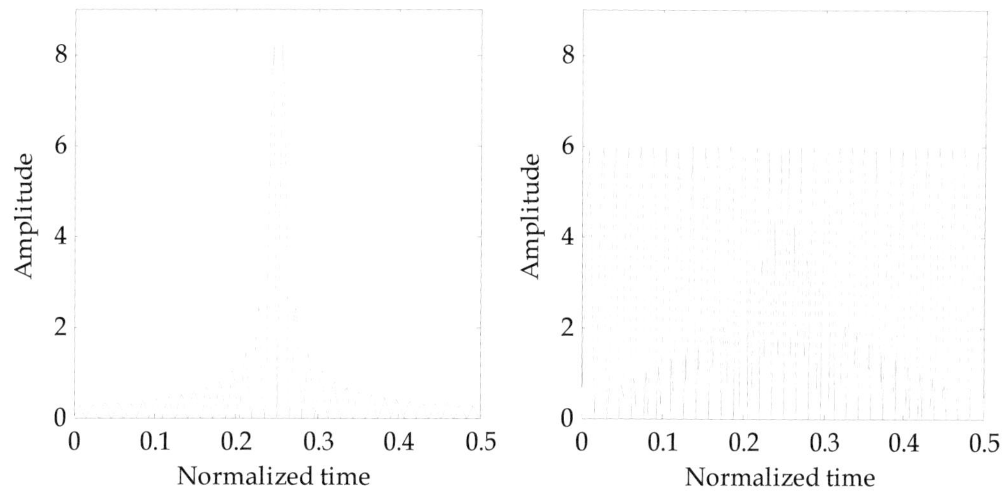

(a) (b)

Fig. 5. Examples of amplitude envelopes in MC-CDMA. (a) Original. (b) With peak values reduced via UR

For comparison purposes, Fig. 6 represents the estimated CCDF of the PAPR, as defined in (4) and (5), obtained under two different conditions for the system load: 8 and 24 active users, respectively. The K_A=8 case represents a "low load" situation (for only 25% of the maximum number of users are active), whereas a system with K_A=24 (75% of the maximum) can be considered as highly loaded. In both cases, we have compared the PAPR of the transmitted signal in the original MC-CDMA system with that obtained when the UR method is applied, using either the exact optimization (11) or the suboptimal active-set approach. For the latter algorithm, we have employed in the clipping procedure, represented by (36) and (37), the approximate impulses given by (22).

It is evident from Fig. 6 that, as it was expected, for an unmodified MC-CDMA system described by (3), the PAPR can become very high if the number of active users is small. Notice also that it is precisely in this case (K_A=8) when the PAPR reduction provided by the UR method is most noticeable. That is because, as K_A decreases, more inactive users are available and the dimensionality of vector $\mathbf{a}^I$ in (7) increases, letting more degrees of freedom to the optimization procedure (11).

We can also see from Fig. 6 that the active-set approach gets close to the optimal if a sufficient number of iterations are allowed. Notice that there is an upper bound for this parameter: the number of iterations cannot exceed the size of vector $\mathbf{a}^I$, because then the matrix involved in the linear system (40) becomes singular.

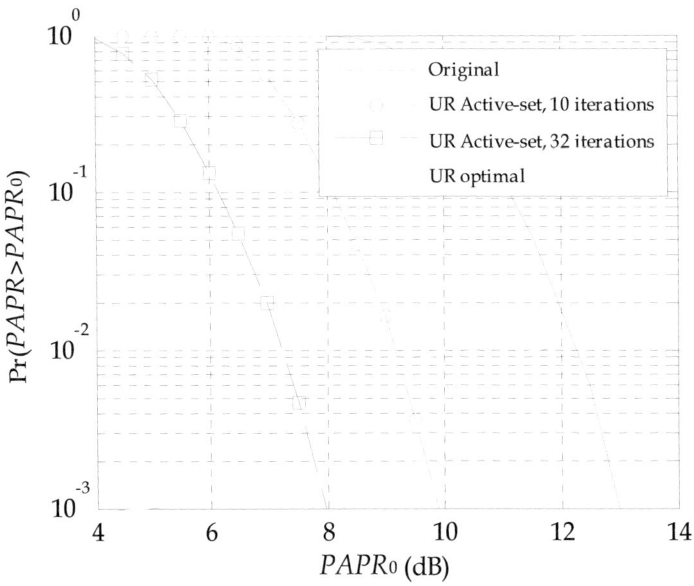

(a)

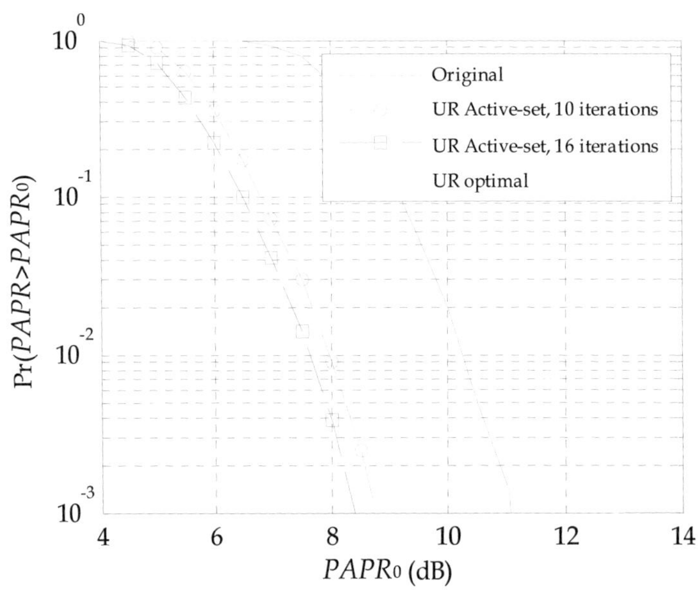

(b)

Fig. 6. CCDF of the PAPR under different MC-CDMA system loads. (a) Low load (8 active users). (b) High load (24 active users)

7. Conclusions

The UR scheme for the reduction of the PAPR of the signal transmitted in an MC-CDMA downlink is explored in this book chapter. This approach does not require any modification

at the receiver side, because it is based on the addition of the spreading codes of users that are inactive. The optimization procedure provides significant improvements in PAPR, especially when the number of active users is relatively low.
The inherent complexity of the SOCP optimization involved in the method can be alleviated if we select only inactive users with WH codes that share the same periods as those of the active users in the system. For further computational savings, suboptimal procedures can be applied to reduce the PAPR; these are based on the idea of iteratively clipping the original signal in the time domain via the addition of impulse-like signals that are synthesized using the WH codes of inactive users.

8. Acknowledgement

The work presented in this book chapter was partially supported by the Spanish Ministry of Science and Innovation under project no. TEC2009-14219-C03-01.

9. References

Alsusa, E. & Yang, L. (2006). MC-CDMA Specific PAPR Reduction Technique Utilising Spreading Code Redistribution. *IEEE Vehicular Technology Conference (VTC Fall '06)*, Montreal, Canada, pp. 1-5.

Bäuml, R. W.; Fischer, R. F. H. & Huber, J. B. (1996). Reducing the Peak-to-Average Power Ratio of Multicarrier Modulation by Selecting Mapping, *IEE Electronics Letters*, Vol. 32, No. 22, pp. 2056-2057.

Breiling, M.; Müller-Weinfurtner, S. H. & Hubber, J. B. (2001). SLM Peak-Power Reduction Without Explicit Side Information, *IEEE Communications Letters*, Vol. 5, No. 6, pp. 239-241.

Cimini, L. J. & Sollenberger, N. R. (2000). Peak-to-Average Power Ratio Reduction of an OFDM Signal using Partial Transmit Sequences, *IEEE Communications Letters*, Vol. 4, No. 3, pp. 86-88.

Erdogan, A. T. (2006). A Low Complexity Multicarrier PAR Reduction Approach Based on Subgradient Optimization, *Signal Processing*, vol. 86, no. 12, pp. 3890-3903.

Fazel, K. & Kaiser, S. (2008). *Multi-Carrier and Spread Spectrum System: From OFDM and MC-CDMA to LTE and WiMAX*, Second ed. The Atrium, Southern Gate, Chichester, West Sussex PO19 8SQ, England: John Wiley & Sons Ltd.

Gatherer, A. & Polley, M. (1997). Controlling Clipping Probability in DMT Transmission, *Proceedings of the 31st Asilomar Conference on Signals, Systems, and Computers*, vol. 1, pp. 578-584, Pacific Grove, Calif, USA, November 1997.

Han, S. H.; Cioffi, J. M. & Lee, J. H. (2006). Tone Injection with Hexagonal Constellation for Peak-to-Average Power Ratio Reduction in OFDM, *IEEE Communications Letters*, Vol. 10, No. 9, pp. 646-648.

Hara, S. & Prasad, R. (1997). Overview of Multicarrier CDMA, *IEEE Communications Magazine*, Vol. 35, No. 12, pp. 126-133.

Haykin, S. (1996). *Adaptive Filter Theory (3rd ed.)*, Prentice-Hall, Inc., Upper Saddle River, NJ.

Jayalath, A. D. S. & Tellambura, C. (2000). Adaptive PTS Approach for Reduction of Peak-to-Average Power Ratio of OFDM Signals, *IEE Electronics Letters*, Vol. 36, No. 14, pp. 1226-1228.

Jones, A. E.; Wilkinson, T. A. & Barton, S. K. (1994). Block Coding Scheme for Reduction of Peak to Mean Envelope Power Ratio of Multicarrier Transmission Schemes, *IEE Electronics Letters*, Vol. 30, No. 25, pp. 2098-2099.

Kang, K.; Choi, K. & Kim, S. (2002). Reduced Search for Optimum Code Sets to Reduce PAPR in MC-CDMA System. *The 5th International Symposium on Wireless Personal Multimedia Communications*, pp. 135-139.

Krongold, B. S. & Jones, D .L. (2004). An Active-Set Approach for OFDM PAR reduction via Tone Reservation, *IEEE Transactions on Signal Processing*, vol. 52, no. 2, pp. 495- 509.

Li, X. & Cimini, L. J. (1997). Effects of Clipping and Filtering on the Performance of OFDM. *IEEE 47th Vehicular Technology Conference (VTC '97)*, pp. 1634-1638.

Ochiai, H. & Imai, H. (1998). OFDM-CDMA with Peak Power Reduction based on the Spreading Sequences. *IEEE International Communications Conference (ICC '98)*, pp. 1299-1303.

Ohkubo, N. & Ohtsuki, T. (2002). A Peak to Average Power Ratio Reduction of Multicarrier CDMA using Selected Mapping. *IEEE 56th Vehicular Technology Conference (VTC '02)*, pp. 2086-2090.

Paredes Hernandez, L. A. P. & García Otero, M. (2009). User Reservation Approach for Peak-to-Average Power Ratio Reduction in MC-CDMA Systems. *IEEE 69th Vehicular Technology Conference (VTC Spring '09)*, Barcelona, Spain, pp. 1-5.

Ruangsurat, N. & Rajatheva, R. M. A. P. (2001). An Investigation of Peak-to-Average Power Ratio in MC-CDMA combined with Partial Transmit Sequence. *IEEE Vehicular Technology Conference (VTC '01)*, pp. 761-765.

Sousa, M.; Vandenberghe, L.; Boyd, S. & Lebret, H. (1998). Applications of Second-Order Cone Programming, *Linear Algebra and its Applications*, vol. 284, pp. 193-228.

Tellado, J. & Cioffi, J. M. (1998). Revisiting DMT's PAR, *ETSI/ANSI TM6, contribution number TD08*.

Väänanen, O.; Vankka, J.; Viero, T. & Halonen, K. (2002). Reducing the Crest Factor of a CDMA Downlink Signal by Adding Unused Channelization Codes, *IEEE Communications Letters*, Vol. 6, No. 10, pp. 443-445.

Wang, M.; Quevedo, D. E.; Goodwin, G. C. & Krongold, B. S. (2008). A Complex-Baseband Active-Set Approach for Tone Reservation PAR Reduction in OFDM Systems, *Australian Communications Theory Workshop, 2008 (AusCTW 2008)*, pp. 113-118, Christchurch, New Zealand, Jan. 30 2008-Feb. 1 2008.

13

Cognitive Radio Communications for Vehicular Technology – Wavelet Applications

Murroni Maurizio[1] and Popescu Vlad[2]
[1]*Department of Electrical and Electronic Engineering, University of Cagliari*
[2]*Department of Electronics and Computers, Transilvania University of Braşov*
[1]*Italy*
[2]*Romania*

1. Introduction

Wireless communications are nowadays a dominant part of our lives: from domotics, through industrial applications and up to infomobility services. The key to the co-existence of wireless systems operating in closely located or even overlapping areas, is sharing of the spectral resource. The optimization of this resource is the main driving force behind the emerging changes in the policies for radio resources allocation. The current approach in spectrum usage specifies fixed frequency bands and transmission power limits for each radio transmitting system. This approach leads to a very low medium utilization factor for some frequency bands, caused by inefficient service allocation over vast geographical areas (radiomobile, radio and TV broadcasting, WiMAX) and also by the usage of large guard bands, obsolete now due to technological progress.

A more flexible use of the spectral resource implies that the radio transceivers have the ability to monitor their radio environment and to adapt at specific transmission conditions. If this concept is supplemented with learning and decision capabilities, we refer to the Cognitive Radio (CR) paradigm. Some of the characteristics of a CR include localization, monitoring of the spectrum usage, frequency changing, transmission power control and, finally, the capacity of dynamically altering all these parameters (Haykin, 2005). This new cognitive approach is expected to have an important impact on the future regulations and spectrum policies.

2. Cognitive radio techniques

The dynamic access at the spectral resource is of extreme interest both for the scientific community as, considering the continuous request for wideband services, for the development of wireless technologies. From this point of view, a fundamental role is played by the Institute of Electrical and Electronic Engineers (IEEE) which in 2007 formed the Standards Coordinating Committee (SCC) 41 on Dynamic Spectrum Access Networks (DySPAN) having as main objective a standard for dynamic access wireless networks. Still within the IEEE frame, the 802.22 initiative defines a new WRAN (Wireless Regional Area Network) interface for wideband access based on cognitive radio techniques in the TV guard bands (the so-called "white spaces").

Coupled with the advantages and flexibility of CR systems and technologies, there is an ever-growing interest around the world in exploiting CR-enabled communications in vehicular and transportation environments. The integration of CR devices and cognitive radio networks into vehicles and associated infrastructures can lead to intelligent interactions with the transportation system, among vehicles, and even among radios within vehicles. Thus, improvements can be achieved in radio resource management and energy efficiency, road traffic management, network management, vehicular diagnostics, road traffic awareness for applications such as route planning, mobile commerce, and much more.

Still open within the framework of dynamic and distributed access to the radio resource are the methods for monitoring the radio environment (the so-called "spectrum sensing") and the transceiver technology to be used on the radio channels.

A CR system works on an opportunistic basis searching for unused frequency bands called "white spaces" within the radio frequency spectrum with the intent to operate invisibly and without disturbing the primary users (PU) holding a license for one or more frequency bands. Spectrum sensing, that is, the fast and reliable detection of the PU's even in the presence of in-band noise, is still a very complex problem with a decisive impact on the functionalities and capabilities of the CRs.

The spectrum sensing techniques can be classified in two types: local and cooperative (distributed). The local techniques are performed by single devices exploiting the spectrum occupancy information in their spatial neighborhood and can be divided into three categories (Budiarjo et al., 2008): "matched filter" (detection of pilot signals, preambles, etc.), "energy detection" (signal strength analysis) and "feature detection" (classification of signals according to their characteristics). Also, a combination of local techniques in a multi-stage design can be used to improve the sensing accuracy (Maleki et al., 2010). Nevertheless, the above-mentioned techniques are mostly inefficient for signals with reduced power or affected by phenomena typical for vehicular technology applications, such as shadowing and multi-path fading. To overcome such problems, cooperatives techniques can be used. Cooperative sensing is based on the aggregation of the spectrum data detected by multiple nodes using cognitive convergence algorithms in order to avoid the channel impairment problems that can lead to false detections.

2.1 Spectrum sensing techniques

Matched filter

The optimal way for any signal detection is a matched filter, since it maximizes received signal-to-noise ratio. However, a matched filter effectively requires demodulation of a primary user signal. This means that cognitive radio has a priori knowledge of primary user signal at both PHY and MAC layers, e.g. modulation type and order, pulse shaping, packet format. Most of the wireless technologies in operation include the transmission of some sort of pilot sequence, to allow channel estimation, to beacon its presence to other terminals and to give a synchronization reference for subsequent messages. Secondary systems therefore can exploit pilot signals in order to detect the presence of transmissions of primary systems in their vicinity.

For example: TV signals have narrowband pilot for audio and video carriers, CDMA systems have dedicated spreading codes for pilot and synchronization channels OFDM packets have preambles for packet acquisition.

If *X[n]* is completely known to the receiver then the optimal detector for this case is

$$T(Y) = \sum_{n=0}^{N-1} Y[n]X[n] \mathop{\lessgtr}_{H0}^{H1} \gamma . \tag{1}$$

If γ is the detection threshold, then the number of samples required for optimal detection is

$$N = [Q^{-1}(P_D - Q^{-1}(P_{FD})]^2 (SNR)^{-1} = O(SNR)^{-1}, \tag{2}$$

where PD and PFD are the probabilities of detection and false detection respectively. Hence, the main advantage of matched filter is that due to coherency it requires less time to achieve high processing gain since only $O(SNR)^{-1}$ samples are needed to meet a given probability of detection constraint. However, a significant drawback of a matched filter is that a cognitive radio would need a dedicated receiver for every primary user class.

Energy detector

One approach to simplify matched filtering approach is to perform non-coherent detection through energy detection. This sub-optimal technique has been extensively used in radiometry. An energy detector can be implemented similar to a spectrum analyzer by averaging frequency bins of a Fast Fourier Transform (FFT), as outlined in figure 1. Processing gain is proportional to FFT size N and observation/averaging time T. Increasing N improves frequency resolution which helps narrowband signal detection. Also, longer averaging time reduces the noise power thus improves SNR.

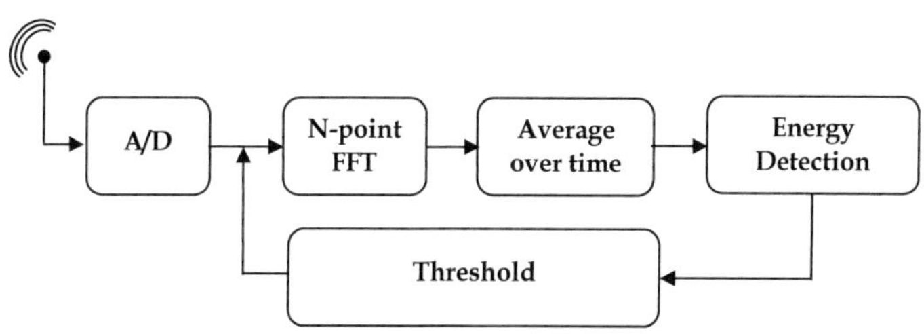

Fig. 1. Block diagram of a matched filter detector

$$T(Y) = \sum_{n=0}^{N-1} Y^2[n] \mathop{\lessgtr}_{H0}^{H1} \gamma \tag{3}$$

$$N = 2[(Q^{-1}(P_{Fd} - Q^{-1}(P_D))(SNR)^{-1} - Q^{-1}(P_D)] = O(SNR)^{-2} \tag{4}$$

Based on the above formula, due to non-coherent processing $O(SNR)^{-2}$ samples are required to meet a probability of detection constraint. There are several drawbacks of energy detectors that might diminish their simplicity in implementation. First, a threshold used for primary user detection is highly susceptible to unknown or changing noise levels. Even if the threshold would be set adaptively, presence of any in-band interference would confuse the energy detector. Furthermore, in frequency selective fading it is not clear how to set the threshold with respect to channel notches. Second, energy detector does not differentiate between modulated signals, noise and interference. Since it cannot recognize the interference, it cannot benefit from adaptive signal processing for canceling the interferer. Furthermore, spectrum policy for using the band is constrained only to primary users, so a cognitive user should treat noise and other secondary users differently. Lastly, an energy detector does not work for spread spectrum signals: direct sequence and frequency hopping signals, for which more sophisticated signal processing algorithms need to be devised (Cabric et al., 2004).

Cyclostationary feature detector

Another method for the detection of primary signals is Cyclostationary Feature Detection in which modulated signals are coupled with sine wave carriers, pulse trains, repeated spreading, hopping sequences, or cyclic prefixes. This results in built-in periodicity. These modulated signals are characterized as cyclostationary because their mean and autocorrelation exhibit periodicity. This periodicity is introduced in the signal format at the receiver so as to exploit it for parameter estimation such as carrier phase, timing or direction of arrival. These features are detected by analyzing a spectral correlation function. The main advantage of this function is that it differentiates the noise from the modulated signal energy. This is due to the fact that noise is a wide-sense stationary signal with no correlation however. Modulated signals are cyclostationary due to embedded redundancy of signal periodicity.
Analogous to autocorrelation function spectral correlation function (SCF) can be defined as:

$$S_x^{\alpha}(f) = lim_{\tau\to\infty} lim_{\Delta t\to\infty} \frac{1}{\Delta t} \int_{-\Delta t/2}^{\Delta t/2} \frac{1}{\tau} X_\tau(t, f+\alpha/2) X_\tau^*(t, f-\alpha/2)dt\,, \tag{5}$$

with the finite time Fourier transform given by

$$X_\tau(t,v) = \int_{t-\frac{\tau}{2}}^{t+\frac{\tau}{2}} x(u)e^{-j2\pi vu}du. \tag{6}$$

Spectral correlation function is also known as cyclic spectrum. While power spectral density (PSD) is a real valued one-dimensional transform, SCF is a complex valued two-dimensional transform. The parameter α is called the cycle frequency. If $a = 0$ then SCF gives the PSD of the signal.

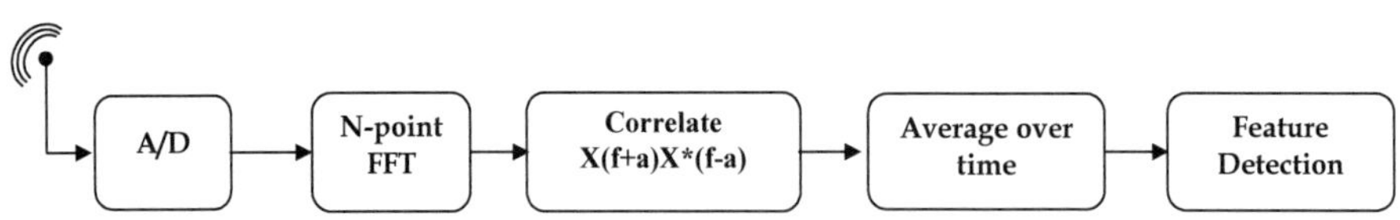

Fig. 2. Block diagram of a cyclostationary feature detector

Because of the inherent spectral redundancy signal selectivity becomes possible. Analysis of signal in this domain retains its phase and frequency information related to timing parameters of modulated signals. Due to this, overlapping features in power spectral

density are non-overlapping features in cyclic spectrum. Hence different types of modulated signals that have identical power spectral density can have different cyclic spectrum.
Implementation of a spectrum correlation function for cyclostationary feature detection is depicted in figure 2. It can be designed as augmentation of the energy detector from figure 1 with a single correlator block. Detected features are number of signals, their modulation types, symbol rates and presence of interferers. Table 1 presents examples of the cyclic frequencies adequate for the most common types of radio signals (Chang, 2006).
The cyclostationary detectors work in two stages. In the first stage the signal *x(k)*, that is transmitted over channel *h(k)*, has to be detected in presence of AWGN *n(k)*. In the second stage, the received cyclic power spectrum is measured at specific cycle frequencies. The signal S_j is declared to be present if a spectral component is detected at corresponding cycle frequencies α_j.

$$S_x^{\alpha}(f)=\begin{cases} S_n^0(f), & \alpha=0,\text{signal absent}\\ |H(f)|^2 S_s^0(f)+S_n^0(f), & \alpha=0,\text{signal present}\\ 0, & \alpha\neq 0,\text{signal absent}\\ H(f+\frac{\alpha}{2})H^*(f-\frac{\alpha}{2})S_s^{\alpha}(f), & \alpha\neq 0,\text{signal present}\end{cases} \tag{7}$$

The advantages of the cyclostationary feature detection are robustness to noise, better detector performance even in low SNR regions, signal classification ability and operation flexibility because it can be used as an energy detector in α = 0 mode. The disadvantage is a more complex processing than energy detection and therefore high speed sensing cannot be achieved. The method cannot be applied for unknown signals because an a priori knowledge of target signal characteristics is needed. Finally, at one time, only one signal can be detected: for multiple signal detection, multiple detectors have to be implemented or slow detection has to be allowed.

Type of Signal	Cyclic Frequencies
Analog Television	Cyclic frequencies at multiples of the TV-signal horizontal line-scan rate
AM signal: $x(t)=a(t)\cos(2\pi f_0 t+\phi_0)$	$\pm 2f_0$
PM and FM signal: $x(t)=\cos(2\pi f_0 t+\phi(t))$	$\pm 2f_0$
Amplitude-Shift Keying: $x(t)=[\sum_{n=-\infty}^{\infty} a_n p(t-nT_0-t_0)]\cos(2\pi f_0 t+\phi_0)$	$k/T_0 (k\neq 0)$ and $\pm 2f_0+k/T_0, k=0,\pm1,\pm2,\ldots$
Phase-Shift Keying: $x(t)=\cos[2\pi f_0 t+\sum_{n=-\infty}^{\infty} a_n p(t-nT_0-t_0)]$.	For QPSK, $k/T_0(k\neq 0)$, and for BPSK $k/T_0(k\neq 0)$ and $\pm 2f_0+k/T_0, k=0,\pm1,\pm2,\ldots$

Table 1. List of cyclic frequencies for various signal types

Besides cyclostationarity, other features of the received signals can be used for detecting the type of signal. One example is the classification of QPSK, 16QAM and 64QAM constellations based on normalized fourth and sixth-order cumulants (Swami et al., 2000). The cumulants are used as features for discriminating different classes of modulation schemes and are calculated

based on the coefficients of the fast Fourier transform. Cumulants are very robust in the presence of Gaussian Noise (higher order cumulants of Gaussian Noise are equal to zero).
When using the Wavelet Transform, the classification of MPSK and MQAM modulations can be done using the normalized histogram of the wavelet coefficients (Prakasam et al., 2008).

Wavelet packets - subband analysis

Within the energy detection method, a particular attention needs to be paid to the properties of the packets wavelet transformation for subband analysis, which, according to the literature, seems to be a feasible alternative to the classical FFT-based energy detection. Vehicular applications are in most cases characterized by the need of coping with fast changes in the radio environment, which lead, in this specific case of cognitive communication, to constrains in terms of short execution time of the spectrum sensing operations. From this point of view, the computational complexity of the wavelet packets method is of the same order of the state-of-the-art FFT algorithms, but the number of mathematical operations is lower using IIR polyphase filters (Murroni et al., 2010).
To define the wavelet packet basis functions we refer to wavelet multiresolution analysis (WMRA). Let $g_0[n]$ be a unit-energy real causal FIR filter of length which is orthogonal to its even translates; i.e., $\sum_n g_0[n]g_0[n-2m]=\delta[m]$, where $\delta[m]$ is the Kroneker delta, and let be the (conjugate) quadrature mirror filter (QMF), $g_1[n]=(-1)^n g_0[N-1-n]$. If $g_0[n]$ satisfies some mild technical conditions, we can use an iterative algorithm to find the function $\phi_{01}(t)=\sqrt{2}\sum_n g_0[n]\phi_{01}(2t-nT_0)$ for an arbitrary interval T_0. Subsequently, we can define the family of functions ϕ_{lm}, $l\geq 0$, $1\leq m\leq 2^l$ in the following (binary) tree-structured manner:

$$\begin{cases}\phi_{l+1,2m-1}(t)=\sum_n g_0[n]\phi_{lm}(t-nT_l)\\ \phi_{l+1,2m}(t)=\sum_n g_1[n]\phi_{lm}(t-nT_l)\end{cases} \qquad (8)$$

where $T_l=2^lT_0$. For any given tree structure, the function at the leafs of the tree form a wavelet packet.
Wavelet packets have a finite duration, $(N-1)T_l$ and are self- and mutually-orthogonal at integer multiples of dyadic intervals. Therefore, they are suitable for subband analysis: a generic signal $x(t)$ can be then decomposed on the wavelet packet basis and represented as a collection of coefficients belonging to orthogonal subbands. Therefore, the total power of $x(t)$ can be evaluated as sum of the contributes of each subband which can be separately computed in the wavelet domain. Let S_k be the k-th subband; if we denote by $\{c_{k,i}\}$ the wavelet coefficients of S_k, the power contribute of S_k is

$$P_k=\frac{2^l}{(N-1)T_l}\sum_i c_{k,i}^2 . \qquad (9)$$

Figure 3 shows an example of a binary tree decomposition (Fig.3a) and the relevant symbolic subband structure (Fig.3b). It is noticeable how for $l>1$ (i.e., packet composed by more than 4 leafs) in the frequency Fourier domain the wavelet packets are not ordered as in the corresponding tree.
A drawback to WMRA as described so far is the higher computational complexity compared to classical Fourier subband analysis. Computational burden is reduced by deploying IIR

polyphase filter banks. It is shown that, whereas the computational complexity of the WMRA based on IIR polyphase filters is of the same order of the state-of-the-art FFT algorithms, the number of mathematical operations is lower. Hence, the wavelet subband division and power calculation is a valid candidate for being used as an alternative to the classic FFT-based energy detection It is therefore suitable for environments where computation time is critical, as the vehicular technology applications are.

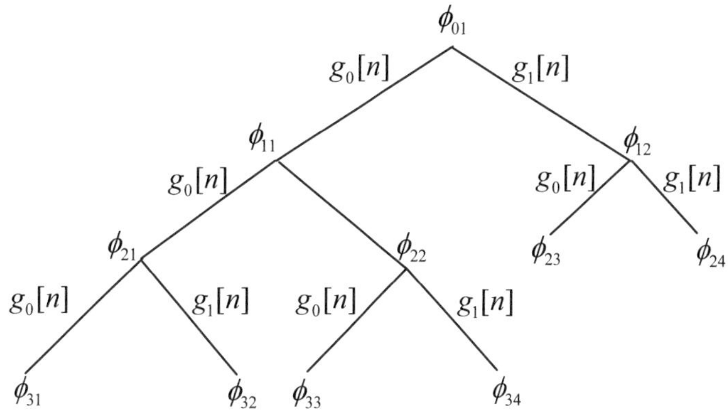

(a)

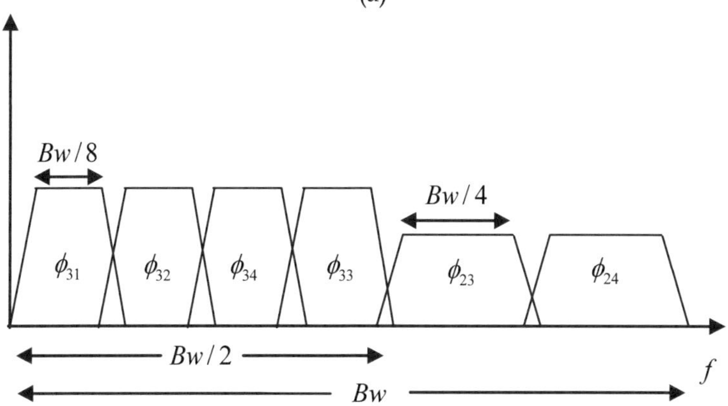

(b)

Fig. 3. (a) Wavelet tree structure (b) Corresponding symbolic subband structure

Combined two-stage detection

Since cyclostationary feature detection is somehow complementary to the energy detection, performing better for narrow bands, a combined approach is suggested in (Maleki et al., 2010), using energy detection for wideband sensing and then, for each detected single channel, a feature detection is applied in order to make the final decision whether the channel is occupied or not. First, a coarse energy detection stage is performed over a wider frequency band. Subsequently the presumed free channels are analyzed with the feature detector in order to take the decision.

We investigated in (Murroni et al., 2010) this combined 2-stage spectrum sensing method for a Wavelet approach, using the wavelet packet transform and the resulting coefficients both

for energy detection as for feature detection. Our research took into consideration a range of frequencies particularly interesting for Vehicular Technologies approaches, the DVB-T channels.
DWPT analysis divides the sensed frequency range into 32 sub-bands in order to contain the all DVB-T channels. The matching process is performed taking into account the known discrepancy between the TV channels order in the Fourier domain and the corresponding sub-bands order in the wavelet domain.
Based on the calculated sub-band power, one or more channels are marked as "white" and can be directly used further by a cognitive transmission system. The list of channels whose signal powers are above a certain threshold is passed to the second stage of the system where a sequential analysis of each of these channels is performed.

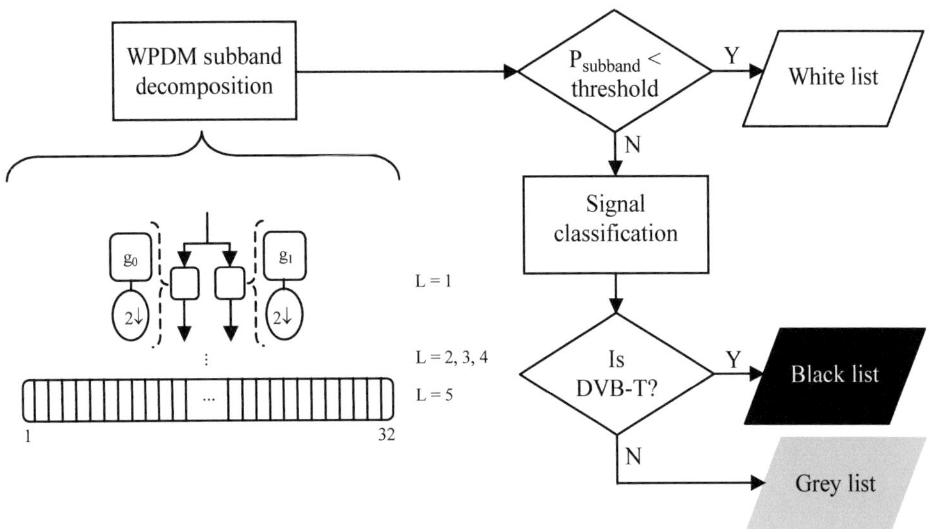

Fig. 4. Flowchart of an application scenario for a two-stage spectrum-sensing scheme in the DVB-T UHF band

The sub-band power computed in the first step is used as a factor to appropriately adjust the values for the thresholds, proportionally with the signal's intensity. Consequently we can compare the corresponding first and second order moments of the normalized histogram in order to distinguish whether the signal is a PU's signal, hence a DVB-T signal with a standard modulation, or not (Figure 4).
If the channel is identified as being occupied by a PU, the corresponding channel is definitively marked as "black", meaning it is undoubtedly used by PUs and therefore not suitable for transmission. If the statistical analysis fails to identify a DVB-T type signal, we can categorize the channel as being "grey", which in our scenario means that there is no PU transmitting, but still the channel is occupied, most probably by another SU. Therefore, the channel is not completely discarded, being a potential candidate to be analyzed again after a certain amount of time in order to be re-evaluated and eventually included in the white list.

2.2 Cooperative issues

Detection of primary user by the secondary system is critical in a cognitive radio environment. However this is rendered difficult due to the challenges in accurate and reliable sensing of the wireless environment. Secondary users might experience losses in the signal which can result in an incorrect judgment of the wireless environment, which can in turn cause interference at the licensed primary user by the secondary transmission. Furthermore, the issues with signal quality are aggravated when secondary users rapidly change location, as it is the case for specific vehicular technology applications. Briefly, as shown in figure 5, unreliable results can be produced based on the following phenomena:

- Multipath: a sensor CR1 under multipath receiving conditions features short term Rayleigh fading. The fluctuations of the power level may cause unreliable detection.
- Shadowing: a sensor CR2 may move behind an obstacle, exhibiting lognormal long term fading. Its covered position may create disturbance for a PRx in its proximity (hidden terminal problem).
- Distance-dependent path loss: a sensor CR3 lies outside the primary transmission range. It receives a low power level due to the distance, but its transmission can produce interference to the primary receiver, which is inside the primary range.

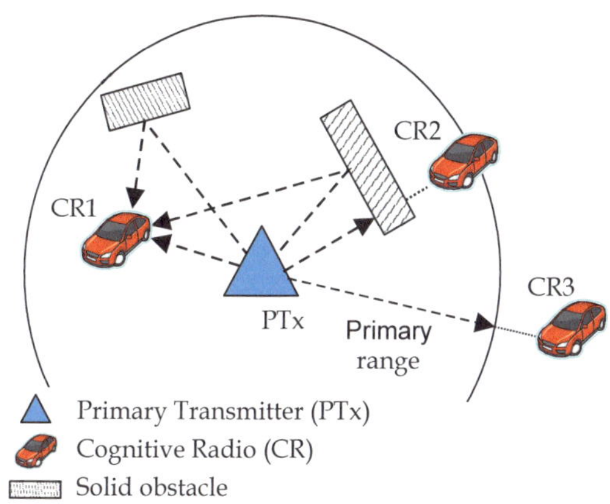

Fig. 5. Layout of a network with moving terminals

This arises the necessity for the cognitive radio to be highly robust to channel impairments and also to be able to detect extremely low power signals. These stringent requirements pose a lot of challenges for the deployment of CR networks.

Channel impairments and low power detection problems in CR can be alleviated if multiple CR users cooperate in sensing the channel. (Thanayankizil & Kailas, 2008) suggest different cooperative topologies that can be broadly classified into three regimes according to their level of cooperation:

Decentralized Uncoordinated Techniques: the cognitive users in the network don't have any kind of cooperation which means that each CR user will independently detect the channel, and if a CR user detects the primary user it would vacate the channel without informing the other users. Uncoordinated techniques are fallible in comparison with coordinated techniques. Therefore, CR users that experience bad channel realizations (shadowed regions) detect the channel incorrectly thereby causing interference at the primary receiver.

Centralized Coordinated Techniques: in these kinds of networks, an infrastructure deployment is assumed for the CR users. CR user that detects the presence of a primary transmitter or receiver informs a CR controller. The CR controller can be a wired immobile device or another CR user. The CR controller notifies all the CR users in its range by means of a broadcast control message. Centralized schemes can be further classified according to their level of cooperation into

- Partially Cooperative: in partially cooperative networks nodes cooperate only in sensing the channel. CR users independently detect the channel inform the CR controller which then notifies all the CR users. One such partially cooperative scheme was considered by (Liu & Shankar, 2006) where a centralized Access Point (CR controller) collected the sensory information from the CR users in its range and allocated spectrum accordingly;
- Totally Cooperative Schemes: in totally cooperative networks nodes cooperate in relaying each other's information in addition to cooperatively sensing the channel. For example, two cognitive users D1 and D2 are assumed to be transmitting to a common receiver and in the first half of the time slot assigned to D1, D1 transmits and in the second half D2 relays D1's transmission. Similarly, in the first half of the second time slot assigned to D2, D2 transmits its information and in the second half D1 relays it.

Decentralized Coordinated Techniques: various algorithms have been proposed for the decentralized techniques, among which the gossiping algorithms, which do cooperative sensing with a significantly lower overhead. Other decentralized techniques rely on clustering schemes where cognitive users form in to clusters and these clusters coordinate amongst themselves, similar to other already known sensor network architecture (i.e. ZigBee).

All these techniques for cooperative spectrum sensing raise the need for a control channel that can be either implemented as a dedicated frequency channel or as an underlay UWB channel. Wideband RF front-end tuners/filters can be shared between the UWB control channel and normal cognitive radio reception/transmission. Furthermore, with multiple cognitive radio groups active simultaneously, the control channel bandwidth needs to be shared. With a dedicated frequency band, a CSMA scheme may be desirable. For a spread spectrum UWB control channel, different spreading sequencing could be allocated to different groups of users.

2.3 Transmission techniques

In a CR environment, terminals are assumed to be able to detect any unoccupied frequencies and to estimate the strength of the received signal of nearby primary users by spectrum sensing, as presented in the previous section. Once a CR user detects free frequency spectrum within the licensed frequency range, he may negotiate with the primary system, or begin data transmission without extra permission, depending on the CR system structure. If any primary users become active in the same frequency band later on, the CR user has to clear this band as soon as possible, giving priority to the primary users. Also, CR users should quit their communication if the estimated SNR levels of the primary users are below an acceptable level. When a CR user operates in a channel adjacent to any active primary users' spectrums, ACI (adjacent channel interference) occurs between the two parties. However, the performance of the primary system should be maintained, whether spectrum sharing is allowed or not. We assume that a minimum SNR requirement is predefined for the primary system so that the maximum allowable ACI at each location can be evaluated by the CR user. The CR user can then determine whether he may use the frequency band or not. At the same time, the CR user needs to avoid the influence of interference from primary users in order to maximize its own data throughput.

Other properties of his type of radio are the ability to operate at variable symbol rates, modulation formats (e.g. low to high order QAM), different channel coding schemes, power

levels and the use of multiple antennas for interference nulling, capacity increase or range extension (beam forming).
The most likely basic strategy will be based on multicarrier OFDM-like modulation across the entire bandwidth in order to most easily resolve the frequency dimension with subsequent spatial and temporal processing.

OFDM Modulation

OFDM has become the modulation of choice in many broadband systems due to its inherent multiple access mechanism and simplicity in channel equalization, plus benefits of frequency diversity and coding. The transmitted OFDM waveform is generated by applying an inverse fast Fourier transform (IFFT) on a vector of data, where number of points N determines the number of sub-carriers for independent channel use, and minimum resolution channel bandwidth is determined by W/N, where W is the entire frequency band accessible by any cognitive user.
The frequency domain characteristics of the transmitted signal are determined by the assignment of non-zero data to IFFT inputs corresponding to sub-carriers to be used by a particular cognitive user. Similarly, the assignment of zeros corresponds to channels not permitted to use due to primary user presence or channels used by other cognitive users. The output of the IFFT processor contains N samples that are passed through a digital-to-analog converter producing the wideband waveform of bandwidth W. A great advantage of this approach is that the entire wideband signal generation is performed in the digital domain, instead of multiple filters and synthesizers required for the signal processing in analog domain.
From the cognitive network perspective, OFDM spectrum access is scalable while keeping users orthogonal and non-interfering, provided the synchronized channel access. However, this conventional OFDM scheme does not provide truly band-limited signals due to spectral leakage caused by sinc-pulse shaped transmission resulted from the IFFT operation. The slow decay of the sinc-pulse waveform, with first side lobe attenuated by only 13.6dB, produces interference to the adjacent band primary users which is proportional to the power allocated to the cognitive user on the corresponding adjacent sub-carrier. Therefore, a conventional OFDM access scheme is not an acceptable candidate for wideband cognitive radio transmission.
To overcome these constraints (Rajbanshi et. al., 2006) suggest non-contiguous OFDM (NC-OFDM) as an alternative, a schematic of an NC-OFDM transceiver being shown in figure 6. The transceiver splits a high data rate input, $x(n)$, into N lower data rate streams. Unlike conventional OFDM, not all the sub carriers are active in order to avoid transmission unoccupied frequency bands. The remaining active sub carriers can either be modulated using M-ary phase shift keying (MPSK), as shown in the figure, or M-ary quadrature amplitude modulation (MQAM). The inverse fast Fourier transform (IFFT) is then used to transform these modulated sub carrier signals into the time domain. Prior to transmission, a guard interval, with a length greater than the channel delay spread, is added to each OFDM symbol using the cyclic prefix (CP) block in order to mitigate the effects of inter-symbol interference (ISI). Following the parallel-to-serial (P/S) conversion, the base band NC-OFDM signal, $s(n)$, is then passed through the transmitter radiofrequency (RF) chain, which amplifies the signal and upconverts it to the desired centre frequency. The receiver performs the reverse operation of the transmitter, mixing the RF signal to base band for processing, yielding the signal $r(n)$. Then the signal is converted into parallel streams, the cyclic prefix is discarded, and the fast Fourier transform (FFT) is applied to transform the time domain data

into the frequency domain. After the distortion from the channel has been compensated via per sub carrier equalization, the data on the sub carriers is demodulated and multiplexed into a reconstructed version of the original high-speed input.

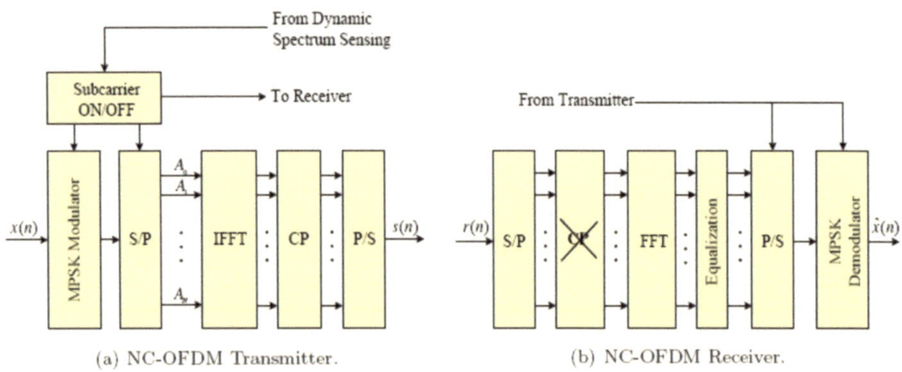

Fig. 6. Schematic of an NC – OFDM transceiver

NC-OFDM was evaluated and compared, both qualitatively and quantitatively with other candidate transmission technologies, such as MC-CDMA and the classic OFDM scheme. The results show that NC-OFDM is sufficiently agile to avoid spectrum occupied by incumbent user transmissions, while not sacrificing its error robustness.

Wavelet packet transmission method

In the last 10 years another multicarrier transmission technique has emerged as a valid alternative to OFDM and its modified versions.
The theoretical background relies on the synthesis of the discrete wavelet packet transform that constructs a signal as the sum of M = 2J waveforms. Those waveforms can be built by J successive iterations each consisting of filtering and upsampling operations. Noting $\langle \cdot,\cdot \rangle$ the convolution operation, the algorithm can be written as:

$$\begin{cases} \varphi_{j,2m}[k] = \langle h_{lo}^{rec}[k], \varphi_{j-1,m}[k/2] \rangle \\ \varphi_{j,2m}[k] = \langle h_{hi}^{rec}[k], \varphi_{j-1,m}[k/2] \rangle \end{cases}, \tag{10}$$

with

$$\varphi_{j,2m}[k] = \begin{cases} 1, \; for \; k = 1 \\ 0, \; otherwise \end{cases} \forall m \,, \tag{11}$$

where j is the iteration index, $1 \le j \le J$ and m the waveform index $0 \le m \le M - 1$.
Using usual notation in discrete signal processing, $\varphi_{j,m}[k/2]$ denotes the upsampled-by-two version of $\varphi_{j,m}[k]$. For the decomposition, the reverse operations are performed, leading to the complementary set of elementary blocks constituting the wavelet packet transform depicted in Figure 7. In orthogonal wavelet systems, the scaling filter h_{lo}^{rec} and dilatation filter h_{hi}^{rec} form a quadrature mirror filter pair. Hence knowledge of the scaling filter and

wavelet tree depth is sufficient to design the wavelet transform. It is also interesting to notice that for orthogonal WPT, the inverse transform (analysis) makes use of waveforms that are time-reversed versions of the forward ones. In communication theory, this is equivalent to using a matched filter to detect the original transmitted waveform.

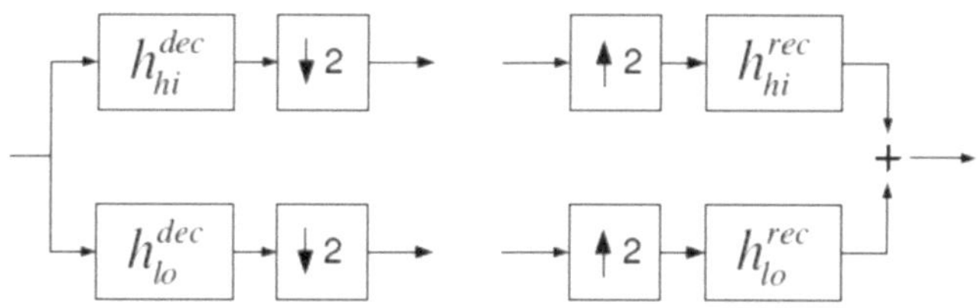

Fig. 7. Wavelet packet elementary block decomposition and reconstruction

A particularity of the waveforms constructed through the WPT is that they are longer than the transform size. Hence, WPM belongs to the family of overlapped transforms, the beginning of a new symbol being transmitted before the previous one(s) ends. The waveforms being M-shift orthogonal, the inter-symbol orthogonality is maintained despite this overlap of consecutive symbols. This allows taking advantage of increased frequency domain localization provided by longer waveforms while avoiding system capacity loss that normally results from time domain spreading. The waveforms length can be derived from a detailed analysis of the tree algorithm. Explicitly, the wavelet filter of length L_0 generates M waveforms of length

The construction of a wavelet packet basis is entirely defined by the wavelet-scaling filter, hence its selection is critical. This filter solely determines the specific characteristics of the transform. In multicarrier systems, the primary characteristic of the waveform composing the multiplex signal is out-of-band energy. Though in an AWGN channel this level of out-of-band energy has no effect on the system performance thanks to the orthogonality condition, this is the most important source of interference when propagation through the channel causes the orthogonality of the transmitted signal to be lost. A waveform with higher frequency domain localization can be obtained with longer time support. On the other hand, it is interesting to use waveforms of short duration to ensure that the symbol duration is far shorter than the channel coherence time. Similarly, short waveforms require less memory, limit the modulation-demodulation delay and require less computation. Those two requirements, corresponding to good localization both in time and frequency domain, cannot be chosen independently. In fact, it has been shown that in the case of wavelets, the bandwidth-duration product is constant. This is usually referred to as the uncertainty principle.

Finally, a minor difference between OFDM and WPM remains to be emphasized. In the former, the set of waveforms is by nature defined in the complex domain. WPM, on the other hand, is generally defined in the *real* domain but can be also defined in the *complex* domain, solely depending of the scaling and dilatation filter coefficients. Since the most commonly encountered WPT are defined in the real domain, it has naturally led the authors to use PAM. It is nevertheless possible to translate the M real waveform directly in the complex domain. The resulting complex WPT is then composed of 2M waveforms forming an orthogonal set.

In WPDM binary messages $x_{lm}[n]$ have polar representation (i.e., $x_{lm}[n] = \pm 1$), waveform coded by pulse amplitude modulation (PAM) of $\phi_{lm}(t - nT_l)$ and then added together to form the composite signal $s(t)$. WPDM can be implemented using a transmultiplexer and a single modulator. For a two level decomposition

$$s(t)=\sum_k x_{01}[k]\phi_{01}(t-kT_0), \tag{12}$$

where $x_{01}[k]=\sum_{(l,m)\in\Gamma}\sum_n f_{lm}[k-2^l n]$, with Γ being the set of terminal index pairs and $f_{lm}[k]$ the equivalent sequence filter from the $(l,m)-th$ terminal to the root of the tree, which can be found recursively from (8). The original message can be recovered from $x_{01}[k]$ using

$$x_{lm}[n]=\sum_k f_{lm}\left[k-2^l n\right]x_{01}[k]. \tag{13}$$

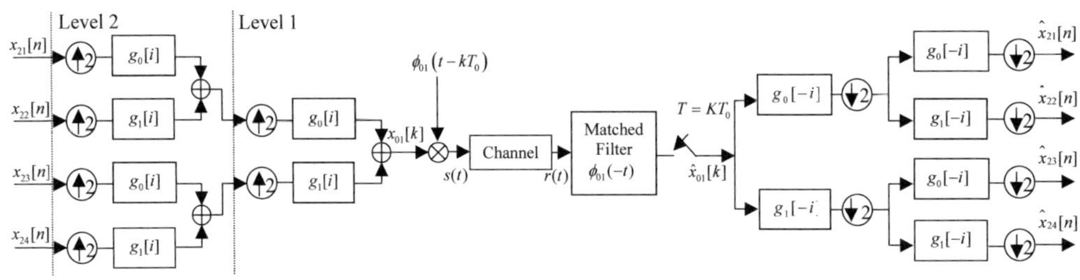

Fig. 8. Transmitter and receiver for a two-level WPDM system

Adaptive modulation

Adaptive modulation is only appropriate for duplex communication between two or more stations because the transmission parameters have to be adapted using some form of a two-way transmission in order to allow channel measurements and signaling to take place. Transmission parameter adaptation is a response of the transmitter to the time-varying channel conditions. In order to efficiently react to the changes in channel quality, the following steps need to be taken:

- Channel quality estimation: to appropriately select the transmission parameters to be employed for the next transmission, a reliable estimation of the channel transfer function during the next active transmission slot is necessary. This is done at the receiver and the information about the channel quality is sent to the transmitter for next transmission through a feedback channel.
- Choice of the appropriate parameters for the next transmission: based on the prediction of the channel conditions for the next time slot, the transmitter has to select the appropriate modulation modes for the sub-carriers.
- Signaling or blind detection of the employed parameters: the receiver has to be informed, as to which demodulator parameters to employ for the received packet.

In a scenario where channel conditions fluctuate dynamically, systems based on fixed modulation schemes do not perform well, as they cannot take into account the difference in channel conditions. In such a situation, a system that adapts to the worst-case scenario would have to be built to offer an acceptable bit-error rate. To achieve a robust and a spectrally efficient communication over multi-path fading channels, adaptive modulation is used, which adapts the transmission scheme to the current channel characteristics. Taking advantage of the time-varying nature of the wireless channels, adaptive modulation based systems alter transmission parameters like power, data rate, coding, and modulation schemes, or any combination of these in accordance with the state of the channel. If the channel can be estimated properly, the transmitter can be easily made to adapt to the current channel conditions by altering the modulation schemes while maintaining a constant BER. This can be typically done by estimating the channel at the receiver and transmitting this estimate back to the transmitter. Thus, with adaptive modulation, high spectral efficiency can be attained at a given BER in good channel conditions, while a reduction in the throughput is experienced in degrading channel conditions. The basic block diagram of an adaptive modulation based cognitive radio system is shown in figure 9. The block diagram provides a detail view of the whole adaptive modulation system with all the necessary feedback paths.

It is assumed that the transmitter has a perfect knowledge of the channel and the channel estimator at the receiver is error-free and there is no time delay. The receiver uses coherent detection methods to detect signal envelopes. The adaptive modulation, M-ary PSK, M-QAM, and M-ary AM schemes with different modes are provided at the transmitter. With the assumption that the estimation of the channel is perfect, for each transmission, the mode is adjusted to maximize the data throughput under average BER constraint, based on the instantaneous channel SNR. Based on the perfect knowledge about the channel state information (CSI), at all instants of time, the modes are adjusted to maximize the data throughput under average BER constraint.

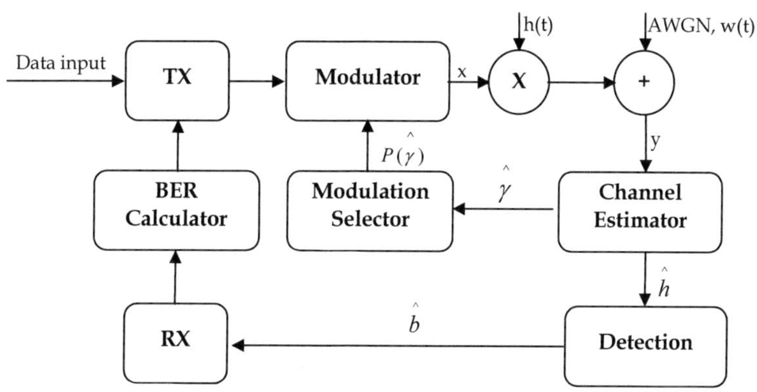

Fig. 9. Basic block diagram of an adaptive modulation - based cognitive radio system

The data stream, *b(t)* is modulated using a modulation scheme given by $P_k(\hat{\gamma})$, the probability of selecting k^{th} modulation mode from K possible modulation schemes available at the transmitter, which is a function of the estimated SNR of the channel. Here, *h(t)* is the fading channel and *w(t)* is the AWGN channel. At the receiver, the signal can be modeled as:

$$y(t) = h(t)\, x(t) + w(t) \tag{14}$$

where *y(t)* is the received signal, *h(t)* is the fading channel impulse response, and *w(t)* is the Additive White Gaussian Noise (AWGN). The estimated current channel information is returned to the transmitter to decide the next modulation scheme. The channel state information $\hat{h}(t)$ is also sent to the detection unit to get the detected stream of data, $\hat{b}(t)$.

3. Conclusion

Our research investigates the use of the Wavelet transform and the Wavelet Packets for cognitive radio purposes. We are applying the wavelet approach both for spectrum sensing, as for adaptive multicarrier transmission, for offering a complete, wavelet-based solution for cognitive application applied on the problematic of vehicular communication (channel impairments, high relative velocity of the communication peers).

4. References

Haykin, S. (2005). Cognitive radio: Brain-empowered wireless communications, *IEEE Journal on Selected Areas in Communications, vol. 25*, pp. 201–22, February 2005.

Budiarjo, I., Lakshmanan, M. K. & H. Nikookar (2008). Cognitive Radio Dynamic Access Techniques. In *Wireless Personal Communications,* Volume 45, Number 3 / May, 2008, Springer Netherlands

Maleki, S., Pandharipande, A. & Leus, G. (2010). Two-Stage Spectrum Sensing for Cognitive Radios. *Proceedings of the ICASSP 2010 Conference*, March 2010, Dallas, USA.

Murroni, M. & Popescu, V. (2010). Cognitive Radio HDTV Multi-Vision System in the 700 MHz UHF TV Band, *IEEE International Symposium on Broadband Multimedia Systems and Broadcasting 2010*, Shanghai, 2010

TWWong, K. M., Wu, J., Davidson, T. N. & Jin, Q. (1997). Wavelet packet division multiplexing and wavelet packet design under timing error effects, *IEEE Trans. Signal Processing, vol. 45*, no. 12, pp- 2877-2890, December 1997.

Cabric, D., Mishra, S.M. & Brodersen, R.W. (2004). Implementation issues in spectrum sensing for cognitive radios. *Signals, Systems and Computers, 2004. Conference Record of the Thirty-Eighth Asilomar Conference*, vol.1, no., pp. 772- 776 Vol.1, 7-10 Nov. 2004

Chang, S (2006). Analysis of Proposed Sensing Schemes: *IEEE 802.22-06/0032r0.* February 2006

Thanayankizil, L. & Kailas, A. (2008). Spectrum Sensing Techniques (II): Receiver Detection and Interference Management

Liu, X. & Shankar, S. (2006). Sensing-based opportunistic channel access, *ACM Journal on Mobile Networks and Applications (MONET)*, Vol. 11, No. 1, Feb. 2006, p. 577-591, 2006.

Rajbanshi R., Chen Q., Wyglinski A. M., Evans J.B. & Minden G. J (2006). Comparative Study of Frequency Agile Data Transmission Schemes for Cognitive RadioTransceivers, *ACM International Conference Proceeding Series; Vol. 222, Proceedings of the first international workshop on Technology and policy for accessing spectrum* Boston, 2006

Maleki, S., Pandharipande A. & Leus, G. (2010). Two-Stage Spectrum Sensing for Cognitive Radios. *Proceedings of the ICASSP 2010 Conference,* March 2010, Dallas, USA.

Swami, A. &Sadler, B.M., (2000) Hierarchical digital modulation classification using cumulants. *IEEE Trans. Communications,* vol. 48, pp. 416-429, March 2000.

Prakasam, P. & Madheswaran, M. (2008). M-ary Shift Keying Modulation Scheme Identification Algorithm Using Wavelet Transform and Higher Order Statistical Moments. *Journal of Applied Science* 8(1), pages 112-119, 2008.

Murroni, M., Fadda, M. & Popescu, V. (2010). Spectrum Sensing in the DVB-T Bands using Combined Energy Detection and Signal Classification in the Wavelet Domain. *Wireless Personal Multimedia Communications Symposium* 2010 (WPMC 2010); Recife, Brazil.

14

Multiple Antenna-Aided Spectrum Sensing Using Energy Detectors for Cognitive Radio

Seung-Hoon Hwang and Jun-Ho Baek
Dongguk University-Seoul,
Korea

1. Introduction

Recently a spectral scarcity has arisen as a serious problem, because current static frequency allocation schemes cannot further accommodate an increasing number of devices requesting higher data rate. Therefore, CR (cognitive radio) technique has been considered as a potential solution to improve spectrum utilization by sharing the spectrum. In this chapter, the main concept of CR was introduced.

1.1 Motivation

The demand for higher data rates is increasing as a result of the transition from voice communications to multimedia applications such as video streaming service, photo mail, and DMB (Digital Multimedia Broadcasting: both satellite or terrestrial type employed in Korea). Generally, wide bandwidth is required to achieve high data rate properly. As you know, most of popular radio spectrums are already assigned. Additional spectrum is not enough to be assigned for new application service. The other problem is that spectrum utilization less than 3GHz concentrates only in several frequency bands, while the majority of frequency bands are inefficiently utilized. According to the FCC's (Federal Communications Commission) report by spectrum policy task force, the usage of allocated spectrum varied around 15% to 85% depending on temporal and geographic situations. Therefore, the new paradigm of using spectrum more efficiently has urged to create a new wireless communication technology.

1.2 Overview of cognitive radio

IEEE 802.22 based WRAN (wireless regional area network) devices sense TV channels and identifies opportunities for transmission. Figure 1-1 shows example of deployment for IEEE 802.22 WRAN. Recently, IEEE 802.22 standards have included cognitive features for the first time. We may say the trend is targeting at this direction, even though it is difficult to expect a wireless standard which is based on wideband spectrum sensing and opportunistic exploitation of the spectrum.

In CR terminology (I. Mitola, J. & J. Maguire, 1999), primary or incumbent user can be defined as the users who have higher priority rights on the usage of a specific part of the spectrum. On the other hand, secondary users with lower priority exploit this spectrum in such away that they do not cause interference to primary users and other secondary users.

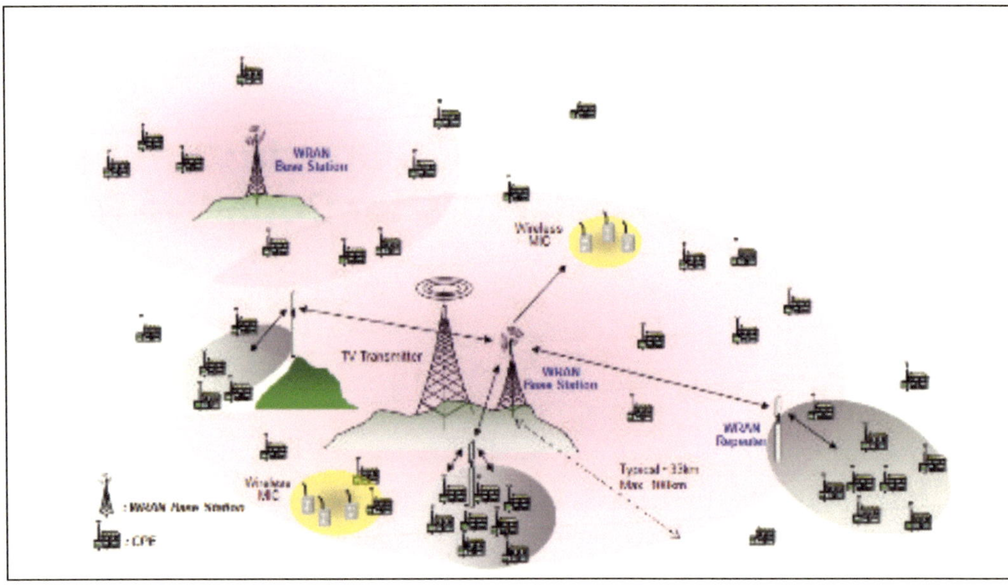

Fig. 1-1. IEEE 802.22 WRAN Deployment Scenario

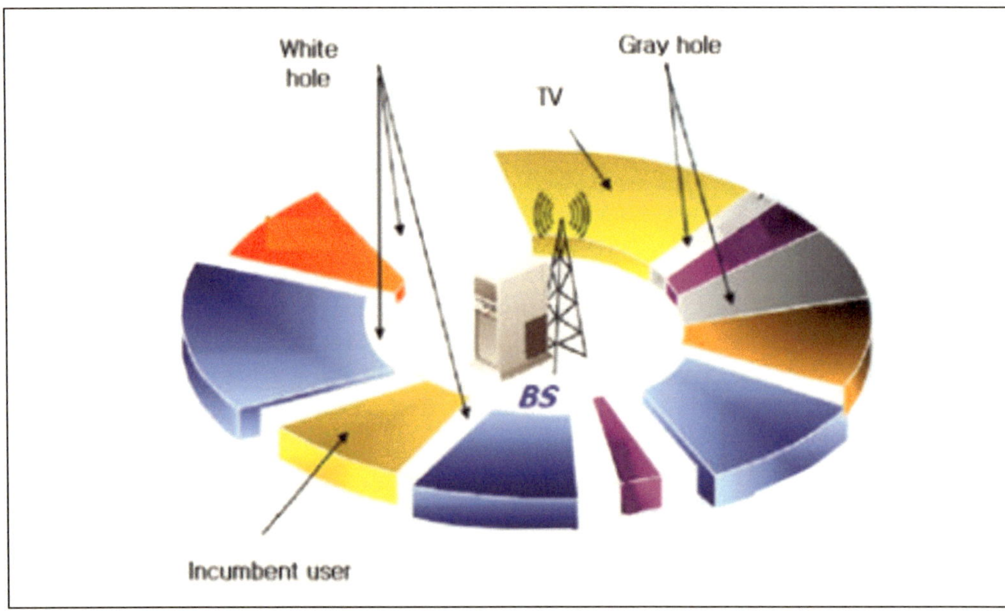

Fig. 1-2. Spectrum Hole Classification (Ian F Akyildiz et al)

In Figure 1-2, spectrum hole can be classified as a black space, white space and gray space. In the white space, there is no interference except noise for the frequency. The Gray space indicated that this spectrum is partially used under acceptable interference. The black space is occupied by incumbent user. The CR technique is aiming at usage of unoccupied spectrum such as the white space or the gray space by adopting the concept of dynamic and autonomous spectrum management, while ensuring the right of privileged primary users.

Therefore, the secondary users should monitor licensed bands and transmit their signals opportunistically whenever no primary signal is detected.
Main functionalities of the CR concept are supported by the ability to measure, sense, learn, and be aware of the parameters related to the radio channel characteristics, the availability of spectrum and power, radio' s operating environment, user requirements and applications, available networks (infrastructures) and nodes, local policies, and the other operating restrictions .
In this chapter, we focus on the spectrum sensing scheme which is the most crucial part to make the CR functionality realized. In Chapter 2, we explain the applicable spectrum sensing methods as well as the cooperative sensing concept. The proposed sensing method was introduced in Chapter 3. Various sensing performances are represented and compared in Chapter 4. Finally, our conclusions were given in Chapter 5.

2. Various spectrum sensing methods

In order to protect the primary user from the secondary users, the spectrum sensing is a key function to decide whether frequency band is empty or not. Generally, various methods such as matched filter detection, cyclostationary feature detection and energy detection have been categorized for the spectrum sensing. In this chapter, we will briefly review several spectrum sensing methods and derive analytical performance of energy detector under AWGN channel environment case.

2.1 Matched filter detection

Matched-filtering is known as the optimum method for detection of the primary users, when the transmitted signal is known. (J. G. Proakis, 2001) The main advantage of the matched filtering is that it takes short time to achieve the spectrum sensing under a certain value of the probability of false alarm or the probability of misdetection, compared to the other methods (R. Tandra & A. Sahai, 2005). However, the matched-filtering requires the perfect knowledge of the primary users' signaling features such as bandwidth, operating frequency, modulation type and order, pulse shaping, and frame format. Moreover, the implementation complexity of sensing unit is impracticably large since the CR needs receivers for all signal types (D. Cabric et al., 2004). Another disadvantage of the match filtering is large power consumption, because various receiver algorithms need to be executed for detection.

2.2 Cyclostationary feature detection

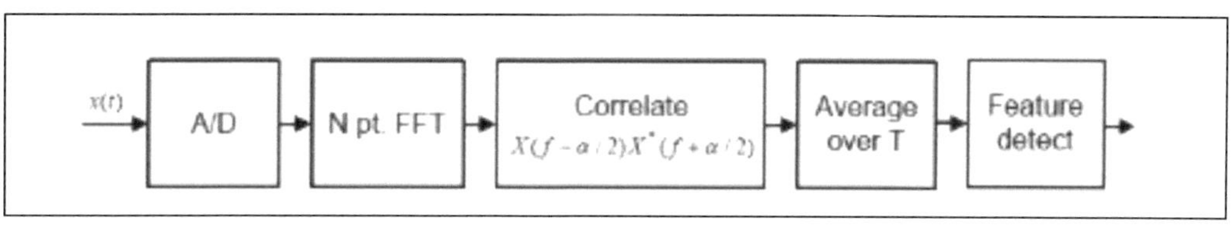

Fig. 2-1. The Structure of Cyclostationary Feature Detector

In the Figure 2-1, cyclostationary features are caused by the periodicity in the signal or its statistics such as mean and autocorrelation or they can be intentionally induced to assist spectrum sensing(O. A. Dobre, et al., 2008), (Mahapatra R & Krusheel, M, 2008). The cyclostationarity-based detection algorithms can differentiate noise from the primary user' signals. This is coming from the fact that the noise is WSS(wide-sense stationary) with no correlation while modulated signals are cyclostationary with spectral correlation due to the redundancy of signal periodicities. Furthermore, the cyclostationarity can be used to distinguish among different types of transmissions. CSD(Cyclic Spectral Density) function of a received signal Eq. (1) can be calculated as follow. (U. Gardner, WA ,1991)

$$s(f,\alpha) = \sum_{\tau=-\infty}^{\infty} R_y^{\alpha}(\tau) e^{-j2\pi f\tau} \tag{1}$$

$$R_y^{\alpha}(\tau) = E[y(n+\tau)y^*(n-\tau)e^{j\pi 2\alpha n}] \tag{2}$$

where $R_y^{\alpha}(\tau)$ in Eq. (2) is the CAF(Cyclic Autocorrelation Function) and is the cyclic frequency. The CSD function outputs peak values when the cyclic frequency is equal to the fundamental frequencies of transmitted signal x(n). The cyclic frequencies can be assumed to be known (M. Ghozzi, et al., 2006).

2.3 Energy detector

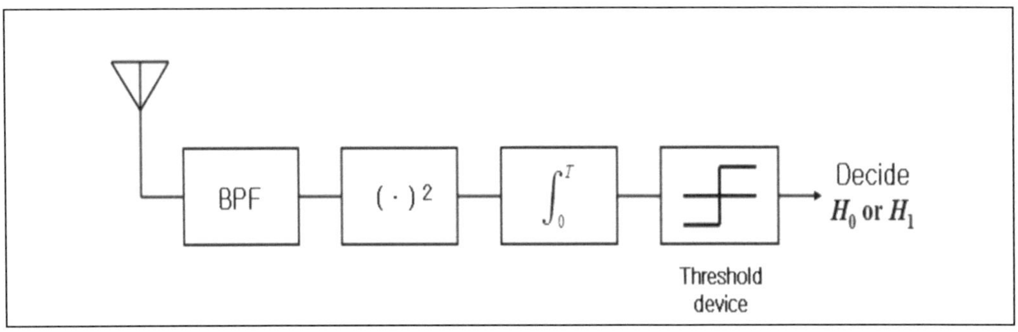

Fig. 2-2. The Structure of Energy Detector

The Figure 2-2 shows the block diagram of a conventional energy detector; the received signal is passed through the BPF(band-pass filter), squared, integrated during the required time, and then compared to a threshold which depends on the noise floor (H. Urkowitz, 1967). In practice, the threshold is chosen to fulfill a certain false alarm rate (D. Cabric et al., 2004). The energy detection method is attractive because it is possible to be applied regardless of the primary signals type and it is quite simple to implement.
Let us assume that the received signal has the following simple form . (H. Urkowitz, 1967)

$$\begin{cases} y(t) = n(t), H_0 \\ y(t) = h(t) * x(t) + n(t), H_1 \end{cases} \tag{3}$$

Where x(t) indicates the transmitted signal by the primary user, n(t) denotes the AWGN and h(t) denotes the amplitude of the channel impulse response. Output y(t) is the received signal from the secondary user. This is equivalent to distinguishing between the following two hypotheses (H_0 or H_1), its property is known as a central Chi-square distribution and a non-central Chi-square distribution, respectively . (H. Urkowitz, 1967)

$$y(t) = \begin{cases} X_{2m}^2, H_0 \\ X_{2m}^2(2\gamma), H_1 \end{cases} \tag{4}$$

where, 2m is degree of freedom, γ is signal-to-noise ratio (SNR), 2γ is non-centrality factor, m is product of time and bandwidth. X_{2m}^2 is a Chi-square distribution and $X_{2m}^2(2\gamma)$ is a non-central Chi-square distribution its probability of density function known as follow (V. I. Kostylev, 2002).

$$f_Y(y) = \begin{cases} \frac{1}{2^m \Gamma(m)} y^{(m-1)} e^{-\frac{y}{2}}, H_0 \\ \frac{1}{2}(\frac{y}{2\gamma})^{\frac{(m-1)}{2}} e^{-\frac{2\gamma+y}{2}} I_{(m-1)}(\sqrt{2\gamma y}), H_1 \end{cases} \tag{5}$$

Commonly, a detector has two types of error events. When channel is really vacant (H_0), the detector can decide that the channel is occupied. The probability of this undesired event is the probability of a false alarm, denoted as P_{FA}. When only noise is existing, the P_{FA} of energy detector can be calculated exactly, refer to Eq. (6). P_{FA} should be kept as small as possible in order to prevent under utilization of transmission opportunities.

$$P_{FA} = P(y > \lambda | H_0) \tag{6}$$

$$= \int_\lambda^\infty \frac{1}{2^m \Gamma(m)} y^{(m-1)} e^{-\frac{y}{2}} dy \tag{7}$$

$$= \frac{1}{2^m \Gamma(m)} \int_\lambda^\infty y^{(m-1)} e^{-\frac{y}{2}} dy \tag{8}$$

Making a change of variables $u = \frac{y}{2}$ in Eq.(8)

$$= \frac{1}{\Gamma(m)} \int_\lambda^\infty u^{(m-1)} e^{-u} du \tag{9}$$

$$= \frac{\Gamma(m, \lambda / 2)}{\Gamma(m)} \tag{10}$$

where, $\Gamma(u)$ gamma function defined in Eq. (11) and (M. Abramowitz and I. Stegun, 1970)

$$\int_0^\infty t^{u-1} e^{(-t)} dt \tag{11}$$

and $\Gamma(\alpha, x)$ is incomplete Gamma function upper bound in Eq. (12)

$$\int_x^\infty t^{\alpha-1} e^{(-t)} dt \tag{12}$$

Meanwhile, the detector may decide that the channel is vacant when channel is actually occupied (H_1). The probability of this undesired event is referred as the probability of misdetection, denoted as P_{MD}.

$$P_{MD} = 1 - P_D \tag{13}$$

And one minus the probability of misdetection is the probability of detection as P_D. When only noise is existing, the P_D of energy detector can be calculated exactly, refer to Eq. (14)

$$P_D = P(y > \lambda | H_1) \tag{14}$$

$$= \int_{\lambda}^{\infty} \frac{1}{2} (\frac{y}{2\gamma})^{\frac{(m-1)}{2}} e^{-\frac{2\gamma + y}{2}} I_{(m-1)}(\sqrt{2\gamma y}) dy \tag{15}$$

Change non-centrality parameter corresponds to $u = m\gamma$ in Eq. (15)

$$= \int_{\lambda}^{\infty} \frac{1}{2} (\frac{y}{2m\gamma})^{\frac{(m-1)}{2}} e^{-\frac{1}{2}(y+2m\gamma)} I_{(m-1)}(\sqrt{2m\gamma y}) dy \tag{16}$$

Change variables $z = \sqrt{y}$ in Eq. (16)

$$= \frac{1}{(\sqrt{2m\gamma})^{m-1}} \int_{\sqrt{\lambda}}^{\infty} z^m e^{-\frac{1}{2}(z^2+2m\gamma)} I_{(m-1)}(z\sqrt{2m\gamma}) dz \tag{17}$$

$$= Q(\sqrt{2m\gamma}, \sqrt{\lambda}) \tag{18}$$

where, $Q_m(a,b)$ is generalized Marcum Q-function, defined as Eq. (19), and modified Bessel function of the first kind of m-1 order is Eq. (20) ~ (21). Refer to (A. H. Nuttall, 1975)

$$Q_m(a,b) = \frac{1}{a^{m-1}} \int_b^{\infty} x^m e^{(-\frac{x^2+a^2}{2})} I_{m-1}(ax) dx \tag{19}$$

$$I_\alpha(x) = i^{-\alpha} J_\alpha(ix) \tag{20}$$

$$J_\alpha(x) = \sum_{m=0}^{\infty} \frac{(-1)^m}{m! \Gamma(m+\alpha)} (\frac{x}{2})^{2m+\alpha} \tag{21}$$

By far, we derived two closed form for probability of detection and misdetection of energy detector under AWGN.

Some of the challenges with energy detector based sensing include selection of the threshold for detecting primary users, inability to differentiate interference from primary users and noise and poor performance under low SNR values. The performance of the energy detection is easily influenced by channel fading, shadowing and interferences (V. I. Kostylev, 2002). As mentioned before, the threshold used in energy detector based sensing algorithms depends on the noise variance. Consequently, a small noise power estimation error causes significant performance loss(A. Sahai et al., 2004). In the following section 2.4, we discussed cooperative sensing approach to improve sensing reliability. (Amir Ghasemi & Elvino S. Sousa, 2005)

2.4 Cooperative spectrum sensing

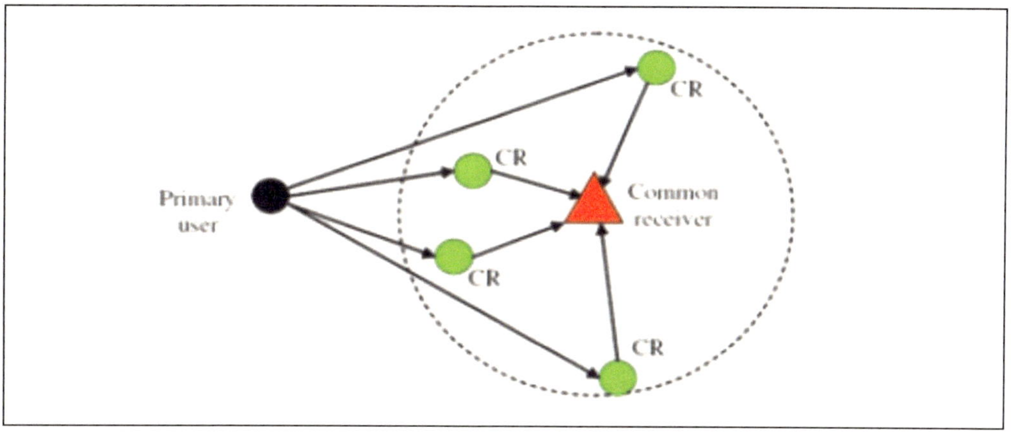

Fig. 2-3. Example of Cooperative Spectrum Sensing (Amir Ghasemi & Elvino S. Sousa, 2005)

Cooperation is proposed in the literature as a solution to problems that arise in spectrum sensing due to noise uncertainty, fading, and shadowing. (Amir Ghasemi & Elvino S. Sousa, 2005) The cooperative sensing scheme share sensing information from the independent number of secondary local user. Cooperative sensing decreases the probabilities of misdetection and false alarm considerably. In addition, cooperation can solve hidden primary user problem and it can decrease sensing time. Cooperative spectrum sensing is most effective when local user observe independent fading or shadowing (D. Cabric et al., 2006). The performance degradation due to correlated shadowing is investigated in terms of missing the opportunities. It is found that it is more advantageous to have the same amount of users cooperating over a large area than over a small area . (A. Ghasemi & E. S. Sousa, 2007)
For example, if we consider the Or-rule as the cooperative decision criterion, where any decision of H1 from the secondary users decides the channel is occupied. For simplicity, we assume that all N users experience independent and identically distributed fading with same average SNR. Then the probability of detection, false alarm and misdetection probabilities for cooperative scheme are shown as follows (P. K. Varshney,1997)

$$Q_D = 1 - (1 - P_D)^N \tag{22}$$

$$Q_{MD} = 1 - Q_D \tag{23}$$

$$Q_{FA} = 1 - (1 - P_{FA})^N \tag{24}$$

where P_D and P_{FA} represents the local detection and false alarm probabilities. In (23) and (24), as N gradually increases, Q_{MD} and Q_{FA} monotonically decrease.
Challenges of cooperative sensing include developing efficient information sharing algorithms and increased complexity (T. Weiss et al., 2003). In cooperative sensing architectures, the control channel (pilot channel) can be implemented using different

methodologies. Depending on the system requirements, one of these methods can be selected. Control channel can be used for sharing spectrum sensing results among cognitive users as well as for sharing channel allocation information.

3. Proposed spectrum sensing method

For such a cooperative approach, it is prerequisite for the interest user to have several cooperative users who may send information independently. Thus, it is necessary to improve the reliability of the energy detector by a standalone approach. In most of the future mobile and wireless communication systems, spatially-separated multiple antennas are considered at both transmitter as well as receiver (Blostein, S.D & Leib, H(2003). Out of their advantages, spatial diversity gain is most attractive, which is resulted from the fact that the signals through the multiple antennas go through different paths and thus experience different channel environments.
Therefore, in this chapter, we introduce a novel spectrum sensing structure combining the multiple antennas with the energy detectors for the standalone approach, where a collaborative decision device is employed to improve the sensing reliability.

3.1 Multiple antenna-aided spectrum sensing method

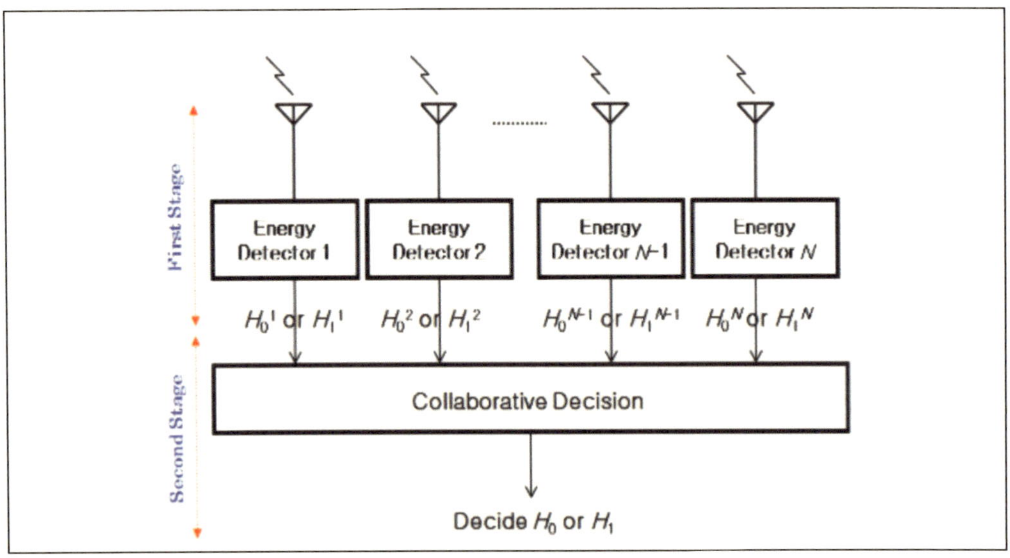

Fig. 3-1. Block Diagram of Multiple Antenna-aided Spectrum Sensing
(Jong-Hwan Lee et al. 2008)

We assume that each antenna are separated more than two over wave length of carrier $(2/\lambda)$. The proposed spectrum sensing scheme is presented in Figure 3-1, and consists of two stages. Tentative decisions are first made by using a conventional energy detector at each antenna. Then, a collaborative decision on whether the band is 'Occupied' (hypothesis H_1) or 'Vacant' (hypothesis H_0) is reached by comparing the number of tentative hard decisions with a reference value K ($1 \le K \le N$), with N as the number of antennas. Of course, the output of energy detector can make decision as a soft decision instead of hard decision. The results presented in (E. Visotsky et al., 2005) show that soft information-combining outperforms hard information-combining method in terms of the probability of missed opportunity. Intuitively, hard-decisions can understand to perform as good as soft decisions when the number of receiver antenna is high.

Other receiver diversity can attained by combining soft output of each square and integrate operation per diversity node. The Square-Law Combine (SLC), Square-Law Selection (SLS) method is well known micro receiver diversity (Digham, F. F et al., 2007). The SLC method can have slightly better under Rayleigh channel without correlation in the (Digham, F. F et al., 2007).

If one of decision fusion the 'Majority rule' is considered here to make a collaborative decision; in this case, the final decision is dependent on the majority of partial decisions, being defined for a number of antennas exceeding three, N ≥3.

In the first stage, the energy detector corresponding to the ith antenna, i= 1,...,N,decides on the hypothesis $H_1^{(i)}$ or $H_0^{(i)}$ by comparing the estimated energy, $y^{(i)}$, with a predefined threshold value, λ . The two hypotheses are defined as $H_1^{(i)}: y^{(i)}(t) = n^{(i)}(t)$, which corresponds to a vacant channel, and , $H_1^{(i)}$: $y^{(i)}(t) = h(t) * x(t) + n^{(i)}(t)$ which corresponds to an occupied channel. Note that $x(t)$ is the signal transmitted by the primary user, $y^{(i)}(t)$, $n^{(i)}(t)$, and $h^{(i)}(t)$ are the signal received by the secondary user, additive white Gaussian noise (AWGN), and channel impulse response corresponding to the ith antenna, respectively, and * represents the convolution operator. Two types of error events can be defined, as mentions before in Chapter 2. Table I describes the example of collaborative decision by Or-Rule criterion.

1st Energy Detector	2nd Energy Detector	3rdEnergy Detector	Collaborative final Decision
H_0	H_0	H_0	H_0
		H_1	H_1
	H_1	H_0	H_1
		H_1	H_1
H_1	H_0	H_0	H_1
		H_1	H_1
	H_1	H_0	H_1
		H_1	H_1

Table 3-1. Example of Or-Rule Criterion for Multiple Antenna-aided Method

4. Performance analysis and simulation results

In this Chapter, the average detection probability of Energy detector over three type of channels Rayleigh, shadowing and Suzuki channel are derived. More importantly, performance of our proposed sensing methods also are represented in terms of ROC curve under various environment. Note that average false alarm probability over fading channel is same as AWGN case, since it' s independent of SNR.

4.1 Performance analysis of energy detector under fading

4.1.1 Rayleigh channel

The requirement that there be many scatterers present means that Rayleigh fading can be a useful model in heavily built-up city centers where there is no line of sight between the transmitter and receiver and many buildings and other objects attenuate, reflect, refract and diffract the signal.

Let received carrier amplitude is modulated by the fading amplitude x, which is random variable with mean square value denote, $\Omega = \overline{x^2}$. PDF of fading amplitude x is dependent on the nature of the radio propagation. The instantaneous signal to noise per symbol, $\gamma = x^2 E_S / N_0$. Average SNR per symbol represents like, $\overline{\gamma} = \Omega E_S / N_0$. Therefore there is relation between fading amplitude and SNR.

$$x = \sqrt{\frac{\gamma}{E_S / N_0}} = \sqrt{\frac{\Omega\gamma}{\overline{\gamma}}} \tag{25}$$

Consequently, PDF of γ is obtained by changing variable in the expression for the fading PDF $P_X(x)$ of x.

$$f_\gamma(\gamma) = \frac{f_X(x)}{2\sqrt{\frac{\gamma\overline{\gamma}}{\Omega}}} \tag{26}$$

If the signal amplitude follows a Rayleigh distribution, then the SNR follows an exponential PDF given by Eq. (27)

$$f_X(x) = \frac{2x}{\Omega} e^{(-\frac{x^2}{\Omega})}, x \geq 0 \tag{27}$$

The Rayleigh PDF of γ is evaluated by substituting (27) into (26).

$$f_\gamma(\gamma) = \frac{f_X(\sqrt{\frac{\Omega\gamma}{\overline{\gamma}}})}{2\sqrt{\frac{\gamma\overline{\gamma}}{\Omega}}} \tag{28}$$

$$= \frac{\frac{2\sqrt{\frac{\Omega\gamma}{\overline{\gamma}}}}{\Omega} e(-\frac{\sqrt{\frac{\Omega\gamma^2}{\overline{\gamma}}}}{\Omega})}{2\sqrt{\frac{\gamma\overline{\gamma}}{\Omega}}} \tag{29}$$

$$= \frac{1}{\overline{\gamma}} e^{(-\frac{\gamma}{\overline{\gamma}})} \tag{30}$$

The average P_D in this case, $P_{D,\ Ray}$, can now be evaluated by averaging Eq.(18) over Eq.(30). refer to (A. H. Nuttall, 1975)

$$\mathrm{P}_{D',\mathrm{Ray}} = \int_x^{\infty} Q_m(\sqrt{2m\gamma},\sqrt{\lambda}) f_\gamma(x)dx \tag{31}$$

$$\int_x^{\infty} Q_m(\sqrt{2m\gamma},\sqrt{\lambda}) \times \frac{1}{\bar{\gamma}} e^{(-\frac{\gamma}{\bar{\gamma}})} d\gamma \tag{32}$$

$$= e^{-\frac{\lambda}{2}} \sum_{i=0}^{N/2-2} \frac{(-\frac{\lambda}{2})^i}{i!} + (\frac{2+2\bar{\gamma}}{2\bar{\gamma}})^{m-1} \times \left[e^{-\frac{\lambda}{2+2\bar{\gamma}}} - e^{-\frac{\lambda}{2}} \left\{ \sum_{i=0}^{m-2} \frac{\frac{2\lambda\bar{\gamma}}{2(2+2\bar{\gamma})}}{i!} \right\} \right] \tag{33}$$

$$\approx \frac{\Gamma(m-1,\frac{\lambda}{2})}{\Gamma(m-1)} + e^{-\frac{\lambda}{2(1+m\bar{\gamma})}} (1+\frac{1}{m_{\bar{\gamma}}})^{m-1} \times \left[1 - \frac{\Gamma(m-1,\frac{\lambda m\bar{\gamma}}{2(1+m\bar{\gamma})})}{\Gamma(m-1)} \right] \tag{34}$$

4.1.2 Shadowing channel

A large obstruction such as a hill or large building obscures the main signal path between the transmitter and the receiver. Normally shadowing is modeled as a log-normal model which is traditionally used to characterize the variation in the local mean power due to shadowing (Kyperountas,. et all, 2007). For shadowing only, $f_\gamma(x)$ in equation (35) is log-normal with mean μ (dB) and a standard deviation σ (dB) of $10\log_{10}\gamma$.

$$f_\gamma(x) = \frac{\xi}{\sqrt{2\pi}\alpha\bar{\gamma}} e^{(-\frac{(10\log_{10}\bar{\gamma}-\mu)^2}{2\alpha^2})} \tag{35}$$

Where, $\xi = 10/\ln 10 = 4.3429$.

The average $\mathrm{P_D}$ in this case, $\mathrm{P_{D,Log}}$ can now be evaluated by averaging (18) over (35). Empirical coefficient values for indoor propagation are shown in Table 4-1.

Building Type	Frequency of Transmission	γ	α [dB]
Vacuum,infinite space		2.0	0
Retail Store	914 MHz	2.2	2.18.7
Grocery Store	914 MHz	1.8	5.2
Office with hard partition	1.5 GHz	3.0	7
Office with soft partition	900 MHz	2.4	9.6
Office with soft partition	1.9 GHz	2.6	14.1
Textile or chemical	1.3 GHz	2.0	3.0
Textile or chemical	4 GHz	2.1	7.0, 9.7
Metalworking	1.3 GHz	1.6	5.8
Metalworking	1.3 GHz	3.3	6.8

Table 4-1. Empirical log-normal mean and standard deviation under indoor propagation (T. S. Rappaport, 2002)

4.1.3 Suzuki channel

The combined effect of large scale and small scale fading may be characterized statistically by a Suzuki distribution (M. Patzold et al., 1994). The Suzuki mixture distribution that characterizes the received power can be expressed in terms of signal power, which is obtained by averaging Rayleigh distribution signal power over the conditional density of log-normal mean signal power in Eq. (36).

$$f_\gamma(x) = \int_0^\infty \frac{1}{\varpi} e^{(-\frac{x}{\varpi})} \frac{\xi}{\sqrt{2\pi}\alpha\varpi} e(-\frac{(10\log_{10}\varpi - \mu)^2}{2\alpha^2}) d\varpi \tag{36}$$

The average P_D in this case, $P_{D,suz}$ can now be evaluated by averaging (18) over (36) refer to (Kyperountas,. et all, 2007).

$$P_{D,Suz} = \int_x^\infty Q_m(\sqrt{2m\gamma}, \sqrt{\lambda}) f_\gamma(x) dx \tag{37}$$

$$= \int_x^\infty \int_\varpi^\infty Q_m(\sqrt{2m\gamma}, \sqrt{\lambda}) \frac{1}{\varpi} w^{\left(-\frac{x}{\varpi}\right)} \frac{\xi}{\sqrt{2\pi}\alpha\varpi} e^{(-\frac{(10\log_{10}\varpi - \mu)^2}{2\alpha^2})} d\varpi dx \tag{38}$$

4.2 Performance analysis of proposed method under fading

The probability of detection for the proposed sensing method is given by Eq.(39) ~ (41). As you can see this formula is exactly same as cooperative sensing refer to Section 2.4.

$$P_D = \sum_{k=n}^{N} \binom{N}{k} P_{D,i}^k (1 - P_{D,i})^{N-k} \tag{39}$$

where $P_{D,i}$ is the probability of detection for each output of antennas. The Or-rule fusion can be evaluated by setting n=1 in Eq. (39).

$$P_{D,or} = 1 - \binom{N}{k} P_{D,i}^k (1 - P_{D,i})^{N-k} |_{k=0} = 1 - (1 - P_{D,k})^N \tag{40}$$

The And-rule corresponds to the case n=N, Eq. (39)

$$P_{D,And} = (P_{D,k})^N \tag{41}$$

Finally, for the case of the majority rule the $n = \lfloor n/2 \rfloor$

$$P_{D,maj} = \sum_{k=\lfloor n/2 \rfloor}^{N} \binom{N}{k} P_{D,i}^k (1 - P_{D,i})^{N-k} \tag{42}$$

4.3 Simulation environment

In this paper, the Suzuki channel was considered by calculating MED (Method of Equal Distance). The characteristic of the MED is such that the difference between two adjacent discrete Doppler frequencies is equidistant. (M. Patzold et al. 1996) We assume independent fading and noise on each branch. We consider from two to four antennas at the receive-side

(N is 1 ~ 4) and apply the several decision criterion for collaborative decision-making in the second stage of the proposed scheme. The parameters used in simulations are listed in Table 4-2. Note that the 0.33ms and 3.3ms sampling durations are chosen according to the IEEE 802.22 draft. The primary user is assumed to transmit a BPSK modulated signal. We will select the threshold so as to get 10% of P_{FA} and then calculate the P_{MD}.

Parameter	Value
Carrier Frequency	900 MHz
Mobile Speed of Secondary User	30 ~ 110 km/h
SNR	3 dB or 6dB
Standard Deviation of shadowing	10dB
Collaborative Decision Criterion	Or-rule, And-rule, Majority-rule
N(number of Receiver Antenna)	1~4
Number of Samples	105
Sample Period	0.33ms
Sensing Duration	3.3 ~ 13.2ms

Table 4-2. The Simulation Parameters

4.4 Performance comparison

We demonstrate simulation results in terms of complementary ROC curve (plot P_{MD} vs. P_{FA}). The ROC (Receiver Operating Characteristics) curve is known to be a suitable tool for visualizing detector performance.

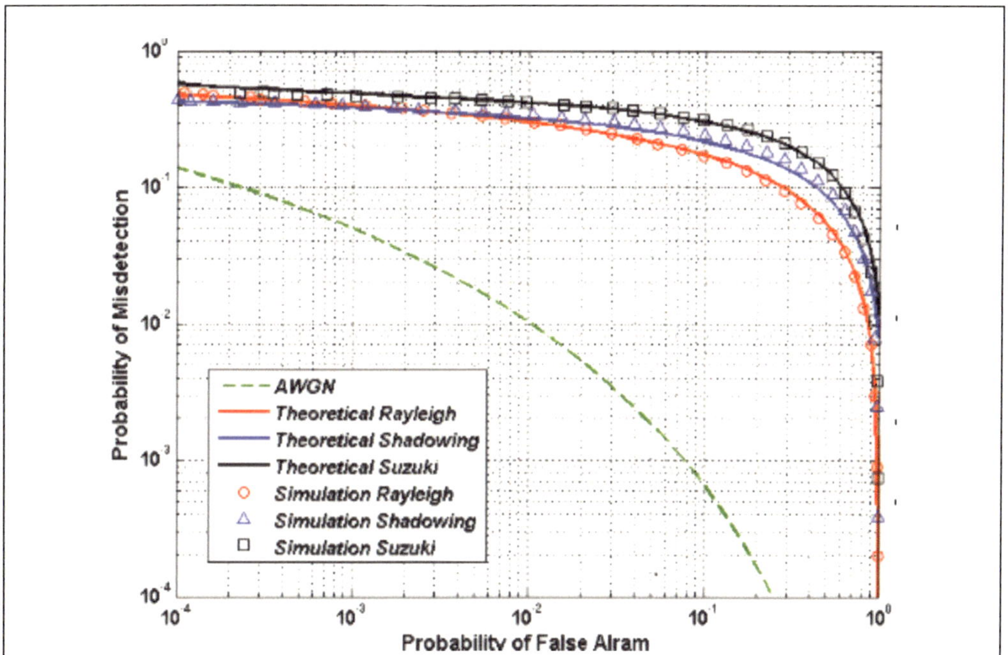

Fig. 4-1. Performance of Conventional Energy Detector under Various Channel (SNR = 6dB, σ =10dB) (Kyperountas,. et all, 2007)

Figure 4-1 shows the simulated and theoretical complementary ROC curves, for a conventional Energy Detector under the various fading environments with average SNR=6dB. As you can see that there is conventional severe performance degradation readily for energy detector.

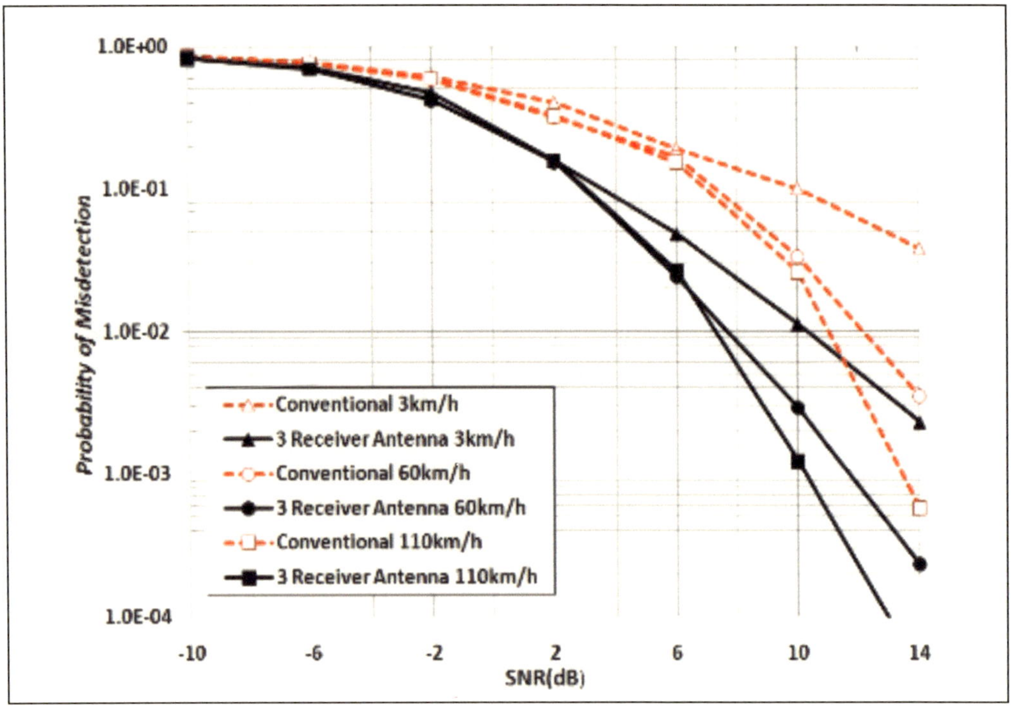

Fig. 4-2. Mobile Speed Effect for Proposed Method in terms of SNR vs. PMD (Mobile Speed = 3, 60, 110km/h) (Jong-Hwan Lee et al. 2008)

At three cases of mobile speed, the performance of the proposed spectrum sensing scheme is much better than conventional energy detector over the whole range of SNR values. Figure 4-2 shows that higher mobile speed may improve the sensing performance in range of high SNR. Because of the rapid channel variation, high mobile speed can overcome deep fading easily. When we require reliable spectrum sensing under the high mobile speed, the proposed scheme may be attractive method even though the complexity of multiple antennas is higher than other sensing methods.

In Figure 4-3 ~ 4-4, the impact of the sensing duration on the performance of the proposed and conventional energy detection schemes is investigated. Figure 4-3 shows the ROC curve (PD vs. PFA) for a mobile speed of 3 km/h, with the sensing duration as a parameter. As expected, as the sensing duration increases, the performances of both proposed and conventional energy detector are improved. However, similar performance is achieved for 6.6 ms and 9.9 ms sensing durations. As such, one can conclude that for sensing durations larger than 6.6 ms there is no substantial enhancement in the sensing performance. In addition, note that the performance of the proposed scheme with the sensing duration of 3.3 ms is similar to that of the conventional scheme for 6.6 ms. This result shows that the use of multiple antennas is equivalent to an increase in the sensing duration with the conventional energy detection.

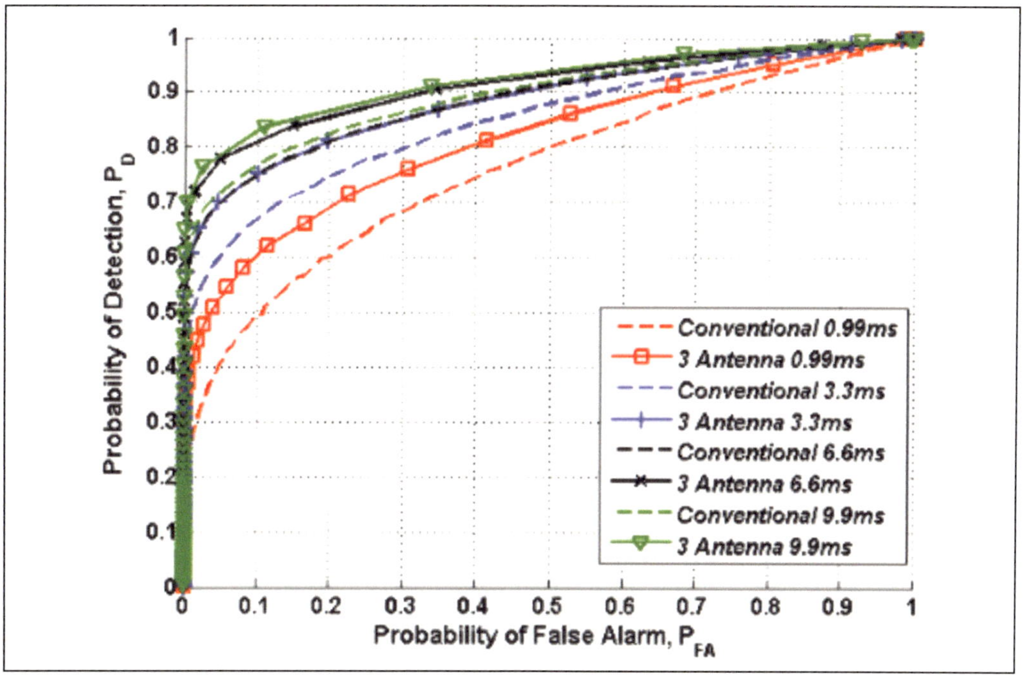

Fig. 4-3. ROC Curve of Proposed Method, for 3 km/h mobile speed and with different sensing durations, PFA vs. PD (S.-H. Hwang et al., 2009)

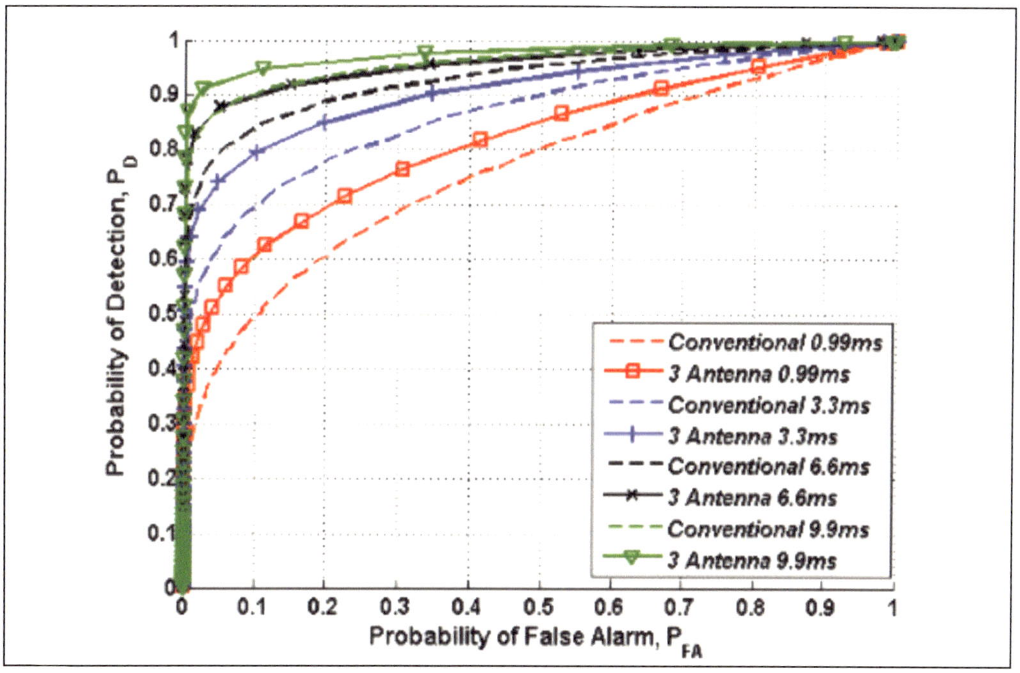

Fig. 4-4. ROC Curve of Proposed Method, for 110 km/h mobile speed and with different sensing durations, P_{FA} vs. P_D (S.-H. Hwang et al., 2009)

In Figure 4-4 results obtained for a mobile speed of 110 km/h are provided, from which one can notice an improvement in performance when compared with Figure 4-3. When increasing the mobile speed, the Doppler spread increases and fast fading will be experienced. This translates into faster variations of the signal at the receive-side. The test statistic used with the energy detector is the received signal energy, which represents the area under the squared signal. For a given sensing duration, this can increase due to faster signal variations. That is, the signal energy can increase for fast fading. Note that the performance of the conventional scheme for the sensing duration of 6.6 ms is better than that of the proposed scheme for 3.3 ms; this result indicates that the effect of the sensing duration is influenced by the mobile speed.

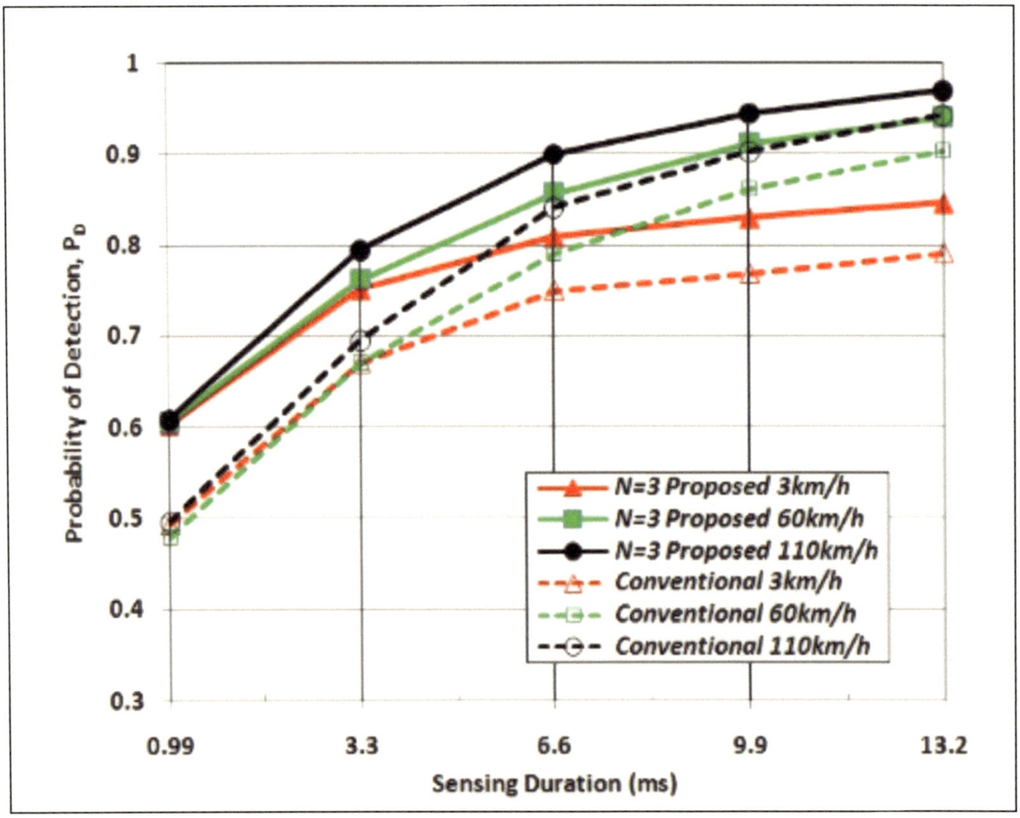

Fig. 4-5. Effect of Sensing Duration for Proposed Method at P_{FA}=0.1, Sensing Duration vs. P_D (S.H.Hwang et al. 2009)

Figure 4-5 summarizes the performances in terms of the P_D versus the sensing duration, for a P_{FA} equal to 0.1. As the sensing duration increases, the performance is improved, regardless the mobile speed. An enhanced performance is attained for an increased speed of the mobile. A P_D above 0.9 is obtained for a sensing duration exceeding 6.6 ms and a mobile speed of 110 km/h, and a sensing duration greater than 9.9 ms and 60 km/h mobile speed. For a mobile speed of 3 km/h, the P_D reaches around 0.85 even above 13.2 ms sensing duration. Note that to guarantee an acceptable sensing performance both P_{FA} and P_{MD} need to be less than 0.1.

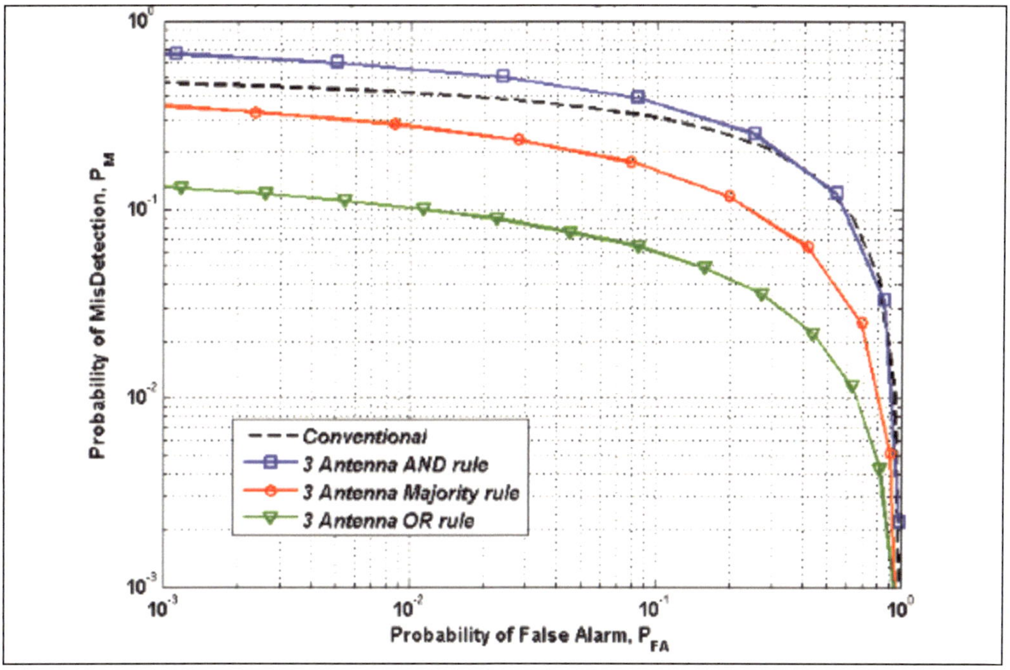

Fig. 4-6. Complementary ROC Curve of Proposed Sensing with Various Decision Rule (N=3, Mobile Speed =3km/h, SNR=3dB), PFA vs. PM (S.-H. Hwang et al., 2009)

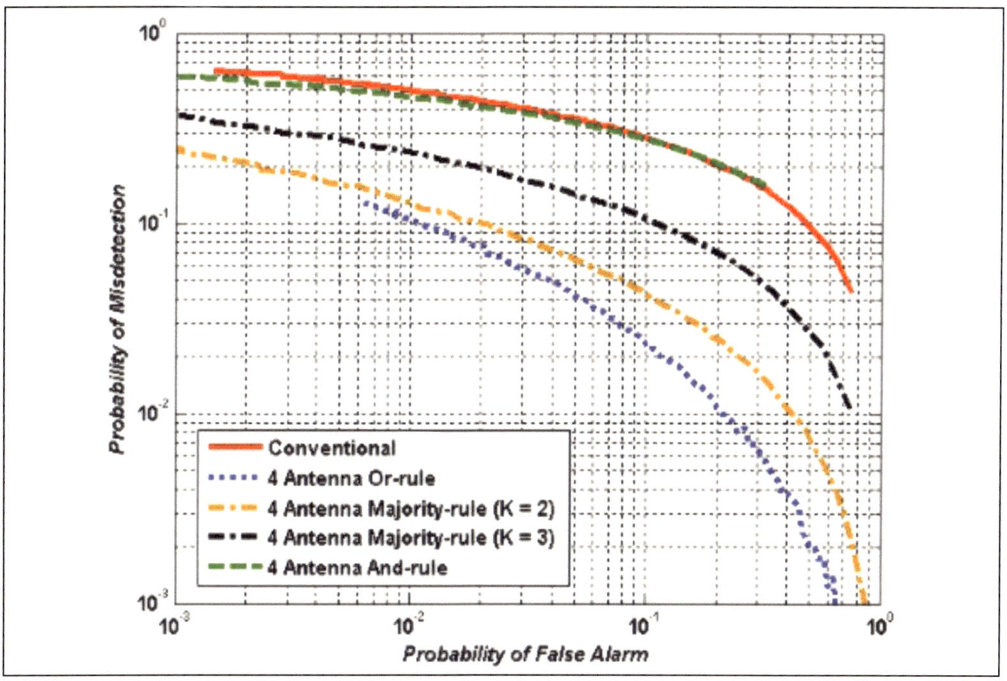

Fig. 4-7. Complementary ROC Curve of Proposed Sensing with Various Decision Rule (N=4, Mobile Speed =3km/h, SNR=3dB), P_{FA} vs. P_M

Figure 4-6 shows only simulative results of three antennas aided energy detector under Suzuki fading channel in terms of Complementary ROC curve, other parameter same as before. The Figure 4-6 presents that the performance of the 'And-rule' is almost same as that of the conventional detector. 'Or-rule' shows the best performances.
Figure 4-7 shows the performance of proposed sensing with four antennas (N = 4). When the number of antenna increases from two to four, the number of applicable criterions increases. For the case of four antennas, it is possible to employ two kinds of 'Majority-rules' where the reference values K can be two or three. The Figure shows that 'Or-rule' -based sensing scheme can achieve the most reliable performance for four antennas as well. Therefore, we can see base on the previous results that 'Or-rule' is very effective to enhance the performance of the proposed scheme with multi-antennas (S.-H. Hwang et al., 2009).

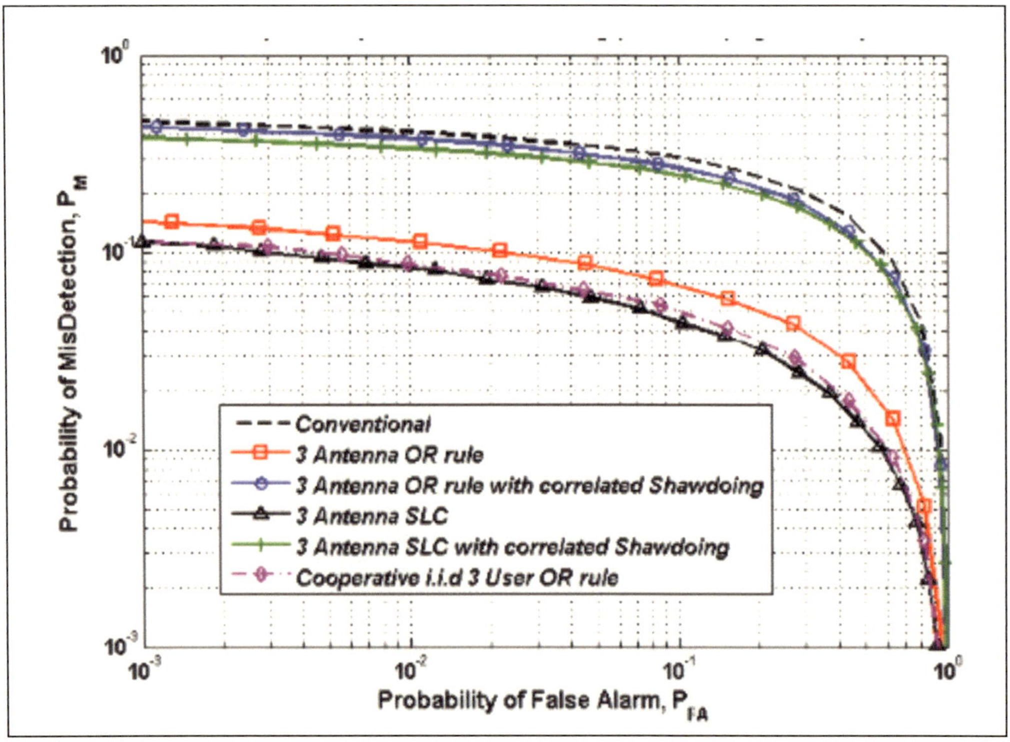

Fig. 4-8. Simulated complement ROC curve for 3 Antenna-aided Sensing under Suzuki channel with heavy correlated and un-correlated shadowing
(SNR = 6dB, Mobile speed = 3km/h), P_{FA} vs. P_M

In the Figure 4-8 non-correlated case, performance of proposed scheme with OR fusion is worse 15% than that of square law combine (SLC). However, the performance degradation is less effective to proposed scheme than SLC scheme due to the correlation. Hence, we may say that proposed scheme is more robust to correlated shadowing effect. Other point of view, SLC gets an advantage for simplicity. Meanwhile proposed is more complex than SLC scheme, since multiple aided proposed scheme has energy detector devices as much as the number of antenna and also need 2nd stage devices.

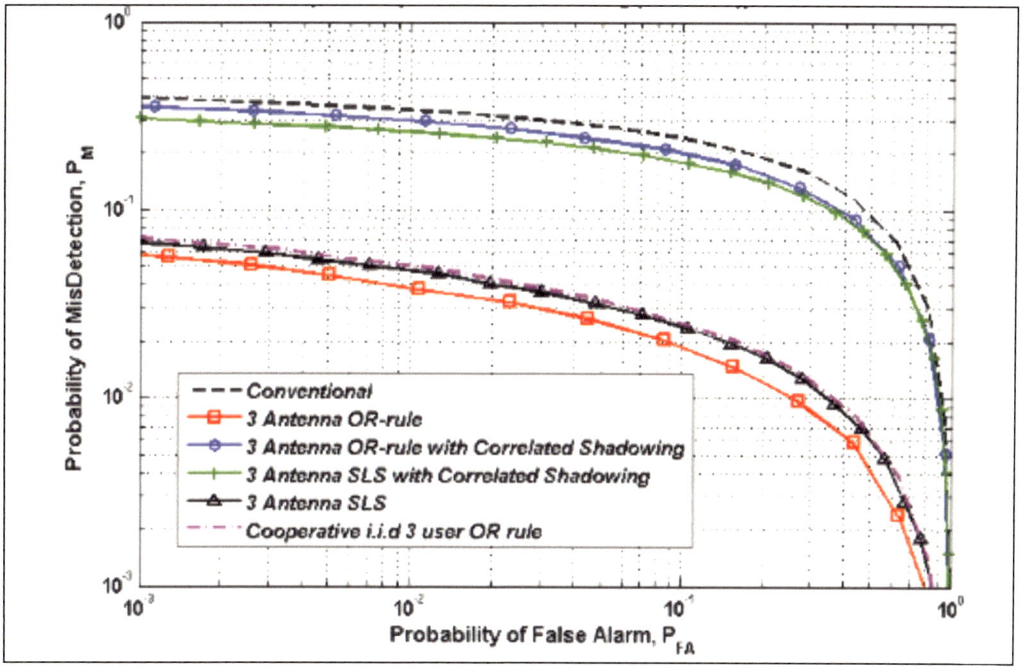

Fig. 4-9. Simulated Complement ROC curve for 3 Antenna-aided Sensing under Suzuki channel with Heavy Correlated and Uncorrelated Shadowing
(SNR = 6dB, Mobile speed = 110km/h), P_{FA} vs. P_M

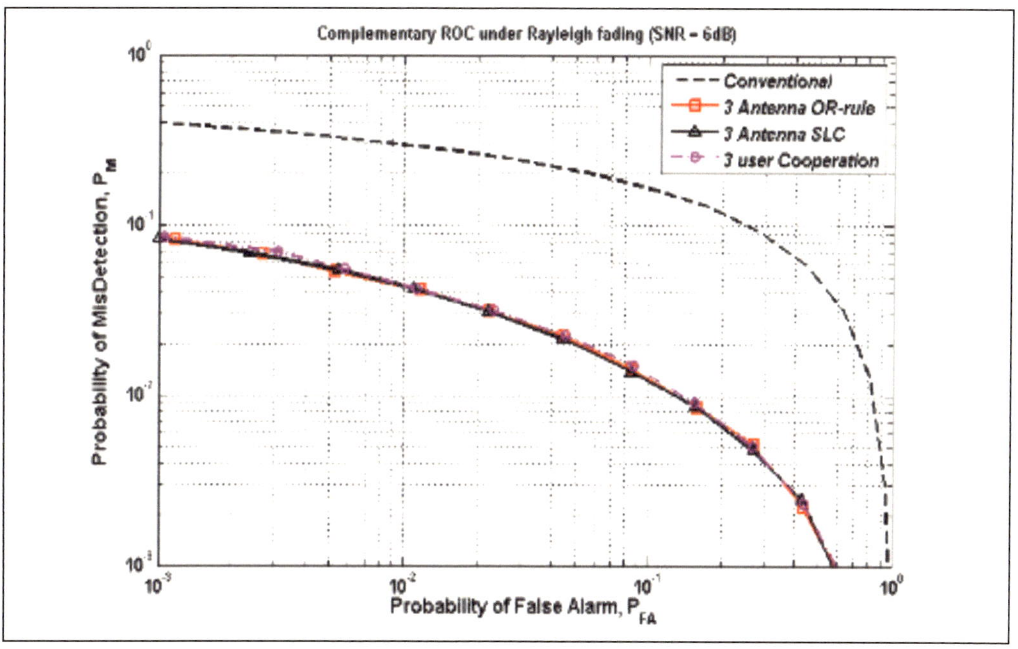

Fig. 4-10. Simulated Complement ROC curve for 3 Antenna-aided Sensing under Rayleigh Channel without Shawdoing Correlation (SNR = 6dB, Mobile speed = 110km/h), P_{FA} vs. P_M

In case of mobile speed up to 110km/h in Figure 4-9, proposed scheme with OR fusion can be achieved the highest performance of than that of other method non-correlated case. When we assume that there is only Rayleigh channel, the proposed scheme, SLC and cooperative sensing method have all same performance in the Figure 4-10.

5. Conclusions and future work

In the paper, we summarized the enabling spectrum sensing methods for cognitive radio. Among the other spectrum sensing method, we focus on the energy detector and derive detection probability under various channels such as AWGN, Rayleigh, Log-normal shadowing and Suzuki channel. We also explained the cooperative sensing concept and introduced novel spectrum sensing scheme which combines the energy detection with multiple receive antennas to improve the sensing reliability.

From the simulated results, our proposed scheme with Or-rule and soft receiver diversity such as SLC are both decreased probability of misdetection and false alarm and they are highly dependent on c and soed shadowing effect. However, the performance degradsoft rby the coand soft ris less effective to proposed scheme than SLC.le nsequently, we may say that proposed is moae robust to c and soed shadowing effect. Now, we can conclude that performance of multiple aided energy detector spectrum sensing scheme can have advantages in some limioed case such as the hur pcoposed varisoft , low correlation.

The large operating bandwidths impose additional requirements on the radio frequencies (RF) components such as antennas and power amplifiers as well. According to resent study, the multiple receiver antenna scheme can have advantages for ultra-wideband spectrum sensing. Because it could manage more flexible wideband sensing. For the future research, we will continue to study spectral sensing method especially for the ultra-wideband frequency band using by multiple receiver antenna.

6. Acknowledgement

This research was supported by the International Research & Development Program of the National Research Foundation of Korea (NRF) funded by the Ministry of Education, Science and Technology(MEST) of Korea(Grant number: F01-2009-000-10188-0)

7. References

A. Ghasemi and Elvino S. Sousa (2005) Collaborative spectrum sensing for opportunistic access in fading environments, *in proc First IEEE International Symposium* on DySPAN, pp.131-136.

A. H. Nuttall, (1975) Some integrals involving the QM function, *in proc IEEE Transactions on Information Theory*, vol. 21, no. 1, pp. 95 - 96.

A. Ghasemi and E. S. Sousa, (2007) Asymptotic performance of collaborative spectrum sensing under correlated log-normal shadowing, *in proc IEEE Commun. Lett.*, vol. 11, no. 1, pp. 34 - 36.

A. Sahai, N. Hoven, and R. Tandra (2004), Some fundamental limits on cognitive radio, *in Proc. Allerton Conf. on Commun., Control, and Computing*, Monticello, Illinois, Oct. 2004.

Blostein, S.D.; Leib, H (2003). Multiple antenna systems: their role and impact in future wireless access, *IEEE Communications Magazine*, Volume 41, Issue 7,(July 2003.) pp. 94 - 101

D. Cabric, S. Mishra, and R. Brodersen (2004), Implementation issues in spectrum sensing for cognitive radios, *in Proc. Asilomar Conf. on Signals, Systems and Computers, vol. 1, Pacific Grove*, California, USA, pp. 772 - 776, Nov. 2004.

D. Cabric, A. Tkachenko, and R. Brodersen (2006), Spectrum sensing measurements of pilot, energy, and collaborative detection, *in Proc. IEEE Military Commun. Conf.*, Washington, D.C., USA, pp. 1 - 7, Oct. 2006.

Digham, F. F.; Alouini, M.-S.; Simon, M. K. (2007), On the Energy Detection of Unknown Signals Over Fading Channels, *IEEE Transactions on Communications*, Volume 55, Issue 1, (Jan. 2007)pp. 21-24

E. Visotsky, S. Kuffner, and R. Peterson (2005), On collaborative detection of TV transmissions in support of dynamic spectrum sharing *in Proc. IEEE Int. Symposium on New Frontiers in Dynamic Spectrum Access Networks*, Baltimore, Maryland, USA,, pp. 338 - 345, Nov. 2005.

H. Urkowitz, (1967) Energy detection of unknown deterministic signals, *in Proc. IEEE*, vol. 55, pp. 523 - 531.

Ian F. Akyildiz, Won-Yeol Lee, Mehmet C. Vuran, Shantidev Mohanty, Next generation/dynamic spectrum access/cognitive radio wireless networks: A survey"

I. Mitola, J. and J. Maguire, G. Q., (1999) Cognitive radio: making software radios more personal, *in proc IEEE Personal Commun* Mag., vol. 6, no. 4, pp. 13 - 18.

J. G. Proakis(2001), *Digital Communications*, 4th ed. McGraw-Hill

J.H Baek, H.J Oh, J.H Lee, S.H Hwang (2008), Performance Improvements of Energy Detector for Spectrum Sensing in Cognitive Radio Environments: Verification using Multiple Antennas, *Journal of IEEK* (Apr. 2008), pp. 38-42,. .

Jong-Hwan Lee, Jun-Ho Baek, Seung-Hoon Hwang,. (2008) Collaborative Spectrum Sensing using Energy Detector in Multiple Antenna System, in *proc ICACT 2008*. 10th International Conference on Volume 1, pp. 427-430

Kyperountas, Spyros, Correal, Neiyer, Shi, Qicai, Ye, Zhuan (2007), Performance Analysis of Cooperative Spectrum Sensing in Suzuki Fading Channels, *in Proc 2nd International Conference on CrownCom*, pp. 428-434.

Mahapatra, R.; Krusheel, M. (2008), Cyclostationary detection for cognitive radio with multiple receivers, *ISWCS '08. IEEE International Symposium*, pp. 493 - 49.

M. Patzold, U. Killat, and F. Laue (1994), A deterministic model for a shadowed Rayleigh land mobile radio channel, *in Proc. PIMRC '94*, The Hague, Netherlands, pp. 1202-1210.

M. Patzold, Killat U., Laue F. (1996), A deterministic digital simulation model for Suzuki processes with application to a shadowed Rayleigh land mobile radio channel, *in proc Vehicular Technology, IEEE Transactions* on Volume 45, Issue 2, pp. 318-331.

M. Ghozzi, F. Marx, M. Dohler, and J. Palicot (2006), Cyclostationarity based test for detection of vacant frequency bands, *in Proc. IEEE Crowncom*,

O. A. Dobre, S. Rajan, and R. Inkol (2008), Exploitation of firstorder cyclostationarity for joint signal detection and classification in cognitive radio, *in Proc. IEEE VTC Fall*, pp. 1-5.

P. K. Varshney (1997), Distributed detection and data fusion. NewYork: Springer-Verlag,

R. Tandra, A. Sahai (2005), Fundamental limits on detection in low SNR under noise uncertainty *inProc. IEEE Int. Conf. Wireless Networks, Commun. and Mobile Computing, vol. 1*, Maui, HI, pp. 464 - 469

S.-H. Hwang, J.-H. Baek, O.A. Dobre (2009), Spectrum sensing using multiple antenna-aided energy detectors for cognitive radio, in *Proc.IEEE CCECE*, pp. 209-212.

T. S. Rappaport (2002), *Wireless communications principles and practices*, Prentice-Hall.

T. Weiss, J. Hillenbrand, A. Krohn, and F. Jondral(2003), Efficient signaling of spectral resources in spectrum pooling systems, in *Proc. IEEE Symposium on Commun. and Veh. Tech nol.*, Eindhoven, Netherlands.

U. Gardner, WA, (1991) Exploitation of spectral redundancy in cyclostationary signals, in *proc IEEE Signal Processing Mag.*, vol. 8, no. 2, pp. 14 - 36.

V. I. Kostylev (2002), Energy Detection of a Signal with Random Amplitude, in *Proc. IEEE International Conference on Communication, vol. 3*, pp. 1606 - 1610.

15

New Method to Generate Balanced 2^n-PSK STTCs

P. Viland, G. Zaharia and J.-F. Hélard
Université européenne de Bretagne (UEB)
Institute of Electronics and Telecommunications of Rennes (IETR)
INSA de Rennes, 20 avenue des Buttes de Coesmes,
F-35708 Rennes
France

1. Introduction

The use of multiple input multiple output (MIMO) is an efficient method to improve the error performance of wireless communications. Tarokh et al. (1998) propose the space-time trellis codes (STTCs) which use trellis-coded modulations (TCM) over MIMO channels. STTCs combine diversity gain and coding gain leading to a reduction of the error probability.

In order to evaluate the performance of STTCs in slow fading channels, the rank and determinant criteria are proposed by Tarokh et al. (1998). In the case of fast fading channels, Tarokh et al. (1998) also present two criteria based on the Hamming distance and the distance product. Ionescu (1999) shows that the Euclidean distance can be used to evaluate the performance of STTCs. Based on the Euclidean distance, Chen et al. (2001) present the trace criterion which governs the performance of STTCs in both slow and fast fading channels, in the case of a great product between the number of transmit and receive antennas. This configuration corresponds to a great number of independent single input single output sub-channels. Liao & Prabhu (2005) explain that the repartition of determinants or Euclidean distances optimizes the performance of STTCs.

Based on these criteria, many codes have been proposed in the previous publications. The main difficulty is a long computing-time to find the best STTCs. Liao & Prabhu (2005) and Hong & Guillen i Fabregas (2007) use an exhaustive search to propose new STTCs, but only for 2 transmit antennas. To reduce the search-time, Chen et al. (2002a;b) advance a sub-optimal method to design STTCs. Thereby, the first STTCs with 3 and 4 transmit antennas are designed. Besides, another method is presented by Abdool-Rassool et al. (2004) where the first STTCs with 5 and 6 transmit antennas are given.

It has been remarked by Ngo et al. (2008; 2007) that the best codes have the same property: the used points of the MIMO constellation are generated with the same probability when the binary input symbols are equiprobable. The codes fulfilling this property are called balanced codes. This concept is also used by set partitioning proposed by Ungerboeck (1987a;b).

Thus, to find the best STTCs, it is sufficient to design and to analyze only the balanced codes. Hence, the time to find the best STTCs is significantly reduced. A first method to design balanced codes is proposed by Ngo et al. (2008; 2007) allowing to find 4-PSK codes with better performance than the previous published STTCs. Nevertheless, this method has been exploited only for the 4-PSK modulation.

The main goal of this chapter is to present a new efficient method to create 2^n-PSK balanced STTCs and thereby to propose new STTCs which outperform the previous published STTCs. The chapter is organized as follows. The next section reminds the representation of STTCs. The existing design criteria is presented in section 3. The properties of the balanced STTCs and the existing method to design these codes are given in section 4. In section 5, the new method is presented and illustrated with examples. In the last section, the performance of new STTCs is compared to the performance of the best published STTCs.

2. System model

In the case of 4-PSK modulation, i.e. $n = 2$, we consider the space-time trellis encoder presented in Fig. 1.

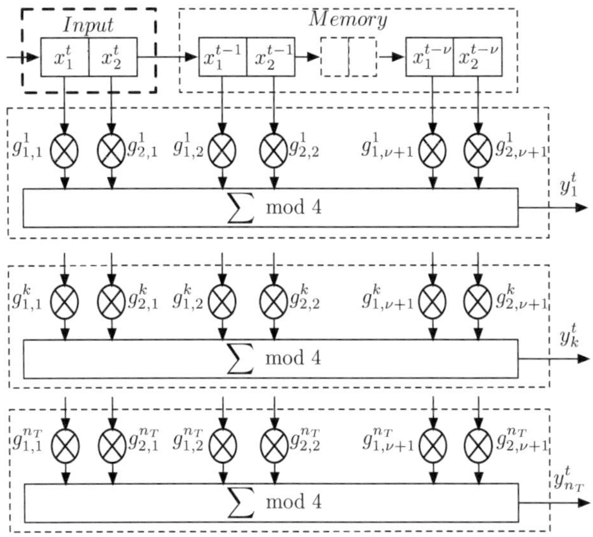

Fig. 1. 4^ν states 4-PSK space-time trellis encoder

In general, a $2^{n\nu}$ states 2^n-PSK space-time trellis encoder with n_T transmit antennas is composed of one input block of n bits and ν memory blocks of n bits. At each time $t \in \mathbb{Z}$, all the bits of a block are replaced by the n bits of the previous block. For each block i, with $i = \overline{1, \nu+1}$ where $\overline{1, \nu+1} = 1, 2, \cdots, \nu+1$, the lth bit with $l = \overline{1, n}$ is associated to n_T coefficients $g_{l,i}^k \in \mathbb{Z}_{2^n}$ with $k = \overline{1, n_T}$. With these $n_T \times n(\nu+1)$ coefficients, the generator matrix $\mathbf{G}$ is obtained and given by

$$\mathbf{G} = \begin{bmatrix} g_{1,1}^1 \cdots g_{n,1}^1 \cdots g_{1,\nu+1}^1 \cdots g_{n,\nu+1}^1 \\ \vdots \qquad \cdots \qquad \vdots \\ g_{1,1}^k \cdots g_{n,1}^k \cdots g_{1,\nu+1}^k \cdots g_{n,\nu+1}^k \\ \vdots \qquad \cdots \qquad \vdots \\ g_{1,1}^{n_T} \cdots g_{n,1}^{n_T} \cdots g_{1,\nu+1}^{n_T} \cdots g_{n,\nu+1}^{n_T} \end{bmatrix}. \tag{1}$$

A state is defined by the binary values of the memory cells corresponding to no-null columns of $\mathbf{G}$. At each time t, the encoder output $Y^t = \left[y_1^t y_2^t \cdots y_{n_T}^t\right]^{\mathrm{T}} \in \mathbb{Z}_{2^n}^{n_T}$ is given by

$$Y^t = \mathbf{G}X^t, \tag{2}$$

where $X^t = [x_1^t \cdots x_n^t \cdots x_1^{t-\nu} \cdots x_n^{t-\nu}]^{\mathrm{T}}$ is the extended-state at time t of the $L_r = n(\nu+1)$ length shift register realized by the input block followed by the ν memory blocks. The matrix $[\cdot]^{\mathrm{T}}$ is the transpose of $[\cdot]$. Thus, STTCs can be defined by a function

$$\Phi : \mathbb{Z}_{2^n}^{L_r} \to \mathbb{Z}_{2^n}^{n_T}. \tag{3}$$

The 2^n-PSK signal sent to the kth transmit antenna at time t is given by $s_k^t = \exp(\mathrm{j}\frac{\pi}{2^{n-1}} y_k^t)$, with $\mathrm{j}^2 = -1$. Thus, the MIMO symbol transmitted over the fading MIMO channel is given by $S^t = \left[s_1^t s_2^t \cdots s_{n_T}^t\right]^{\mathrm{T}}$.
For the transmission of an input binary frame of $L_b \in \mathbb{N}^*$ bits, where L_b is a multiple of n, the first and the last state of the encoder are the null state. At each time t, n bits of the input binary frame feed into the encoder. Hence, $L = \frac{L_b}{n} + \nu$ MIMO symbols regrouped in the codeword

$$\mathbf{S} = \left[S^1 \cdots S^t \cdots S^L\right] \tag{4}$$

$$= \begin{bmatrix} s_1^1 & \cdots & s_1^t & \cdots & s_1^L \\ \vdots & \ddots & \vdots & \ddots & \vdots \\ s_k^1 & \cdots & s_k^t & \cdots & s_k^L \\ \vdots & \ddots & \vdots & \ddots & \vdots \\ s_{n_T}^1 & \cdots & s_{n_T}^t & \cdots & s_{n_T}^L \end{bmatrix} \tag{5}$$

are sent to the MIMO channel.
The vector of the signals received at time t by the n_R receive antennas $R^t = [r_1^t \cdots r_{n_R}^t]^{\mathrm{T}}$ can be written as

$$R^t = \mathbf{H}^t S^t + N^t, \tag{6}$$

where $N^t = [n_t^1 \cdots n_{n_R}^t]^{\mathrm{T}}$ is the vector of complex additive white gaussian noises (AWGN) at time t. The $n_R \times n_T$ matrix $\mathbf{H}^t$ representing the complex path gains of the SISO channels between the transmit and receive antennas at time t is given by

$$\mathbf{H}^t = \begin{bmatrix} h_{1,1}^t & \cdots & h_{1,n_T}^t \\ \vdots & \ddots & \vdots \\ h_{n_R,1}^t & \cdots & h_{n_R,n_T}^t \end{bmatrix}. \tag{7}$$

In this chapter, only the case of Rayleigh fading channels is considered. The path gain $h_{k',k}^t$ of the SISO channel between the kth transmit antenna and $(k')^{th}$ receive antenna is a complex random variable. The real and the imaginary parts of $h_{k',k}^t$ are zero-mean Gaussian random variables with the same variance. Two types of Rayleigh fading channels can be considered:

- Slow Rayleigh fading channels: the complex path gains of the channels do not change during the transmission of the symbols of the same codeword.
- Fast Rayleigh fading channels: the complex path gains of the channels change independently at each time t.

3. Performance criteria

The main goal of this design is to reduce the pairwise error probability (PEP) which is the probability that the decoder selects an erroneous codeword **E** while a different codeword **S** was transmitted. We consider a codeword of L MIMO signals starting at $t = 1$ by a $n_T \times L$ matrix $\mathbf{S} = [S^1 S^2 \cdots S^L]$ where S^t is the tth MIMO signal. An error occurs if the decoder decides that another codeword $\mathbf{E} = [E^1 E^2 \cdots E^L]$ is transmitted. Let us define the $n_T \times L$ difference matrix

$$\mathbf{B} = \mathbf{E} - \mathbf{S} = \begin{bmatrix} e_1^1 - s_1^1 & \dots & e_1^L - s_1^L \\ \vdots & \ddots & \vdots \\ e_{n_T}^1 - s_{n_T}^1 & \dots & e_{n_T}^L - s_{n_T}^L \end{bmatrix}. \tag{8}$$

The $n_T \times n_T$ product matrix $\mathbf{A} = \mathbf{B}\mathbf{B}^*$ is introduced, where $\mathbf{B}^*$ denotes the hermitian of $\mathbf{B}$. The minimum rank of $\mathbf{A}$ $r = \min(\text{rank}(\mathbf{A}))$, computed for all pairs $(\mathbf{E}, \mathbf{S})$ of different codewords is defined. The design criteria depend on the value of the product rn_R.

First case: $rn_R \leq 3$:

In this case, for slow Rayleigh fading channels, two criteria have been proposed by Tarokh et al. (1998) and Liao & Prabhu (2005) to reduce the PEP:

- **A** has to be a full rank matrix for any pair $(\mathbf{E}, \mathbf{S})$. Since the maximal value of r is n_T, the achievable spatial diversity order is $n_T n_R$.
- The coding gain is related to the inverse of $\eta = \sum_d N(d) d^{-n_R}$, where $N(d)$ is defined as the average number of error events with a determinant d equal to

$$\begin{aligned} d = \det(\mathbf{A}) &= \prod_{k=1}^{n_T} \lambda_k \\ &= \prod_{k=1}^{n_T} \left(\sum_{t=1}^{L} \left| e_k^t - s_k^t \right|^2 \right). \end{aligned} \tag{9}$$

  The best codes must have the minimum value of η.

In the case of fast Rayleigh fading channels, different criteria have been obtained by Tarokh et al. (1998). They define the Hamming distance $d_H(\mathbf{E}, \mathbf{S})$ between two codewords **E** and **S** as the number of time intervals for which $|E^t - S^t| \neq 0$. To maximize the diversity advantage, the minimal Hamming distance must be maximized for all pairs of codewords $(\mathbf{E}, \mathbf{S})$. In this case, the achieved spatial diversity order is equal to $d_H(\mathbf{E}, \mathbf{S}) n_R$. In the same way, Tarokh *et al.* introduce the product distance $d_p^2(\mathbf{E}, \mathbf{S})$ given by

$$d_p^2(\mathbf{E}, \mathbf{S}) = \prod_{\substack{t=1 \\ E^t \neq S^t}}^{L} d_E^2(E^t, S^t), \tag{10}$$

where $d_E^2(E^t, S^t) = \sum_{k=1}^{n_T} \left| e_k^t - s_k^t \right|^2$ is the squared Euclidean distance between the MIMO signals E^t and S^t at time t. In order to reduce the number of error events, $\min \left\{ d_p^2(\mathbf{E}, \mathbf{S}) \right\}$ must be maximized for all pairs $(\mathbf{E}, \mathbf{S})$.

Second case: $rn_R \geq 4$:

Chen et al. (2001) show that for a large value of rn_R, which corresponds to a large number of independent SISO channels, the PEP is minimized if the sum of all the eigenvalues of the matrices **A** is maximized. Since **A** is a square matrix, the sum of all the eigenvalues is equal to its trace

$$\mathrm{tr}(\mathbf{A}) = \sum_{k=1}^{n_T} \lambda_k = \sum_{t=1}^{L} d_E^2(E^t, S^t). \quad (11)$$

For each pair of codewords, $\mathrm{tr}(\mathbf{A})$ is computed. The minimum trace (which is the minimum value of the squared Euclidean distance between two codewords) is the minimum of all these values $\mathrm{tr}(\mathbf{A})$. The concept of Euclidean distance for STTCs has been previously introduced by Ionescu (1999). The minimization of the PEP amounts to using a code which has the maximum value of the minimum Euclidean distance between two codewords. Liao & Prabhu (2005) state also that to minimize the frame error rate (FER), the number of error events with the minimum squared Euclidean distance between codewords has to be minimized.
In this paper, we consider only the case $rn_R \geq 4$ which is obtained when the rank of the STTCs is greater than or equal to 2 and there are at least 2 receive antennas.

4. Balanced STTCs

4.1 Definitions

The concept of "balanced codes" proposed by Ngo et al. (2008; 2007) is based on the observation that each good code has the same property given by the following definition.

Definition 1 (Number of occurrences). *The number of occurrences of a MIMO symbol is the number of times where the MIMO symbol is generated when we consider the entire set of the extended-states.*

Definition 2 (Balanced codes). *A code is balanced if and only if the generated MIMO symbols have the same number of occurrences $n_0 \in \mathbb{N}^*$, if the binary input symbols are equiprobable.*

Definition 3 (Fully balanced code). *A code is fully balanced if and only if the code is balanced and the set of generated MIMO symbols is $\Lambda = \mathbb{Z}_{2^n}^{n_T}$.*

Definition 4 (Minimal length code). *A code is a minimal length code if and only if the code is fully balanced and the number of occurrences of each MIMO symbol is $n_0 = 1$.*

To check that a code is balanced, the MIMO symbols generated by all the extended-states must be computed. Then, the number of occurrences of each MIMO symbol can be obtained.

For example, let us consider two generator matrices

$$\mathbf{G}_1 = \begin{bmatrix} 0 & 0 & 2 & 1 \\ 2 & 1 & 0 & 0 \end{bmatrix} \quad (12)$$

and

$$\mathbf{G}_2 = \begin{bmatrix} 0 & 0 & 2 & 1 \\ 3 & 1 & 0 & 0 \end{bmatrix}. \quad (13)$$

Remark: $\mathbf{G}_1$ is the generator matrix of the code proposed by Tarokh et al. (1998).
The repartition of MIMO symbols in function of extended-states is given in Tables 1 et 2 for the generator matrices $\mathbf{G}_1$ and $\mathbf{G}_2$ respectively. The decimal value of the extended-state $X^t = [x_1^t\ x_2^t\ x_1^{t-1}\ x_2^{t-1}]^\mathrm{T} \in \mathbb{Z}_2^4$ is computed by considering x_1^t the most significant bit. The number

of occurrences of each generated MIMO symbol is also given in these tables. The generator matrix $\mathbf{G}_1$ corresponds to a minimal length code, whereas $\mathbf{G}_2$ corresponds to a no balanced code.

X^t	0	1	2	3	4	5	6	7	8	9	10	11	12	13	14	15
Y^t	$\begin{bmatrix}0\\0\end{bmatrix}$	$\begin{bmatrix}1\\0\end{bmatrix}$	$\begin{bmatrix}2\\0\end{bmatrix}$	$\begin{bmatrix}3\\0\end{bmatrix}$	$\begin{bmatrix}0\\1\end{bmatrix}$	$\begin{bmatrix}1\\1\end{bmatrix}$	$\begin{bmatrix}2\\1\end{bmatrix}$	$\begin{bmatrix}3\\1\end{bmatrix}$	$\begin{bmatrix}0\\2\end{bmatrix}$	$\begin{bmatrix}1\\2\end{bmatrix}$	$\begin{bmatrix}2\\2\end{bmatrix}$	$\begin{bmatrix}3\\2\end{bmatrix}$	$\begin{bmatrix}0\\3\end{bmatrix}$	$\begin{bmatrix}1\\3\end{bmatrix}$	$\begin{bmatrix}2\\3\end{bmatrix}$	$\begin{bmatrix}3\\3\end{bmatrix}$
occurrences	1	1	1	1	1	1	1	1	1	1	1	1	1	1	1	1

Table 1. MIMO symbols generated by $\mathbf{G}_1$

X^t	0 12	1 13	2 14	3 15	4	5	6	7	8	9	10	11
Y	$\begin{bmatrix}0\\0\end{bmatrix}$	$\begin{bmatrix}1\\0\end{bmatrix}$	$\begin{bmatrix}2\\0\end{bmatrix}$	$\begin{bmatrix}3\\0\end{bmatrix}$	$\begin{bmatrix}0\\1\end{bmatrix}$	$\begin{bmatrix}1\\1\end{bmatrix}$	$\begin{bmatrix}2\\1\end{bmatrix}$	$\begin{bmatrix}3\\1\end{bmatrix}$	$\begin{bmatrix}0\\3\end{bmatrix}$	$\begin{bmatrix}1\\3\end{bmatrix}$	$\begin{bmatrix}2\\3\end{bmatrix}$	$\begin{bmatrix}3\\3\end{bmatrix}$
Nb occurrences	2	2	2	2	1	1	1	1	1	1	1	1

Table 2. MIMO symbols generated by $\mathbf{G}_2$

4.2 Properties of $\mathbb{Z}_{2^n}^{n_T}$

We define the subgroup $\mathcal{C}_0$ of $\mathbb{Z}_{2^n}^{n_T}$ by

$$\mathcal{C}_0 = 2^{n-1}\mathbb{Z}_2^{n_T}. \tag{14}$$

Property 1. $\forall V \in \mathcal{C}_0$, $V = -V$ *with* $V + (-V) = [0 \cdots 0]^{\mathrm{T}}$.

Proof. Let us consider $V \in \mathcal{C}_0$. As $\mathcal{C}_0 = 2^{n-1}\mathbb{Z}_2^{n_T}$,

$$V = 2^{n-1}q \in \mathcal{C}_0 \text{ with } q \in \mathbb{Z}_2^{n_T}. \tag{15}$$

Therefore

$$V + V = 2^n q = 0 \in \mathbb{Z}_{2^n}^{n_T} \pmod{2^n}. \tag{16}$$

Thus, it exists $V' = V$ such as $V + V' = 0 \in \mathbb{Z}_{2^n}^{n_T}$. Hence, $-V = V, \forall V \in \mathcal{C}_0$. □

Besides, it is possible to make a partition of the group $\mathbb{Z}_{2^n}^{n_T}$ into $2^{n_T(n-1)}$ cosets, as presented by Coleman (2002) such as

$$\mathbb{Z}_{2^n}^{n_T} = \bigcup_{P \in \mathbb{Z}_{2^n}^{n_T}} \mathcal{C}_P, \tag{17}$$

where P is a coset representative of the coset $\mathcal{C}_P = P + \mathcal{C}_0$. Based on these cosets, another partition can be created and given by

$$\mathbb{Z}_{2^n}^{n_T} = \bigcup_{q=0}^{n-1} \mathcal{E}_q, \tag{18}$$

where $\mathcal{E}_0 = \mathcal{C}_0$. For $q = \overline{1, n-1}$, the other $\mathcal{E}_q$ are defined by

$$\mathcal{E}_q = \bigcup_{P_q} (P_q + \mathcal{C}_0) = \bigcup_{P_q} \mathcal{C}_{P_q}, \tag{19}$$

where $P_q \in 2^{n-q-1}\mathbb{Z}_{2^q}^{n_T} \backslash 2^{n-q}\mathbb{Z}_{2^{q-1}}^{n_T}$. The set $\mathbb{Z}_1^{n_T}$ contains only the nul element of $\mathbb{Z}_{2^n}^{n_T}$.

Definition 5. *Let us consider a subgroup Λ of $\mathbb{Z}_{2^n}^{n_T}$. A coset $C_P = P + \Lambda$ with $P \in \mathbb{Z}_{2^n}^{n_T}$ is relative to $Q \in \Lambda$ if and only if $2P = Q$.*

Thus, for $q = \overline{2, n-1}$, each coset $\mathcal{C}_{P_q} = P_q + \mathcal{C}_0 \subset \mathcal{E}_q$ is relative to $R = 2P_q \in \mathcal{E}_{q-1}$.

For example, it is possible to make a partition of the group $\mathbb{Z}_{2^n}^{n_T}$ which is the set of 4-PSK MIMO symbols with 2 transmit antennas. This partition is represented in Table 3.

$\mathcal{E}_0$:	$\mathcal{C}_0$	$\begin{bmatrix}0\\0\end{bmatrix}$	$\begin{bmatrix}0\\2\end{bmatrix}$	$\begin{bmatrix}2\\0\end{bmatrix}$	$\begin{bmatrix}2\\2\end{bmatrix}$
$\mathcal{E}_1$:	$\mathcal{C}_{\begin{bmatrix}0\\1\end{bmatrix}}$	$\begin{bmatrix}0\\1\end{bmatrix}$	$\begin{bmatrix}0\\3\end{bmatrix}$	$\begin{bmatrix}2\\1\end{bmatrix}$	$\begin{bmatrix}2\\3\end{bmatrix}$
	$\mathcal{C}_{\begin{bmatrix}1\\0\end{bmatrix}}$	$\begin{bmatrix}1\\0\end{bmatrix}$	$\begin{bmatrix}1\\2\end{bmatrix}$	$\begin{bmatrix}3\\0\end{bmatrix}$	$\begin{bmatrix}3\\2\end{bmatrix}$
	$\mathcal{C}_{\begin{bmatrix}1\\1\end{bmatrix}}$	$\begin{bmatrix}1\\1\end{bmatrix}$	$\begin{bmatrix}1\\3\end{bmatrix}$	$\begin{bmatrix}3\\1\end{bmatrix}$	$\begin{bmatrix}3\\3\end{bmatrix}$

Table 3. Partition of $\mathbb{Z}_4^2$

The red coset $\mathcal{C}_{\begin{bmatrix}0\\1\end{bmatrix}} = \begin{bmatrix}0\\1\end{bmatrix} + \mathcal{C}_0$ is relative to the red element $\begin{bmatrix}0\\2\end{bmatrix} \in \mathcal{C}_0$. The green coset $\mathcal{C}_{\begin{bmatrix}1\\0\end{bmatrix}} = \begin{bmatrix}1\\0\end{bmatrix} + \mathcal{C}_0$ is relative to the green element $\begin{bmatrix}2\\0\end{bmatrix} \in \mathcal{C}_0$. The yellow coset $\mathcal{C}_{\begin{bmatrix}1\\1\end{bmatrix}} = \begin{bmatrix}1\\1\end{bmatrix} + \mathcal{C}_0$ is relative to the yellow element $\begin{bmatrix}2\\2\end{bmatrix} \in \mathcal{C}_0$.

Property 2. *If Λ_l is a subgroup of $\mathbb{Z}_{2^n}^{n_T}$ given by*

$$\Lambda_l = \left\{ \sum_{m=1}^{l} x_m V_m \ mod\ 2^n \,/\, x_m \in \{0,1\} \right\},\ l \in \{1,2,\cdots,nn_T\} \tag{20}$$

with $V_m \in \mathbb{Z}_{2^n}^{n_T}$ and if the number of occurrences of each MIMO symbol $V \in \Lambda_l$ is $n(V) = n_0 = 1$ i.e. $\text{card}(\Lambda_l) = 2^l$, then there is at least one element V_m which belongs to $\mathcal{C}_0^*$.

Proof. The Lagrange's theorem states that for a finite group Λ, the order of each subgroup Λ_l of Λ divides the order of Λ. In the case of 2^n-PSK, $\text{card}(\Lambda) = \text{card}(\mathbb{Z}_{2^n}^{n_T}) = 2^{nn_T}$, then $\text{card}(\Lambda_l) = 2^l$. Hence, $\text{card}(\Lambda_l)$ is a even number. The null element belongs to Λ_l and the opposite of each element is included in Λ_l. Thus, in order to obtain an even number for $\text{card}(\Lambda_l)$, there are at least one element $V_m \neq 0$ which respects $V_m = -V_m$. Only the elements of $\mathcal{C}_0$ respect $V_m = -V_m$. Therefore, there is at least one element $V_m \in \mathcal{C}_0^*$. □

Definition 6 (Linearly independent vectors). *If $\text{card}(\Lambda_l) = 2^l$, the vectors $V_1, V_2, \cdots V_l$ are linearly independent. Hence, they form a base of Λ_l.*

Remarks :

1. $\min\left\{l \in \mathbb{N} \backslash \Lambda_l = \mathbb{Z}_{2^n}^{n_T}\right\} = nn_T = L_{min} = \dim\left(\mathbb{Z}_{2^n}^{n_T}\right)$ i.e. nn_T is a minimal number of vectors giving a fully balanced code.
2. nn_T is the maximal number of vectors to obtain a number of occurrences $n(V) = n_0 = 1$, $\forall V \in \Lambda_l$.

Property 3. *To generate a subgroup* $\Lambda_l = \left\{\sum_{m=1}^{l} x_m V_m \bmod 2^n / x_m \in \{0,1\}\right\}$ *with* $V_m \in \mathbb{Z}_{2^n}^{n_T}$, $l \in \{1, 2, \cdots, nn_T\}$ *and* $\mathrm{card}(\Lambda_l) = 2^l$, *the elements* V_m *must be selected as follows:*

- *The first element* V_1 *must belong to* $\mathcal{C}_0^*$.
- *If* $m-1$ *elements* $V_1, V_2, \cdots V_{m-1}$ *have been already selected with* $m \in \{2 \cdots l\}$, *the* m^{th} *element* V_m *must not belong to*

$$\Lambda_{m-1} = \left\{\sum_{m'=1}^{m-1} x_{m'} V_{m'} \bmod 2^n / x_{m'} \in \{0,1\}\right\} \tag{21}$$

and must belong to $\mathcal{C}_0^*$ *or to the cosets relative to an element of* Λ_{m-1}.

Proof. As shown by property 2, there is at least one element which belongs to $\mathcal{C}_0^*$. Thus, if $V_1 \in \mathcal{C}_0^*$, $\Lambda_1 = \{0, V_1\}$ is a subgroup of $\mathbb{Z}_{2^n}^{n_T}$.
We consider that $m-1$ elements have been selected to generate a subgroup Λ_{m-1} with $m = \{2, \cdots nn_T\}$.
If we select $V_m \in \mathbb{Z}_{2^n}^{n_T} \backslash \Lambda_{m-1}$ such as $2V_m = Q \in \Lambda_{m-1}$, a set Λ_m is defined by

$$\Lambda_m = \Lambda_{m-1} \bigcup \mathcal{C}_{V_m} \tag{22}$$

where $\mathcal{C}_{V_m}$ is the coset defined by

$$\mathcal{C}_{V_m} = V_m + \Lambda_{m-1}. \tag{23}$$

In order to show that Λ_m is a subgroup of $\mathbb{Z}_{2^n}^{n_T}$, the following properties must be proved:

1. $0 \in \Lambda_m$. *Proof:* As $\Lambda_m = \Lambda_{m-1} \bigcup (\Lambda_{m-1} + V_m)$ and $0 \in \Lambda_{m-1}$, we have $0 \in \Lambda_m$.
2. $\forall V_1, V_2 \in \Lambda_m$, $V_1 + V_2 \in \Lambda_m$. Three cases must be considered.
   1rd case: $V_1, V_2 \in \Lambda_{m-1}$. In this case, as Λ_{m-1} is a subgroup $V_1 + V_2 \in \Lambda_{m-1} \subset \Lambda_m$.
   2st case: $V_1, V_2 \in \mathcal{C}_{V_m}$. In this case, $V_1 = V_m + Q_1$ and $V_2 = V_m + Q_2$ with $Q_1, Q_2 \in \Lambda_{m-1}$. Thus, $V_1 + V_2 = 2V_m + Q_1 + Q_2$. As $2V_m \in \Lambda_{m-1}$ and Λ_{m-1} is a subgroup, $V_1 + V_2 \in \Lambda_{m-1} \subset \Lambda_m$.
   3nd case: $V_1 \in \Lambda_{m-1}$ and $V_2 \in \mathcal{C}_{V_m}$, In this case, $V_2 = V_m + Q_2$, with $Q_2 \in \Lambda_{m-1}$. We have $V_1 + V_2 = V_1 + V_m + Q_2 = V_m + (V_1 + Q_2) \in \mathcal{C}_{V_m} \subset \Lambda_m$ because $V_1 + Q_2 \in \Lambda_{m-1}$ (Λ_{m-1} is a subgroup).
   Thus, Λ_m is a closed set under addition.
3. $\forall V \in \Lambda_m, \exists -V \in \Lambda_m$ such as $V + (-V) = 0$.
   Proof : Two cases must be considered.
   1st case: $V \in \Lambda_{m-1}$.
   In this case, as Λ_{m-1} is a subgroup, so $-V \in \Lambda_{m-1} \subset \Lambda_m$.

2nd case: $V \in \mathcal{C}_{m-1}$.

In this case, $V = V_m + Q$ with $Q \in \Lambda_{m-1}$. Because Λ_{m-1} is a subgroup, $-Q = -V_m + (-V_m) = -2V_m \in \Lambda_{m-1}$ with $V_m + (-V_m) = 0$ and $Q + (-Q) = 0$. Thus, we have

$$-V_m = V_m + (-2V_m) = V_m + (-Q) \in \mathcal{C}_{V_m} \subset \Lambda_m. \tag{24}$$

Hence $-V = -V_m + (-Q) = V_m + (-Q) + (-Q) \in \mathcal{C}_m \subset \Lambda_m$ because $(-Q) + (-Q) \in \Lambda_{m-1}$.

Thus, the opposite of each element of Λ_m belongs to Λ_m.

In conclusion, if each new element is selected within a coset relative to a generated element, the created set Λ_m is also a subgroup. □

4.3 Properties of balanced STTCs

Each MIMO symbol belongs to $\mathbb{Z}_{2^n}^{n_T}$. At each time t, the generated MIMO symbol is given by the value of extended-state X^t and the generator matrix $\mathbf{G}$. In this section, several properties of the generator matrix are given in order to reduce the search-time.

Property 4. *For a fully balanced code, the number of columns of the generator matrix* $\mathbf{G}$ *is* $L_r \geq L_{min} = nn_T$. *As shown by the definition (4), if* $\mathbf{G}$ *has* L_{min} *columns, then the code is a minimal length code.*

Proof. Let us consider the generator matrix $\mathbf{G}$ of a 2^n-PSK 2^{L_r-n} states STTC with n_T transmit antennas. The extended-state can use 2^{L_r} binary value. The maximal number of generated MIMO symbols is given by

$$\sum_{Y \in \mathbb{Z}_{2^n}^{n_T}} n(Y) = 2^{L_r}. \tag{25}$$

$\mathbf{G}$ is the generator matrix of a fully balanced code, $\forall Y \in \mathbb{Z}_{2^n}^{n_T}$, $n(Y) = n_0$. The number of possible MIMO symbols is $\text{card}(\mathbb{Z}_{2^n}^{n_T}) = 2^{nn_T}$. Thus, the previous expression is

$$n_0 2^{nn_T} = 2^{L_r}. \tag{26}$$

The number of occurrences of each $Y \in \mathbb{Z}_{2^n}^{n_T}$, $n(Y) \geq 1$, so $2^{L_r} \geq \text{card}(\mathbb{Z}_{2^n}^{n_T}) = nn_T$. Therefore $L_r \geq L_{min} = nn_T$. □

Property 5. *If* $\mathbf{G}$ *is the generator matrix with* L_r *columns of a fully balanced code, for any additional column* $G_{L_r+1} \in \mathbb{Z}_{2^n}^{n_T}$, *the resulting generator matrix* $\mathbf{G}' = [\mathbf{G}\, G_{L_r+1}]$ *corresponds to a new fully balanced code.*

Proof. For a fully balanced code, the generated MIMO symbols belong to $\Lambda_{L_r} = \mathbb{Z}_{2^n}^{n_T}$. If a new column $G_{L_r+1} \in \mathbb{Z}_{2^n}^{n_T}$ is added to $\mathbf{G}$, the new set of columns generates $\Lambda_{L_r+1} = \Lambda_{L_r} \bigcup (\Lambda_{L_r} + G_{L_r+1})$. As $G_{L_r+1} \in \mathbb{Z}_{2^n}^{n_T}$ and $\mathbb{Z}_{2^n}^{n_T}$ is a group, $\Lambda_{L_r} + G_{L_r+1} = \Lambda_{L_r}$. Thereby, $\Lambda_{L_r+1} = \mathbb{Z}_{2^n}^{n_T}$. The number of occurrences of the elements belonging to Λ_{L_r} is n_0. The number of occurrences of the elements belonging to $\Lambda_{L_r} + G_{L_r+1}$ is also n_0. Thus, for each new column of the fully balanced code, the number of occurrences of the elements of the new code is $2n_0$. □

Property 6. *If* $\mathbf{G}$ *is the matrix of a balanced code, each permutation of columns or/and lines generates the generator matrix of a new balanced code.*

Proof. The set of MIMO symbols belongs to a subgroup of the commutative group $\mathbb{Z}_{2^n}^{n_T}$. Therefore, the permutations between the columns of the generator matrix generate the same MIMO symbols. So, the new codes are also balanced.
A permutation of lines of the generator matrix corresponds to a permutation of the transmit antennas. If the initial code is balanced, the code will be balanced.

□

5. New method to generate the balanced codes

5.1 Generation of the fully balanced codes

The property 5 states that the minimal length code is the first step to create any fully balanced code. In fact, for each column added to the generator matrix of a fully balanced code, the new code is also fully balanced. Thus, only the generation of minimal length codes is presented.
To create a minimal length STTC, the columns G_i with $i = \overline{1, nn_T}$ must be selected with the following rules.

Rule 1. *The first column G_1 must be selected within C_0^*. This first selection creates the subgroup $\Lambda_1 = \{0, G_1\}$.*

Rule 2. *If the first $m \in \{1, 2, \cdots, L_{min} - 1\}$ columns of the generator matrix have been selected to generate a subgroup $\Lambda_m = \left\{ \sum_{m'=1}^{m} x_{m'} G_{m'} \bmod 2^n / x_{m'} \in \{0, 1\} \right\}$ of $\mathbb{Z}_{2^n}^{n_T}$, the add of next column G_{m+1} create a new subgroup Λ_{m+1} of $\mathbb{Z}_{2^n}^{n_T}$ with $\mathrm{card}(\Lambda_{m+1}) = 2\mathrm{card}(\Lambda_m)$. Hence, the column G_{m+1} of $\mathbf{G}$ must be selected in $\mathbb{Z}_{2^n}^{n_T} \backslash \Lambda_m$ in a coset relative to an element of Λ_m or in C_0^*.*

If this algorithm is respected, the new generated set

$$\Lambda_{m+1} = \Lambda_m \bigcup (\Lambda_m + G_m) \tag{27}$$

is a subgroup of $\mathbb{Z}_{2^n}^{n_T}$.
Thereby, as presented by Forney (1988), a chain partition of $\mathbb{Z}_{2^n}^{n_T}$

$$\Lambda_1 / \Lambda_2 \cdots / \Lambda_m / \cdots / \Lambda_{L_{min}} \tag{28}$$

is obtained with $\mathrm{card}(\Lambda_{m+1}) = 2\mathrm{card}(\Lambda_m)$. Since $\mathrm{card}(\Lambda_1) = 2$, then the L_{min} columns generate a subgroup $\Lambda_{L_{min}}$ with $\mathrm{card}(\Lambda_{L_{min}}) = 2^{L_{min}}$.
To obtain a fully balanced code with a generator matrix with $L_r > L_{min}$, it is sufficient to add new columns which belong to $\mathbb{Z}_{2^n}^{n_T}$.
The main advantage of this method is to define distinctly the set where each new column of $\mathbf{G}$ must be selected. In the first method presented by Ngo et al. (2008; 2007), the linear combination of the generated and selected elements and their opposites must be blocked, i.e. these elements must not be further selected. Due to the new method, there are no blocked elements. This is a significant simplification of the first method.

5.2 Generation of the balanced codes

This section treats the generation of balanced STTCs (not fully balanced STTCs) i.e. the set of the generated MIMO symbols is a subset of the group $\mathbb{Z}_{2^n}^{n_T}$ and not the entire set $\mathbb{Z}_{2^n}^{n_T}$.
For the generation of these STTCs with the generator matrix constituted by L_r columns, the algorithm which is presented in the previous section must be used for the selection of the first $m_0 \leq \min(L_{min} - 2, L_r)$ columns. Thus,

- The m_0 first columns $G_1, G_2, \cdots G_{m_0}$ must be selected with the Rules 1 and 2. Hence, the set generated by the first m_0 columns is the subgroup

$$\Lambda_{m_0} = \left\{ \sum_{m=1}^{m_0} x_m G_m \bmod 2^n / x_m \in \{0,1\} \right\}. \tag{29}$$

  The number of occurrences of each MIMO symbol $V \in \Lambda_{m_0}$ is 1.

- The columns $G_{m_0+1} \cdots G_{L_r-1}$ must be selected in the subgroup Λ_{m_0}. Thus, the subgroup created by the $L_r - 1$ first columns is $\Lambda_{L_r-1} = \Lambda_{m_0}$, but the number of occurrences of each MIMO symbol is $2^{L_r-m_0-1}, \forall Y \in \Lambda_{m_0}$.
- The last column must be selected everywhere in $\mathbb{Z}_{2^n}^{n_T}$. Two cases can be analyzed :
  - If $G_{L_r} \in \Lambda_{L_r-1}$, then the number of occurrences of each element is multiplied by two. The resulting code is also balanced. In this case, the set of generated MIMO symbols is the subgroupΛ_{L_r-1}.
  - If $G_{L_r} \in \mathbb{Z}_{2^n}^{n_T} \backslash \Lambda_{L_r-1}$, the generated set $\Lambda_{L_r} = \Lambda_{L_r-1} \bigcup (\Lambda_{L_r-1} + G_{L_r})$ is not necessarily a subgroup of $\mathbb{Z}_{2^n}^{n_T}$. Each element of the generated set has $2^{L_r-m_0-1}, \forall Y \in \Lambda_{L_r}$, but $\text{card}(\Lambda_{L_r}) = 2\text{card}(\Lambda_{L_r-1})$.

5.3 Example of the generation of a fully balanced 64 states 4-PSK STTC with 3 transmit antennas

The generator matrix of the 64 states 4-PSK STTCs with 3 transmit antennas is

$$\mathbf{G} = [G_1 G_2 | G_3 G_4 | G_5 G_6 | G_7 G_8] \tag{30}$$

with $G_i \in \left[G_i^1 G_i^2 G_i^3\right]^{\mathrm{T}} \in \mathbb{Z}_4^3$ for $i = \overline{1,8}$. In order to create a fully balanced 64 states 4-PSK STTC with 3 transmit antennas, 8 elements must be selected in $\mathbb{Z}_4^3$. The first element must belong to $\mathcal{C}_0^* = 2\mathbb{Z}_2^3 \backslash \{[0\ 0\ 0]^{\mathrm{T}}\}$. The element $G_1 = \begin{bmatrix}0\\2\\2\end{bmatrix}$ can be selected. Thereby, the second column of $\mathbf{G}$ must be selected either in $\mathcal{C}_0^*$ or in the coset relative to G_1. In Table 4, the green element represents the selected element and the blue element represent the generated element.

$\mathcal{C}_0$	$\begin{bmatrix}0\\0\\0\end{bmatrix}$ (blue)	$\begin{bmatrix}0\\0\\2\end{bmatrix}$	$\begin{bmatrix}0\\2\\0\end{bmatrix}$	$\begin{bmatrix}0\\2\\2\end{bmatrix}$ (green)	$\begin{bmatrix}2\\0\\0\end{bmatrix}$	$\begin{bmatrix}2\\0\\2\end{bmatrix}$	$\begin{bmatrix}2\\2\\0\end{bmatrix}$	$\begin{bmatrix}2\\2\\2\end{bmatrix}$
$\mathcal{C}_{\begin{bmatrix}0\\1\\1\end{bmatrix}}$	$\begin{bmatrix}0\\1\\1\end{bmatrix}$	$\begin{bmatrix}0\\1\\3\end{bmatrix}$	$\begin{bmatrix}0\\3\\1\end{bmatrix}$	$\begin{bmatrix}0\\3\\3\end{bmatrix}$	$\begin{bmatrix}2\\1\\1\end{bmatrix}$	$\begin{bmatrix}2\\1\\3\end{bmatrix}$	$\begin{bmatrix}2\\3\\1\end{bmatrix}$	$\begin{bmatrix}2\\3\\3\end{bmatrix}$

Table 4. Selection of G_1 of $\mathbf{G}$

The next column G_2 of $\mathbf{G}$ must be selected in the white set. If G_2 is selected into $\mathcal{C}_{\begin{bmatrix}0\\1\\1\end{bmatrix}}$, for example $\begin{bmatrix}2\\1\\3\end{bmatrix}$, the generated subgroup Λ_2 is represented by the colored elements of Table 5. No new element of $\mathcal{C}_0$ is generated. Thereby, G_3 must belong to $\mathcal{C}_0^*$ or $\mathcal{C}_{\begin{bmatrix}0\\1\\1\end{bmatrix}}$ and must not belong to Λ_2.

$\mathcal{C}_0$	$\begin{bmatrix}0\\0\\0\end{bmatrix}$	$\begin{bmatrix}0\\0\\2\end{bmatrix}$	$\begin{bmatrix}0\\2\\0\end{bmatrix}$	$\begin{bmatrix}0\\2\\2\end{bmatrix}$	$\begin{bmatrix}2\\0\\0\end{bmatrix}$	$\begin{bmatrix}2\\0\\2\end{bmatrix}$	$\begin{bmatrix}2\\2\\0\end{bmatrix}$	$\begin{bmatrix}2\\2\\2\end{bmatrix}$
$\mathcal{C}_{\begin{bmatrix}0\\1\\1\end{bmatrix}}$	$\begin{bmatrix}0\\1\\1\end{bmatrix}$	$\begin{bmatrix}0\\1\\3\end{bmatrix}$	$\begin{bmatrix}0\\3\\1\end{bmatrix}$	$\begin{bmatrix}0\\3\\3\end{bmatrix}$	$\begin{bmatrix}2\\1\\1\end{bmatrix}$	$\begin{bmatrix}2\\1\\3\end{bmatrix}$	$\begin{bmatrix}2\\3\\1\end{bmatrix}$	$\begin{bmatrix}2\\3\\3\end{bmatrix}$

Table 5. Selection of G_2 of **G**

The 3rd element can be $G_3 = \begin{bmatrix}0\\0\\2\end{bmatrix}$. As presented in Table 6, two new elements of $\mathcal{C}_0$ are generated (or selected). Thus, G_4 must be selected among the white elements of Table 6. If $G_1 = 2P_1$ and $G_3 = 2P_2$, the set of white elements is $\mathcal{C}_0 \bigcup(\mathcal{C}_0 + P_1) \bigcup(\mathcal{C}_0 + P_2) \bigcup(\mathcal{C}_0 + P_1 + P_2) \backslash \Lambda_2$.

$\mathcal{C}_0$	$\begin{bmatrix}0\\0\\0\end{bmatrix}$	$\begin{bmatrix}0\\0\\2\end{bmatrix}$	$\begin{bmatrix}0\\2\\0\end{bmatrix}$	$\begin{bmatrix}0\\2\\2\end{bmatrix}$	$\begin{bmatrix}2\\0\\0\end{bmatrix}$	$\begin{bmatrix}2\\0\\2\end{bmatrix}$	$\begin{bmatrix}2\\2\\0\end{bmatrix}$	$\begin{bmatrix}2\\2\\2\end{bmatrix}$
$\mathcal{C}_{\begin{bmatrix}0\\1\\1\end{bmatrix}}$	$\begin{bmatrix}0\\1\\1\end{bmatrix}$	$\begin{bmatrix}0\\1\\3\end{bmatrix}$	$\begin{bmatrix}0\\3\\1\end{bmatrix}$	$\begin{bmatrix}0\\3\\3\end{bmatrix}$	$\begin{bmatrix}2\\1\\1\end{bmatrix}$	$\begin{bmatrix}2\\1\\3\end{bmatrix}$	$\begin{bmatrix}2\\3\\1\end{bmatrix}$	$\begin{bmatrix}2\\3\\3\end{bmatrix}$
$\mathcal{C}_{\begin{bmatrix}0\\0\\1\end{bmatrix}}$	$\begin{bmatrix}0\\0\\1\end{bmatrix}$	$\begin{bmatrix}0\\0\\3\end{bmatrix}$	$\begin{bmatrix}0\\2\\1\end{bmatrix}$	$\begin{bmatrix}0\\2\\3\end{bmatrix}$	$\begin{bmatrix}2\\0\\1\end{bmatrix}$	$\begin{bmatrix}2\\0\\3\end{bmatrix}$	$\begin{bmatrix}2\\2\\1\end{bmatrix}$	$\begin{bmatrix}2\\2\\3\end{bmatrix}$
$\mathcal{C}_{\begin{bmatrix}0\\1\\0\end{bmatrix}}$	$\begin{bmatrix}0\\1\\0\end{bmatrix}$	$\begin{bmatrix}0\\1\\2\end{bmatrix}$	$\begin{bmatrix}0\\3\\0\end{bmatrix}$	$\begin{bmatrix}0\\3\\2\end{bmatrix}$	$\begin{bmatrix}2\\1\\0\end{bmatrix}$	$\begin{bmatrix}2\\1\\2\end{bmatrix}$	$\begin{bmatrix}2\\3\\0\end{bmatrix}$	$\begin{bmatrix}2\\3\\2\end{bmatrix}$

Table 6. Selection of G_3 of **G**

Now, we select $G_4 = \begin{bmatrix}2\\2\\3\end{bmatrix}$. A new subgroup is created, as shown by the colored elements in Table 7.

$\mathcal{C}_0$	$\begin{bmatrix}0\\0\\0\end{bmatrix}$	$\begin{bmatrix}0\\0\\2\end{bmatrix}$	$\begin{bmatrix}0\\2\\0\end{bmatrix}$	$\begin{bmatrix}0\\2\\2\end{bmatrix}$	$\begin{bmatrix}2\\0\\0\end{bmatrix}$	$\begin{bmatrix}2\\0\\2\end{bmatrix}$	$\begin{bmatrix}2\\2\\0\end{bmatrix}$	$\begin{bmatrix}2\\2\\2\end{bmatrix}$
$\mathcal{C}_{\begin{bmatrix}0\\1\\1\end{bmatrix}}$	$\begin{bmatrix}0\\1\\1\end{bmatrix}$	$\begin{bmatrix}0\\1\\3\end{bmatrix}$	$\begin{bmatrix}0\\3\\1\end{bmatrix}$	$\begin{bmatrix}0\\3\\3\end{bmatrix}$	$\begin{bmatrix}2\\1\\1\end{bmatrix}$	$\begin{bmatrix}2\\1\\3\end{bmatrix}$	$\begin{bmatrix}2\\3\\1\end{bmatrix}$	$\begin{bmatrix}2\\3\\3\end{bmatrix}$
$\mathcal{C}_{\begin{bmatrix}0\\0\\1\end{bmatrix}}$	$\begin{bmatrix}0\\0\\1\end{bmatrix}$	$\begin{bmatrix}0\\0\\3\end{bmatrix}$	$\begin{bmatrix}0\\2\\1\end{bmatrix}$	$\begin{bmatrix}0\\2\\3\end{bmatrix}$	$\begin{bmatrix}2\\0\\1\end{bmatrix}$	$\begin{bmatrix}2\\0\\3\end{bmatrix}$	$\begin{bmatrix}2\\2\\1\end{bmatrix}$	$\begin{bmatrix}2\\2\\3\end{bmatrix}$
$\mathcal{C}_{\begin{bmatrix}0\\1\\0\end{bmatrix}}$	$\begin{bmatrix}0\\1\\0\end{bmatrix}$	$\begin{bmatrix}0\\1\\2\end{bmatrix}$	$\begin{bmatrix}0\\3\\0\end{bmatrix}$	$\begin{bmatrix}0\\3\\2\end{bmatrix}$	$\begin{bmatrix}2\\1\\0\end{bmatrix}$	$\begin{bmatrix}2\\1\\2\end{bmatrix}$	$\begin{bmatrix}2\\3\\0\end{bmatrix}$	$\begin{bmatrix}2\\3\\2\end{bmatrix}$

Table 7. Selection of G_4 of **G**

If G_5 is selected among to the white elements of Table 7, for example $G_5 = \begin{bmatrix} 2 \\ 2 \\ 2 \end{bmatrix}$, $\mathcal{C}_0$ is totally generated. Hence, a new set to select the last element, which is represented by the white elements of Table 8 is created.

$\mathcal{C}_0$	$\begin{bmatrix} 0 \\ 0 \\ 0 \end{bmatrix}$	$\begin{bmatrix} 0 \\ 0 \\ 2 \end{bmatrix}$	$\begin{bmatrix} 0 \\ 2 \\ 0 \end{bmatrix}$	$\begin{bmatrix} 0 \\ 2 \\ 2 \end{bmatrix}$	$\begin{bmatrix} 2 \\ 0 \\ 0 \end{bmatrix}$	$\begin{bmatrix} 2 \\ 0 \\ 2 \end{bmatrix}$	$\begin{bmatrix} 2 \\ 2 \\ 0 \end{bmatrix}$	$\begin{bmatrix} 2 \\ 2 \\ 2 \end{bmatrix}$
$\mathcal{C}_{\begin{bmatrix} 0 \\ 1 \\ 1 \end{bmatrix}}$	$\begin{bmatrix} 0 \\ 1 \\ 1 \end{bmatrix}$	$\begin{bmatrix} 0 \\ 1 \\ 3 \end{bmatrix}$	$\begin{bmatrix} 0 \\ 3 \\ 1 \end{bmatrix}$	$\begin{bmatrix} 0 \\ 3 \\ 3 \end{bmatrix}$	$\begin{bmatrix} 2 \\ 1 \\ 1 \end{bmatrix}$	$\begin{bmatrix} 2 \\ 1 \\ 3 \end{bmatrix}$	$\begin{bmatrix} 2 \\ 3 \\ 1 \end{bmatrix}$	$\begin{bmatrix} 2 \\ 3 \\ 3 \end{bmatrix}$
$\mathcal{C}_{\begin{bmatrix} 0 \\ 0 \\ 1 \end{bmatrix}}$	$\begin{bmatrix} 0 \\ 0 \\ 1 \end{bmatrix}$	$\begin{bmatrix} 0 \\ 0 \\ 3 \end{bmatrix}$	$\begin{bmatrix} 0 \\ 2 \\ 1 \end{bmatrix}$	$\begin{bmatrix} 0 \\ 2 \\ 3 \end{bmatrix}$	$\begin{bmatrix} 2 \\ 0 \\ 1 \end{bmatrix}$	$\begin{bmatrix} 2 \\ 0 \\ 3 \end{bmatrix}$	$\begin{bmatrix} 2 \\ 2 \\ 1 \end{bmatrix}$	$\begin{bmatrix} 2 \\ 2 \\ 3 \end{bmatrix}$
$\mathcal{C}_{\begin{bmatrix} 0 \\ 1 \\ 0 \end{bmatrix}}$	$\begin{bmatrix} 0 \\ 1 \\ 0 \end{bmatrix}$	$\begin{bmatrix} 0 \\ 1 \\ 2 \end{bmatrix}$	$\begin{bmatrix} 0 \\ 3 \\ 0 \end{bmatrix}$	$\begin{bmatrix} 0 \\ 3 \\ 2 \end{bmatrix}$	$\begin{bmatrix} 2 \\ 1 \\ 0 \end{bmatrix}$	$\begin{bmatrix} 2 \\ 1 \\ 2 \end{bmatrix}$	$\begin{bmatrix} 2 \\ 3 \\ 0 \end{bmatrix}$	$\begin{bmatrix} 2 \\ 3 \\ 2 \end{bmatrix}$
$\mathcal{C}_{\begin{bmatrix} 1 \\ 0 \\ 0 \end{bmatrix}}$	$\begin{bmatrix} 1 \\ 0 \\ 0 \end{bmatrix}$	$\begin{bmatrix} 1 \\ 0 \\ 2 \end{bmatrix}$	$\begin{bmatrix} 1 \\ 2 \\ 0 \end{bmatrix}$	$\begin{bmatrix} 1 \\ 2 \\ 2 \end{bmatrix}$	$\begin{bmatrix} 3 \\ 0 \\ 0 \end{bmatrix}$	$\begin{bmatrix} 3 \\ 0 \\ 2 \end{bmatrix}$	$\begin{bmatrix} 3 \\ 2 \\ 0 \end{bmatrix}$	$\begin{bmatrix} 3 \\ 2 \\ 2 \end{bmatrix}$
$\mathcal{C}_{\begin{bmatrix} 1 \\ 0 \\ 1 \end{bmatrix}}$	$\begin{bmatrix} 1 \\ 0 \\ 1 \end{bmatrix}$	$\begin{bmatrix} 1 \\ 0 \\ 3 \end{bmatrix}$	$\begin{bmatrix} 1 \\ 2 \\ 1 \end{bmatrix}$	$\begin{bmatrix} 1 \\ 2 \\ 3 \end{bmatrix}$	$\begin{bmatrix} 3 \\ 0 \\ 1 \end{bmatrix}$	$\begin{bmatrix} 3 \\ 0 \\ 3 \end{bmatrix}$	$\begin{bmatrix} 3 \\ 2 \\ 1 \end{bmatrix}$	$\begin{bmatrix} 3 \\ 2 \\ 3 \end{bmatrix}$
$\mathcal{C}_{\begin{bmatrix} 1 \\ 1 \\ 0 \end{bmatrix}}$	$\begin{bmatrix} 1 \\ 1 \\ 0 \end{bmatrix}$	$\begin{bmatrix} 1 \\ 1 \\ 2 \end{bmatrix}$	$\begin{bmatrix} 1 \\ 3 \\ 0 \end{bmatrix}$	$\begin{bmatrix} 1 \\ 3 \\ 2 \end{bmatrix}$	$\begin{bmatrix} 3 \\ 1 \\ 0 \end{bmatrix}$	$\begin{bmatrix} 3 \\ 1 \\ 2 \end{bmatrix}$	$\begin{bmatrix} 3 \\ 3 \\ 0 \end{bmatrix}$	$\begin{bmatrix} 3 \\ 3 \\ 2 \end{bmatrix}$
$\mathcal{C}_{\begin{bmatrix} 1 \\ 1 \\ 1 \end{bmatrix}}$	$\begin{bmatrix} 1 \\ 1 \\ 1 \end{bmatrix}$	$\begin{bmatrix} 1 \\ 1 \\ 3 \end{bmatrix}$	$\begin{bmatrix} 1 \\ 3 \\ 1 \end{bmatrix}$	$\begin{bmatrix} 1 \\ 3 \\ 3 \end{bmatrix}$	$\begin{bmatrix} 3 \\ 1 \\ 1 \end{bmatrix}$	$\begin{bmatrix} 3 \\ 1 \\ 3 \end{bmatrix}$	$\begin{bmatrix} 3 \\ 3 \\ 1 \end{bmatrix}$	$\begin{bmatrix} 3 \\ 3 \\ 3 \end{bmatrix}$

Table 8. Selection of G_5 of $\mathbf{G}$

If G_6 is chosen among belong to the white elements of Table 8, the totality of $\mathbb{Z}_{2^n}^{n_T}$ is generated. For example, we can select $G_6 = \begin{bmatrix} 3 \\ 3 \\ 1 \end{bmatrix}$. The created code is a minimal length code with the generator matrix

$$\mathbf{G} = \begin{bmatrix} 0 & 2 & 0 & 2 & 2 & 3 \\ 2 & 1 & 0 & 2 & 2 & 3 \\ 2 & 3 & 2 & 3 & 2 & 1 \end{bmatrix}. \tag{31}$$

As stated by the property 5, if an additional column is added to the generator matrix, the resulting code is also fully balanced. The number of occurrences of each MIMO symbol is given by $2^{L_r - L_{min}}$, where $L_r - L_{min}$ is the number of additional columns. Thus, the columns $G_7 = \begin{bmatrix} 0 \\ 2 \\ 2 \end{bmatrix}$ and $G_8 = \begin{bmatrix} 2 \\ 1 \\ 1 \end{bmatrix}$ are added to the generator matrix. We obtain

$$\mathbf{G} = \left[\begin{array}{cc|cc|cc|cc} 0 & 2 & 0 & 2 & 2 & 3 & 0 & 2 \\ 2 & 1 & 0 & 2 & 2 & 3 & 2 & 1 \\ 2 & 3 & 2 & 3 & 2 & 1 & 2 & 1 \end{array}\right]. \tag{32}$$

Table 9 shows the code proposed by Chen et al. (2002b) and the new generated code which are both fully balanced. The minimal rank and the minimal trace of each code are also given. The new code has a better rank and trace than the corresponding Chen's code. The FER and bit error rate (BER) of these two STTCs are presented respectively in Figs. 2 and 3. For the simulation, the channel fading coefficients are independent samples of a complex Gaussian process with zero mean and variance 0.5 per dimension. These channel coefficients are assumed to be known by the decoder. Each codeword consists of 130 MIMO symbols. For the simulation, 2 and 4 receive antennas are considered. The decoding is performed by the Viterbi's algorithm. We remark that the new code slightly outperforms the Chen's code.

Name	G	Rank	d_E^2 min
Chen et al. (2002b)	$\left[\begin{array}{cc\|cc\|cc\|cc} 0 & 2 & 3 & 2 & 3 & 0 & 3 & 2 \\ 2 & 2 & 1 & 2 & 3 & 0 & 2 & 0 \\ 2 & 0 & 0 & 2 & 2 & 3 & 1 & 1 \end{array}\right]$	2	28
New	$\left[\begin{array}{cc\|cc\|cc\|cc} 0 & 2 & 0 & 2 & 2 & 3 & 0 & 2 \\ 2 & 1 & 0 & 2 & 2 & 3 & 2 & 1 \\ 2 & 3 & 2 & 3 & 2 & 1 & 2 & 1 \end{array}\right]$	3	32

Table 9. 64 states 4-PSK STTCs with 3 transmit antennas

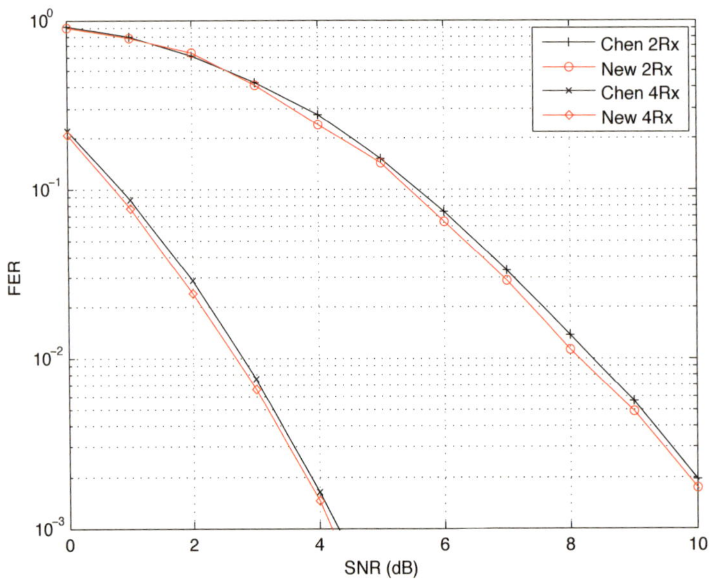

Fig. 2. FER of 64 states 4-PSK STTCs with 3 transmit antennas

5.4 Example of the generation of a balanced 8 states 8-PSK STTC with 4 transmit antennas

This section presents a design of balanced 8 states 8-PSK STTCs with 4 transmit antennas. The generator matrix is

$$\mathbf{G} = [G_1 G_2 G_3 | G_4 G_5 G_6], \tag{33}$$

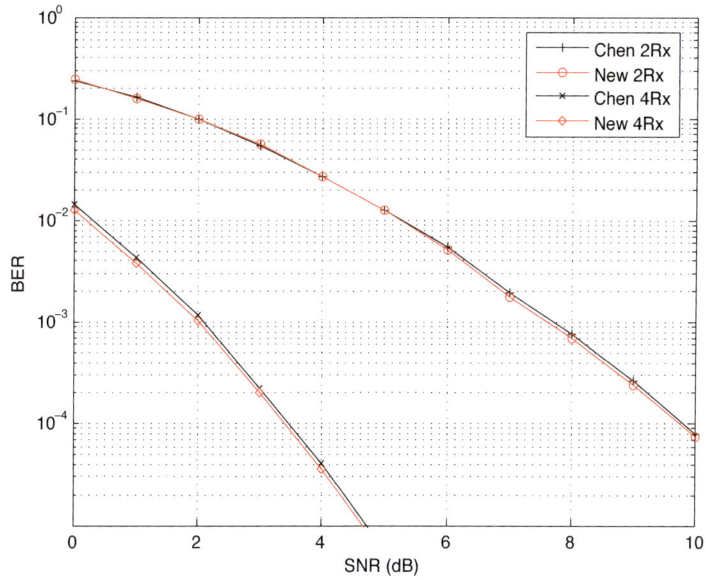

Fig. 3. BER of 64 states 4-PSK STTCs with 3 transmit antennas

with $G_i = \left[G_i^1 G_i^2 G_i^3 G_i^4\right]^{\mathrm{T}} \in \mathbb{Z}_8^4$ for $i = \overline{1,6}$. This code can not be fully balanced but just balanced. In fact, a fully balanced 8-PSK STTC with 4 transmit antennas must have 12 columns to be fully balanced (c.f. property 4).
The group $\mathbb{Z}_8^4$ is divided into 3 sets $\mathcal{E}_0$, $\mathcal{E}_1$ and $\mathcal{E}_2$. Each set $\mathcal{E}_i$ is the union of several cosets. Each coset included in $\mathcal{E}_i$ is relative to one vector of $\mathcal{E}_{i-1}$ with i={ 1,2}.
In general, the first no null column of the generator matrix must belong to

$$\mathcal{C}_0 = 2^{n-1}\mathbb{Z}_2^{n_T}. \tag{34}$$

Thereby, for the design of 8 states 8-PSK STTCs with 4 transmit antennas, $G_1 \in 2\mathbb{Z}_2^4$ for example $G_1 = \begin{bmatrix} 4\\0\\4\\4 \end{bmatrix} = 2P_1$, with $P_1 = \begin{bmatrix} 2\\0\\2\\2 \end{bmatrix}$. The first generated subgroup is $\Lambda_1 = \{0, G_1\}$. The next column of **G** must belong to

$$\mathcal{S}_2 = \left(\mathcal{C}_0 \bigcup \mathcal{C}_{P_1}\right) \backslash \Lambda_1, \tag{35}$$

where $\mathcal{C}_{P_1} \subset \mathcal{E}_1$ is the coset relative to G_1. The second column of **G** can be $G_2 = \begin{bmatrix} 2\\4\\2\\2 \end{bmatrix}$. The subgroup generated by the first two columns is

$$\Lambda_2 = \{0, G_1, G_2, G_1 + G_2\} \tag{36}$$

$$= \left\{ \begin{bmatrix} 0\\0\\0\\0 \end{bmatrix}, \begin{bmatrix} 4\\0\\4\\4 \end{bmatrix}, \begin{bmatrix} 2\\4\\2\\2 \end{bmatrix}, \begin{bmatrix} 6\\4\\6\\6 \end{bmatrix} \right\}, \tag{37}$$

with $G_1 + G_2 = -G_2$. As G_2 and $-G_2 \in \mathcal{E}_1$, the next column must belong to the set

$$\mathcal{S}_3 = \left(\mathcal{C}_0 \bigcup \mathcal{C}_{P_1} \bigcup \mathcal{C}_{P_2} \bigcup \mathcal{C}_{P_1+P_2}\right) \backslash \Lambda_2, \tag{38}$$

with $G_3 = \begin{bmatrix} 7\\6\\3\\3 \end{bmatrix} \in \mathcal{C}_{P_1+P_2}$. The next generated subgroup is

$$\Lambda_3 = \Lambda_2 \bigcup (\Lambda_3 + G_3) = \left\{ \begin{bmatrix} 0\\0\\0\\0 \end{bmatrix}, \begin{bmatrix} 4\\0\\4\\4 \end{bmatrix}, \begin{bmatrix} 2\\4\\2\\2 \end{bmatrix}, \begin{bmatrix} 6\\4\\6\\6 \end{bmatrix}, \begin{bmatrix} 7\\6\\3\\3 \end{bmatrix}, \begin{bmatrix} 3\\6\\7\\7 \end{bmatrix}, \begin{bmatrix} 1\\2\\5\\5 \end{bmatrix}, \begin{bmatrix} 5\\2\\1\\1 \end{bmatrix} \right\}. \tag{39}$$

The new generated elements $\Lambda_3 \backslash \Lambda_2$ belong to $\mathcal{E}_2$. Since no coset is relative to the elements of $\mathcal{E}_2$, the set using to select G_4 is

$$\mathcal{S}_4 = \mathcal{S}_3 \backslash \Lambda_3 = \left(\mathcal{C}_0 \bigcup \mathcal{C}_{P_1} \bigcup \mathcal{C}_{P_2} \bigcup \mathcal{C}_{(P_1+P_2)} \right) \backslash \Lambda_3, \tag{40}$$

It is possible to select $G_4 = \begin{bmatrix} 0\\4\\4\\0 \end{bmatrix}$. The new generated subgroup is

$$\Lambda_4 = \Lambda_3 \bigcup (\Lambda_3 + G_4) \tag{41}$$

$$= \left\{ \begin{array}{l} \begin{bmatrix} 0\\0\\0\\0 \end{bmatrix}, \begin{bmatrix} 4\\0\\4\\4 \end{bmatrix}, \begin{bmatrix} 2\\4\\2\\2 \end{bmatrix}, \begin{bmatrix} 6\\4\\6\\6 \end{bmatrix}, \begin{bmatrix} 7\\6\\3\\3 \end{bmatrix}, \begin{bmatrix} 3\\6\\7\\7 \end{bmatrix}, \begin{bmatrix} 1\\2\\5\\5 \end{bmatrix}, \begin{bmatrix} 5\\2\\1\\1 \end{bmatrix}, \\ \begin{bmatrix} 0\\4\\4\\0 \end{bmatrix}, \begin{bmatrix} 4\\4\\4\\4 \end{bmatrix}, \begin{bmatrix} 2\\0\\6\\2 \end{bmatrix}, \begin{bmatrix} 6\\0\\2\\6 \end{bmatrix}, \begin{bmatrix} 7\\2\\7\\3 \end{bmatrix}, \begin{bmatrix} 3\\2\\3\\7 \end{bmatrix}, \begin{bmatrix} 1\\6\\1\\5 \end{bmatrix}, \begin{bmatrix} 5\\6\\5\\1 \end{bmatrix} \end{array} \right\}. \tag{42}$$

The 5th column of $\mathbf{G}$ must be selected either in $\mathcal{C}_0^*$ or in a coset relative to an element of the set

$$\Lambda_4 \bigcap \left(\mathcal{E}_0^* \bigcup \mathcal{E}_1 \right) = \left\{ \begin{bmatrix} 4\\0\\4\\4 \end{bmatrix}, \begin{bmatrix} 2\\4\\2\\2 \end{bmatrix}, \begin{bmatrix} 6\\4\\6\\6 \end{bmatrix}, \begin{bmatrix} 0\\4\\4\\0 \end{bmatrix}, \begin{bmatrix} 4\\4\\4\\4 \end{bmatrix}, \begin{bmatrix} 2\\0\\6\\2 \end{bmatrix}, \begin{bmatrix} 6\\0\\2\\6 \end{bmatrix} \right\}. \tag{43}$$

Thus, it is possible to select $G_5 = \begin{bmatrix} 0\\6\\6\\4 \end{bmatrix} \in \mathcal{C}_{\begin{bmatrix} 0\\2\\2\\0 \end{bmatrix}}$. The new generated subgroup is

$$\Lambda_5 = \Lambda_4 \bigcup (\Lambda_4 + G_5) \tag{44}$$

$$= \left\{ \begin{array}{l} \begin{bmatrix} 0\\0\\0\\0 \end{bmatrix}, \begin{bmatrix} 4\\0\\4\\4 \end{bmatrix}, \begin{bmatrix} 2\\4\\2\\2 \end{bmatrix}, \begin{bmatrix} 6\\4\\6\\6 \end{bmatrix}, \begin{bmatrix} 7\\6\\3\\3 \end{bmatrix}, \begin{bmatrix} 3\\6\\7\\7 \end{bmatrix}, \begin{bmatrix} 1\\2\\5\\5 \end{bmatrix}, \begin{bmatrix} 5\\2\\1\\1 \end{bmatrix}, \\ \begin{bmatrix} 0\\4\\4\\0 \end{bmatrix}, \begin{bmatrix} 4\\4\\4\\4 \end{bmatrix}, \begin{bmatrix} 2\\0\\6\\2 \end{bmatrix}, \begin{bmatrix} 6\\0\\2\\6 \end{bmatrix}, \begin{bmatrix} 7\\2\\7\\3 \end{bmatrix}, \begin{bmatrix} 3\\2\\3\\7 \end{bmatrix}, \begin{bmatrix} 1\\6\\1\\5 \end{bmatrix}, \begin{bmatrix} 5\\6\\5\\1 \end{bmatrix}, \\ \begin{bmatrix} 0\\6\\6\\4 \end{bmatrix}, \begin{bmatrix} 4\\6\\2\\0 \end{bmatrix}, \begin{bmatrix} 2\\2\\0\\6 \end{bmatrix}, \begin{bmatrix} 6\\2\\4\\2 \end{bmatrix}, \begin{bmatrix} 7\\4\\1\\7 \end{bmatrix}, \begin{bmatrix} 3\\4\\5\\3 \end{bmatrix}, \begin{bmatrix} 1\\0\\3\\3 \end{bmatrix}, \begin{bmatrix} 5\\0\\7\\5 \end{bmatrix}, \\ \begin{bmatrix} 0\\2\\2\\4 \end{bmatrix}, \begin{bmatrix} 4\\2\\2\\0 \end{bmatrix}, \begin{bmatrix} 2\\6\\4\\6 \end{bmatrix}, \begin{bmatrix} 6\\6\\0\\2 \end{bmatrix}, \begin{bmatrix} 7\\0\\5\\7 \end{bmatrix}, \begin{bmatrix} 3\\0\\1\\3 \end{bmatrix}, \begin{bmatrix} 1\\4\\7\\1 \end{bmatrix}, \begin{bmatrix} 5\\4\\3\\5 \end{bmatrix} \end{array} \right\}. \tag{45}$$

To generate a balanced STTC, the last column of the generator matrix must be selected anywhere in $\mathbb{Z}_{2^n}^{n_T}$. Thus, G_6 can be $\begin{bmatrix} 4\\3\\0\\2 \end{bmatrix}$. The generator matrix is

$$\mathbf{G} = \left[\begin{array}{ccc|ccc} 4 & 2 & 7 & 0 & 0 & 4 \\ 0 & 4 & 6 & 4 & 6 & 3 \\ 4 & 2 & 3 & 4 & 6 & 0 \\ 4 & 2 & 3 & 0 & 4 & 2 \end{array} \right]. \tag{46}$$

After the selection of the last column G_6, the generated set Λ_6 is not a subgroup.
Table 10 shows the code 8 states 8-PSK STTC with 4 transmit antennas presented by Chen et al. (2002a) and the new generated code which are both balanced. The minimal rank and the minimal trace $d^2_{E_{\min}}$ of each code are also presented. The new code has a better trace than the corresponding Chen's code. The FER and BER of these two STTCs with 2 and 4 receive antennas are presented respectively in Figs. 4 and 5. We remark that the new code slightly outperforms the Chen's code.

Code	G	Rank	$d^2_{E_{\min}}$
Chen et al. (2002a)	$\begin{bmatrix} 2 & 4 & 0 & \vert & 3 & 2 & 4 \\ 1 & 6 & 4 & \vert & 4 & 0 & 0 \\ 3 & 2 & 4 & \vert & 0 & 4 & 2 \\ 7 & 2 & 4 & \vert & 5 & 4 & 0 \end{bmatrix}$	2	16.58
New	$\begin{bmatrix} 4 & 2 & 7 & \vert & 0 & 0 & 4 \\ 0 & 4 & 6 & \vert & 4 & 6 & 3 \\ 4 & 2 & 3 & \vert & 4 & 6 & 0 \\ 4 & 2 & 3 & \vert & 0 & 4 & 2 \end{bmatrix}$	2	17.17

Table 10. 8 states 8-PSK STTCs with 4 transmit antennas

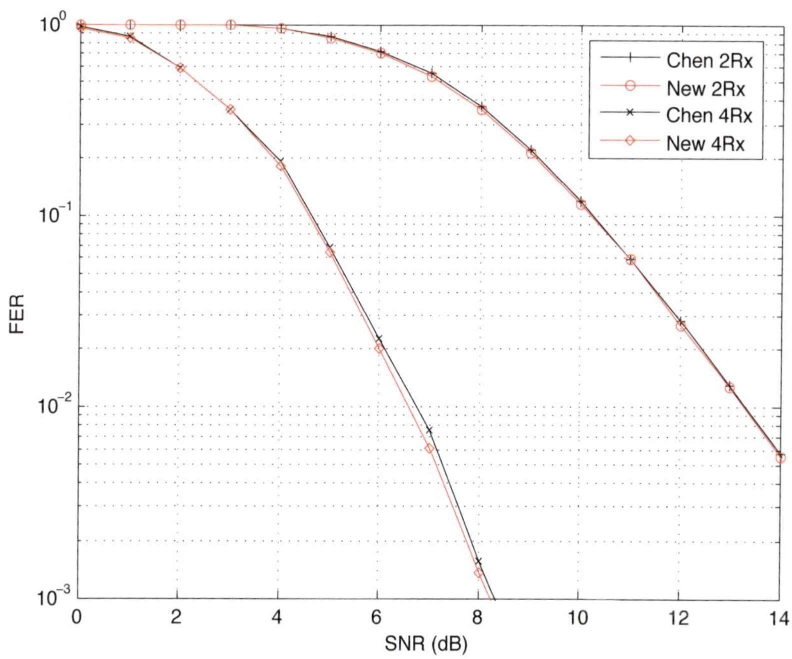

Fig. 4. FER of 8 states 8-PSK STTCs with 4 transmit antennas

6. Other new 4-PSK STTCs

Via this new method, other STTCs have been generated. Example of balanced 4-PSK STTCs are presented in Table 11. The performance of these STTCs is shown in Fig. 6. The codes noted by 'B' are balanced, those by 'FB' are fully balanced and those noted by 'NB' are not balanced.

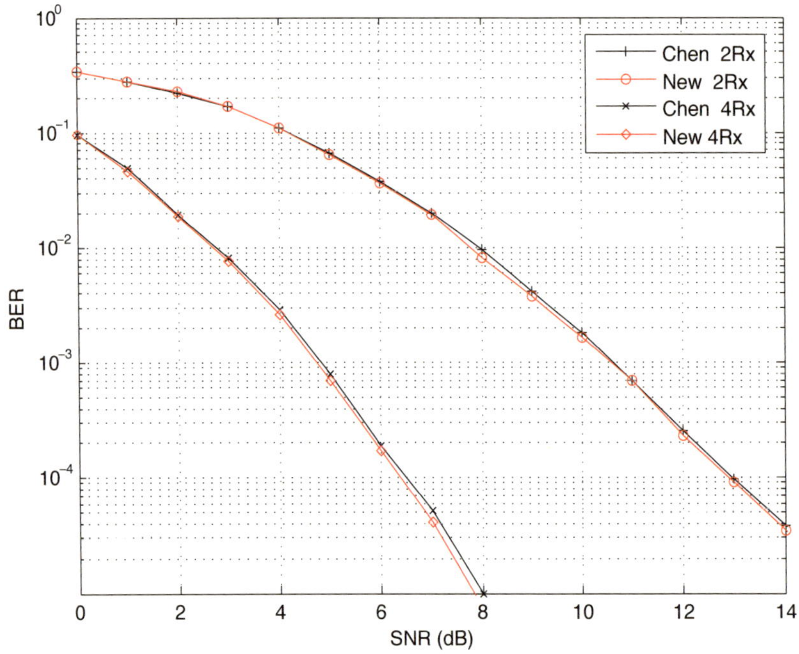

Fig. 5. BER of 8 states 8-PSK STTCs with 4 transmit antennas

n_T	States	Code	G		$d^2_{E\,min}$
3	32	Chen *et al.* Chen et al. (2002a)	$\begin{bmatrix} 0&2&2&1&1&2&0&2\\ 2&2&3&2&2&3&0&0\\ 2&0&2&2&2&1&0&0 \end{bmatrix}$	FB	24
		New	$\begin{bmatrix} 2&1&2&3&2&3&0&2\\ 2&3&0&2&2&1&0&0\\ 2&1&2&1&0&0&0&2 \end{bmatrix}$	FB	26
4	32	Chen *et al.* Chen et al. (2002a)	$\begin{bmatrix} 0&2&2&1&1&2&0&2\\ 2&2&3&2&2&3&0&0\\ 2&0&3&2&2&1&0&0\\ 2&1&2&0&1&0&0&2 \end{bmatrix}$	NB	36
		New	$\begin{bmatrix} 2&3&2&1&2&1&0&2\\ 0&2&2&1&2&3&0&3\\ 2&3&2&3&0&0&0&2\\ 2&1&0&2&2&1&0&0 \end{bmatrix}$	B	36
	64	Chen *et al.* Chen et al. (2002a)	$\begin{bmatrix} 0&2&3&2&''&0&3&2\\ 2&2&1&2&3&0&2&0\\ 2&0&0&2&2&3&1&1\\ 1&2&2&0&2&1&3&3 \end{bmatrix}$	NB	38
		New	$\begin{bmatrix} 1&2&2&0&3&2&1&2\\ 3&2&3&2&2&0&3&2\\ 2&0&1&2&3&2&3&2\\ 1&2&2&0&2&0&2&0 \end{bmatrix}$	B	40

Table 11. 4-PSK STTCs

7. Conclusion

The use of STTCs is an efficient solution to improve the performance of wireless MIMO systems. However, difficulties arise in terms of computational time to find the best codes

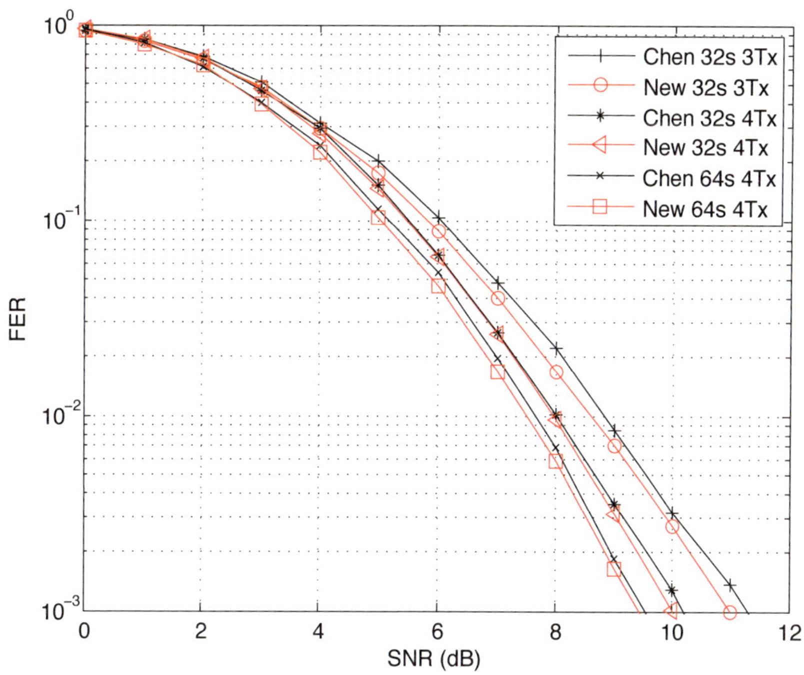

Fig. 6. FER of 4-PSK STTCs with 2 receive antennas

especially for a great number of transmit antennas. In order to reduce the search-time, this chapter presents a new and simple method to design balanced STTCs. A balanced STTC is a STTC which generates the MIMO symbols with the same probability when the binary input symbols are equiprobable. Each best STTC belongs to this class. Thereby, it is sufficient to generate only the balanced STTCs to find those with the best performance. Consequently, the search-time is considerably reduced.

The new method proposed in this chapter is simpler than the first method proposed by Ngo et al. (2008; 2007) and used only for 4-PSK modulation. Besides, the new method is given for 2^n-PSK STTCs and n_T transmit antennas. It is based on the generation of the subgroup of $\mathbb{Z}_{2^n}^{n_T}$ which determines a partition of $\mathbb{Z}_{2^n}^{n_T}$ in cosets.

Furthermore, several new 4-PSK and 8-PSK balanced STTCs have been proposed. These new STTCs outperform slightly the best corresponding published codes.

8. References

Abdool-Rassool, B., Nakhai, M., Heliot, F., Revelly, L. & Aghvami, H. (2004). Search for space - time trellis codes: novel codes for rayleigh fading channels, *Communications, IEE Proceedings-* 151(1): 25 – 31.

Chen, Z., Vucetic, B., Yuan, J. & Lo, K. L. (2002a). Space-time trellis codes for 4-PSK with three and four transmit antennas in quasi-static flat fading channels, *Communications Letters, IEEE* 6(2): 67 –69.

Chen, Z., Vucetic, B., Yuan, J. & Lo, K. L. (2002b). Space-time trellis codes with two, three and four transmit antennas in quasi-static flat fading channels, *Communications, 2002. ICC 2002. IEEE International Conference on*, Vol. 3, pp. 1589 – 1595 vol.3.

Chen, Z., Yuan, J. & Vucetic, B. (2001). Improved space-time trellis coded modulation scheme on slow rayleigh fading channels, *Electronics Letters* 37(7): 440 –441.

Coleman, J. (2002). Coset decomposition in lattices yields sample-block number systems, *Proc. IEEE ISCAS 2002*, Vol. 2, pp. 688–691 vol.2.

Forney, G.D., J. (1988). Coset codes. I. Introduction and geometrical classification, *IEEE Transactions on Information Theory* 34(5): 1123–1151.

Hong, Y. & Guillen i Fabregas, A. (2007). New space-time trellis codes for two-antenna quasi-static channels, *Vehicular Technology, IEEE Transactions on* 56(6): 3581 –3587.

Ionescu, D. (1999). New results on space-time code design criteria, *Wireless Communications and Networking Conference, 1999. WCNC. 1999 IEEE*, pp. 684 –687 vol.2.

Liao, C. & Prabhu, V. (2005). Improved code design criteria for space-time codes over quasi-static flat fading channels, *Signal Processing Advances in Wireless Communications, 2005 IEEE 6th Workshop on*, pp. 7 – 11.

Ngo, T. M. H., Viland, P., Zaharia, G. & Helard, J.-F. (2008). Balanced QPSK space-time trellis codes, *Electronics Letters* 44(16): 983 –985.

Ngo, T. M. H., Zaharia, G., Bougeard, S. & Helard, J. (2007). A new class of balanced 4-PSK STTC for two and three transmit antennas, pp. 1 –5.

Tarokh, V., Seshadri, N. & Calderbank, A. (1998). Space-time codes for high data rate wireless communication: performance criterion and code construction, *Information Theory, IEEE Transactions on* 44(2): 744 –765.

Ungerboeck, G. (1987a). Trellis-coded modulation with redundant signal sets part I: Introduction, *IEEE Communication Magazine* 25: 5–11.

Ungerboeck, G. (1987b). Trellis-coded modulation with redundant signal sets part II: State of the art, *IEEE Communication Magazine* 25: 12–21.

16

Correlation Coefficients of Received Signal I and Q Components in a Domain with Time and Frequency Axes under Multipath Mobile Channel with LOS and NLOS

Shigeru Kozono, Kenji Ookubo, Takeshi Kozima
and Tomohiro Hamashima
Department of Electrical, Electronics and Computer Engineering,
Chiba Institute of Technology
2-17-1 Tsudanuma, Narashino-shi, 275-0016,
Japan

1. Introduction

To support realtime multimedia communication, future mobile communications will require a high-bit-rate transmission system with high utilization of the frequency spectrum in very rich multipath channels [1]. Systems capable of fulfilling this requirement, with features such as orthogonal frequency division multiplexing (OFDM) [2], [3] and multiple-input multiple-output (MIMO) [4]–[7], have been studied extensively. Furthermore, the use of M-ary quadrature amplitude modulation (M-ary QAM) with coherent detection helps to meet the transmission requirements. Such systems overcome frequency selective fading by using a narrowband multiple-carrier signal called a sub-channel. Those systems use sophisticated and refined techniques and several of those techniques are usually combined in one system. To make full use of the techniques in the system design and achieve high-quality transmission, we must design a system taking into consideration wide and detailed multipath properties in a domain with time and frequency axes. OFDM works best with a lot of narrow bands for the sub-channel. MIMO also requires multipath properties between the antennas composing the MIMO antenna, and it requires a low correlation coefficient. M-ary QAM detection also requires the device to compensate accurately for both the in-phase and quadrature (I and Q) components of the sub-channel for the ever-changing multipath channel associated with terminal movement, and this is achieved by putting a pilot signal in the information data [8].

Figure 1 shows an example of the sub-channel state of the I and Q components and phase ψ ($\tan^{-1}(Q/I)$) as movement in a multipath channel, and Fig. 2 shows the signal state diagram at the moment. With movement, the I, Q, and the phase ψ change irregularly as shown by solid lines in Fig.1, so we cannot use the sub-channel as a transmission line. Therefore, the sub-channel needs to be a stable state with compensation by a pilot signal. The states of the sub-channel compensated every $10T_{sy}$ (T_{sy}: symbol length) in information data, which corresponds to a 0.1λ [m] running distance, are shown by broken lines in Fig. 1, the I and Q components become stable, and the sub-channel can then be used as a transmission line. This phenomenon in Fig. 2 appears as concentrated points in a small broken circle, so it is possible that a received signal with M-ary QAM is detected. The pilot signal is inserted at a fixed interval in information data: in particular the pilot signal in a combined system with OFDM and M-ary QAM is allocated in a domain with time and frequency axes and there must be a high correlation coefficient between pilot signals. This signal allocation is very important because it influences transmission efficiency and bit error rate (BER).

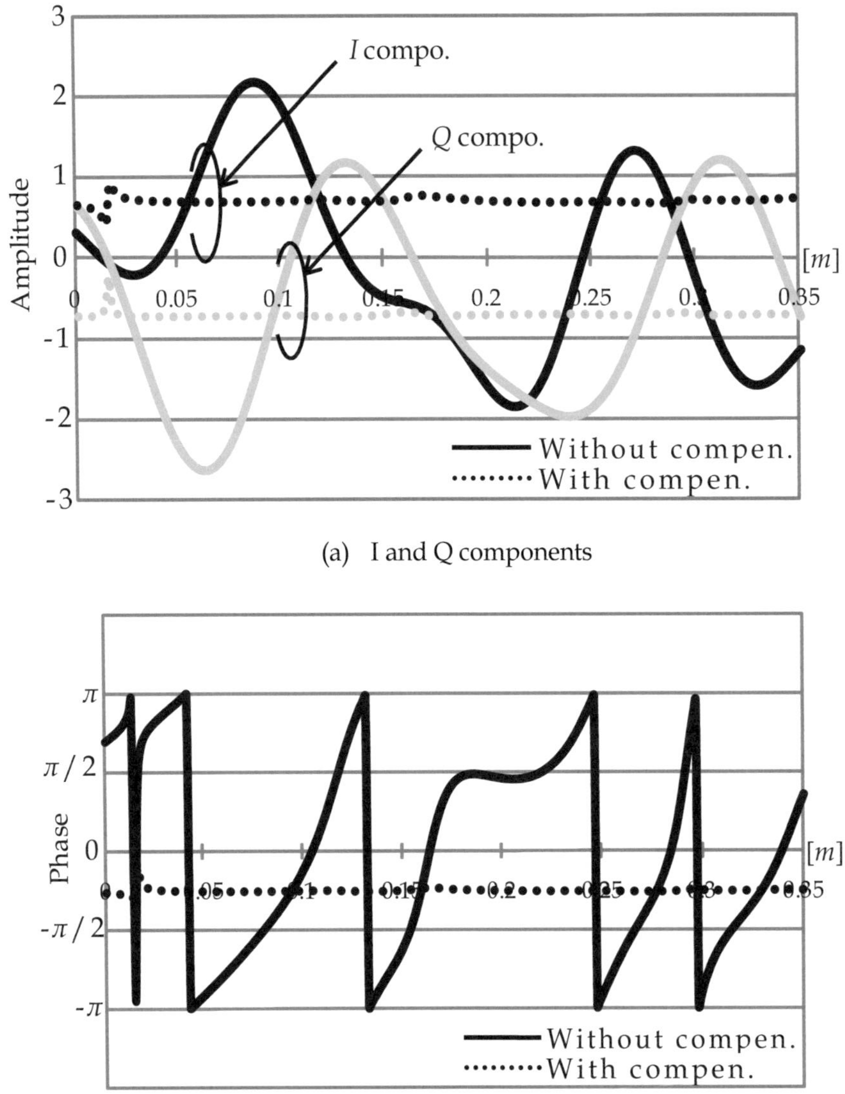

(a) I and Q components

(b) Phase ψ

Fig. 1. Example of sub-channel variations with movement

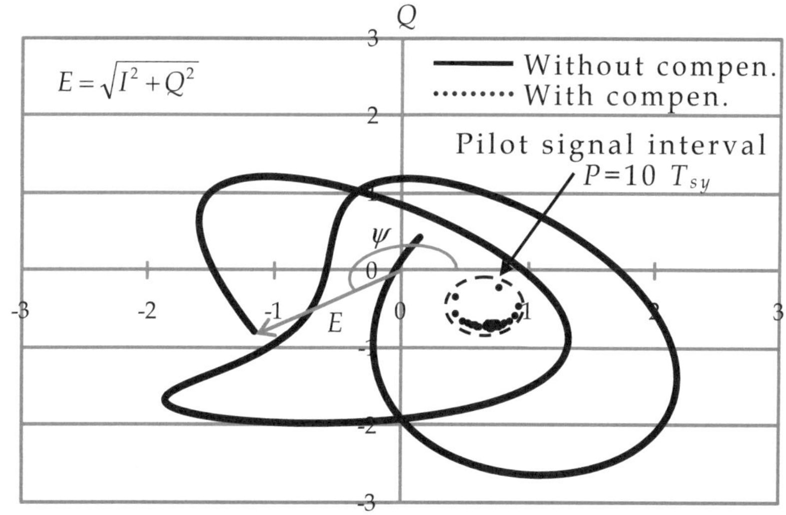

Fig. 2. Signal state diagram with I and Q components in Fig. 1

What the wireless system design requires in terms of basic multipath properties seems to be the correlation of the sub-channel's I and Q components in a domain with time and frequency axes. Many theoretical and experimental studies have been performed on the correlation of received signal amplitudes [9]–[12], but the correlation of the I and Q components has hardly been studied at all [9]. Furthermore, the ordinary formula for amplitude correlation given by Jakes [10] is derived on the basis of the I and Q components with a random Gaussian variable in the non-line-of-sight (NLOS) condition. With this in mind, we paid attention to the delay profile with actual multipath data in a mobile channel and developed an analysis model with a domain having time, frequency, and space (delay time) axes to investigate the correlation using the profile [13]. Furthermore, we derived a correlation formula for various profile types for the line-of-sight (LOS) and NLOS conditions and also verified the theory by computer simulation.

This chapter is organized as follows. Section 2 describes a propagation model and analysis model. Section 3 covers the theoretical study. First, we derive the general correlation formula for the I and Q components using a delay profile for LOS or NLOS. Next as examples, we derived formulas for delay profiles when the arriving wave has an exponentially decreasing amplitude and a random amplitude. Section 4 covers simulation. The simulation method and parameters are described and the simulation results are presented. The simulation was done for autocorrelation and for frequency correlation on time and frequency axes for LOS and NLOS and also done in the domain for those profiles. Then, the correlation formulas were evaluated. Finally, section 5 summarizes the results.

2. Propagation model and analysis model

A. Propagation model

In a mobile radio channel, numerous waves arrive at a receiving point via multiple paths, with each wave having a certain amplitude, a certain delay time, and a random arrival angle. Consequently, the arriving waves interfere with each other and standing waves are produced. Thus, the received signal amplitude varies with terminal movement and has a Rayleigh or Nakagami-Rice distribution. An example of the delay profile is shown in Fig. 3 for the following conditions.

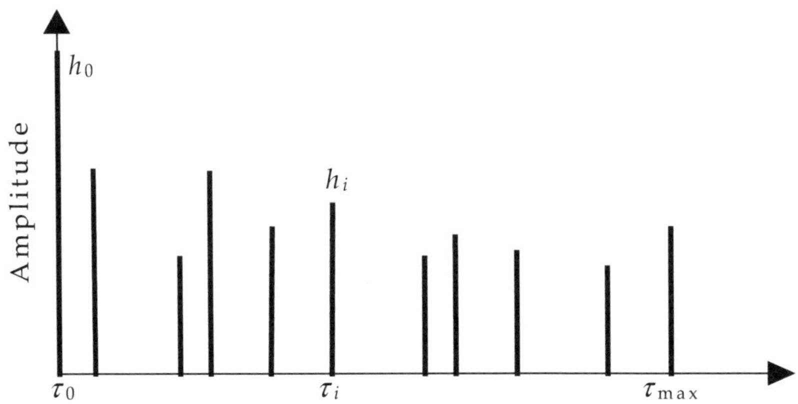

Fig. 3. Example of delay profile

i. The number of arriving waves is N+1, the waves are independent of each other, and the ith arriving wave is denoted by subscript i, where i=0 means a directive wave and i≥1 means no directive waves.
ii. The waves have an excess delay time τ_i relative to the directive wave and a maximum excess delay time τ_{max}. The τ_i values are random from τ_0 to τ_{max}.
iii. The amplitude is h_i, the power ratio of the directive and nondirective waves is denoted by k (K: in dB), and k=0 (K=-∞ dB) means NLOS. Furthermore, the nondirective wave's power is normalized to 1.
iv. The arrival angle is ξ_i, the initial phase is ϕ_i, and the values are random from 0 to 2π rad.

B. Multipath channel

Under the above conditions, the envelope $E(t, f_c)$ of a received signal is given by (1), where $I(t, f_c)$ and $Q(t, f_c)$ express the in-phase (real) and quadrature (imaginary) components of $E(t, f_c)$, respectively.

$$E(t, f_c) = \sum_{i=0}^{N} h_i e^{j\theta_i}$$

$$= I(t, f_c) + jQ(t, f_c) \quad (1)$$

$$I(t, f_c) = \sum_{i=0}^{N} h_i \cos\theta_i \quad (2\text{-a})$$

$$Q(t, f_c) = \sum_{i=0}^{N} h_i \sin\theta_i \quad (2\text{-b})$$

$$\theta_i = 2\pi(f_c\tau_i - f_m t\cos\xi_i) + \phi_i \tag{3}$$

Here, θ_i is the path phase of the ith arriving wave, f_c is the radio frequency, and f_m is the maximum fading frequency calculated as v/λ, where v is the moving speed and λ is the wavelength. As shown by (2) and (3), the I and Q components are functions of variable θ_i, which is a function of f_c, t, and v, for a given delay profile. When the sub-channel bandwidth of the system, such as an OFDM system, is f_0, we have $f_c >> f_0$ since we treat a narrowband channel in a multipath environment. Therefore, we treat the frequency separately from f_c as $\Delta f = nf_0$ (n: integer). Figure 4 shows an example of an OFDM channel with a lot of sub-channels in a domain with time (vt) and frequency (nf_0): each sub-channel amplitude $|E(t, f_c)|$ changes severely and deeply in the domain.

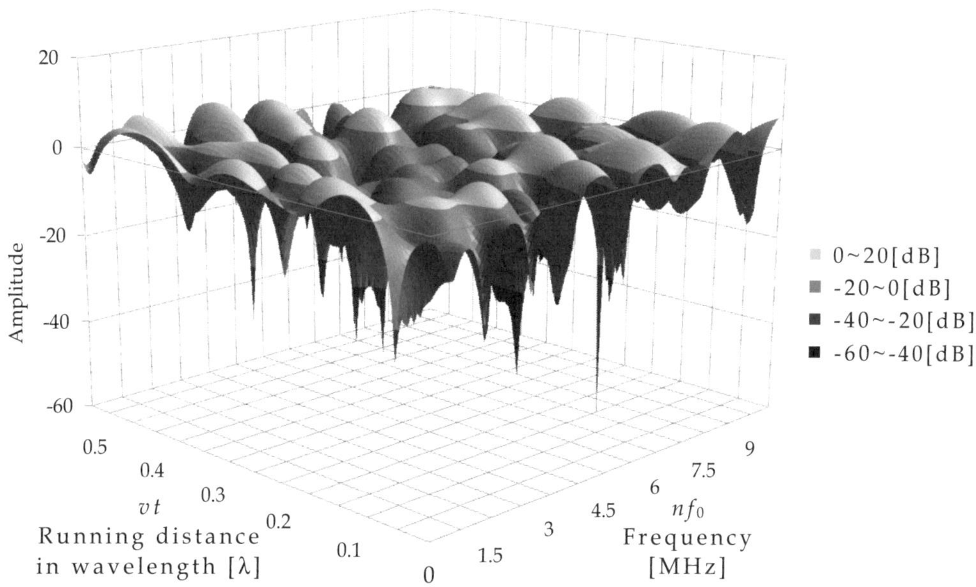

Fig. 4. Example of OFDM channel in multipath condition

C. Analysis model

As shown (2), the $I(t, f_c)$ and $Q(t, f_c)$ depend on θ_i by (3) for a given delay profile and the θ_i is a function of $f_c\tau_i$ and $f_m t$. Therefore, we study the correlation of $I(t, f_c)$ and $Q(t, f_c)$ by using the variables of $f_m T_s$ (normalized maximum Doppler frequency) on the time axis and $\Delta f\sigma$ (σ: delay spread) on the frequency axis. For example, T_s is absolute time t minus relative time T_0 for a directive wave arriving at the receiving point, i.e., $T_s = t - T_0$; Δf is the frequency separation from the radio frequency f_c, i.e., $\Delta f = f - f_c$. Furthermore, to simplify the expression of those variables, we introduce $x = f_m T_s + \Delta f\sigma$ as a variable of time and frequency. Moreover, we choose the origin $P_0(0,0)$ and the point under consideration for the correlation $P_1(f_m T_s, \Delta f\sigma)$ to correspond to $x=0$ and $x=\Delta x$, respectively. Consequently, those points are expressed by $P_0(0)$ and $P_1(\Delta x)$, and we can calculate the correlation coefficient $\rho(\Delta x)$ as the correlation between $P_0(0)$ and $P_1(\Delta x)$. We denote the correlation coefficients for the I and Q components using the symbols $\rho_I(\Delta x)$ and $\rho_Q(\Delta x)$, respectively. The path phases of the ith arriving wave θ_i given by (3) at $P_0(0)$ and $P_1(\Delta x)$ are also represented by (4-a) and (4-b), respectively.

$$\theta_i(0) = 2\pi f_c \tau_i + \phi_i, \tag{4-a}$$

$$\theta_i(\Delta x) = 2\pi((f_c + \Delta f)\tau_i - f_m T_s \cos\xi_i) + \phi_i \tag{4-b}$$

3. Derivation of correlation coefficients

A. General formula derivation

We start by studying the correlation coefficient of the I component. With the variable x, the correlation coefficient of the I component is generally expressed by

$$\rho_I(\Delta x) = \frac{\langle (I(0)I(\Delta x)\rangle - \langle I(0)\rangle\langle I(\Delta x)\rangle}{[\{<(I(0))^2> - <I(0)>^2\}\{<(I(\Delta x))^2> - <I(\Delta x)>^2\}]^{1/2}}. \tag{5}$$

Here, the symbol < > means an ensemble average. Under the conditions of the propagation model in 2-*A* and assuming that N is a large of number, we can start to calculate each term in (5).

First, we calculate the terms $<I(0)>$ and $<I(\Delta x)>$. Because τ_i, ξ_i, and ϕ_i have random values, the ensemble averages at x=0 and Δx become zero when (4-a) and (4-b) are used to substitute for θ_i in (2-a), so these terms become

$$\langle I(0)\rangle = \langle I(\Delta x)\rangle = 0. \tag{6}$$

Second, we calculate $<I(0)^2>$ and $<I(\Delta x)^2>$. Though the ensemble averages for the product of different ith arriving waves vanish since τ_i, ξ_i, and ϕ_i have random values, those of the same ith arriving waves remain as the power, and we get

$$\langle I(0)^2\rangle = \langle I(\Delta x)^2\rangle = \frac{1}{2}\langle\sum_{i=0}^{N} h_i^2\rangle. \tag{7}$$

Next, we calculate the term $<I(0)I(\Delta x)>$. When we consider τ_i, ξ_i, and ϕ_i to be random values and take an odd function of sine, this term becomes as follows (see appendix A).

$$\langle I(0)I(\Delta x)\rangle = \frac{1}{2}\langle\sum_{i=0}^{N} h_i^2 \cos(2\pi\Delta f\tau_i)\cos(-2\pi f_m' T_s \cos\xi_i)\rangle \tag{8}$$

Through (6) to (8), we can calculate the denominator and numerator in (5). Both $\langle I(0)^2\rangle - \langle I(0)\rangle^2$ and $\langle I(\Delta x)^2\rangle - \langle I(\Delta x)\rangle^2$ in the denominator are $\frac{1}{2}\langle\sum_{i=0}^{N} h_i^2\rangle$ by (7) minus (6), so the denominator is $\frac{1}{2}\langle\sum_{i=0}^{N} h_i^2\rangle$. On the other hand, the numerator becomes $\frac{1}{2}\langle\sum_{i=0}^{N} h_i^2 \cos(2\pi\Delta f\tau_i)\cos(-2\pi f_m' T_s \cos\xi_i)\rangle$ by (6) and (8). Finally, we can rewrite correlation coefficient $\rho_I(\Delta x)$ in (5) as (9).

$$\rho_I(\Delta x) = \frac{\frac{1}{2}\langle \sum_{i=0}^{N} h_i^2 \cos(2\pi\Delta f \tau_i)\cos(-2\pi f_m^{'} T_s \cos\xi_i) \rangle}{\frac{1}{2}\langle \sum_{i=0}^{N} h_i^2 \rangle} \tag{9}$$

Equation (9) represents a general correlation coefficient formula for the I component, and it is applicable to a delay profile for both LOS and NLOS conditions.
Next, we study the correlation coefficient of the Q component. It is easy to understand that the correlation coefficient of the Q component is given by the same formula (9) as that for the I component. The reason is as follows. Since (9) does not contain ϕ_i, we can replace ϕ_i in (3) by $\phi_i + \pi/2$ and substitute $\phi_i + \pi/2$ for ϕ_i in θ_i for $\sin\theta_i$ in (2-b). Consequently, the term $\sin\theta_i$ becomes $\cos\theta_i$ analytically, and the properties of $Q(t, f_c)$ then become statistically similar to those of $I(t, f_c)$. Thus, since we can calculate the correlation coefficients of both the I and Q components by using the I component, from here on we treat only the I component, which we denote simply as $\rho(\Delta x)$.

B. Correlation coefficient in given delay profile

i) Type 1: Delay profile with exponential distribution

We first calculate correlation coefficient $\rho(\Delta x)$ in (9) for a delay profile with an exponential distribution, as shown in Fig. 5(a). It has a directive wave with amplitude h_0 and excess delay time τ_0=0 and also has a lot of nondirective waves for which h_i decreases exponentially according to (10) as excess delay time τ_i increases.

$$h_i = h\exp(-\tau_i/\tau_{av}) \;, \tag{10}$$

where h is the amplitude at τ_i=0 and τ_{av} is the average excess delay time. Under this condition and for the propagation model in 2-*A*, we analytically derive the correlation coefficient formula on the basis of $\rho(\Delta x)$ in (9). The denominator in (9) is k+1 because the power of the nondirective wave was normalized to 1. Furthermore, since τ_i and ξ_i are independent, the numerator in (9) can be separated into two parts, A_1 and B_1, as in (11).

$$A_1 B_1 = \langle \frac{1}{2}\sum_{i=0}^{N} h_i^2 \cos(2\pi\Delta f \tau_i)\rangle \langle \sum_{i=0}^{N} \cos(-2\pi f_m^{'} T_s \cos\xi_i)\rangle \tag{11}$$

When N is a large number, $\tau_{\max}$ is large, and h_i decreases according to (10), the first term A_1 can be calculated by integrating with respect to τ. As shown in appendix *B*, it becomes

$$\begin{aligned} A_1 &= \langle \frac{1}{2} h_0^2 \rangle + \frac{1}{2}\int_{\tau_{\min}}^{\tau_{\max}} h^2 \exp(-2\tau/\tau_{av})\cos(2\pi\Delta f \tau) d\tau \\ &= k + \frac{1}{1 + (2\pi\Delta f)^2 (\tau_{av}/2)^2}. \end{aligned} \tag{12}$$

Concerning the second part, B_1, it is well known that $B_1 = J_0(2\pi f'_m T_s)$, where $J_0(\bullet)$ is a zeroth-order Bessel function of the first kind and f_m' is the maximum fading frequency at $f_c + \Delta f$. Moreover, the value of $\tau_{av}/2$ in (12) is equal to delay spread σ when $\tau_{\max}$ is very large and h_i becomes very small at $\tau_{\max}$. Replacing $2\pi\Delta f$ by s, we finally get the correlation coefficient $\rho(\Delta x)$.

$$\rho(\Delta x) = \frac{k + \frac{1}{(1 + s^2\sigma^2)}}{k+1} J_0(2\pi f_m^{'} T_s) \tag{13}$$

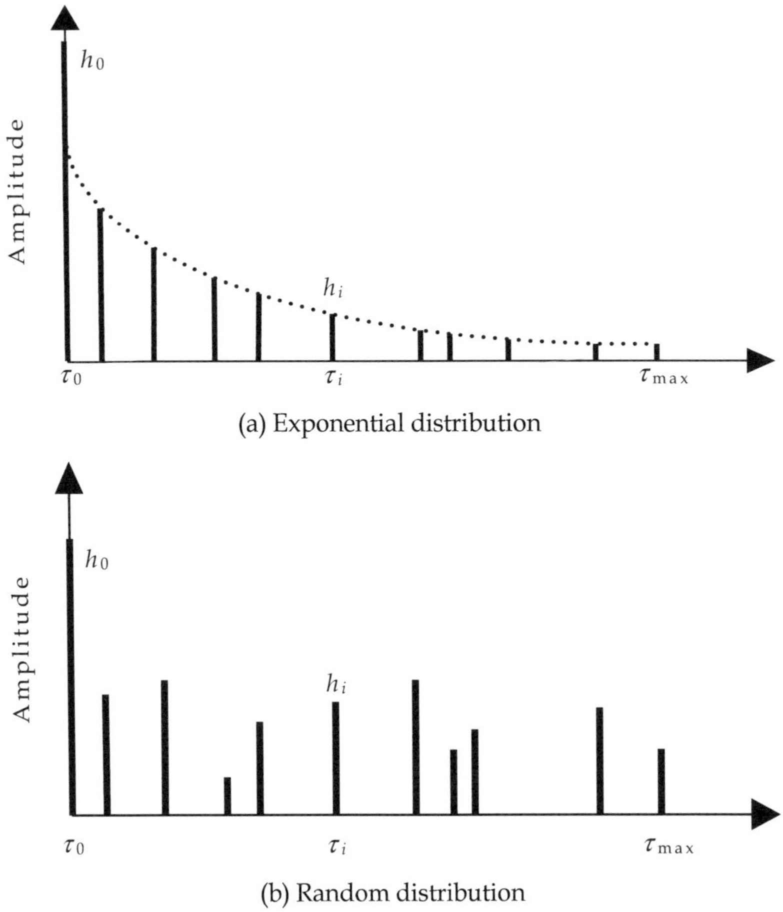

Fig. 5. Delay profile used formula derivation

ii) Type 2: *Delay profile with random distribution*

Next, we calculate the correlation $\rho(\Delta x)$ for a delay profile with a random distribution, as shown in Fig. 5(b). It has a directive wave and a lot of nondirective waves like Fig. 5(a), but h_i and τ_i of the nondirective waves are independent. Excess delay time τ_i has maximum value τ_{max} and is random between τ_{min} (close to 0) and τ_{max}. In a similar way to that for type 1, the denominator in (9) is also k+1, and the numerator in (9) can also be separated into A_2 and B_2; moreover, A_2 separates into directive and nondirective waves, as shown in (14).

$$A_2B_2 = \langle \frac{1}{2}h_0^2 + \frac{1}{2}\sum_{i=1}^{N} h_i^2 \cos(2\pi\Delta f \tau_i)\rangle \langle \sum_{i=0}^{N} \cos(-2\pi f_m^{'} T_s \cos\xi_i)\rangle \tag{14}$$

In the first ensemble average in (14), the power of a directive wave is k and that of a nondirective one is 1; moreover, h_i and τ_i are independent. Therefore, A_2 becomes (15) by a similar integration to that for type 1 (see appendix C).

$$A_2 = \langle k + \sum_{i=1}^{N} \cos(2\pi\Delta f \tau_i)\rangle \quad = k + \frac{\sin(2\pi\Delta f \tau_{max})}{2\pi\Delta f \tau_{max}} \tag{15}$$

B_2 in (14) also becomes $J_0(2\pi f_m'T_s)$. From the above description, the correlation coefficient $\rho(\Delta x)$ becomes

$$\rho(\Delta x)=\frac{k+\sin(2\pi\Delta f\tau_{\max})\big/2\pi\Delta f\tau_{\max}}{k+1}J_0(2\pi f_m^{'}T_s). \tag{16}$$

From (13) and (16), the autocorrelation on the time axis is independent of the delay profile and always seems to be a zeroth-order Bessel function of the first kind, $J_0(\bullet)$, whenever arriving angle ξ_i has a uniform distribution over 0 to 2π. However, the frequency correlation on the frequency axis depends on whether the delay profile is for LOS or NLOS and whether it is dependent on or independent of h_i and τ_i.

4. Simulation

A. Simulation method

A computer simulation was performed to verify the correlation coefficient formula derived in 3-*B*. The simulation parameters are listed in Table 1.

Frequency f_c	2 GHz
Number of arriving waves N +1	10+1
Nakagami-Rice factor K	$-\infty$, 5 dB
Average excess delay time τ_{av} Maximum excess delay time $\tau_{\max}$	1 μs (type 1) 2 μs (type 2)
Effective amplitude h_i	-20 dB

Table 1. Simulation parameters

Simulation was done to confirm (13) for type 1 with an exponential distribution and (16) for type 2 with a random distribution, and it mainly simulated correlations such as autocorrelation on the time axis, frequency correlation on the frequency axis, and correlation in a domain with time and frequency axes. The correlation was simulated by using the delay profile in actually and frequently encountered environments within the propagation model in 2-*A*, such that N was 10, the effective amplitude of arriving wave was -20 dB relative to the maximum for a nondirective wave, and the K factor took values of $-\infty$ and 5 dB. For type 1, the average delay time τ_{av} was 1 μs when the statistical delay spread σ was about 0.42 μs at K= $-\infty$ dB. For type 2, the maximum delay time $\tau_{\max}$ was 2 μs and σ was 0.50 μs. The simulation was performed on the basis of (2) and (5). Each simulated value of $\rho(\Delta x)$ was calculated from an ensemble average for more than 10^6 delay profiles. Thus, we performed a statistical analysis in the simulation.

B. Simulation results and considerations

i) Type 1: Delay profile with exponential distribution

Figure 6 shows the correlation coefficient simulated using a delay profile with exponential distributions of K=$-\infty$ and 5 dB denoted by symbols × and ▲, respectively. Figures 6(a) and

(b) are the correlations on the time and frequency axes, i.e., the well known autocorrelation and frequency correlation, respectively. In Fig. 6(a), the fine line is the theoretical value of $\rho(\Delta x)$ obtained from (13) by putting Δf =0 or $\rho(\Delta x)=J_0(2\pi f'_m T_s)$. The simulated autocorrelations for both K=-∞ and 5 dB have almost the same features and agree with the theoretical value. This shows that the autocorrelation is independent of the K factor and is expressed by $J_0(2\pi f'_m T_s)$. In Fig. 6(b), the fine and broken lines are theoretical values of $\rho(\Delta x)$ obtained from (13) by putting $f'_m T_s$=0 or $\frac{k + 1/(1+s^2\sigma^2)}{k+1}$. Here, we note that delay spread σ was calculated for NLOS without any directive waves. The frequency correlations for K=-∞ and 5 dB have different features, so the correlation depends on the K factor. The simulated

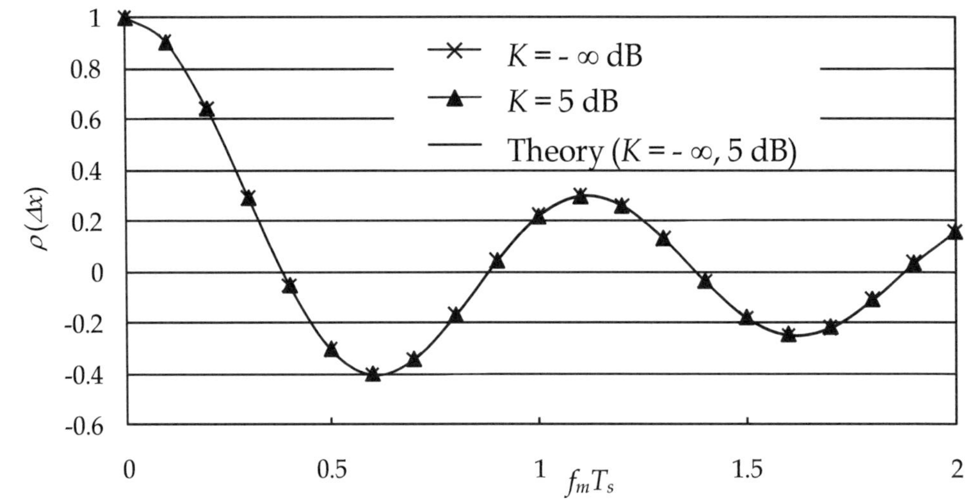

(a) Autocorrelation

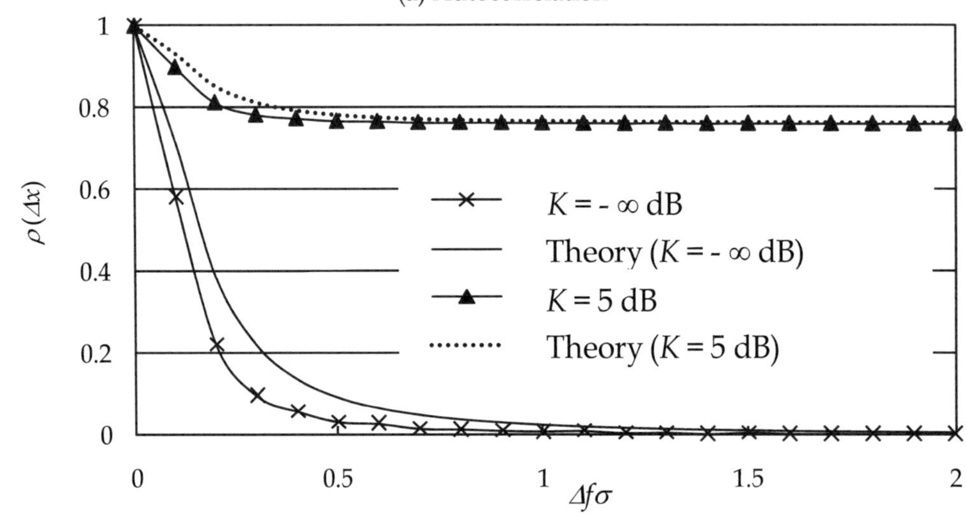

(b) Frequency correlation

Fig. 6. Correlation coefficient (Type 1: exponential distribution)

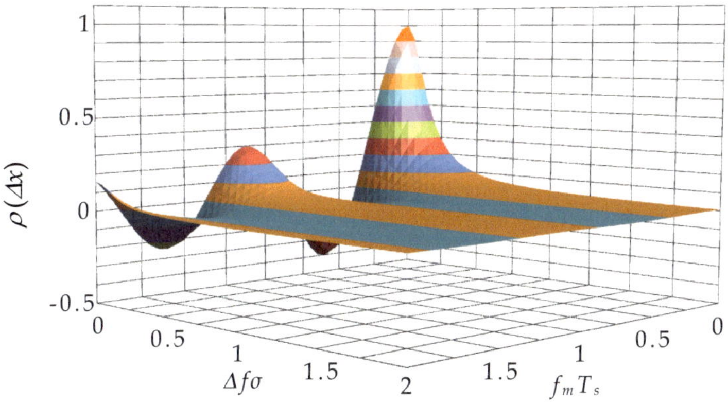

(a) Bird's-eye view (theory)

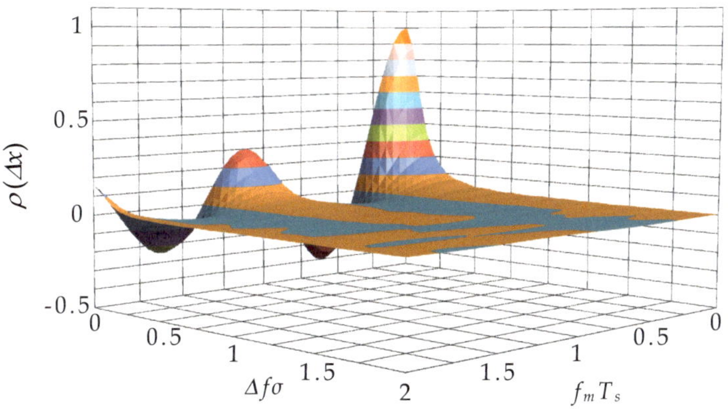

(b) Bird's-eye view (simulation)

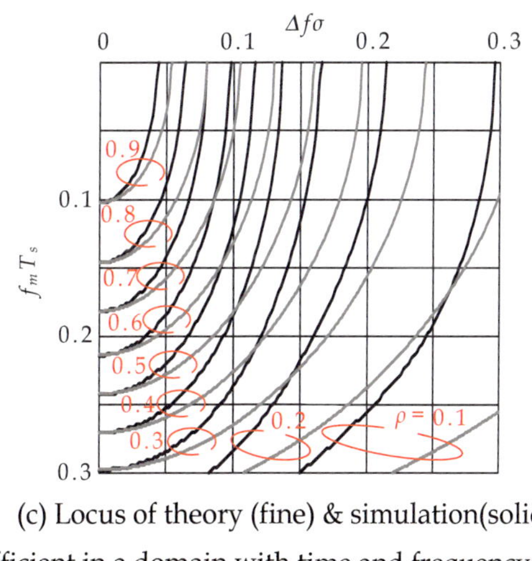

(c) Locus of theory (fine) & simulation(solid)

Fig. 7. Correlation coefficient in a domain with time and frequency axes (Type 1: exponential distribution , $K = -\infty$ dB) correlation for K= -∞ dB is slightly lower than the theoretical one. We suppose that the difference is due to different conditions between the theoretical and simulation models for the delay profile, such as the number of waves N and the counted effective amplitude in h_i. Therefore, we repeated the simulation, changing from N=10 and effective amplitude of -20 dB to N>100 and effective amplitude of less than -50 dB. As a result, the correlation became close to the theoretical one. The above results show that the autocorrelation is independent of the K factor, while the frequency correlation depends on the K factor. The correlation coefficient in a domain with time and frequency axes for NLOS is shown in Fig. 7. Figure 7(a) is a bird's-eye view of $\rho(\Delta x)$ obtained from (13), which makes it easy to comprehend the overall $\rho(\Delta x)$. We can see that $\rho(\Delta x)$ has a peak $\rho(0)=1$ at the origin $\Delta x = 0$ and that $\rho(\Delta x)$ becomes smaller with increasing Δx or as $f'_m T_s$ and $\Delta f\sigma$ become larger. Furthermore, $\rho(\Delta x)$ decreases in a fluctuating manner on the time axis, but decreases monotonically on the frequency axis. The high area of $\rho(\Delta x)$ needed to allocate the pilot signal in coherent detection, such as $\rho(\Delta x)$ >0.8, is very small in the domain, whereas the low area needed to design the diversity antenna, such as $\rho(\Delta x)$ <0.5, is spread out widely. Figure 7(b) also shows a bird's-eye view of the simulated correlation; Figs. 7(a) and (b) both exhibit almost the same features. Figure 7(c) shows the loci of the theoretical and simulated correlations in a small area up to 0.3 on both axes, with contour lines of $\rho(\Delta x)$. It is easy to compare them. The theoretical value on the time axis agrees well with the simulated one, but the theoretical value on the frequency axis is slightly higher than the simulated one. The reason for this is a different model for the delay profile, as described earlier. We can see from Fig. 7(c) that the theoretical value agrees roughly with the simulated one. Furthermore, we note that in Fig. 7(c), the locus of the correlation coefficient $\rho(\Delta x)$ seems to be an ellipse with its major axis on the time axis and its origin Δx=0 in the domain.

ii) Type 2: Delay profile with random distribution

Figure 8 shows the correlation coefficient for a delay profile with random distributions of K=-∞ and 5 dB. The values plotted by symbols of × and ▲ in Figs. 8(a) and (b) were simulated in a similar way to Fig. 6, and the fine and broken lines show the theoretical value of $\rho(\Delta x)$ obtained from (16) by putting Δf =0 for autocorrelation or $\rho(\Delta x)=J_0(2\pi f'_m T_s)$ and by putting $f'_m T_s$=0 for frequency correlation or $\dfrac{k+\sin(2\pi\Delta f\tau_{\max})/2\pi\Delta f\tau_{\max}}{k+1}$. As shown in Fig. 8(a), the simulated autocorrelations for both K=-∞ and 5 [dB] have almost the same features and agree well with the theoretical values, so the correlation seems to be independent of the K factor. On the other hand, the simulated frequency correlations for K=-∞ and 5 dB in Fig. 8(b) have different features that also depend on the K factor. The simulated and theoretical values agree well.

The correlation coefficient in a domain with time and frequency axes for NLOS is shown in Fig. 9 by a similar method to that for Fig. 7. Figure 9(a) is a bird's-eye view of the theoretical value of $\rho(\Delta x)$ obtained from (16). We can also see that $\rho(\Delta x)$ has a peak $\rho(0)$=1 at the origin Δx = 0, and $\rho(\Delta x)$ becomes smaller with increasing Δx. However, in this case, $\rho(\Delta x)$ decreases in a fluctuating manner not only on the time axis, but also on the frequency axis. The high area of $\rho(\Delta x)$ in the domain, such as $\rho(\Delta x)$ >0.8, is larger than that for an exponential distribution; the low area, such as $\rho(\Delta x)$ <0.5, is spread out widely.

Figure 9(b) is the simulated correlation, which exhibits similar features to the theoretical one in Fig. 9(a). Figure 9(c) shows the loci of the theoretical and simulated correlations. The theoretical value in the domain agrees well with the simulated ones. For a delay profile with random distribution, the locus of correlation coefficient $\rho(\Delta x)$ is also an ellipse in the domain.

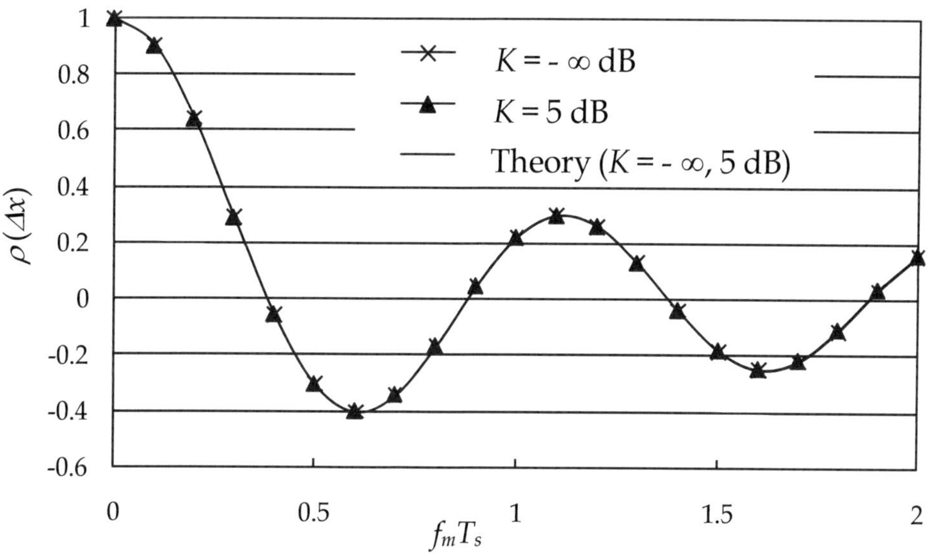

(a) Autocorrelation

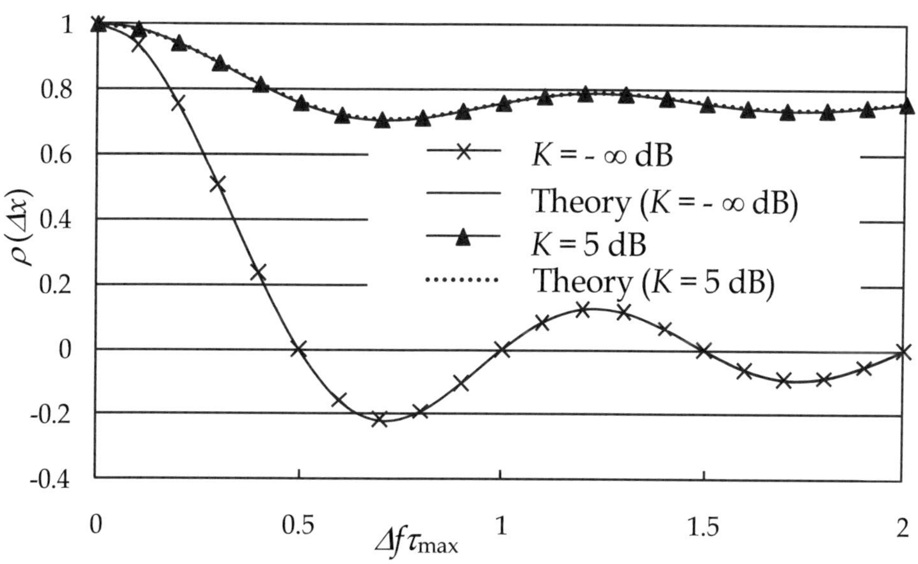

(b) Frequency correlation

Fig. 8. Correlation coefficient (Type 2 : random distribution)

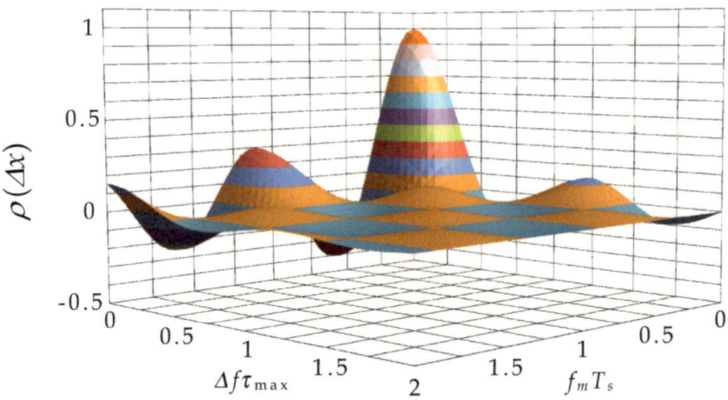

(a) Bird's-eye view (theory)

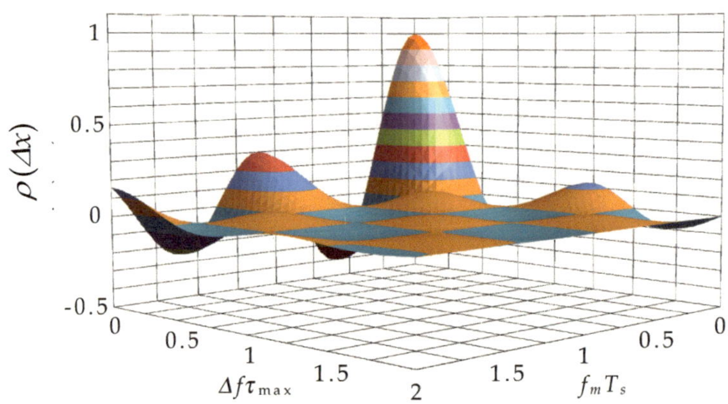

(b) Bird's-eye view (simulation)

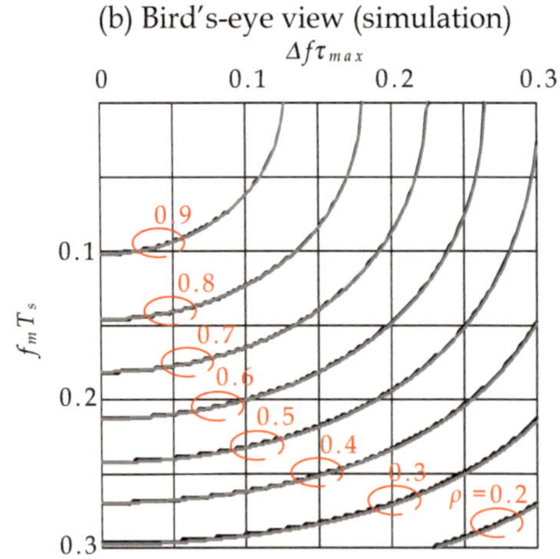

(c) Locus of theory (fine) & simulation (solid)

Fig. 9. Correlation coefficient in a domain with time and frequency axes (Type 2: random distribution , K = -∞ dB)

5. Conclusion

An analysis model having a domain with time, frequency, and space axes was prepared to study the correlation coefficients of the I and Q components in a mobile channel, which are needed in order to allocate a pilot signal with M-ary QAM detection, such as in an OFDM system, and to compose antennas in the MIMO technique. For a multipath environment, the general correlation coefficient formula was derived on the basis of a delay profile with and without a directive wave, and as examples, the formulas for delay profiles with exponential and random distributions were then derived. The formulas exhibit some interesting features: the autocorrelation on the time axis is independent of the K factor and is expressed by $J_0(2\pi f'_m T_s)$, but the frequency correlation depends on the K factor and delay profile type. The locus of a fixed value correlation is an ellipse in the domain with time and frequency axes. The correlations were also shown in the domain using bird's-eye views for easy comprehension. Furthermore, computer simulation was performed to verify the derived formulas and the theoretical and simulated values agree well. Therefore, it is possible to estimate logically the pilot signal allocation of M-ary QAM in OFDM and antenna construction for MIMO in the domain.

6. Appendix

6.1 Appendix A derivation of (8)

Under the conditions of the propagation model in 2-A and assuming that N is a large number, $\langle I(0)I(\Delta x)\rangle$ is analyzed as follows. It is separated into terms of the same ith and other ith arriving waves, as shown in (17).

$$\begin{aligned}\langle I(0)I(\Delta x)\rangle &= \langle[\sum_{i=0}^{N} h_i\cos(2\pi f_c\tau_i+\phi_i)][\sum_{i=0}^{N} h_i\cos\{2\pi((f_c+\Delta f)\tau_i - f'_m T_s\cos\xi_i)+\phi_i\}]\rangle \\ &= \langle\sum_{i=0}^{N} h_i^2\cos(2\pi f_c\tau_i+\phi_i)\cos\{2\pi((f_c+\Delta f)\tau_i - f'_m T_s\cos\xi_i)+\phi_i\} \\ &\quad + \sum_{i=0}^{N}{}'\sum_{j=0}^{N}{}' h_i h_j\cos(2\pi f_c\tau_i+\phi_i)\cos\{2\pi((f_c+\Delta f)\tau_j - f'_m T_s\cos\xi_j)+\phi_j\}\rangle, \end{aligned} \tag{17}$$

$$where\ i \neq j\ on\ \sum_{i=0}^{N}{}'\sum_{j=0}^{N}{}'$$

Furthermore, the second term in (17) vanishes because the values of τ_i, ξ_i, and ϕ_i for the ith wave and τ_j, ξ_j, and ϕ_j for the jth wave are independent of each other and are also random values, and the sum of the products of $\cos\theta_i$ and $\cos\theta_j$ then becomes zero. With this in mind and considering the small values of $\Delta f\tau_i$ and $f_m' T_s$, the first term of (17) is modified as (18) to (20) by using, for example, a transforming trigonometric function, τ_i, ξ_i, and ϕ_i with random values, and an odd function of sine. In this procedure, the second terms in (18) and (19) also vanish, so we finally get (20) as the result for $\langle I(0)I(\Delta x)\rangle$.

$$Eq.\ (17) = \langle\frac{1}{2}\sum_{i=0}^{N} h_i^2[\cos(2\pi(\Delta f\tau_i - f'_m T_s\cos\xi_i)) + \cos\{2\pi((2f_c+\Delta f)\tau_i - f'_m T_s\cos\xi_i)+2\phi_i\}]\rangle \tag{18}$$

$$=\langle\frac{1}{2}\sum_{i=0}^{N}h_i^2[\cos(2\pi\Delta f\tau_i)\cos(-2\pi f_m^{'}T_s\cos\xi_i)-\sin(2\pi\Delta f\tau_i)\sin(-2\pi f_m^{'}T_s\cos\xi_i)]\rangle \quad (19)$$

$$=\langle\frac{1}{2}\sum_{i=0}^{N}h_i^2\cos(2\pi\Delta f\tau_i)\cos(-2\pi f_m^{'}T_s\cos\xi_i)\rangle \quad (20)$$

6.2 Appendix B derivation of (12)

In the first term A_1 in (11), we change A_1 to (21) because the directive wave's amplitude h_0 is usually much larger than that of the nondirective one and because τ_0=0. Moreover, the amplitude h_i of the ith arriving wave depends on excess delay time τ_i according to (10). So by substituting (10) for h_i, we can rewrite (21) as (22).

$$A_1=\langle\frac{1}{2}[h_0^2+\sum_{i=1}^{N}h_i^2\cos(2\pi\Delta f\tau_i)]\ \rangle \quad (21)$$

$$=\langle\frac{1}{2}h_0^2\rangle+\langle\frac{1}{2}\sum_{i=1}^{N}[h\exp(-\tau_i/\tau_{av})]^2\cos(2\pi\Delta f\tau_i)\rangle \quad (22)$$

We try to calculate by replacing the ensemble average of the second term in (22) by integration with respect to τ_i assuming a large N. As a result, we get (23) assuming $\tau_{\min}$ is close to 0 and $\tau_{\max}$ is large.

$$A_1=\langle\frac{1}{2}h_0^2\rangle+\frac{h^2}{2}\int_{\tau_{\min}}^{\tau_{\max}}\exp(-2\tau/\tau_{av})\cos(2\pi\Delta f\tau)d\tau$$

$$=\langle\frac{1}{2}h_0^2\rangle+\frac{h^2\tau_{av}/4}{1+(\tau_{av}/2)^2(2\pi\Delta f)^2} \quad (23)$$

Furthermore, we need to adjust h in (23) to a normalized power of 1 when N and $\tau_{\max}$ are large, so we get

$$\sum_{i=1}^{N}\frac{1}{2}h_i^2=\frac{1}{2}\sum_{i=1}^{N}[h\exp(-\tau_i/\tau_{av})]^2$$

$$=\frac{h^2}{2}\int_{\tau_{\min}}^{\tau_{\max}}\exp(-2\tau/\tau_{av})d\tau$$

$$=\frac{1}{4}h^2\tau_{av}=1. \quad (24)$$

Consequently, the numerator of the second term in (23) should be 1. Furthermore, considering $\left\langle\frac{1}{2}h_0^2\right\rangle$, we get (25).

$$A_1=k+\frac{1}{1+(\tau_{av}/2)^2(2\pi\Delta f)^2} \quad (25)$$

6.3 Appendix C derivation of (15)

We calculate the ensemble average of A_2 in (14) in a similar manner to that for (23) by integrating with the provability density function $1/\tau_{max}$ of τ. Considering the independence of h_i and τ_i and the power of the directive and nondirective waves, or k and 1, we get (26).

$$A_2 = \langle \ \frac{1}{2}h_0^2 + \frac{1}{2}\sum_{i=1}^{N} h_i^2 \cos(2\pi\Delta f \tau_i) \ \rangle$$

$$= k + \langle \frac{1}{2}\sum_{i=1}^{N} h_i^2 \rangle \langle \sum_{i=1}^{N} \cos(2\pi\Delta f \tau_i \rangle$$

$$= k + \frac{1}{\tau_{max}} \int_{\tau_{min}}^{\tau_{max}} \cos(2\pi\Delta f \tau) d\tau$$

$$= k + \frac{\sin(2\pi\Delta f \tau_{max})}{2\pi\Delta f \tau_{max}} \tag{26}$$

7. References

[1] ITU Circular Letter 5/LCCE/2, Radiocommunication Bureau, 7 March 2008.

[2] Richard van Nee and Ramjee Prasad, OFDM FOR WIRELESS MULTIMEDIA COMMUNICATIONS, Artech House, 1999.

[3] T. Hwang, C. Yang, G. Wu, S. Li, and G. Y. Li, OFDM and Its Wireless Applications: A Survey, *IEEE Trans. Veh. Technol.*, Vol. 58, No. 4, pp. 1673-1694, May 2009.

[4] G. J. Foschini and M. J. Gans, On limits of wireless communications in fading environments when using multiple antennas, *Wireless Personal Commun.* Vol. 6, pp. 311-335, 1998.

[5] D. Shiu, G. Foschini, M. J. Gans, and J. Kahn, Fading Correlation and Its Effect on the Capacity of Multielement Antenna Systems, *IEEE Trans. Commun.*, Vol. 48, No. 3, pp. 502-513, March 2000.

[6] Andreas F. Molisch, Martin Steinbauer, Martin Toeltsch, Ernst Bonek, and Reiner S. Thoma, Capacity of MIMO Systems Based on Measured Wireless Channel, *IEEE JSAC*, Vol. 20, No. 3, pp. 561-569, April 2002.

[7] H. Nishimoto, Y. Ogawa, T. Nishimura, and T. Ohgana, Measurement-Based Performance Evaluation of MIMO Spatial Multiplexing in Multipath-Rich Indoor Environment, *IEEE Trans. Antennas and Propag.*, Vol. 55, No. 12, pp. 3677-3689, Dec. 2007.

[8] Seiichi Sampei and Terumi Sunaga, Rayleigh Fading Compensation for QAM in Land Mobile Radio Communications, *IEEE Trans. Veh. Technol.*, Vol. 42, No. 2, pp. 137-147, May 1993.

[9] R. H. Clarke, A statistical theory of mobile-radio reception, *Bell Syst. Tech. J.*, Vol. 47, No. 6, pp. 957-1000, 1968.

[10] William C. Jakes, MICROWAVE MOBILE COMMUNICATIONS, John Wiley & Sons, Inc., 1974.

[11] S. Kozono, T. Tsuruhara, and M. Sakamoto, Base Station Polarization Diversity Reception for Mobile Radio, *IEEE Trans. Veh. Technol.*, Vol. VT33, No. 4, pp. 301-306, Nov. 1984.
[12] H. Nakabayashi and S. Kozono, Theoretical Analysis of Frequency-Correlation Coefficient for Received Signal Level in Mobile Communications, *IEEE Trans. Veh. Technol.*, Vol. 51, No. 4, pp. 729-737, July 2002.
[13] S. Kozono, K. Ookubo, and K. Yoshida, Study of Correlation Coefficients of Complex Envelope and Phase in a Domain with Time and Frequency Axes in Narrowband Multipath Channel, in *69th IEEE Veh. Technol. Conf.*, Barcelona, Spain, April 2009.

17

Multimodulus Blind Equalization Algorithm Using Oblong QAM Constellations for Fast Carrier Phase Recovery

Jenq-Tay Yuan[1] and Tzu-Chao Lin[2]

Graduate Institute of Applied Science and Engineering, Fu Jen Catholic University Taipei 24205, Taiwan, R.O.C.

1. Introduction

Adaptive channel equalization without a training sequence is known as blind equalization [1]-[11]. Consider a complex baseband model with a channel impulse response of c_n. The channel input, additive white Gaussian noise, and equalizer input are denoted by s_n, w_n and u_n, respectively, as shown in Fig. 1. The transmitted data symbols, s_n, are assumed to consist of stationary independently and identically distributed (i.i.d.) complex non-Gaussian random variables. The channel is possibly a non-minimum phase linear time-invariant filter. The equalizer input, $u_n = s_n * c_n + w_n$ is then sent to a tap-delay-line blind equalizer, f_n, intended to equalize the distortion caused by inter-symbol interference (ISI) without a training signal, where $*$ denotes the convolution operation. The output of the blind equalizer, $y_n = f_n^\star * u_n = s_n * h_n + f_n^\star * w_n$, can be used to recover the transmitted data symbols, s_n, where $\star$ denotes complex conjugation and $h_n = f_n^\star * c_n$ denotes the impulse response of the combined channel-equalizer system whose parameter vector can be written as the time-varying vector $\mathbf{h}_n = [h_n(1), h_n(2), \ldots]^T$ with M arbitrarily located non-zero components at a particular instant, n, during the blind equalization process, where $M = 1,2,3,\ldots$. For example, if $M = 3$ and $I_M = \{1,2,5\}$ is any M-element subset of the integers, then $\mathbf{h}_n = [h_n(1), h_n(2), 0, 0, h_n(5), 0\ldots 0]^T$ is a representative value of $\mathbf{h}_n$.

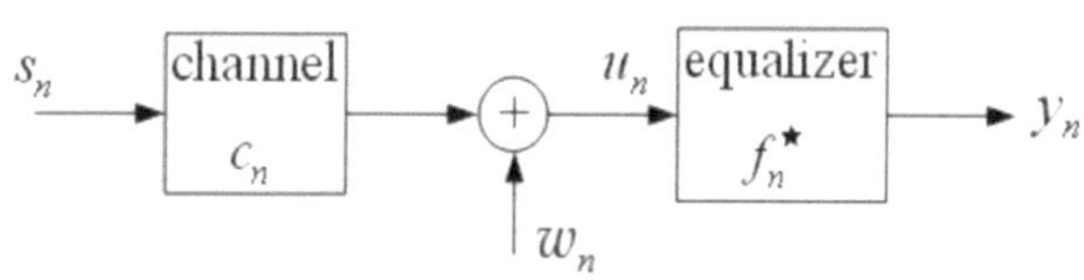

Fig. 1. A complex basedband-equivalent model.

The constant modulus algorithm (CMA) is one of the most widely used blind equalization algorithms [1]-[3]. CMA is known to be phase-independent and one way to deal with its phase ambiguity is through the use of a carrier phase rotator to produce the correct constellation orientation, which increases the complexity of the implementation of the receiver. Moreover, for high-order quadrature amplitude modulation (QAM) signal constellations (especially for cross constellations such as 128-QAM, in which the corner points containing significant phase information are not available), both the large adaptation noise and the increased sensitivity to phase jitter may make the phase rotator spin due to the crowded signal constellations [10]-[13]. Wesolowski [4], [5] Oh and Chin [6], and Yang, Werner and Dumont [7], proposed the multimodulus algorithm (MMA), whose cost function is given by

$$J_{MMA} = E\{[y_{R,n}^2 - R_{2,R}]^2\} + E\{[y_{I,n}^2 - R_{2,I}]^2\} \tag{1}$$

where $y_{R,n}$ and $y_{I,n}$ are the real and imaginary parts of the equalizer output, respectively, while $R_{2,R}$ and $R_{2,I}$ are given by $R_{2,R} = E[s_{R,n}^4]/E[s_{R,n}^2]$ and $R_{2,I} = E[s_{I,n}^4]/E[s_{I,n}^2]$, in which $s_{R,n}$ and $s_{I,n}$ denote the real and imaginary parts of s_n, respectively. Decomposing the cost function of the MMA into real and imaginary parts thus allows both the *modulus* and the *phase* of the equalizer output to be considered; therefore, joint *blind equalization* and *carrier phase recovery* may be simultaneously accomplished, eliminating the need for an adaptive phase rotator to perform separate constellation phase recovery in steady-state operation. The tap-weight vector of the MMA, $\mathbf{f}_n$, is updated according to the stochastic gradient descent (SGD) to obtain the blind equalizer output $y_n = \mathbf{f}_n^H \mathbf{u}_n$

$$\mathbf{f}_{n+1} = \mathbf{f}_n - \mu \cdot \nabla J_{MMA} = \mathbf{f}_n - \mu \cdot \frac{\partial J_{MMA}}{\partial \mathbf{f}_n^\star} = \mathbf{f}_n - \mu \cdot e_n^\star \cdot \mathbf{u}_n \tag{2}$$

where $\mathbf{u}_n = [u_{n+l}, \ldots, u_{n-l}]^T$ and $e_n = e_{R,n} + je_{I,n}$ in which $e_{R,n} = y_{R,n} \cdot [y_{R,n}^2 - R_{2,R}]$, $e_{I,n} = y_{I,n} \cdot [y_{I,n}^2 - R_{2,I}]$ and $L = 2l + 1$ is the tap length of the equalizer.
The analysis in [9], which concerns only the square constellations, indicates that the MMA can remove inter-symbol interference (ISI) and simultaneously correct the phase error. However, when the transmitted symbols are drawn from a QAM constellation having an odd number of bits per symbol ($N = 2^{2i+1}, i = 2, 3, \ldots$), the N-points constellations can be arranged into an *oblong* constellation [14], [15] so long as $E[s_{R,n}^2] \neq E[s_{I,n}^2]$ is satisfied. For example, Fig. 2 illustrates a 128-QAM arranged by oblong (8×16)-QAM with the required average energy of 82. The conventional 128-cross QAM constellations with the required average energy of 82 can be obtained from a square constellation of $12 \times 12 = 144$ points by removing the four outer points in each corner as illustrated in Fig. 3 [7]. Notably, the distance between two adjacent message points in the oblong constellations illustrated in Fig. 2 has been modified to be 1.759 instead of 2 as in the conventional cross 128-QAM, so that the average energies required by both cross and oblong constellations are almost identical. In this chapter, the oblong constellation illustrated in Fig. 2 is used as an example to demonstrate that the MMA using asymmetric oblong QAM constellations with an odd number of bits per symbol may significantly outperform its cross counterpart in the recovery of the carrier phase introduced by channels, without requiring additional average transmitted power. We use the term *asymmetric* because the oblong QAM is not quadrantally symmetric, i.e., $E[s_{R,n}^2] \neq E[s_{I,n}^2]$, and as a consequence $E[s_n^2] \neq 0$. Although reducing the distance between adjacent message points in the proposed oblong constellation in Fig. 2 may increase the steady-state symbol-error rate (SER) or mean-squared error (MSE) of the adaptive equalizer, this chapter is concerned with the unique feature of fast carrier phase recovery associated with the MMA using oblong constellations during blind equalization process owing to its non-identical nature of the real and imaginary parts of the source statistics.

2. Analysis of MMA using oblong constellations

This section presents an analysis of the MMA using oblong QAM constellations from the perspective of its stationary points. Our analytical results demonstrate that the four saddle

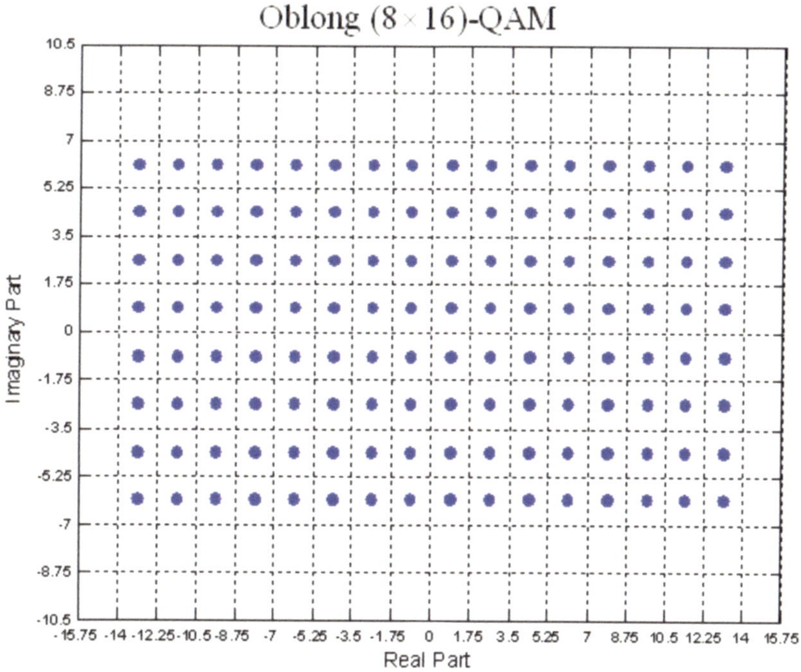

Fig. 2. Oblong constellations for 128-QAM sources.

points existing in the square and cross constellations along $\theta(k) = \pi/4, 3\pi/4, 5\pi/4, 7\pi/4$ are absent when using oblong constellations. Consequently, the frequency of being attracted towards the vicinity of the saddle points, around which the MMA exhibits slow convergence, before converging to the desired minimum, is significantly reduced when using oblong constellations. The use of oblong constellations may thus accelerate the magnitude equalization process during the transient operation.

2.1 MMA Cost function of oblong constellations

After some algebraic manipulation, the expansion of the MMA cost function in (1) for a complex i.i.d. zero-mean QAM source (for *square*, *cross*, and *oblong* constellations) with each member of the symbol alphabet being equiprobable, and a complex baseband channel excluding additive channel noise, can be written in the combined channel-equalizer space $\mathbf{h}_n$ as

$$\begin{aligned} J_{MMA} = {} & \frac{1}{4} \cdot \Re\{E[s^4] \cdot \sum_i h^4(i) + 3E^2[s^2] \cdot \sum_i \sum_{l \neq i} h^2(i) h^2(l)\} + \frac{3}{4}(k_s \sigma_s^4 \cdot \sum_i |h(i)|^4 \\ & + 2\sigma_s^4 \sum_i \sum_{l \neq i} |h(i)|^2 |h(l)|^2 + E^2[s^2] \cdot \sum_i \sum_{l \neq i} h^2(i)(h^2(l))^*) - R_{2,R} \cdot (\Re\{E[s^2] \cdot \sum_i h^2(i)\} \\ & + \sigma_s^2 \cdot \sum_i |h(i)|^2) - R_{2,I} \cdot (-\Re\{E[s^2] \cdot \sum_i h^2(i)\} + \sigma_s^2 \cdot \sum_i |h(i)|^2) + R_{2,R}^2 + R_{2,I}^2 \end{aligned} \quad (3)$$

where $\sigma_s^2 = E[|s_n|^2] = E[s_{R,n}^2] + E[s_{I,n}^2]$ and $k_s = E[|s_n|^4]/\sigma_s^4$ is the source kurtosis. The first term of (3), $(1/4) \cdot \Re\{E[s^4] \cdot \sum_i h^4(i)\}$, which is related to the *fourth-order statistics* (or

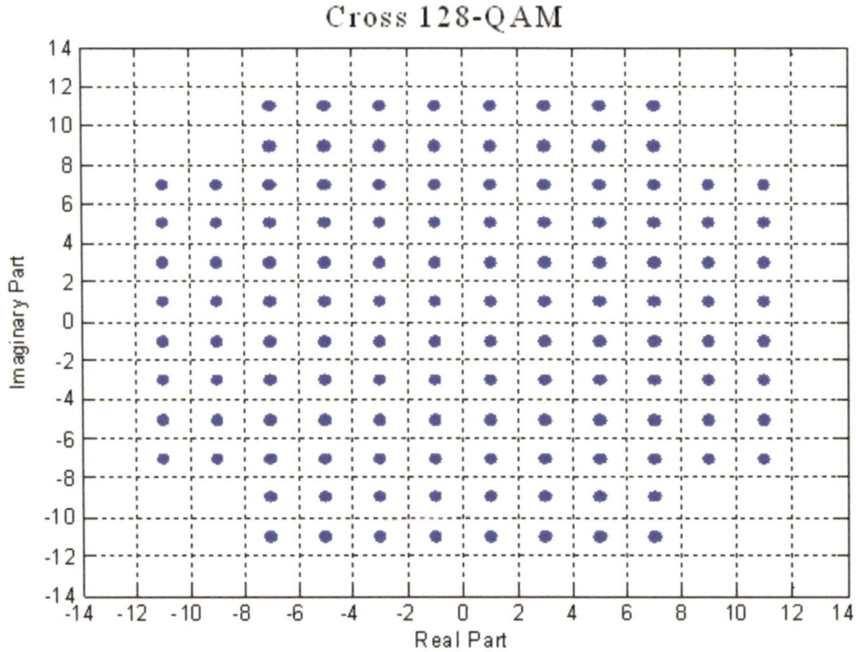

Fig. 3. cross constellations for 128-QAM sources.

fourth-power phase estimator) in [12], [13], [16]-[18] containing the phase information and is absent from the CMA cost function, allows the MMA to recover a possible phase rotation of the equalizer output. Note that when there is no possibility of confusion, the notation is simplified by suppressing the time index n, so for example, $s_n \triangleq s = s_R + js_I$ and $h_n(k) \triangleq h(k)$. Terms s_R and s_I are assumed to be uncorrelated, and both are zero-mean, sub-Gaussian (such that $E[|s|^4] - 2E^2[|s|^2] - |E[s^2]|^2 < 0$). The values of the source statistics of, for example, the oblong (8×16)-QAM can be computed to be $E[s_R^2] = 65.08$ and $E[s_I^2] = 16.08$. For convenience of mathematical analysis, (3) can also be expressed in the following polar space with $[r(k), \theta(k)] = [\sqrt{h_R^2(k) + h_I^2(k)}, \tan^{-1}(h_I(k)/h_R(k))]$, where $h(k) = h_R(k) + jh_I(k)$,

$$J_{MMA} = \frac{1}{4}E[s^4]\cdot\sum_i r^4(i)\cos 4\theta(i) + \frac{3}{2}\cdot E^2[s^2]\cdot\Big(\sum_i\sum_{l\neq i} r^2(i)r^2(l)\cos[2\theta(i)]\cos[2\theta(l)]\Big)$$

$$+\frac{3}{4}\cdot k_s\sigma_s^4\sum_i r^4(i) + \frac{3}{2}\cdot\sigma_s^4\sum_i\sum_{l\neq i} r^2(i)r^2(l) - R_{2,R}\cdot[(E[s_R^2] - E[s_I^2])\cdot\sum_i r^2(i)\cos 2\theta(i) \quad (4)$$

$$+\sigma_s^2\cdot\sum_i r^2(i)] - R_{2,I}\cdot[-(E[s_R^2] - E[s_I^2])\cdot\sum_i r^2(i)\cos 2\theta(i) + \sigma_s^2\cdot\sum_i r^2(i)] + R_{2,R}^2 + R_{2,I}^2$$

Notably, for oblong constellations, $R_{2,R} \neq R_{2,I}$, $E[s_R^4] \neq E[s_I^4]$, $E[s_R^2] \neq E[s_I^2]$ and $E^2[s^2] = E^2[s_R^2] + E^2[s_I^2] - 2E[s_R^2]E[s_I^2] \neq 0$. This asymmetric nature makes the major difference between an oblong constellation and a square (or cross) constellation, since the shape of their resulting MMA cost surfaces would be significantly different, as is revealed later in this chapter.

2.2 Stationary points of MMA using oblong constellations

The equalizer is assumed to be either doubly-infinite in length as in [8] or of finite-length fractionally spaced as in [3] under the equalizability conditions. The set of stationary points of the MMA for oblong constellations considering $M \geq 1$ can be obtained by setting the gradient of J_{MMA} in (4) to zero, such that $\nabla J_{MMA} = \mathbf{r}\frac{\partial J_{MMA}}{\partial r(k)} + \frac{\Theta}{r(k)}\frac{\partial J_{MMA}}{\partial \theta(k)} = 0$. The components of the gradients are

$$\frac{\partial J_{MMA}}{\partial r(k)} = E[s^4]\cdot r^3(k)\cdot \cos 4\theta(k) + 6E^2[s^2]\cdot r(k)\cdot \cos 2\theta(k)\cdot (\sum_{l\neq k} r^2(l)\cdot \cos 2\theta(l))$$

$$+ 3k_s\cdot \sigma_s^4\cdot r^3(k) + 6r(k)\cdot \sigma_s^4\cdot \sum_{l\neq k} r^2(l) - 2R_{2,R}\cdot r(k)\cdot \{[(E[s_R^2] - E[s_I^2]) \tag{5}$$

$$\cdot \cos 2\theta(k)] + \sigma_s^2\} - 2R_{2,I}\cdot r(k)\cdot \{-[(E[s_R^2] - E[s_I^2])\cdot \cos 2\theta(k)] + \sigma_s^2\}$$

$$\frac{\partial J_{MMA}}{\partial \theta(k)} = -E[s^4]\cdot [\sin 4\theta(k)]\cdot r^4(k) - 6^2[s^2]\cdot r^2(k)\cdot [\sin 2\theta(k)]\cdot (\sum_{l\neq k} r^2(l)\cdot \cos 2\theta(l)) \tag{6}$$

$$+2\sin 2\theta(k)\cdot (E[s_R^2] - E[s_I^2])\cdot (R_{2,R} - R_{2,I})\cdot r^2(k)$$

where we set each to zero. Clearly, one stationary point is at $r(k) = 0$. For $r(k) > 0$, (6) yields both $\sin 4\theta(k) = 0$ and $\sin 2\theta(k) = 0$ or

$$\theta(k) = \frac{n\pi}{2}, (n = 0, 1, 2, 3) \tag{7}$$

which reveals the location of the four stationary points of using the oblong constellations. This is in contrast to the square (or cross) constellations, for which $R_{2,R} = R_{2,I}$ and $E[s_R^2] = E[s_I^2]$ such that the last two terms of (6) are both zero. Therefore, only $\sin 4\theta(k) = 0$ is required and, consequently,

$$\theta(k) = \frac{n\pi}{4}, (n = 0, \ldots, 7) \tag{8}$$

The four stationary points given in (7) are now located. Substituting $\theta(k) \in \{0, \pi\}$ and $\theta(l) \in \{0, \pi\}$ (such that $\cos 2\theta(k) = \cos 2\theta(l) = 1$) into (5) yields

$$4r_+^2(k)(E[s_R^4] + E[s_I^4]) + 12(E^2[s_R^2] + E^2[s_I^2])\cdot \sum_{l\neq k} r_+^2(l) - 4E[s_R^4] - 4E[s_I^4] = 0, k = 1, \ldots, M \tag{9}$$

where we have used $E[s^4] = E[s_R^4] + E[s_I^4] - 6E[s_R^2]\cdot E[s_I^2]$. Similarly, substituting $\theta(k) \in \{\pi/2, 3\pi/2\}$ and $\theta(l) \in \{\pi/2, 3\pi/2\}$ (such that $\cos 2\theta(k) = \cos 2\theta(l) = -1$) into (5) yields

$$4r_-^2(k)(E[s_R^4] + E[s_I^4]) + 12(E^2[s_R^2] + E^2[s_I^2])\cdot \sum_{l\neq k} r_-^2(l) - 4R_{2,R}\cdot E[s_I^2] - 4R_{2,I}\cdot E[s_R^2] = 0, k = 1, \ldots, M \tag{10}$$

Clearly, (9) and (10) give $r_+^2(1) = r_+^2(2) = \ldots = r_+^2(M)$ and $r_-^2(1) = r_-^2(2) = \ldots = r_-^2(M)$, respectively. Consequently, (9) and (10) given $k = M$ suffice to determine $r_+^2(M)$ and $r_-^2(M)$:

$$r_+^2(M) = \frac{E[s_R^4] + E[s_I^4]}{E[s_R^4] + E[s_I^4] + [3(M-1)]\cdot (E^2[s_R^2] + E^2[s_I^2])} \tag{11}$$

$$r_-^2(M) = \frac{R_{2,R}\cdot E[s_I^2] + R_{2,I}\cdot E[s_R^2]}{E[s_R^4] + E[s_I^4] + [3(M-1)]\cdot (E^2[s_R^2] + E^2[s_I^2])} \tag{12}$$

The following form for the four stationary points along $\theta(k) = 0, \pi/2, \pi, 3\pi/2$ for each of the M non-zero components of $\mathbf{h}_n$ (for *oblong constellations* only) can thus be derived

$$|h(k)|^2 = [r_+^2(M) \sum_{i \in I_n, \theta(k) \in \{0,\pi\}} \delta(k-i)] + [r_-^2(M) \sum_{i \in I_n, \theta(k) \in \{\frac{\pi}{2}, \frac{3\pi}{2}\}} \delta(k-i)] \tag{13}$$

Notably, since the four stationary points for each element of $\mathbf{h}_n$ have precisely the same location in the $[h_R(k), h_I(k)]$ plane for a given M, the four stationary points given in (13) may be viewed as the stationary points in terms of the overall vector $\mathbf{h}_n$ in the MMA cost in (3) under the common $h_R(k)$ and $h_I(k)$ space denoting the real and imaginary parts of $h(k)$.

2.3 Two special cases: Square and cross constellations

For the special cases of both square and cross constellations, i.e., $R_{2,R} = R_{2,I}$, $E[s_R^2] = E[s_I^2]$ and $E[s_R^4] = E[s_I^4]$, (4) reduces to the following cost function

$$J_{MMA} = \frac{1}{4} E[s^4] \cdot \sum_i r^4(i) \cos 4\theta(i) + \frac{3}{4} [k_s \sigma_s^4 \sum_i r^4(i) + 2\sigma_s^4 \sum_i \sum_{l \neq i} r^2(i) r^2(l)] \tag{14}$$

$$-2\sigma_s^2 \cdot R_{2,R} \cdot \sum_i r^2(i) + 2(R_{2,R})^2$$

It has been shown in [9] that the following form for all possible stationary points of the MMA (for *square and cross constellations*), except for $r(k) = 0$, can be expressed as

$$|h_M(k)|^2 = [r_\pm^2(M) \sum_{i \in I_n, \theta(k) \in \{0, \frac{\pi}{2}, \pi, \frac{3\pi}{2}\}} \delta(k-i)] + [r_\times^2(M) \sum_{i \in I_n, \theta(k) \in \{\frac{\pi}{4}, \frac{3\pi}{4}, \frac{5\pi}{4}, \frac{7\pi}{4}\}} \delta(k-i)] \tag{15}$$

where $r_\pm^2(M) = E[s_R^4]/(E[s_R^4] + [3(M-1)] \cdot E^2[s_R^2])$ and $r_\times^2(M) = 2E[s_R^4]/(E[s_R^4] + [3(2M-1)] \cdot E^2[s_R^2])$. Figure 4 depicts the MMA cost surface for a cross $128 - QAM$ input for $M = 1$. Notably, both $r_+^2(M)$ in (11) and $r_-^2(M)$ in (12) reduce to $r_\pm^2(M)$ when $E[s_R^2] = E[s_I^2]$ and $E[s_R^4] = E[s_I^4]$.

2.4 Unstable equilibria of MMA when $M \geq 2$

If the distribution of s_n is sub-Gaussian, then all the pre-specified $\mathbf{h}_n$ (with the associated I_M) with the stationary points shown in (13), for $M \geq 2$, can be shown to be unstable equilibria (saddle points) by applying the concept proposed by Foschini [8]. Consequently, all the vectors $\mathbf{h}_n$, $M = 2, 3, \ldots$, are saddle points. The locations of the four stationary points at $[\pm r_+(M), 0]$ and $[0, \pm r_-(M)]$ for $\mathbf{h}_n$ with the $(8 \times 16) - QAM$ source for different M can be computed by using (11) and (12). For example, $r_+(10) = 0.249$, $r_-(10) = 0.17$; $r_+(5) = 0.36$, $r_-(5) = 0.245$; $r_+(2) = 0.611$, $r_-(2) = 0.416$; $r_+(1) = 1$, $r_-(1) = 0.68$. Clearly, the four saddle points are located nearer the origin as the number of non-zero components of $\mathbf{h}_n$ rises. As M decreases during the blind equalization process, the locations of the four saddle points for $\mathbf{h}_n$ move dynamically away from the origin in four mutually perpendicular directions. The four saddle points eventually converge as $M \to 1$ to $[\pm 1, 0]$ and $[0, \pm r_-(1)]$, where the former two stationary points become the only two global minima, and the latter two are still two saddle points.

When compared with (8) and (15), the result in (7) and (13) is significant, since it implies that, for $M \geq 2$ (during the transient (or startup) mode operation), the number of saddle points is only half those of the square and cross constellations when the oblong constellations

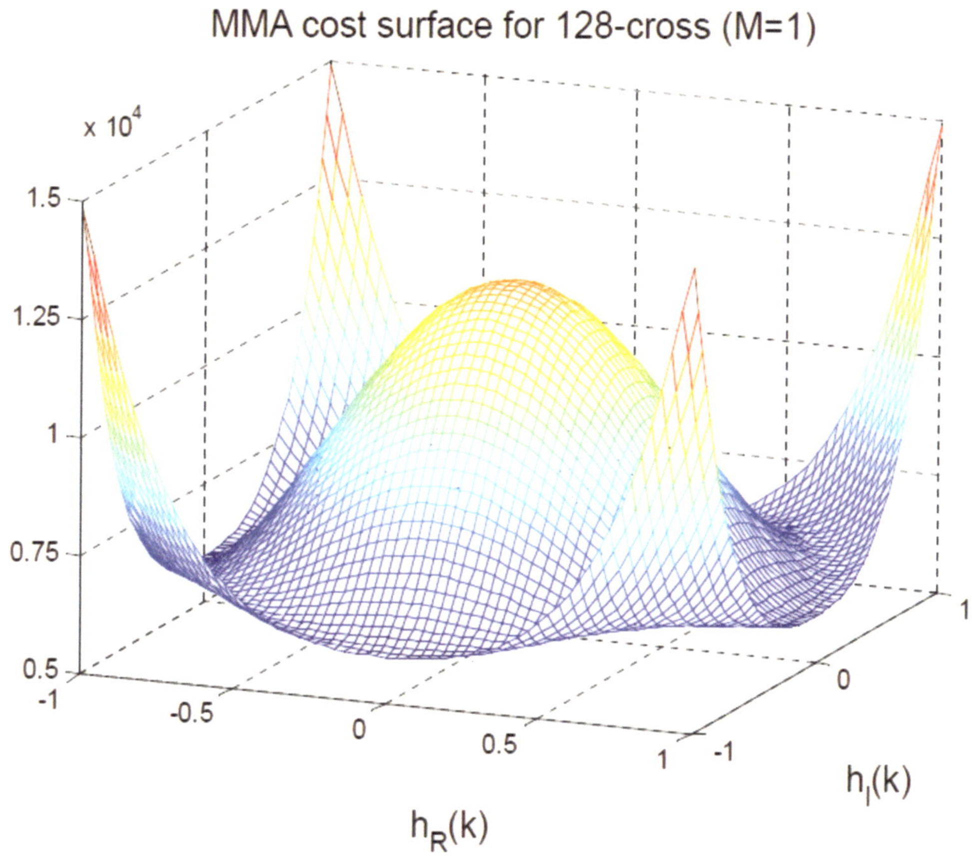

Fig. 4. MMA cost surface for 128-cross input as a function of $h_R(k)$ and $h_I(k)$ for $M = 1$

are adopted (i.e., the four saddle points existing in square and cross constellations along $\theta(k) = \pi/4, 3\pi/4, 5\pi/4, 7\pi/4$ disappear). Consequently, the frequency of attraction toward the vicinity of the saddle points, around which it exhibits slow convergence, before converging to the desired minimum, is significantly diminished when using oblong constellations. Accordingly, using oblong constellations may accelerate the *magnitude equalization* (or residual ISI removing) process during the transient mode operation.

2.5 Local maximum and local minimum of MMA when $M \geq 2$

Computer simulations performed in this study demonstrate that once blind equalization started, a non-zero component of $\mathbf{h}_n$ with maximum magnitude square, $r^2(k)$, rose, and the sum of the magnitude squared of the remaining $M-1$ nonzero components, $\sum_{l \neq k} r^2(l)$, fell rapidly, eventually diminishing to zero. The non-zero component of $\mathbf{h}_n$ with maximum magnitude square is called " $h(k)$ *with maximum modulus*" in the remainder of this chapter. This chapter focuses on the MMA cost in terms of $h(k)$ with maximum modulus, which indicates the performance of the MMA during the transient mode operation ($M \geq 2$). To discover how $h(k)$ with maximum modulus evolves during the transient operation when the MMA adopts the oblong QAM based on the SGD, the MMA cost can be considered in terms of $h(k)$ with maximum modulus alone by substituting the approximations $\sum_k \sum_{l \neq k} r^2(k) r^2(l) \cong 0$,

$\sum_l r^4(l) \cong r^4(k)$, and $\sum_l r^2(l) \cong r^2(k)$ (i.e., the sum of magnitude square of the remaining $(M-1)$ non-zero components of $\mathbf{h}_n$ are very small once the MMA begins functioning) into (4).

From (5) and (6), one stationary point is located at the origin. The second derivative can be shown that when $M \geq 2$, $\nabla^2 J_{MMA} \cong 12 \cdot k_s \sigma_s^4 \cdot r^2(k) - 4 \cdot \sigma_s^2 \cdot [R_{2,R} + R_{2,I}]$. Substituting $r^2(k) = 0$ into the above equation, yields $\nabla^2 J_{MMA} \cong -4 \cdot \sigma_s^2 \cdot [R_{2,R} + R_{2,I}] < 0$, because the situation involving $h(k)$ with maximum modulus at the origin is our concern. The origin thus corresponds to a local maximum in the MMA cost function in terms of $h(k)$ with maximum modulus alone during the transient mode operation. However, the origin corresponds to a local minimum in the MMA cost function in terms of $h(l), l \neq k$ (i.e., the remaining $M-1$ nonzero components of $\mathbf{h}_n$), during the transient mode operation, provided that $\nabla^2 J_{MMA} > 0$ at the origin (or $r^2(k) > [R_{2,R} + R_{2,I}] \cdot \sigma_s^2 / 3k_s\sigma_s^4 \cong 0.3868$) for oblong (8×16)-QAM input.

2.6 Desired global minima of MMA when $M = 1$

When $M \to 1$ (as in steady-state mode), the combined channel-equalizer impulse response vector $\mathbf{h}_n = [0, \ldots, 0, h(k), 0, \ldots, 0]^T$ can be shown to be the only set of minima. The set of all possible stationary points of vector $\mathbf{h}_n$ for oblong constellations given $M = 1$ can be summarized as (i) $r(k) = 0$; (ii) $r_+(k) = 1, \theta(k) = 0$ or π; (iii) $r_-^2(k) = \frac{R_{2,R} \cdot E[s_I^2] + R_{2,I} \cdot E[s_R^2]}{E[s_R^4] + E[s_I^4]}, \theta(k) = \pi/2$ or $3\pi/2$ by substituting $M = 1$ into (11) and (12). Examining the second derivative of the cost function in (4) indicates that for a sub-Gaussian input, $[h_R(k), h_I(k)] = [0, 0], [h_R(k), h_I(k)] = [\pm 1, 0]$ and $[h_R(k), h_I(k)] = [0, \pm\sqrt{\frac{R_{2,R} \cdot E[s_I^2] + R_{2,I} \cdot E[s_R^2]}{E[s_R^4] + E[s_I^4]}}]$ correspond to a local maximum, local (hence global) minima and two saddle points, respectively. Notably, the finding that $[1, 0]$ and $[-1, 0]$ are the only two global minima reveals that any phase error within 180° can be correctly detected when using oblong constellations. This finding contrasts with equalizer output with a 90° phase ambiguity when using the quadrantally symmetric QAM square and cross constellations because phase errors that are multiples of 90° are undetectable. Figure 5 depicts the MMA cost surface for oblong $(8 \times 16) - QAM$ input with $M = 1$ in (3) such that $\mathbf{h}_n = [0, \ldots, 0, h_R(k) + jh_I(k), 0, \ldots, 0]^T$.

Even for $M \geq 2$, the MMA cost function in (4) can be visualized in terms of each non-zero element of $\mathbf{h}_n$ having four saddle points along $\theta(k) = 0, \pi/2, \pi, 3\pi/2$. More specifically, there is one local maximum at the origin in terms of $h(k)$ with maximum modulus and one local minimum at the origin in terms of $h(l), l \neq k$, provided that $r^2(k) > 0.3868$, explaining why $h(k)$ with maximum modulus moves toward $[\pm r_+(M), 0]$, rather than the origin, while the remaining $M-1$ nonzero components of $\mathbf{h}_n$ move toward the origin, based on the SGD method, during the transient operation.

3. Carrier phase tracking capability

This section presents two superior phase tracking capabilities of the oblong constellation. First, when using the MMA with the oblong QAM based on the SGD, $h(k)$ with maximum modulus automatically moves toward the stationary points at $[r_+(M), 0]$ and $[-r_+(M), 0]$, respectively, when $-90° < \theta(k) < 90°$ and $90° < \theta(k) < 270°$, once the magnitude of $h(k)$ with maximum modulus is large enough during the transient mode operation. Meanwhile, the remaining $M-1$ nonzero components of $\mathbf{h}_n$ tend to diminish, and the two stationary points at $[r_+(M), 0] \to [1, 0]$ and $[-r_+(M), 0] \to [-1, 0]$ automatically become the only two global minima when $M \to 1$. Second, an oblong constellation permits much faster phase

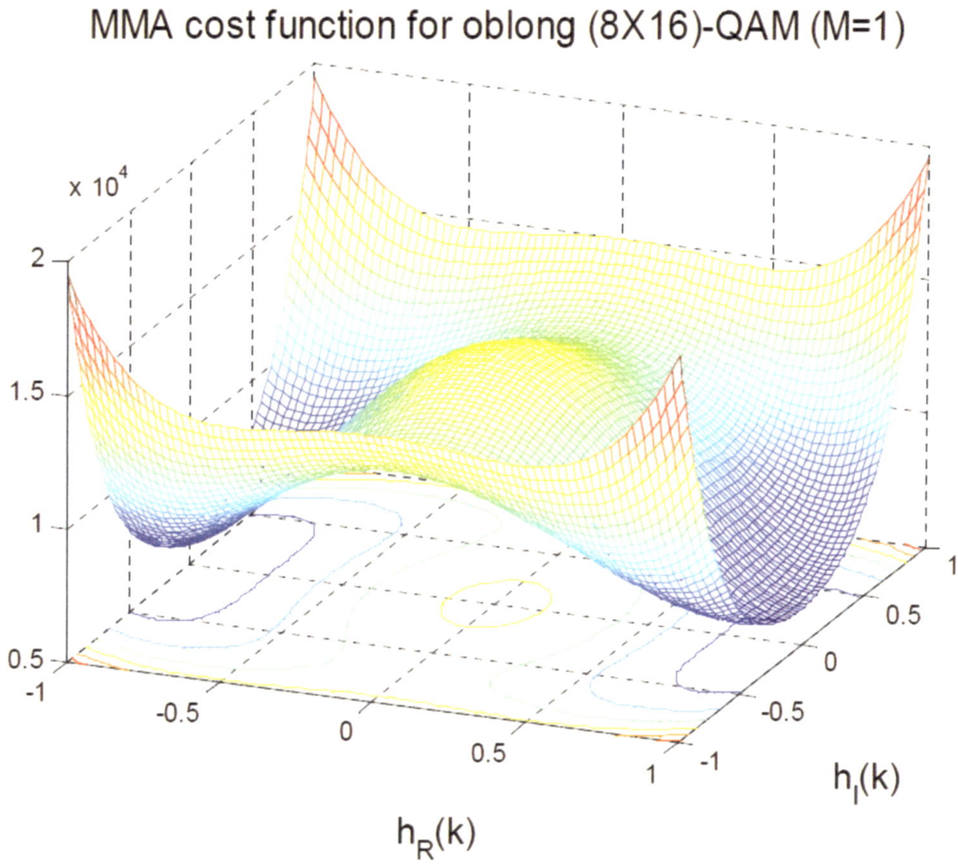

Fig. 5. MMA cost surface for $(8 \times 16) - QAM$ input as a function of $h_R(k)$ and $h_I(k)$ for $M = 1$

acquisition and tracking rates than a cross constellation, because the shape of the cost surface of the oblong constellation inherently has a much steeper average gradient toward the two global minima than the cross constellation.

3.1 Dynamic phase tracking using $-\nabla_\theta J_{MMA}$

We start by rewriting (6) as $\frac{\partial J_{MMA}}{\partial \theta} = 2\sin 2\theta(k)\{F \cdot r^2(k) - A \cdot r^4(k)\cos 2\theta(k) - 3B \cdot r^2(k) \cdot \sum_{l \neq k} r^2(l) \cdot \cos 2\theta(l)\}$, where, for notational simplicity, $A = E[s^4]$, $B = E^2[s^2]$, and $F = \{E[s_R^2] - E[s_I^2]\} \cdot [R_{2,R} - R_{2,I}]$. For the special cases of square and cross constellations, $F = 0$ and $B = 0$ leading to $\nabla_\theta = \frac{\Theta}{r(k)} \frac{\partial J_{MMA}}{\partial \theta(k)} = -E[s^4] \cdot r^3(k) \cdot \sin 4\theta(k)$, which is exactly the result shown in [9] for the square QAM during the transient operation. This section is mainly concerned with the magnitude of $h(k)$ with maximum modulus [i.e., $r(k) = max(|h(k)|)$] and its corresponding phase $\theta(k) = tan^{-1}\frac{h_I(k)}{h_R(k)}$. If $C = \{F \cdot r^2(k) - A \cdot r^4(k)\cos 2\theta(k) - 3B \cdot r^2(k) \cdot \sum_{l \neq k} r^2(l) \cdot \cos 2\theta(l)\} > 0$ can be proved, then the sign of $\frac{\partial J_{MMA}}{\partial \theta(k)} = 2\sin 2\theta(k) \cdot C$ is determined entirely by $\sin 2\theta(k)$. Hence, the non-zero component of $\mathbf{h}_n$ with maximum modulus based on the SGD in (2) can be readily shown to move toward the stationary points along $\theta(k) = 0$ and

$\theta(k) = 180°$, respectively, when $-90° < \theta(k) < 90°$ and $90° < \theta(k) < 270°$, as shown in Fig. 6(a), provided that $C > 0$.
We consider the strictest condition when $\cos 2\theta(l) = 1$, for $l \neq k$ and, consequently, C may be rewritten as $C = \{F \cdot r^2(k) - A \cdot r^4(k)\cos 2\theta(k) - 3B \cdot r^2(k) \cdot \sum_{l \neq k} r^2(l)\}$. In order to obtain the condition of the size of $r^2(k)$ required for $C > 0$ [see (16) below], $E[|y|^2] \cong E[|s|^2]$ was assumed to be satisfied (i.e., magnitude equalization was successfully accomplished with the constant power constraint). Consequently, $\sum_{l \neq k} r^2(l) \cong 1 - r^2(k)$ was obtained by using $E[|y|^2] = E[|s|^2] \cdot \sum_l r^2(l)$ [30]. This assumption is justifiable because one unique feature of the MMA is that its magnitude equalization and carrier phase recovery remain simultaneously effective once it starts functioning. However, this carrier phase recovery takes much longer than the magnitude equalization, which might be almost accomplished in fewer than 1500 iterations, in contrast to 10000 iterations or even more (e.g., for 128-QAM) for the carrier phase recovery. Consequently, the MMA might still produce a high SER, because the equalizer output still exhibits a large phase rotation even once the magnitude equalization has almost been accomplished (ISI $\leq -20dB$), where ISI is defined as $ISI(dB) = 10\log\{\frac{\sum_k |h(k)|^2 - max(|h(k)|^2)}{max(|h(k)|^2)}\}$. By using the approximation, $\sum_{l \neq k} r^2(l) \cong 1 - r^2(k)$, it is straightforward to show that the condition for $C > 0$ is

$$r^2(k) = h_R^2(k) + h_I^2(k) > \frac{3B - F}{3B - A \cdot \cos 2\theta(k)} \tag{16}$$

Consequently, the MMA using the oblong QAM based on the SGD moves toward $\theta(k) = 0°$ and $\theta(k) = 180°$ when $-90° < \theta(k) < 90°$ and $90° < \theta(k) < 270°$, respectively. Equation (16) reveals that the carrier phase recovery, and the tracking of any frequency offset of the MMA using the oblong QAM, hinges on the success of magnitude equalization. To give a rough idea of the size of $r^2(k)$ required to satisfy (16), thus allowing carrier phase recovery to be undertaken and allowing the equalizer to be switched to the decision-directed mode, let $\theta(k) \rightarrow 0$ (or $\theta(k) \rightarrow \pi$) such that the blind equalizer is transitioning to phase recovery regime. Then $\cos 2\theta(k)$ approximates unity and (16) becomes $r^2(k) \cong h_R^2(k) > (3B - F)/(3B - A)$, which equals around 0.54 for $(8 \times 16) - QAM$ input. The condition in (16) becomes increasingly strict as $\theta(k) \rightarrow 0$ (or $\theta(k) \rightarrow \pi$). For example, if $\theta(k) = \pi/6$, then $r^2(k) > 0.459$ for the $(8 \times 16) - QAM$ input. Simulation results confirm that the condition in (16) is almost immediately satisfied as long as the channel is well equalized by choosing an appropriate step-size parameter. By contrast, for the square and cross constellations, $h(k)$ with maximum modulus *unconditionally* moves toward the four stationary points at $[\pm r_{\pm}(M), 0]$ or $[0, \pm r_{\pm}(M)]$, as long as $E[s^4] < 0$, which is valid for most sub-Gaussian source (other than 8-PSK, 4-PSK, and V.29 modem [28]), as shown in Fig. 6(b), since for these two special cases, $F = 0$, $B = 0$ and $A < 0$, (16) becomes $r^2(k) > 0$, which is naturally satisfied. Also note that the narrower the constellations (e.g., $(4 \times 32) - QAM$ input) is used, the condition in (16) becomes stricter. As an example, for the $(4 \times 32) - QAM$ input, (16) becomes $r^2(k) > 0.95$ when $\theta(k) \rightarrow 0$. This result reveals that the use of a very narrow oblong constellation may not yield satisfactory phase tracking capability.

3.2 Fast carrier phase recovery

The *average slope* (i.e., the average gradient) of the MMA cost surface using oblong constellation, in terms of $h(k)$ with maximum modulus alone, from saddle point $[0, r_-(M)]$ toward $[r_+(M), 0]$, was compared with that of the cross constellation from saddle point

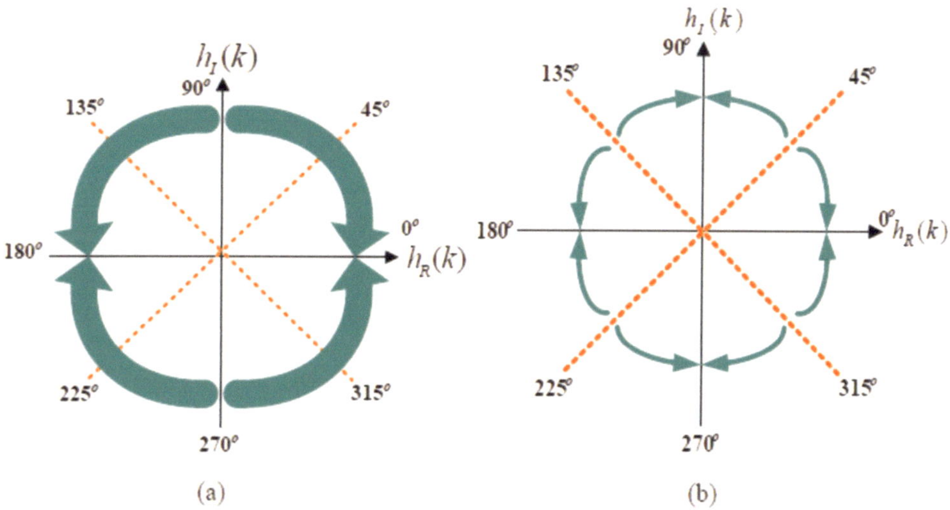

Fig. 6. Trajectory of $h(k)$ with maximum modulus using SGD for (a) oblong QAM and (b) square and cross QAM.

$(1/\sqrt{2}) \cdot [r_\times(M), r_\times(M)]$ toward $[r_\pm(M), 0]$ in order to compare their phase tracking speeds. This is because the rate of phase tracking hinges on the diminishing rate of the imaginary part of $h(k)$ with maximum modulus, which eventually diminishes to zero when the phase rotation is fully recovered. Notably, due to the 180° symmetric QAM oblong constellation and the quadrantally symmetric QAM cross constellations, exactly the same results would have been obtained using the slope from the other saddle point $[0, -r_-(M)]$ to $[\pm r_+(M), 0]$ for the oblong constellation, and the slope from $(1/\sqrt{2}) \cdot [\pm r_\times(M), \pm r_\times(M)]$ to $[\pm r_\pm(M), 0]$ or $[0, \pm r_\pm(M)]$ for the cross constellation.

The average slope between two points at $[a, b]$ and $[c, d]$ in the MMA cost surface in terms of $h(k)$ with maximum modulus alone is defined as $slope = \frac{J_{MMA}[a,b] - J_{MMA}[c,d]}{\sqrt{(a-c)^2 + (b-d)^2}}$, where $J_{MMA}[a, b]$ is calculated by substituting $[a, b]$ into the MMA cost function in (4) (for oblong constellation) or in (14) (for cross constellation) by using the approximations $\sum_l r^2(l) \cong r^2(k)$, $\sum_l r^4(l) \cong r^4(k)$, and $\sum_k \sum_{l \neq k} r^2(k) r^2(l) \cong 0$ (i.e., the sum of magnitude square of the remaining $(M-1)$ non-zero components of $\mathbf{h}_n$ are very small once the MMA begins functioning). The average slope of the MMA cost from $[0, r_-(M)]$ to $[r_+(M), 0]$ for the oblong constellation and the average slope of the MMA cost from $(1/\sqrt{2}) \cdot [r_\times(M), r_\times(M)]$ to $[r_\pm(M), 0]$ for the cross constellation can be approximated as, respectively,

$$slope_{obl} = \frac{J_{MMA}[0, r_-(M)] - J_{MMA}[r_+(M), 0]}{\sqrt{r_+^2(M) + r_-^2(M)}} = \frac{A}{\sqrt{r_+^2(M) + r_-^2(M)}} \tag{17}$$

and

$$slope_{cross} = \frac{J_{MMA}[\frac{r_\times(M)}{\sqrt{2}}, \frac{r_\times(M)}{\sqrt{2}}] - J_{MMA}[r_\pm(M), 0]}{\sqrt{[\frac{r_\times(M)}{\sqrt{2}} - r_\pm(M)]^2 + [\frac{r_\times(M)}{\sqrt{2}} - 0]^2}} \tag{18}$$

$$= \frac{B}{\sqrt{r_\times^2(M) + r_\pm^2(M) - \sqrt{2}\cdot r_\pm(M)\cdot r_\times(M)}}$$

where $A = (-1/4)\cdot E[s^4]\cdot[r_+^4(M) - r_-^4(M)] + (3/4)\cdot k_s\sigma_s^4\cdot[r_-^4(M) - r_+^4(M)] - [R_{2,R} + R_{2,I}]\cdot \sigma_s^2\cdot[r_-^2(M) - r_+^2(M)] + [E[s_R^2] - E[s_I^2]]\cdot[R_{2,R} - R_{2,I}]\cdot[r_-^2(M) + r_+^2(M)]$ and $B = (-1/4)\cdot E[s^4]\cdot[r_\times^4(M) + r_\pm^4(M)] + (3/4)\cdot k_s\sigma_s^4\cdot[r_\times^4(M) - r_\pm^4(M)] - 2\cdot R_{2,R}\cdot\sigma_s^2\cdot[r_\times^2(M) - r_\pm^2(M)]$. Comparison of A with B term by term indicates that the first three terms of both A and B are similar. However, the dominating term, which is the last term in A, contributes to the large average slope of the MMA cost surface of using the oblong constellation, owing to its asymmetric nature of $E[s_R^2] > E[s_I^2]$ and $R_{2,R} > R_{2,I}$ inherent in the oblong constellation. The ratio between the slopes of the two MMA cost surfaces can then be defined as $r = \frac{slope_{obl}}{slope_{cross}} = \frac{A\cdot\sqrt{r_\times^2(M)+r_\pm^2(M)-\sqrt{2}\cdot r_\pm(M)\cdot r_\times(M)}}{B\cdot\sqrt{r_+^2(M)+r_-^2(M)}}$, which is calculated as 3.1, 7.6, 12.56, 17.54 and 25.02 for $M = 1, 3, 5, 7$ and 10, respectively, when comparing oblong $(8\times 16) - QAM$ with cross $128 - QAM$. The imaginary part of $h(k)$ with maximum modulus using oblong constellation would therefore diminish to zero much faster than that using cross constellation. Consequently, when using oblong QAM based on the SGD, $h(k)$ with maximum modulus moves toward the two global minima at $[\pm 1, 0]$ much more rapidly than that of using the cross constellation. Notably, the MMA cost surface of oblong constellation slopes down more sharply than its cross counterpart as the number of non-zero components of $\mathbf{h}_n$ rises. This finding reveals that the imaginary part of $h(k)$ with maximum modulus of oblong constellation diminishes much faster than that of using the cross constellation, especially during the startup operation of blind equalization. Also note that the larger the difference between $E[s_R^2]$ and $E[s_I^2]$ for a particular oblong constellation, the steeper the average gradient of the cost surface of oblong constellation toward global minima than that of cross constellation.

4. Computer simulations

Some blind carrier phase tracking algorithms [20]-[23] and some multimodulus-based algorithms [24], [25] for cross-QAM signals have been developed recently. Computer simulations were performed to compare the performance of the MMA using both oblong and cross constellations, and the performance of CMA using cross constellations followed by either a *dispersion minimization derotator (DMD)* [20] or a *phase tracker (PT)* [21], [22] [29] to correct the carrier phase offset of the CMA output. The computer simulations also included a sliced constant cross algorithm (SCXA) for joint blind equalization and phase recovery of odd-bit cross QAM proposed by Abrar and Qureshi [24]. The transmitted data symbol s_n is an i.i.d., and may be one of the following four possible sources: conventional 128-cross, oblong (8×16)-QAM , and two modified 128-QAM constellations proposed by Cartwright [23], known as the 128A-COB and 128C-COB, whose variance of the fourth power phase estimator is much smaller than that of the 128-cross. The average energies of the conventional 128-cross, (8×16)-oblong , 128A-COB, and 128C-COB, are all 82, while the peak-to-average energy ratios for the four constellations are 2.073, 2.585, 2.645, and 2.881, respectively. Notably, the distance between two adjacent message points in 128A-COB and 128C-COB has been reduced to 1.893 and 1.976, respectively, instead of 2. The channel used for the simulations was the Brazil Ensemble C [26]. Computer experiments were carried out by using the baseband-equivalent channel model developed in [27] with fractionally spaced equalizers (FSE), and the sampling frequency was chosen to be four times the symbol rate. The input to each sub-equalizer is a 40° phase-rotated version of $u_n^{(j)} = \sum_i c_i^{(j)} s_{n-i} + w_n^{(j)}, j = 1,2,3,4$. The real and imaginary parts of the complex-valued additive white Gaussian noise $w_n^{(j)}$ were assumed to be independent, and had equal variance such that the signal-noise-ratios (SNRs) were 30 dB. Here, $SNR = P_{avg}/2\sigma_w^2$ in which P_{avg} is the average power of the signal constellation and σ_w^2 is the variance of each component of the complex valued noise source. In the FSE, each of the four sub-equalizers had 31 complex tap weights with 15 units of time delay. The tap weights of the fourth sub-equalizer were initialized by setting the central tap weight to 1 and the others to zero, while all the tap weights of the remaining three sub-equalizers were set to zero. The ensemble-averaged SER over 10 independent runs was used as a performance index shown in Fig. 7 in which the symbol duration is $0.093\mu s$ as in the ATSC DTV standard.

The MMA using $(8 \times 16) - oblong$ displayed a faster convergence rate than that of $CMA + DMD$, $CMA + PT$, $SCXA$, MMA, using the conventional 128-cross and $CMA + PT$ using both 128A-COB and 128C-COB, according to Fig. 7. However, the MMA using (8×16)-oblong yielded a higher steady-state SER than that of $CMA + PT$ and $CMA + DMD$ using the conventional 128-cross because the use of oblong constellations generates a higher steady-state MSE than its conventional cross counterpart. Notably, the use of MMA without requiring a phase rotator such as DMD and PT reduces the complexity of the implementation of the receiver. Moreover, both DMD and PT yield results with a possible ambiguity of 90° when 128-cross, 128A-COB, and 128C-COB are used. Therefore, the use of MMA with oblong constellations reduces the number of possible phase ambiguities by half. The $CMA + PT$ using the conventional 128-cross outperformed the $CMA + PT$ using 128A-COB and 128C-COB because the latter two constellations whose kurtosis $k_s = 1.462$ and $k_s = 1.399$, respectively, are larger than that of the 128-cross (with $k_s = 1.34$), sacrifice some equalizer convergence rate (i.e., residual ISI removal speed) in order to gain some phase tracking speed, especially when the ISI of the channel is severe. Simulation results similar to those of Fig. 7 were also obtained when the same channel with different phase rotations were adopted because the carrier phase offset was overshadowed by the severity of the ISI of the channel. However, the simulation results also show that the SCXA proposed in [24] outperformed all the other schemes when the phase rotations were less than 30° because the SCXA sometimes became trapped in some undesirable local minima when the phase rotations were greater than 30°.

5. Conclusion

The analysis in this chapter demonstrates that the MMA using oblong constellations is less likely to be attracted to saddle points than the MMA using both square and cross constellations because the number of saddle points is only half of those of the square and cross constellations when adopting oblong constellations during the transient mode operation. Moreover, the finding that $[1,0]$ and $[-1,0]$ are the only two global minima reveals that any phase error within 180° can be correctly detected when using oblong constellations. The 180° phase ambiguity (and the 90° phase ambiguity for 128-cross, 128A-COB, and 128C-COB) may be overcome by using differential encoding when necessary. This chapter also shows that the oblong constellations to be used with the MMA may outperform the conventional cross constellations in terms of carrier phase recovery owing to $E[s_R^2] \neq E[s_I^2]$. Intuitively, this may be because the MMA cost function given by (1) reduces to $J_{MMA} = 2E\{[y_{R,n}^2 - R_{2,R}]^2\}$, $J_{MMA} = 2E\{[y_{R,n}^2 - R_{2,R}]\}$ which essentially considers only the real part of the equalizer output, when the symmetric constellations are such that $R_{2,R} = R_{2,I}$, $E[s_R^4] = E[s_I^4]$ and

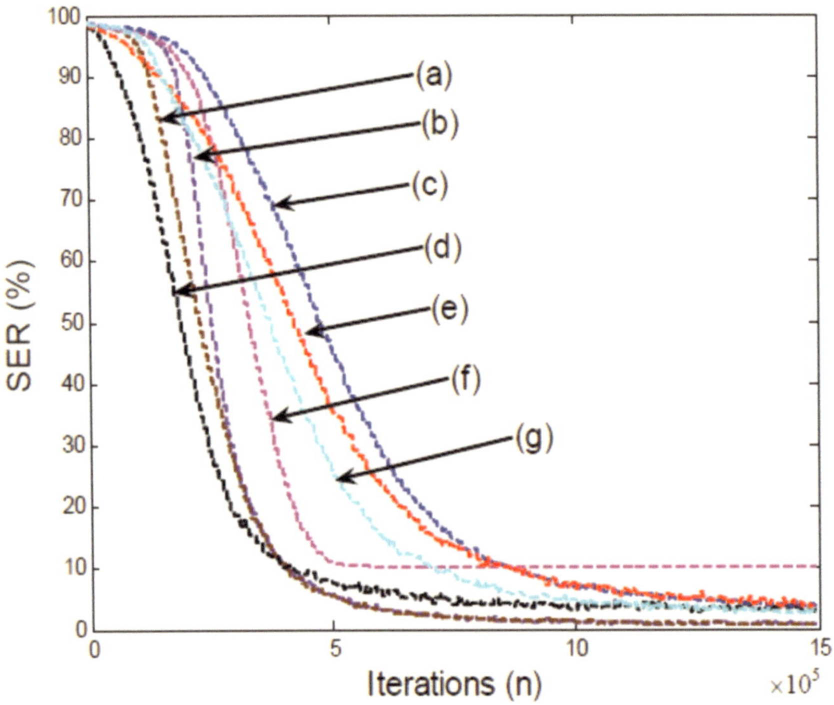

Fig. 7. Ensemble-averaged SER performance using the Brazil Ensemble C for $SNR = 30dB$ and phase rotation 40°. Traces: (a) $CMA + PT$ using 128-cross ($\mu = 9 \times 10^{-9}, \mu_{PT} = 2 \times 10^{-8}$), (b) $CMA + DMD$ using 128-cross ($\mu = 9 \times 10^{-9}, \mu_{DMD} = 4 \times 10^{-8}$), (c) MMA using 128-cross ($\mu = 6 \times 10^{-9}$),(d) MMA using oblong (8 × 16)-QAM ($\mu = 10^{-8}$), (e) $CMA + PT$ using 128A-COB ($\mu = 6 \times 10^{-9}, \mu_{PT} = 5 \times 10^{-8}$), (f) SCXA using 128-cross ($\mu = 3 \times 10^{-9}$), and (g) $CMA + PT$ using 128C-COB ($\mu = 6 \times 10^{-9}, \mu_{PT} = 10^{-8}$).

$E[s_R^2] = E[s_I^2]$. From a statistical perspective, the MMA cost function in (1) can be exploited fully for blind equalization only when the *asymmetric* nature of the constellations is utilized. However, the oblong constellation has a slightly higher peak-to-average power ratio and a higher steady-state SER than the conventional cross constellation at the same average energy level. Other asymmetric constellations may be designed (such as by adjusting the symbol probabilities without reducing the distance between adjacent message points) to achieve the best compromise among fast carrier phase recovery, low steady-state SER and low average transmitted power, as long as $E[s_R^2] \neq E[s_I^2]$.

6. References

[1] D. N. Godard (1980). "Self-recovering equalization and carrier tracking in two-dimensional data communication system," *IEEE Trans. Commun.*, vol. COM-28, pp.1867-1875, Nov. 1980.

[2] J. R. Treichler and M. G. Larimore (1985). "New Processing Techniques Based on the Constant Modulus Algorithm," *IEEE Trans. Acoust., Speech, Signal Process.*, vol. ASSP-33, pp.420-431, Apr. 1985.

[3] C. R. Johnson, Jr., P. Schniter, T. J. Endres, J. D. Behm, D. R. Brown, and R. A. Casas (1998). "Blind equalization using the constant modulus criterion: A review," *Proc. IEEE*, vol. 86, pp.1927-1950, Oct. 1998.

[4] K. Wesolowski (1987). "Self-recovering adaptive equalization algorithms for digital radio and voiceband data modems," *Proc. European Conf. Circuit Theory and Design*, 1987, pp. 19-24.

[5] K. Wesolowski (1992). "Analysis and properties of the modified constant modulus algorithm for blind equalization," *European Trans. Telecommun.*, vol. 3, pp. 225-230, May-June 1992.

[6] K. N. Oh and Y. O. Chin (1995). "Modified constant modulus algorithm: blind equalization and carrier phase recovery algorithm," *Proc. IEEE Int. Conf. Commun.*, 1995, vol. 1, pp. 498-502.

[7] J. Yang, J.-J. Werner, and G. A. Dumont (2002). "The multimodulus blind equalization and its generalized algorithms," *IEEE J. Sel. Areas Commun.*, vol. 20, no. 5, pp. 997-1015, Jun. 2002.

[8] G. J. Foschini (1985). "Equalization without altering or detecting data," *AT&T Tech. J.*, vol. 64, pp. 1885-1911, Oct. 1985.

[9] J.-T. Yuan and K.-D. Tsai (2005). "Analysis of the Multimodulus Blind equalization Algorithm in QAM Communication Systems," *IEEE Trans. Commun.*, Vol. 53, No. 9, pp. 1427-1431, Sep. 2005.

[10] N. K. Jablon (1992). "Joint blind equalization, carrier recovery, and timing recovery for high-order QAM signal constellations," *IEEE Trans. Signal Process.*, vol.40, no. 6, pp.1383-1397, Jun. 1992.

[11] J.-T. Yuan and L.-W. Chang (2007). "Carrier phase tracking of multimodulus blind equalization algorithm using QAM oblong constellations," Proc. *IEEE Int. Conf. Commun.*, Glasgow, Scotland, UK, 2007, pp. 2991-2998.

[12] C. N. Georghiades (1997). "Blind carrier phase acquisition for QAM constellations," *IEEE Trans. Commun.*, vol. 45, pp. 1477-1486, Nov. 1997.

[13] E. Serpedin, P. Ciblat, G.. B. Giannakis, and P. Loubaton (2001). "Performance analysis of blind carrier phase estimators for general QAM constellations," *IEEE Trans. Signal Process.*, vol. 49, no. 8, pp. 1816-1823, Aug. 2001.

[14] J. G. Smith (1975). "Odd-bit quadrature amplitude-shift keying," *IEEE Trans. Commun.*, vol. 23, pp. 385-389, Mar. 1975.

[15] K. Cho and D. Yoon (2002). "On the general BER expression of one- and two-dimensional amplitude modulations," *IEEE Trans. Commun.*, vol. 50, pp. 1074-1080, Jul. 2002.

[16] M. Moeneclaey and G. de Jonghe (1994). "ML-oriented NDA carrier synchronization for general rotationally symmetric signal constellations," *IEEE Trans., Commun.*, vol. 42, Aug. 1994, pp. 2531-2533.

[17] K. V. Cartwright (1999). "Blind phase recovery in general QAM communication systems using alternative higher order statistics," *IEEE Signal Processing Lett.*, vol. 6, no. 12, Dec. 1999, pp. 327-329.

[18] R. Lopez-Valcarce (2004). "Cost minimization interpretation of fourth power phase estimator and links to multimodulus algorithm," *Electron. Lett.*, vol. 40, no. 4, Feb. 2004.

[19] J. P. LeBlanc, I. Fijalkow, and C. R. Johnson, Jr. (1998). "CMA fractionally spaced equalizers: stationary points and stability under iid and temporally correlated sources," *Int. J. Adapt. Control Signal Process.*, vol. 12, pp. 135-155, 1998.

[20] W. Chung, W. A. Sethares, and C. R. Johnson, Jr. (2004). "Performance analysis of blind adaptive phase offset correction based on dispersion minimization," *IEEE Trans. Signal Process.*, vol.52, no. 6, pp.1750-1759, Jun. 2004.

[21] H. Mathis (2001). "Blind phase synchronization for VSB signals," *IEEE Trans. on Broadcast.*, vol.47, no. 4, pp.340-347, Dec. 2001.

[22] S. Abrar (2007). "An adaptive method for blind carrier phase recovery in a QAM receiver," in *Proc. Int. Conf. on Information Emerging Technologies*, 2007, pp. 1-6.

[23] K. V. Cartwright (2000). "Fourth power phase estimation with alternative two-dimensional odd-bit constellations," *IEEE Commun. Lett.*, vol. 4, pp. 199-201, Jun. 2000.

[24] S. Abrar and I. M. Qureshi (2006). "Blind equalization of cross-QAM signals," *IEEE Signal Processing Lett*, vol. 13, no. 12, pp. 745-748, Dec. 2006.

[25] S. Chanmanon and T. Thaiupathump (2006). "Performance analysis of double-ellipse algorithm for QAM equalization and phase recovery," *2006 Int. Symposium on Commun. and Inf. Technologies*, pp. 657-662.

[26] C. W. Rhodes (2002). "Some recent improvements in the design of DTV receivers for the ATSC standard," *IEEE Trans. Consumer Electron.*, vol. 48, pp. 938-945, Nov. 2002.

[27] H. -N. Kim, Y. -T. Lee, and S. W. Kim (2003). "Mathematical modeling of VSB-based digital television systems," *ETRI J.*, vol. 25, pp. 9-18, Feb. 2003.

[28] J.-T. Yuan and T.-C. Lin (2010). "Equalization and Carrier Phase Recovery of CMA and MMA in Blind Adaptive Receivers," *IEEE Trans. Signal Process.*, vol. 58, no.6, pp. 3206-3217, Jun. 2010.

[29] J.-T. Yuan and Y.-F. Huang (2010). "Blind Carrier Phase Acquisition and Tracking for 8-VSB Signals," *IEEE Trans. Commun.*, vol. 58, no. 3, pp. 769-774, Mar. 2010.

[30] O. Shalvi and E. Weinstein (1990). "New criteria for blind deconvolution of nonminimum phase systems (channels)," *IEEE Trans. Inf. Theory*, vol. 36, pp. 312-321, Mar. 1990.

18

Peak-to-Average Power Ratio Reduction for Wavelet Packet Modulation Schemes via Basis Function Design

Ngon Thanh Le[1], Siva D. Muruganathan[2] and Abu B. Sesay[1]
[1]*Department of Electrical & Computer Engineering, University of Calgary,*
[2]*Department of Electrical & Computer Engineering, University of Alberta*
Canada

1. Introduction

Orthogonal frequency division multiplexing (OFDM) is a well known and widely employed high-data rate transmission technology. By dividing the wideband channel into multiple narrowband sub-channels, OFDM mitigates the detrimental effects of the multi-path fading channel. A main drawback of OFDM is that the transmitted signal possesses a high peak-to-average power ratio (PAPR). Since the OFDM modulated signal is the superposition of many constituent narrowband signals, it is possible for these constituent narrowband signals to align in phase and produce high peak values. This results in high PAPRs where the peak power of the OFDM modulated signal is very large compared to the average power. For applications employing highly power-efficient power amplifiers, the input signal with a high PAPR causes the power amplifier to be operated in the non-linear region. The non-linear characteristic of the high power amplifier (HPA) will introduce in-band distortions which in turn will degrade system performance. Additionally, a HPA operated in the non-linear region will also produce out-of-band power emissions which will generate spectral regrowth and produce unwanted interference to the adjacent channel users. To alleviate the high PAPR problem in OFDM systems, several PAPR reduction schemes have been proposed in the literature (see Han & Lee, 2005; Jiang & Wu, 2008, and references therein).

Wavelet packet modulation (WPM) is an alternative multi-carrier technology that has received recent research attention (see Lakshmana & Nikookar, 2006 and references therein). WPM systems have been shown to have better immunity to impulse and narrowband noises than OFDM (Lindsey, 1995). Furthermore, WPM systems attain better bandwidth efficiency than OFDM (Sandberg & Tzannes, 1995), and they do not require cyclic prefix extension unlike OFDM (Lakshmana & Nikookar, 2006). To benefit from these advantages, WPM has been recently applied in the various areas including multi-carrier multi-code code division multiple access (CDMA) (Akho-Zhieh & Ugweje, 2008), cognitive radio systems (Lakshmanan, Budiarjo, & Nikookar, 2007), and multiple-input multiple-output systems (Lakshmanan, Budiarjo, & Nikookar, 2008), to name a few. However, similar to OFDM, WPM also suffers from the PAPR problem.

The objectives of this chapter are twofold. Firstly, we provide an overview of the PAPR reduction methods already published in the literature for WPM systems. As part of the overview, we discuss the merits and limitations associated with the existing PAPR reduction schemes. Secondly, we formulate the design criteria for a set of PAPR minimizing orthogonal basis functions for WPM systems. To evaluate the merits of the PAPR minimizing orthogonal basis functions, performance comparisons are made with the conventional Daubechies basis functions and OFDM. Furthermore, we also provide a qualitative comparison between the PAPR minimizing orthogonal basis functions and the existing PAPR reduction techniques for WPM systems.

The rest of the chapter is organized as follows. First, an overview of WPM based multi-carrier systems is provided in Section 2. An overview of some PAPR reduction methods already proposed in the literature for WPM systems is then presented in Section 3. Next, in Section 4, we present the design criteria of the PAPR minimizing orthogonal basis functions. This is followed by simulation results and discussions in Section 5. Finally, the chapter is concluded in Section 6.

The following notations are used throughout the chapter: $E\{\theta\}$ denotes the statistical average of random variable θ; $\Pr\{\theta > \theta_0\}$ represents the probability of the event $\theta > \theta_0$; $max_n\{\tilde{\theta}(n)\}$ represents the maximum value of $\tilde{\theta}(n)$ over all instances of time index n; $\text{sign}\{\bullet\}$ is the signum function; $\delta(p)$ represents the Kronecker delta function; $\lceil \theta \rceil$ denotes the smallest integer greater than or equal to θ; $\forall$ represents universal quantification.

Parts of this work is based on "An Efficient PAPR Reduction Method for Wavelet Packet Modulation Schemes", by N. T. Le, S. D. Muruganathan, and A. B. Sesay which appeared in IEEE Vehicular Technology Conference, Spring-2009. The previously published material is reused with permission from IEEE. Copyright [2009] Institute of Electrical and Electronics Engineers.

2. Overview of WPM based multi-carrier systems

In general, the WPM based multi-carrier systems can be characterized using sets of orthogonal basis functions called wavelet packets (Lakshmanan & Nikookar, 2006). These orthogonal basis functions are defined at multiple levels with the mth level consisting of 2^m distinct basis functions. Let us denote the orthogonal basis functions corresponding to the mth level by the set

$$\left\{\varphi_{m,0}(t),\ \varphi_{m,1}(t),\ \ldots,\ \varphi_{m,k}(t)\ \ldots,\ \varphi_{m,2^m-2}(t),\ \varphi_{m,2^m-1}(t)\right\}, \tag{1}$$

where k $(k = 0, 1, \ldots, 2^m - 1)$ denotes the orthogonal basis function index. Given (1), the orthogonal basis functions corresponding to the $(m+1)^{th}$ level are defined as

$$\varphi_{m+1,\,2k}(t) = \sum_n \tilde{h}(n)\,\varphi_{m,k}(2t-n), \tag{2}$$

$$\varphi_{m+1,\,2k+1}(t) = \sum_n \tilde{g}(n)\,\varphi_{m,k}(2t-n), \tag{3}$$

where $\tilde{h}(n)$ and $\tilde{g}(n)$ respectively denote low-pass and high-pass filter impulse responses forming a quadrature mirror filter (QMF) pair at the receiver. Hence, $\tilde{h}(n)$ and $\tilde{g}(n)$ are related through (Lakshmanan & Nikookar, 2006)

$$\tilde{g}(n) = (-1)^n \tilde{h}(2N - 1 - n), \tag{4}$$

where $2N$ denotes length of both filters. In addition, the low-pass filter impulse response $\tilde{h}(n)$ is also required to satisfy the following conditions (Daubechies, 1992):

$$\sum_n \tilde{h}(n)\tilde{h}(n + 2p) = \delta(p), \tag{5}$$

$$\sum_n \tilde{h}(n) = \sqrt{2}. \tag{6}$$

Let us next define the time-reversed version of the QMF pair at the transmitter as follows:

$$h(n) = \tilde{h}(-n), \tag{7}$$

$$g(n) = \tilde{g}(-n), \tag{8}$$

where $h(n)$ and $g(n)$ represent the time-reversed low-pass and high-pass filter impulse responses, respectively.

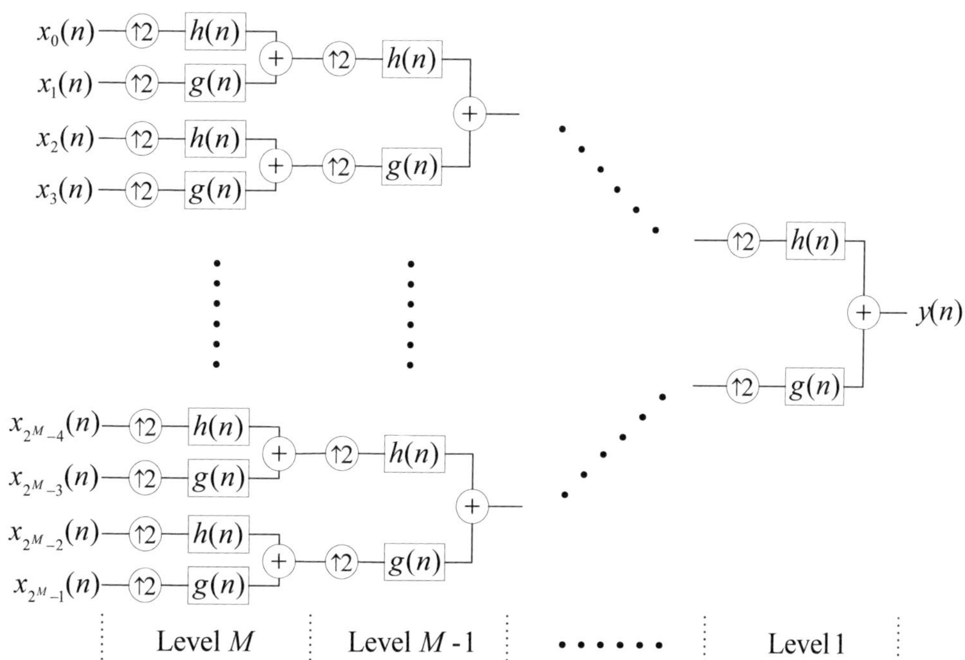

Fig. 1. A 2^M sub-channel WPM transmitter structure

Now, the transmitter structure for a WPM multi-carrier system with 2^M sub-channels can be constructed as shown in Figure 1. In Figure 1, the input data symbol streams $x_0(n)$, $x_1(n)$, ..., $x_{2^M-2}(n)$, $x_{2^M-1}(n)$ are multiplexed onto 2^M sub-channels via the successive application of time-reversed QMF pairs (i.e., $h(n)$ and $g(n)$). For a system with 2^M sub-channels, we require M levels of QMF pairs at the transmitter, wherein the mth

level consists of 2^{m-1} QMF pairs. Generally, the successive application of the M levels of QMF pairs yields the output $y(n)$ of the WPM transmitter. It should be noted that the WPM transmitter structure of Figure 1 is also referred to as the wavelet packet tree (WPT) in the literature. Additionally, the combination of the up-sampler and the filter (i.e., either $h(n)$ or $g(n)$) is sometimes called a node.

Alternatively, the WPM transmitter structure of Figure 1 can also be represented as in Figure 2 (Daly et al., 2002). In Figure 2, the equivalent impulse response corresponding to the kth $(k = 0, 1, \ldots, 2^M - 1)$ sub-channel is denoted as $w_k(n)$. Next, let us represent the Z-transform of the filter impulse response corresponding to the kth sub-channel and the mth level (refer to Figure 1) as

$$T_{k,m}(z) \in \{H(z), G(z)\}, \tag{9}$$

where $H(z)$ and $G(z)$ denote the Z-transforms of $h(n)$ and $g(n)$, respectively. Then, the Z-transform of the equivalent impulse response $w_k(n)$ is given by (Daly et al., 2002)

$$W_k(z) = \prod_{m=1}^{M} T_{k,m}(z^{2^{m-1}}). \tag{10}$$

3. PAPR reduction techniques for WPM systems

Although PAPR reduction techniques have been extensively studied for OFDM systems (see Han & Lee, 2005, Jiang & Wu, 2008, and references therein), PAPR reduction methods for WPM systems have only recently gained the attention of the Communications and Signal Processing research communities. In this section, we provide an overview of some PAPR reduction methods already proposed in the literature for WPM systems. In doing so, we discuss the merits and limitations of the existing PAPR reduction schemes for WPM systems. This section also helps set the stage for later sections by emphasizing the motivation for proposing a basis function design based PAPR reduction scheme for WPM systems.

3.1 Tree pruning based methods

In the tree pruning approach, adjacent nodes of the wavelet packet tree (WPT) are selectively combined (or split) to form a single node (or two separate nodes) in order to reduce the PAPR. Baro and Ilow recently proposed a tree pruning based PAPR reduction scheme where a sequence of modulated symbols are first passed through N_T distinct pruned WPTs to produce N_T different output sequences (Baro & Ilow, 2007a). The N_T distinct pruned WPTs are attained by performing a single adjacent node-pair joining at different locations of the unpruned WPT (i.e., a single adjacent node-pair joining is performed at N_T different locations of the unpruned WPT). Then, the sequence with the lowest PAPR among the N_T different output sequences is chosen for transmission. Additionally, to facilitate recovery of the information symbols, details about the pruned WPT chosen for transmission is sent to the receiver as side information via additional sub-carriers.

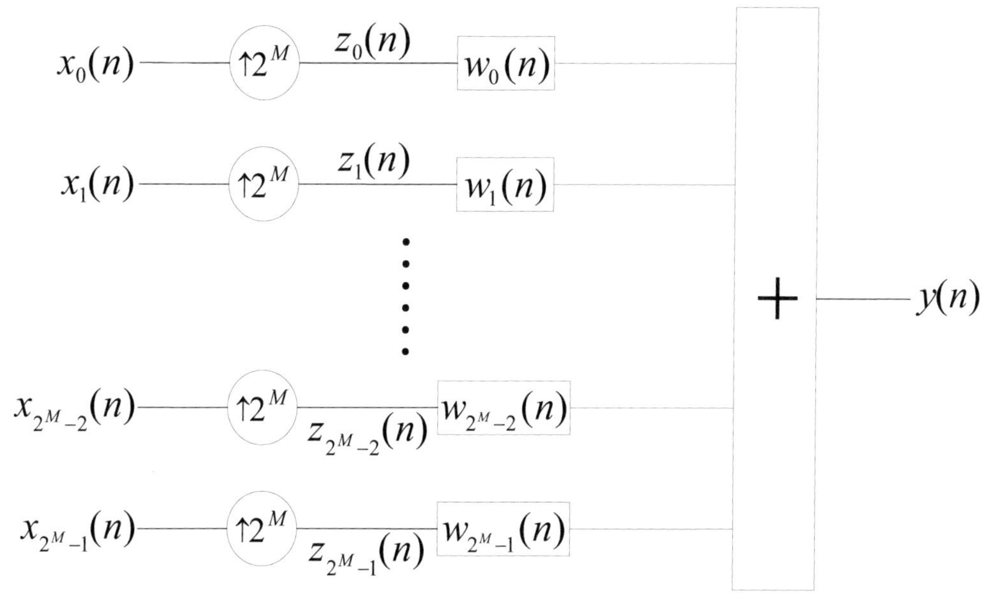

Fig. 2. An equivalent representation of the 2^M sub-channel WPM transmitter structure

Since the scheme proposed in (Baro & Ilow, 2007a) requires an exhaustive search for the output sequence with the lowest PAPR, the number N_T of pruned WPTs determines the computational complexity associated with it. Furthermore, the number N_T also affects the amount of side information required since a minimum of $\lceil \log_2(N_T) \rceil$ bits are necessary in order for the receiver to identify the pruned WPT that was used by the transmitter. To alleviate the complexity and side information overhead issues, the authors in (Baro & Ilow, 2007a) also propose two low-complexity pruning algorithms. In the first low-complexity algorithm, the authors reduce the number N_T of pruned WPTs by choosing only the most frequently chosen N'_T pruned WPTs (i.e., the ones that generate the output sequence with the lowest PAPR most frequently) obtained through several random simulation trials. This scheme not only reduces the complexity associated with the original tree pruning algorithm but also decreases the side information overhead. The second low-complexity algorithm involves setting a PAPR threshold $PAPR_{th}$, where the first pruned WPT that produces an output PAPR below $PAPR_{th}$ is chosen for transmission. Depending on the values of N'_T and $PAPR_{th}$ chosen, the two low-complexity versions are shown to slightly reduced PAPR reduction performance (Baro & Ilow, 2007a).

In (Baro & Ilow, 2007b), an alternative tree pruning is proposed for PAPR reduction in WPM systems. As opposed to the scheme in (Baro & Ilow, 2007a) which is based on a single adjacent node-pair joining approach, the scheme proposed in (Baro & Ilow, 2007b) allows multiple joins on multiple non-leaf nodes. In the later scheme, a set $P^{(D)}$ of allowable non-leaf nodes are first defined, and the joining of multiple nodes is performed iteratively. Since this scheme involves searching for a pruned WPT that yields the least PAPR via multiple iterations (or passes), the authors refer to it as the multi-pass tree pruning method. In reducing the processing time and computational complexity, it is shown in (Baro & Ilow, 2007b) that the multi-pass tree pruning method provides good PAPR reduction capability

when the number of iterations (or passes) is set to 5. Furthermore, a computationally simpler version of the multi-pass tree pruning algorithm based on setting a PAPR threshold $PAPR_{th}$ is also proposed in (Baro & Ilow, 2007b). In the simplified version, the iterative search procedure is terminated once a pruned WPT that yields a PAPR value below $PAPR_{th}$ is found. For the receiver to identify the pruned WPT that was used by the transmitter, the multi-pass pruning scheme requires a minimum of $\left\lceil \log_2\left(size\left\{P^{(D)}\right\}\times N_i\right)\right\rceil$ bits as side information. Here, $size\left\{P^{(D)}\right\}$ and N_i respectively denote the size of the allowable non-leaf node set and the maximum number of iterations. It should be noted that with increasing $size\left\{P^{(D)}\right\}$, the multi-pass pruning scheme yields improved PAPR reduction capability and requires more side information overhead.
The major advantage of tree pruning based methods is their capability to provide high PAPR reduction. The results presented in (Baro & Ilow, 2007a) show that the single node-pair joining based tree pruning method can yield up to 3.5 dB of PAPR reduction when compared to the WPM scheme using the unpruned WPT. Likewise, the multi-pass tree pruning method of (Baro & Ilow, 2007b) can provide up to 5 dB PAPR reduction over the unpruned WPM scheme. Furthermore, the tree pruning based PAPR reduction methods are distortionless and do not introduce spectral regrowth. The major drawbacks of the tree pruning based PAPR reduction methods are twofold. Firstly, the tree pruning approach requires side information to be sent to the receiver which reduces the bandwidth efficiency of the system. Moreover, when the side information is received in error, the receiver will not be able to recover the transmitted data sequence. Hence, some form of protection such as the employment of channel encoding may be necessary to reliably receive the side information. However, the employment of channel encoding in the transmission of side information will result in further loss of bandwidth efficiency. The second major disadvantage of tree pruning based PAPR reduction methods is that the pruned WPTs result in sub-channels with different bandwidths. Hence, the application of tree pruning based approach does not guarantee that all sub-channels undergo frequency-flat fading. This will reduce the multipath resilience of the WPM system under broadband communication environments.

3.2 Clipping and amplitude threshold based methods

The clipping method is the simplest and one of the widely used PAPR reduction methods in multicarrier communication systems (Jiang & Wu, 2008). Using the clipping method, any desired amount of PAPR reduction can be achieved by presetting the clipping level at the transmitter. In (Rostamzadeh & Vakily, 2008), a clipping based PAPR reduction scheme is investigated with application to WPM. To reduce the effect of nonlinear distortion introduced by the clipping process, the authors in (Rostamzadeh & Vakily, 2008) adopt an iterative maximum likelihood (ML) based approach at the receiver. In this approach, estimates of the transmitted symbols are first attained via the ML detector. These transmitted symbol estimates are then used to compute an estimate of the nonlinear distortion component. Next, the nonlinear component estimate is removed from the received signal, and revised ML estimates of the transmitted symbols are obtained. The revised ML estimates of the transmitted symbols are once again used to attain a revised estimate of the nonlinear distortion component. The processes of revised nonlinear distortion component estimation and revised ML estimation of the transmitted symbols are repeated iteratively until a desired level of performance is attained. Bit error rate (BER) results presented in (Rostamzadeh & Vakily, 2008) show that the iterative ML receiver based clipping approach can nearly mitigate the in-band distortion introduced by the clipping process with 3 iterations in an additive white Gaussian noise (AWGN) channel. However, in a multipath fading channel, it is generally difficult to estimate the nonlinear distortion component at the receiver (Jiang & Wu, 2008). Thus, the iterative ML receiver based clipping approach proposed in (Rostamzadeh & Vakily, 2008) may suffer performance degradation due to in-band distortion in a multipath fading environment. Other major disadvantages of the iterative ML receiver based clipping approach include increased out-of-band radiation and a higher receiver complexity.

In (Zhang, Yuan, & Zhao, 2005), an amplitude threshold based method is proposed for PAPR reduction in WPM systems. In this method, the signal samples whose amplitudes are below a threshold T are set to zero, and the samples with amplitudes exceeding T are unaltered. Since setting the low amplitude samples to zero is a nonlinear process, this method also suffers from in-band distortion and introduces out-of-band power emissions. Another major disadvantage with the amplitude threshold based PAPR reduction scheme proposed in (Zhang, Yuan, & Zhao, 2005) is that it results in an increase in the average power of the modified signal. Although the PAPR is reduced due to an increased average power, this method will result in BER degradation when the transmitted signal is normalized back to its original signal power level (Han & Lee, 2005). Furthermore, the amplitude threshold based PAPR reduction scheme also requires HPAs with large linear operation regions (Jiang & Wu, 2008). Lastly, the criterion for choosing the threshold value T is not defined in (Zhang, Yuan, & Zhao, 2005), and hence, may depend on the characteristics of the HPA.

An alternative amplitude threshold based scheme called adaptive threshold companding transform is proposed in (Rostamzadeh, Vakily, & Moshfegh, 2008). Generally, the application of nonlinear companding transforms to multicarrier communication systems are very useful since these transforms yield good PAPR reduction capability with low implementation complexity (Jiang & Wu, 2008). In the adaptive threshold companding scheme proposed (Rostamzadeh, Vakily, & Moshfegh, 2008), signal samples with amplitudes higher than a threshold T are compressed at the transmitter via a nonlinear companding function; the signal samples with amplitudes below T are unaltered. To undo the nonlinear companding transform, received signal samples corresponding to signal samples that underwent compression at the transmitter are nonlinearly expanded at the receiver. The threshold value T is determined adaptively at the transmitter and sent to the receiver as side information. Specifically, T is determined adaptively to be a function of the median and the standard deviation of the signal. By compressing the signal samples with high amplitudes (i.e., amplitudes exceeding T), the adaptive threshold companding scheme achieves notable PAPR reductions in WPM systems. Results presented in (Rostamzadeh, Vakily, & Moshfegh, 2008) show that the adaptive threshold companding scheme yields a significantly enhanced symbol error rate performance over the clipping method in an AWGN channel. However, the authors in (Rostamzadeh, Vakily, & Moshfegh, 2008) do not provide performance results corresponding to the multipath fading environment. The received signal samples to be nonlinearly expanded are identified by comparing the received signal amplitudes to the threshold value T at the receiver. Although this approach works reasonably well in the AWGN channel, it may not be practical in a multipath fading channel due to imperfections associated with fading mitigation techniques such as non-ideal channel estimation, equalization, interference cancellation, etc. Another major disadvantage associated with the adaptive threshold companding scheme is that it is very sensitive to channel noise. For instance, the larger the amplitude compression at the transmitter, the higher the BER (or symbol error rate) at the receiver due to noise amplification. Furthermore, the adaptive threshold companding scheme suffers a slight loss in bandwidth efficiency due to side information being transmitted from the transmitter to the receiver. The application of this scheme will also result in additional performance loss if the side information is received in error.

3.3 Other methods

In (Gautier et al., 2008), PAPR reduction for WPM based multicarrier systems is studied using different pulse shapes based on the conventional Daubechies wavelet family. Simulation results presented in this work show that the employment of wavelet packets can yield notable PAPR reductions when the number of subchannels is low. However, to improve PAPR, the authors in (Gautier et al., 2008) increase the wavelet index of the conventional Daubechies basis functions which results in an increased modulation complexity.

In (Rostamzadeh & Vakily, 2008), two types of partial transmit sequences (PTS) methods are applied to reduce PAPR in WPM systems. The first method is a conventional PTS scheme where the input signal block is partitioned into Φ disjoint sub-blocks. Each of the Φ disjoint sub-blocks then undergoes inverse discrete wavelet packet transformations to produce Φ different output signals. Next, the transformed output signals are rotated by different phase factors b_ϕ $(\phi = 1, 2, \ldots, \Phi)$. The rotated and transformed output signals are lastly combined to form the transmitted signal. In the conventional PTS scheme, the phase factors $\{b_\phi\}$ are chosen such that the PAPR of the combined signal (i.e., the signal to be transmitted) is minimized. The second PTS method applied to WPM systems in (Rostamzadeh & Vakily, 2008) is the sub-optimal iterative flipping technique which was originally proposed in (Cimini & Sollenburger, 2000). In the iterative flipping technique, the phase factors $\{b_\phi\}$ are restricted to the values ± 1. The iterative flipping PTS scheme first starts off with phase factor initializations of $b_\phi = +1$ for $\forall\phi$ and a calculation of the corresponding PAPR (i.e., the PAPR corresponding to the case $b_\phi = +1$, $\forall\phi$). Next, the phase factor b_1 is flipped to -1, and the resulting PAPR value is calculated again. If the new PAPR is lower than the original PAPR, the phase factor $b_1 = -1$ is retained. Otherwise, the phase factor b_1 is reset to its original value of $+1$. This phase factor flipping procedure is then applied to the other phase factors $b_2, b_3, \ldots, b_\Phi$ to progressively reduce the PAPR. Furthermore, if a desired PAPR is attained after applying the phase factor flipping procedure to b_ϕ $(2 \le \phi < \Phi)$, the algorithm can be terminated in the middle to reduce the computational complexity associated with the iterative flipping PTS technique.

In general, the PTS based methods are considered important for their distortionless PAPR reduction capability in multi-carrier systems (Jiang & Wu, 2008; Han & Lee, 2005). The amount of PAPR reduction achieved by PTS based methods depends on the number Φ of disjoint sub-blocks and the number W of allowed phase factor values. However, the PAPR reduction achieved by PTS based methods come with an increased computational complexity. When applied to WPM systems, PTS based methods require Φ inverse discrete wavelet packet transformations at the transmitter. Moreover, the conventional PTS scheme incurs a high computational complexity in the search for the optimal phase factors. Another disadvantage associated with the PTS based methods is the loss in bandwidth efficiency due to the need to transmit side information about the phase factors from the transmitter to the receiver. The minimum number of side information bits required for the conventional PTS scheme and the iterative flipping PTS scheme are $\lceil \Phi \log_2(W) \rceil$ and Φ, respectively (Jiang & Wu, 2008; Han & Lee, 2005). It should also be noted that the PTS schemes will yield degraded system performance if the side-information bits are received in error at the receiver.

Another method considered for reducing the PAPR of WPM systems in (Rostamzadeh & Vakily, 2008) is the selective mapping (SLM) approach. In the SLM method, the input sequence is first multiplied by U different phase sequences to generate U alternative sequences. Then, the U alternative sequences are inverse wavelet packet transformed to produce U different output sequences. This is followed by a comparison of the PAPRs corresponding to the U output sequences. The output sequence with the lowest PAPR is lastly selected for transmission. To recover the original input sequence at the receiver, side information about the phase sequence that generated the output sequence with the lowest PAPR must be transmitted to the receiver. The PAPR reduction capability of the SLM method depends on the number U of phase sequences considered and the design of the phase sequences (Han & Lee, 2005). Similar to the PTS method, the SLM approach is distortionless and does not introduce spectral regrowth. The major disadvantages of the SLM method are its high implementation complexity and the bandwidth efficiency loss it incurs due to the requirement to transmit side information. When applied to WPM systems, the SLM method requires U inverse discrete wavelet packet transformations at the transmitter. Furthermore, a minimum of $\lceil \log_2(U) \rceil$ side-information bits are required to facilitate recovery of the original input sequence at the receiver (Jiang & Wu, 2008; Han & Lee, 2005). Similar to the PTS based methods, the SLM approach will also degrade system performance if the side-information bits are erroneously received at the receiver.

4. Orthogonal basis function design approach for PAPR reduction

In this section, we present a set of orthogonal basis functions for WPM-based multi-carrier systems that reduce the PAPR without the abovementioned disadvantages of previously proposed techniques. Given the WPM transmitter output signal $y(n)$, the PAPR is defined as

$$PAPR \triangleq \frac{max_n\{|y(n)|^2\}}{E\{|y(n)|^2\}}, \tag{11}$$

where $max_n\{\bullet\}$ represents the maximum value over all instances of time index n. The PAPR reduction method presented here is based on the derivation of an upper bound for the PAPR. With regards to (11), we showed in (Le, Muruganathan, & Sesay, 2008) that

$$E\{|y(n)|^2\} = \sigma_x^2, \tag{12}$$

where σ_x^2 is the average power of any one of the input data symbol streams $x_0(n),\ x_1(n),\ \ldots,\ x_{2^M-2}(n),\ x_{2^M-1}(n)$. In Section 4.1, we complete the derivation of the

PAPR upper bound by deriving an upper bound for the peak power $max_n\{|y(n)|^2\}$. The design criteria for the orthogonal basis functions that minimize the PAPR upper bound are then presented in Section 4.2.

4.1 Upper bound for PAPR

Let us first consider the derivation of an upper bound for the peak power $max_n\{|y(n)|^2\}$. Using the notation introduced in Figure 2, the WPM transmitter output signal $y(n)$ can be expressed as

$$y(n)=\sum_{k=0}^{2^M-1}\sum_{p} z_k(p)w_k(n-p), \tag{13}$$

where $z_k(n)$ is the up-sampled version of $x_k(n)$ which is defined as

$$z_k(n)=\begin{cases} x_k\left(\frac{n}{2^M}\right), & \text{if } \mathrm{mod}(n,2^M)=0, \\ 0, & \text{otherwise.}\end{cases} \tag{14}$$

Now, substituting (14) into (13), it can be shown that

$$y(n)=\sum_{k=0}^{2^M-1}\sum_{p} x_k(p)w_k(n-2^Mp). \tag{15}$$

We next apply the triangular inequality to (15) and obtain the following upper bound for $|y(n)|$:

$$|y(n)| \le \max_{n,k}\{|x_k(n)|\}\left[\sum_{k=0}^{2^M-1}\sum_{p}\left|w_k(n-2^Mp)\right|\right]. \tag{16}$$

In (16), $\max_{n,k}\{|x_k(n)|\}$ denotes the peak amplitude of the input data symbol stream $x_k(n)$ over all sub-channels (i.e., $\forall k$) and all instances of time index n. Hence, from (16), the peak value of $|y(n)|$ over all instances of n can be upper bounded as

$$\max_n\{|y(n)|\} \le \max_{n,k}\{|x_k(n)|\}\times\max_n\left\{\sum_{k=0}^{2^M-1}\sum_{p}\left|w_k(n-2^Mp)\right|\right\}, \tag{17}$$

where the notation $max_n\{\bullet\}$ is as defined in (11). It can be shown that in (17), equality holds if and only if the input data symbol streams in (15) (i.e., $x_k(p)$ for $\forall k$) satisfy the condition

$$x_k(p)=\max_{n,k}\{|x_k(n)|\}\times\mathrm{sign}\left\{w_k(n-2^Mp)\right\}\times e^{j\alpha}, \tag{18}$$

where α denotes an arbitrary phase value.

Now, using the result in (17), the upper bound for the peak power $max_n\{|y(n)|^2\}$ is attained as

$$\max_n \left\{ |y(n)|^2 \right\} \le \max_{n,k} \left\{ |x_k(n)|^2 \right\} \times \left[\max_n \left\{ \sum_{k=0}^{2^M-1} \sum_p \left| w_k\left(n - 2^M p\right) \right| \right\} \right]^2. \tag{19}$$

Lastly, substitution of (12) and (19) into (11) yields the PAPR upper bound as

$$PAPR \le PAPR_{UB} = \frac{\max_{n,k}\left\{ |x_k(n)|^2 \right\}}{\sigma_x^2} \left[\max_n \left\{ \sum_{k=0}^{2^M-1} \sum_p \left| w_k\left(n - 2^M p\right) \right| \right\} \right]^2. \tag{20}$$

4.2 Design criteria for PAPR minimizing orthogonal basis functions

Considering (20), we first note that the new PAPR upper bound is the product of two factors. The first factor

$$\frac{\max_{n,k}\left\{ |x_k(n)|^2 \right\}}{\sigma_x^2}$$

is only dependent on the input data symbol streams $x_0(n),\ x_1(n),\ \ldots,\ x_{2^M-2}(n),\ x_{2^M-1}(n)$. By virtue of (9)-(10), the second factor

$$\left[\max_n \left\{ \sum_{k=0}^{2^M-1} \sum_p \left| w_k\left(n - 2^M p\right) \right| \right\} \right]^2$$

is solely determined by the reversed QMF pair, $h(n)$ and $g(n)$. Furthermore, since $h(n)$ and $g(n)$ are closely related through (7)-(8) and (4), the abovementioned second factor can be expressed entirely in terms of $h(n)$. Recalling from (2)-(8) that the orthogonal basis functions are characterized by the time-reversed low-pass filter impulse response $h(n)$, we now strive to minimize the PAPR upper bound in (20) by minimizing the cost function

$$CF_M = \max_n \left\{ \sum_{k=0}^{2^M-1} \sum_p \left| w_k\left(n - 2^M p\right) \right| \right\} \tag{21}$$

by appropriately designing $h(n)$.

Firstly, let $H(\omega)$ denote the Fourier transform of the low-pass filter impulse response $h(n)$ with length $2N$. Given a set of orthogonal basis functions with regularity L $(1 \le L \le N)$, the magnitude response corresponding to $h(n)$ can be written as (Burrus, Gopinath, & Guo, 1998)

$$|H(\omega)|^2 = \left[\cos^2(\omega/2)\right]^L P\left(\sin^2(\omega/2)\right), \tag{22}$$

where

$$P\left(\sin^2(\omega/2)\right) = \sum_{\ell=0}^{L-1} \binom{L-1+\ell}{\ell} \left[\sin^2(\omega/2)\right]^\ell + \left[\sin^2(\omega/2)\right]^L R(\cos(\omega)). \tag{23}$$

In (23), $R(\cos(\omega))$ denotes an odd polynomial defined as

$$R(\cos(\omega)) = \begin{cases} 0, & if\ L = N, \\ \sum_{i=1}^{N-L} a_i (\cos(\omega))^{2i-1}, & if\ 1 \le L < N. \end{cases} \tag{24}$$

It should be noted that the first case (i.e., $L = N$) of (24) corresponds to the case of the conventional Daubechies basis functions. In the second case of (24) where $1 \le L < N$, the coefficients $\{a_i\}$ are chosen such that

$$P(\sin^2(\omega/2)) \ge 0, \tag{25}$$

for $0 \le \sin^2(\omega/2) \le 1$. Now, substituting (24) into (23) yields

$$P(\sin^2(\omega/2)) = \sum_{\ell=0}^{L-1} \binom{L-1+\ell}{\ell} \left[\sin^2(\omega/2)\right]^{\ell} + \left[\sin^2(\omega/2)\right]^{L} \sum_{i=1}^{N-L} a_i (\cos(\omega))^{2i-1}. \tag{26}$$

Since $0 \le \sin^2(\omega/2) \le 1$, we note that

$$\left[\sin^2(\omega/2)\right]^{\ell} \ge \left[\sin^2(\omega/2)\right]^{L} \tag{27}$$

for $\ell = 0, 1, ..., (L-1)$. Then, using (27), the first term on the right hand side of (26) can be lower bounded as

$$\sum_{\ell=0}^{L-1} \binom{L-1+\ell}{\ell} \left[\sin^2(\omega/2)\right]^{\ell} \ge \left[\sin^2(\omega/2)\right]^{L} \sum_{\ell=0}^{L-1} \binom{L-1+\ell}{\ell}. \tag{28}$$

Next, noting that $\left| a_i (\cos(\omega))^{2i-1} \right| \le |a_i|$, we have

$$a_i (\cos(\omega))^{2i-1} \ge -|a_i| \tag{29}$$

for $i = 1, 2, ..., (N-L)$. Using (29), the second term on the right hand side of (26) can be lower bounded as

$$\left[\sin^2(\omega/2)\right]^{L} \sum_{i=1}^{N-L} a_i (\cos(\omega))^{2i-1} \ge -\left[\sin^2(\omega/2)\right]^{L} \sum_{i=1}^{N-L} |a_i|. \tag{30}$$

Now, combining (28) and (30) with (26) yields

$$\begin{aligned} P(\sin^2(\omega/2)) &= \sum_{\ell=0}^{L-1} \binom{L-1+\ell}{\ell} \left[\sin^2(\omega/2)\right]^{\ell} + \left[\sin^2(\omega/2)\right]^{L} \sum_{i=1}^{N-L} a_i (\cos(\omega))^{2i-1} \\ &\ge \left[\sin^2(\omega/2)\right]^{L} \sum_{\ell=0}^{L-1} \binom{L-1+\ell}{\ell} - \left[\sin^2(\omega/2)\right]^{L} \sum_{i=1}^{N-L} |a_i|. \end{aligned} \tag{31}$$

Recalling the constraint $P\left(\sin^2(\omega/2)\right) \geq 0$ from (25), we can further lower bound (31) as

$$P\left(\sin^2(\omega/2)\right) \geq \left[\sin^2(\omega/2)\right]^L \sum_{\ell=0}^{L-1} \binom{L-1+\ell}{\ell} - \left[\sin^2(\omega/2)\right]^L \sum_{i=1}^{N-L} |a_i| \geq 0. \tag{32}$$

From the last inequality of (32), we have

$$\sum_{i=1}^{N-L} |a_i| \leq \sum_{\ell=0}^{L-1} \binom{L-1+k}{k}. \tag{33}$$

Next, using (33), the range of values for coefficient a_1 is chosen as $-A_1 \leq a_1 \leq A_1$, where

$$A_1 = \sum_{\ell=0}^{L-1} \binom{L-1+k}{k}. \tag{34}$$

Likewise, the ranges of the remaining coefficients a_i $(i = 2, 3, \ldots, N-L)$ are set as $-A_i \leq a_i \leq A_i$, wherein

$$A_i = \sum_{\ell=0}^{L-1} \binom{L-1+k}{k} - \sum_{k=1}^{i-1} |a_k|. \tag{35}$$

Then, each coefficient a_i is searched within its respective range in predefined intervals. For each given set of coefficients $\{a_i\}$, the associated cost function CF_M is computed using (4), (7)-(10), and (21). Lastly, the time-reversed low-pass filter impulse response $h(n)$ that minimizes the PAPR upper bound of (20) is determined by choosing the set of coefficients $\{a_i\}$ that minimizes the cost function CF_M of (21). It should be noted that the cost function CF_M of (21) is independent of the input data symbol streams. Hence, using the presented method the impulse response $h(n)$ can first be designed offline and then be employed even in real-time applications.

5. Simulation results and discussions

In this section, we present simulation results to evaluate the performance of the PAPR reduction method of Section 4. Throughout this section, we compare the performance of the PAPR minimizing orthogonal basis functions (which are determined by the time-reversed impulse response $h(n)$ designed in Section 4.2) to the performance of the conventional Daubechies basis functions. Additionally, we also make performance comparisons with multi-carrier systems employing OFDM. Throughout the simulations, the number of sub-channels in all three multi-carrier systems is set to 64 (i.e., $M = 6$), and the channel bandwidth is assumed to be 22 MHz. Furthermore, the input data symbol streams $x_0(n),\ x_1(n),\ \ldots,\ x_{2^M-2}(n),\ x_{2^M-1}(n)$ are drawn from a 4-QAM symbol constellation. In the cases of the proposed orthogonal basis functions and conventional Daubechies basis functions, we set $N = 6$. Furthermore, for the PAPR minimizing orthogonal basis functions, the regularity L is chosen to be 3 (recall that for the conventional Daubechies basis functions $L = N = 6$).

The first performance metric we consider is the complementary cumulative distribution function (CCDF) which is defined as

$$CCDF(PAPR_0) = \Pr\{PAPR > PAPR_0\}.$$

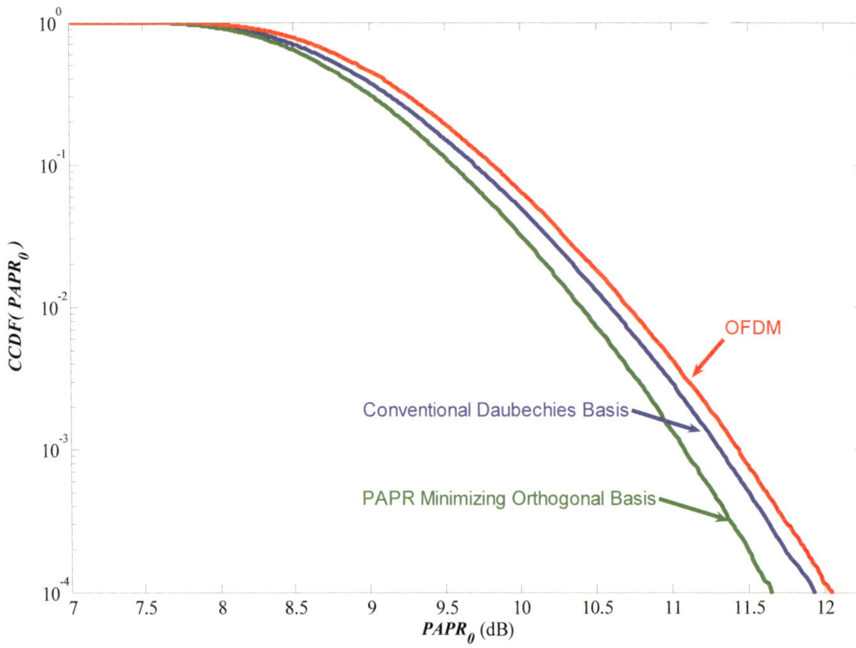

Fig. 3. CCDF performance comparison

The CCDF performance comparison between the three multi-carrier systems is presented in Figure 3. From Figure 3, we note that the PAPR minimizing orthogonal basis functions achieve PAPR reductions of 0.3 dB over the conventional Daubechies basis functions and 0.4 dB over OFDM. It should be emphasized that these performance gains are attained with no need for side information to be sent to the receiver, no distortion, and no loss in bandwidth efficiency. Furthermore, if additional PAPR reduction is desired, the PAPR minimizing orthogonal basis functions can also be combined with some of the PAPR reduction methods surveyed in (Han & Lee, 2005) and (Jiang & Wu, 2008).

Next, we compare the bit error rate (BER) performances of the three multi-carrier systems under consideration. In the BER comparisons, we utilize the Rapp's model to characterize the high power amplifier with the non-linear characteristic parameter chosen as 2 and the saturation amplitude set to 3.75 (van Nee & Prasad, 2000). Furthermore, a 10-path channel with an exponentially decaying power delay profile and a root mean square delay spread of 50 ns is assumed. The BER results as a function of the normalized signal-to-noise ratio (SNR) are shown in Figure 4. From the figure, it is noted that at a target BER of 3×10^{-4}, the PAPR minimizing orthogonal basis functions achieve an SNR gain of 2.9 dB over the conventional Daubechies basis functions. The corresponding SNR gain over OFDM is 6.5 dB.

We next quantify the out-of-band power emissions associated with the PAPR minimizing orthogonal basis functions, the conventional Daubechies basis functions, and the OFDM system. This is done by analyzing the adjacent channel power ratio (ACPR)-CCDF corresponding to the three different schemes. The ACPR-CCDF is defined as

$$CCDF\left(ACPR_0\right)=\Pr\left\{ACPR>ACPR_0\right\}.$$

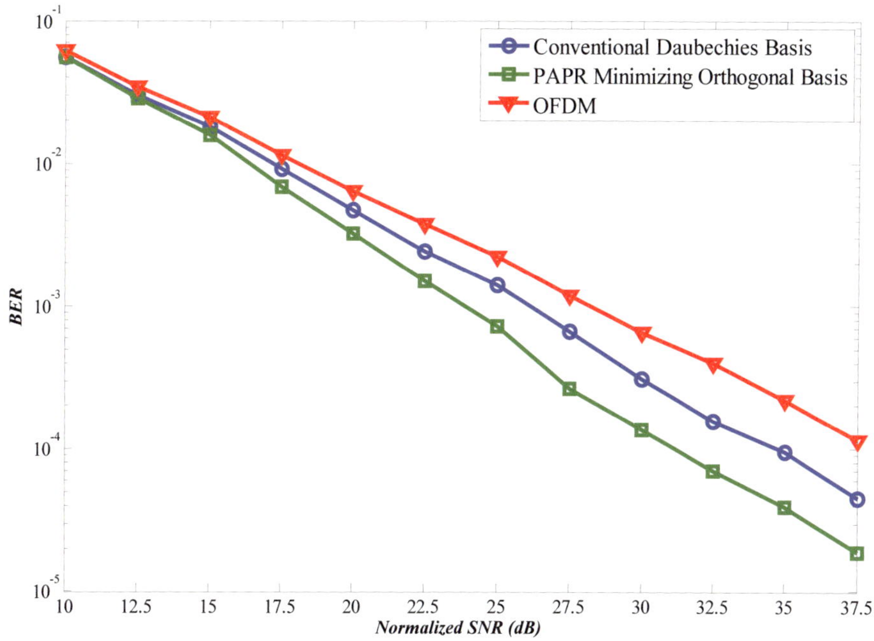

Fig. 4. BER performance comparison

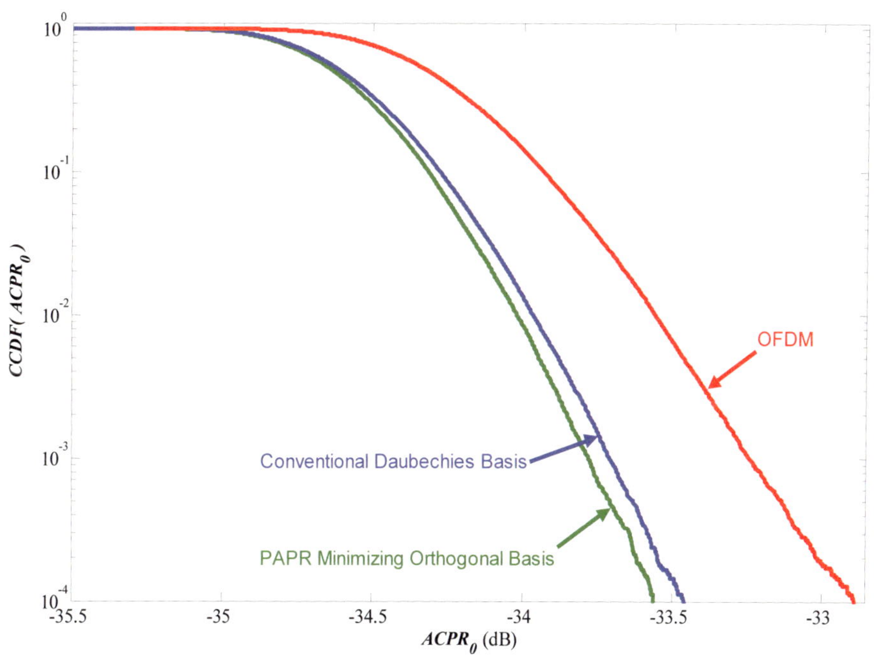

Fig. 5. ACPR-CCDF performance comparison

Method	PAPR minimizing orthogonal basis functions	PTS and SLM	Using different Daubechies pulse shapes	Adaptive threshold companding	Lower amplitude threshold	Clipping	Tree pruning
PAPR Reduction Capability	Moderate	High	High	High	Moderate	Variable	High
Implementation Complexity	Low (since the PAPR minimizing orthogonal basis functions are designed offline)	High	High (due to an increase in the wavelet index)	Low	Low	Low	High
Spectral Regrowth	No	No	No	Yes	Yes	Yes	No
BER Degradation in AWGN	No	No	No	Yes	Yes	Yes	No
Applicability in Multipath Fading	Performs well	Performs well	May not be feasible due to a high complexity	May not be practical in a multipath environment	Not practical in a multipath environment	Yields poor performance	Reduced multipath resilience
Distortionless	Yes	Yes	Yes	Yes (for AWGN)	No	No	Yes
Side Information	Not required	Required	Not required	Required	Not required	Not required	Required

Table 1. Comparison of different PAPR reduction methods for WPM systems

Figure 5 shows the ACPR-CCDF results which are generated using the same power amplifier model used to generate the results of Figure 4. For a CCDF probability of 10^{-4}, we note from Figure 5 that the PAPR minimizing orthogonal basis functions achieve an ACPR reduction of approximately 0.67 dB over the OFDM system. It should be noted that when compared to OFDM, the PAPR minimizing orthogonal basis functions reduce the ACPR by reducing the out-of-band power emissions introduced by the non-linear power amplifier. Furthermore, it is also noted from Figure 5 that the PAPR minimizing orthogonal basis functions yield a 0.1 dB ACPR reduction over the conventional Daubechies basis functions. This ACPR reduction is achieved mainly due to the superior PAPR reduction performance associated with the PAPR minimizing orthogonal basis functions when compared to the conventional Daubechies basis functions.

Lastly, in Table 1, we provide a qualitative comparison of the PAPR minimizing orthogonal basis functions presented in Section 4 with the other PAPR reduction techniques overviewed in Section 3. It should be noted that although the PAPR minimizing orthogonal basis functions yield a marginal PAPR reduction, this PAPR reduction is achieved without the disadvantages associated with the other techniques. Moreover, if a high PAPR reduction is desired, the PAPR reducing orthogonal basis functions also offer the flexibility to be combined with other PAPR reduction methods such as PTS, SLM, etc.

6. Conclusions

In this chapter, a PAPR reduction method for WPM multicarrier systems based on the orthogonal basis function design approach is presented. Firstly, we provide an overview of the WPM system and survey some PAPR reduction methods already proposed in the literature for WPM systems. Next, we derive a new PAPR upper bound that applies the triangular inequality to the WPM transmitter output signal. Using the new PAPR upper bound derived, design criteria for PAPR minimizing orthogonal basis functions are next formulated. The performance of the PAPR minimizing orthogonal basis functions is compared to those of the conventional Daubechies basis functions and OFDM. These comparisons show that the PAPR minimizing orthogonal basis functions outperform both the conventional Daubechies basis functions and OFDM. Furthermore, we also provide a qualitative comparison between the PAPR minimizing orthogonal basis functions and the existing PAPR reduction techniques. Through this comparison, it is shown that the PAPR minimizing orthogonal basis functions reduce the PAPR without the disadvantages associated with the other techniques.

7. References

Akho-Zhieh, M. M. & Ugweje, O. C. (2008). Diversity performance of a wavelet-packet-based multicarrier multicode CDMA communication system. *IEEE Transactions on Vehicular Technology*, vol. 57, no. 2, pp. 787-797.

Baro, M. & Ilow, J. (2007a). PAPR reduction in wavelet packet modulation using tree pruning. *IEEE 65th Vehicular Technology Conference (VTC-Spring)*, pp. 1756-1760.

Baro, M. & Ilow, J. (2007b). Improved PAPR reduction for wavelet packet modulation using multi-pass tree pruning. *IEEE 18th International Symposium on Personal, Indoor and Mobile Radio Communications (PIMRC)*, pp. 1-5.

Burrus, C. S., Gopinath, R. A., & Guo, H. (1998). *Introduction to Wavelets and Wavelet Transforms: A Primer,* Prentice Hall.

Cimini, L. J. & Sollenburger, N. R. (2000). Peak-to-average power ratio reduction of an OFDM signal using partial transmit sequences. *IEEE Communications Letters,* vol. 4, no. 3, pp. 86-88.

Daly, D., Heneghan, C., Fagan, A., & Vetterli, M. (2002). Optimal wavelet packet modulation under finite complexity constraint. *IEEE Int'l Conf. Acoustics, Speech, and Signal Processing (ICASSP),* vol. 3, pp. 2789-2792.

Daubechies, I. (1992). *Ten lectures on wavelets,* SIAM.

Gautier, M., Lereau, C., Arndt, M., & Lienard, J. (2008). PAPR analysis in wavelet packet modulation. *IEEE International Symposium on Communications, Control, and Signal Processing (ISCCSP),* pp. 799-803.

Han, S. H. & Lee, J. H. (2005). An overview of peak-to-average power ratio reduction techniques for multicarrier transmission. *IEEE Wireless Communications,* vol. 12, no. 2, pp. 56-65.

Jiang, T. & Wu, Y. (2008). An overview: peak-to-average power ratio reduction techniques for OFDM signals. *IEEE Transactions on Broadcasting,* vol. 54, no. 2, pp. 257-268.

Lakshmanan, M. K. & Nikookar, H. (2006). A review of wavelets for digital communication. *Wireless Personal Communications,* vol. 37, no. 3-4, pp. 387-420.

Lakshmanan, M. K., Budiarjo, I., & Nikookar, H. (2007). Maximally frequency selective wavelet packets based multi-carrier modulation scheme for cognitive radio systems. *IEEE Global Telecommunications Conference,* pp. 4185-4189.

Lakshmanan, M. K., Budiarjo, I., & Nikookar, H. (2008). Wavelet packet multi-carrier modulation MIMO based cognitive radio systems with VBLAST receiver architecture. *IEEE Wireless Communications and Networking Conference,* pp. 705-710.

Le, N. T., Muruganathan, S. D., & Sesay, A. B. (2008). Peak-to-average power ratio reduction for wavelet packet modulation schemes via basis function design. *IEEE Vehicular Technology Conference (VTC-Fall),* pp. 1-5.

Lindsey, A. R. (1995). Generalized orthogonal multiplexed communications via wavelet packet bases. *Ph.D. Dissertation,* Ohio University.

Rostamzadeh, M. & Vakily, V. T. (2008). PAPR reduction in wavelet packet modulation. *IEEE 5th International Multi-Conference on Systems, Signals and Devices (SSD),* pp. 1-6.

Rostamzadeh, M., Vakily, V. T., & Moshfegh, M. (2008). PAPR reduction in WPDM and OFDM systems using an adaptive threshold companding scheme. *IEEE 5th International Multi-Conference on Systems, Signals and Devices (SSD),* pp. 1-6.

Sandberg, S. D. & Tzannes, M. A. (1995). Overlapped discrete multitone modulation for high speed copper wire communications. *IEEE Journal of Selected Areas in Communications,* vol. 13, no. 9, pp. 1571-1585.

van Nee, R. & Prasad, R. (2000). *OFDM for Wireless Multimedia Communications,* Artech House.

Zhang, H., Yuan, D., & Zhao, F. (2005). Research of PAPR reduction method in multicarrier modulation system. *IEEE International Conference on Communications, Circuits and Systems (ICCCAS),* vol. 1, pp. 91-94.

19

Outage Performance and Symbol Error Rate Analysis of L-Branch Maximal-Ratio Combiner for κ-μ and η-μ Fading

Mirza Milišić, Mirza Hamza and Mesud Hadžialić
Faculty of Electrical Engineering, University of Sarajevo
Bosnia and Herzegovina

1. Introduction

This chapter treats performances of Maximal-Ratio Combiner (MRC) in presence of two general fading distributions, the κ-μ distribution and the η-μ distribution (Yacoub, 2007.). Namely, performances of Maximal-Ratio Combiner in fading channels have been of interest for a long time, which can be seen by a numerous publications concerning this topic. Most of these papers are concerned by Rayleigh, Nakagami-*m*, Hoyt (Nakagami-*q*), Rice (Nakagami-*n*) and Weibull fading (Kim et al., 2003), (Annamalai et al., 2002), (da Costa et al., 2005), (Fraidenraich et al., a, 2005), and (Fraidenraich et al., b, 2005). Beside MRC, performances of selection combining, equal-gain combining, hybrid combining and switched combining in fading channels have also been studied. Most of the papers treating diversity combining have examined only dual-branch combining because of the inability to obtain closed-form expressions for evaluated parameters of diversity system. Scenarios of correlated fading in combiner's branches have also been examined in numerous papers. Nevertheless, depending on system used and combiner's implementation, one must take care of resources available at the receiver, such as: space, frequency, complexity, etc. Moreover, fading statistic doesn't necessary have to be the same in each branch, e.g. probability density function (PDF) can be the same, but with different parameters (Nakagami-*m* fading in *i*-th and *j*-th branches, with $m_i \neq m_j$), or probability density functions (PDF) in different branches are different (Nakagami-*m* fading in *i*-th branch, and Rice fading in *j*-th branch). This chapter treats MRC outage performances in presence of κ-μ and η-μ distributed fading (Milišić et al., a, 2008), (Milišić et al., b, 2008), (Milišić et al., a, 2009) and (Milišić et al., b, 2009). This types of fading have been chosen because they include, as special cases, Nakagami-*m* and Nakagami-*n* (Rice) fading, and their entire special cases as well (e.g. Rayleigh and one-sided Gaussian fading). It will be shown that the sum of κ-μ squares is a κ-μ square as well (but with different parameters), which is an ideal choice for MRC analysis. This also applies to η-μ distribution. Throughout this chapter probability of outage and average symbol error rate, at the L-branch Maximal-Ratio Combiner's output, will be analyzed. Chapter will be organized as follows.

In the first part of the chapter we will present κ-μ and η-μ distributions, their importance, physical models, derivation of the probability density function, and relationships to other commonly used distributions. Namely, these distributions are fully characterized in terms of

measurable physical parameters. The κ-μ distribution includes Rice (Nakagami-n), Nakagami-m, Rayleigh, and One-Sided Gaussian distributions as special cases. The η-μ distribution includes the Hoyt (Nakagami-q), Nakagami-m, Rayleigh, and One-Sided Gaussian distributions as special cases. In particular, κ-μ distribution better suites line-of-sight scenarios, whereas η-μ distribution gives better results for non-line-of-sight scenarios.
Second part of this chapter will treat L-branch Maximal-Ratio Combiner and it's operational characteristics. We treat Maximal-Ratio Combiner because it has been shown that MRC receiver is the optimal multichannel receiver, regardless of fading statistics in various diversity branches since it results in a ML receiver. In this part of the chapter we will use the same framework used for derivation of κ-μ and η-μ probability density functions to derive probability density functions at combiner's output for κ-μ and η-μ fading. Derived probability density function will be used to obtain outage probability at combiner's output.
In third part of the chapter analysis of symbol error rate, at combiner's output, will be conduced. This analysis will be carried out for coherent and non-coherent detection. Although coherent detection results in smaller error probability than corresponding non-coherent detection for the same average signal-to-noise ratio, sometimes it is suitable to perform non-coherent detection depending on receiver structure complexity. In this part of the chapter we will derive and analyze average symbol error probability for κ-μ and η-μ fading at combiner's output, based upon two generic expressions for symbol error probability for coherent and non-coherent detection types for various modulation techniques.
In fourth part we will discuss Maximal-Ratio Combiner's performances obtained by Monte Carlo simulations. Theoretical expressions for outage probability and average symbol error rate probability for κ-μ and η-μ fading will be compared to results of the simulations. We will also draw some conclusions, and some suggestions for future work that needs to be done in this field of engineering.

2. The κ-μ distribution and the η-μ distribution

2.1 The κ-μ distribution

The κ-μ distribution is a general fading distribution that can be used to better represent the small-scale variations of the fading signal in a Line-of-Sight (LoS) conditions.The fading model for the κ-μ distribution considers a signal composed of clusters of multipath waves, propagating in a nonhomogenous environment. Within single cluster, the phases of the scattered waves are random and have similar delay times, with delay-time spreads of different clusters being relatively large. It is assumed that the clusters of multipath waves have scattered waves with identical powers, and that each cluster has a dominant component with arbitrary power. Given the physical model for the κ-μ distribution, the envelope R, and instantaneous power P_R can be written in terms of the in-phase and quadrature components of the fading signal as:

$$R^2 = P_R = \sum_{i=1}^{n}(X_i + p_i)^2 + \sum_{i=1}^{n}(Y_i + q_i)^2 \tag{1}$$

, where X_i and Y_i are mutually independent Gaussian processes with $\overline{X_i} = \overline{Y_i} = 0$ and $\overline{X_i^2} = \overline{Y_i^2} = \sigma^2$. p_i and q_i are respectively the mean values of the in-phase and quadrature components of the multipath waves of cluster i, and n is the number of clusters of multipath.

Total power of the i-th cluster is $P_{R,i} = R_i^2 = (X_i + p_i)^2 + (Y_i + q_i)^2$. Since $P_{R,i}$ equals to the sum of two non-central Gaussian random variables (RVs), its Moment Generating Function (MGF), in accordance to (Abramowitz and Stegun, 1972, eq. 29.3.81), yields in:

$$M_{P_{R,i}}(s) = \frac{\exp\left(\dfrac{d_i^2 \cdot s}{1 - 2\sigma^2 \cdot s}\right)}{1 - 2\sigma^2 \cdot s} \tag{2}$$

, where $d_i^2 = p_i^2 + q_i^2$, and s is the complex frequency. Knowing that the $P_{R,i}$, $i=1,2,\ldots,n$, are independent RVs, the MGF of the $f_{P_R}(P_R)$, where $P_R = \sum_{i=1}^{n} P_{R,i}$, is found to be:

$$M_{P_R}(s) = \prod_{i=1}^{n} M_{P_{R,i}}(s) = \frac{\exp\left(\dfrac{sd^2}{1 - 2s\sigma^2}\right)}{\left(1 - 2s\sigma^2\right)^n} \tag{3}$$

, where $d^2 = \sum_{i=1}^{n} d_i^2$. The inverse of (3) is given by (Abramowitz and Stegun, 1972, eq. 29.3.81):

$$f_{P_R}(P_R) = \frac{1}{2\sigma^2} \cdot \left(\frac{P_R}{d^2}\right)^{\frac{n-1}{2}} \cdot \exp\left(-\frac{P_R + d^2}{2\sigma^2}\right) \cdot I_{n-1}\left(\frac{d\sqrt{P_R}}{\sigma^2}\right) \tag{4}$$

It can be seen that $\Omega = \overline{R^2} = \overline{P_R} = 2n\sigma^2 + d^2$, and $\overline{R^4} = \overline{P_R^{\,2}} = 4n\sigma^4 + 4\sigma^2 d^2 + \left(2n\sigma^2 + d^2\right)^2$. We define $\kappa = \dfrac{d^2}{2n\sigma^2}$ as the ratio between the total power of the dominant components and the total power of the scattered waves. Therefore we obtain:

$$\frac{\left(\overline{R^2}\right)^2}{\overline{R^4} - \left(\overline{R^2}\right)^2} = \frac{\left(\overline{\gamma}\right)^2}{\overline{P_R^2} - \left(\overline{P_R}\right)^2} = n \cdot \frac{(1+\kappa)^2}{(1+2\kappa)}. \tag{5}$$

From (5), note that n may be expressed in terms of physical parameters, such as mean-squared value of the power, the variance of the power, and the ratio of the total power of the dominant components and the total power of the scattered waves. Note also, that whereas these physical parameters are of a continuous nature, n is of a discrete nature. It is plausible to presume that if these parameters are to be obtained by field measurements, their ratios, as defined in (5), will certainly lead to figures that may depart from the exact n. Several reasons exist for this. One of them, and probably the most meaningful, is that although the model proposed here is general, it is in fact an approximate solution to the so-called random phase problem, as are all the other well-known fading models approximate solutions to the random phase problem. The limitation of the model can be made less stringent by defining μ to be:

$$\mu = \frac{\left(\overline{R^2}\right)^2}{\overline{R^4} - \left(\overline{R^2}\right)^2} \cdot \frac{(1+2\kappa)}{(1+\kappa)^2} = \frac{\left(\overline{\gamma}\right)^2}{\overline{P_R^2} - \left(\overline{P_R}\right)^2} \cdot \frac{(1+2\kappa)}{(1+\kappa)^2} \tag{6}$$

with μ being the real extension of *n*. Non-integer values of the parameter μ may account for:
- non-Gaussian nature of the in-phase and quadrature components of each cluster of the fading signal,
- non-zero correlation among the clusters of multipath components,
- non-zero correlation between in-phase and quadrature components within each cluster, etc.

Non-integer values of clusters have been found in practice, and are extensively reported in the literature, e.g. (Asplund et al., 2002.). Using the definitions for parameters κ and μ, and the considerations as given above, the κ-μ power PDF can be written from (4) as:

$$f_{P_R}(P_R) = \frac{\mu(1+\kappa)^{\frac{\mu+1}{2}}}{\kappa^{\frac{\mu-1}{2}} \cdot \exp(\mu\kappa) \cdot \Omega^{\frac{\mu+1}{2}}} \cdot P_R^{\frac{\mu-1}{2}} \cdot \exp\left[-\frac{\mu(1+\kappa)P_R}{\Omega}\right] \cdot I_{\mu-1}\left[2\mu\sqrt{\frac{\kappa(1+\kappa)P_R}{\Omega}}\right]. \tag{7}$$

Therefore, the κ-μ envelope PDF can be obtained from (7) as:

$$f_R(R) = \frac{2\mu \cdot (1+\kappa)^{\frac{\mu+1}{2}} \cdot R^{\mu}}{\kappa^{\frac{\mu-1}{2}} \cdot \exp(\mu\kappa) \cdot \Omega^{\frac{\mu+1}{2}}} \cdot \exp\left[-\frac{\mu \cdot (1+\kappa) \cdot R^2}{\Omega}\right] \cdot I_{\mu-1}\left[2\mu \cdot \sqrt{\frac{\kappa \cdot (1+\kappa)}{\Omega}} \cdot R\right] \tag{8}$$

Instantaneous and average signal-to-noise ratio (SNR) are given by: $\gamma = \frac{P_R}{N_0} = \frac{R^2}{N_0}$, $\bar{\gamma} = \frac{\overline{P_R}}{N_0} = \frac{\Omega}{N_0}$, and therefore the κ-μ SNR PDF can be obtained from (8) as:

$$f_\gamma(\gamma) = \frac{\mu \cdot (1+\kappa)^{\frac{\mu+1}{2}} \cdot \gamma^{\frac{\mu-1}{2}}}{\kappa^{\frac{\mu-1}{2}} \cdot \exp(\mu\kappa) \cdot \left(\bar{\gamma}\right)^{\frac{\mu+1}{2}}} \cdot \exp\left[-\frac{\mu \cdot (1+\kappa) \cdot \gamma}{\bar{\gamma}}\right] \cdot I_{\mu-1}\left[2\mu \cdot \sqrt{\frac{\kappa \cdot (1+\kappa) \cdot \gamma}{\bar{\gamma}}}\right] \tag{9}$$

N_0 represents single-sided power spectral density of additive white Gaussian nosie. From (9) we can derive Cumulative Distribution Function (CDF) of instantaneous SNR as:

$$F_\gamma(\gamma) = \int_0^\gamma f_\gamma(t)dt = 1 - Q_\mu\left[\sqrt{2\kappa\mu}, \sqrt{\frac{2(\kappa+1)\mu\gamma}{\Omega}}\right] \tag{10}$$

, where Q stands for generalized Marcum-Q function defined in (Marcum, 1947.) as: $Q_\nu(a,b) = \frac{1}{a^{\nu-1}} \int_b^\infty x^\nu \cdot \exp\left(-\frac{x^2+a^2}{2}\right) \cdot I_{\nu-1}(ax) \cdot dx$. Now we can obtain closed-form expression for the n-th order moment of RV γ as:

$$\overline{\gamma^n} = \frac{\Gamma(\mu+n)}{\Gamma(\mu)\exp(\kappa\mu)} \cdot \left(\frac{\bar{\gamma}}{(1+\kappa)\mu}\right)^n \cdot {}_1F_1(\mu+n;\mu;\kappa\mu) \tag{11}$$

, where Γ(•) stands for gamma function (Abramowitz and Stegun, 1972, eq. 6.1.1), and ${}_1F_1(\bullet;\bullet;\bullet)$ represents confluent hypergeometric function (Abramowitz and Stegun, 1972, eq. 13.1.2).

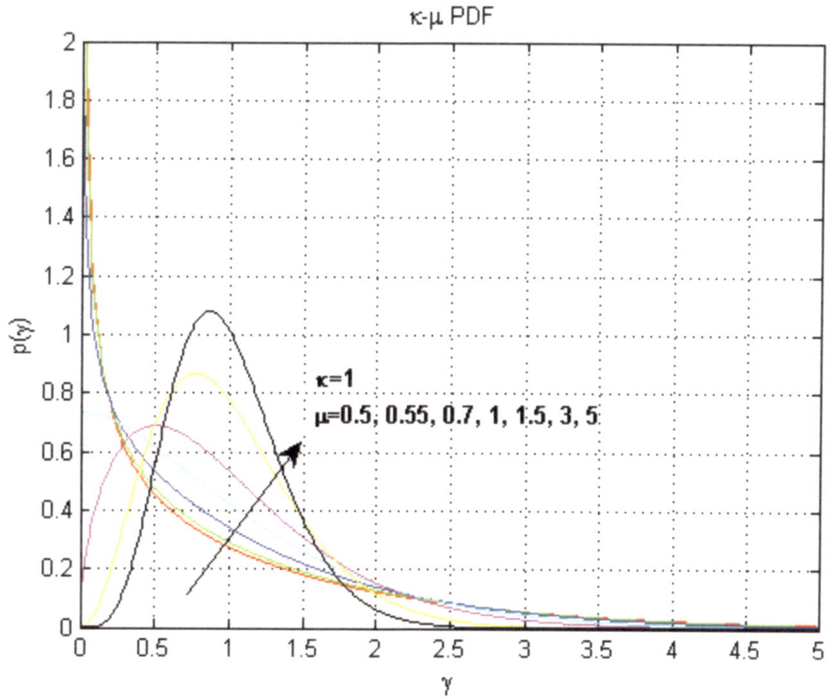

Fig. 1. PDF of SNR for κ=1 and various values of μ

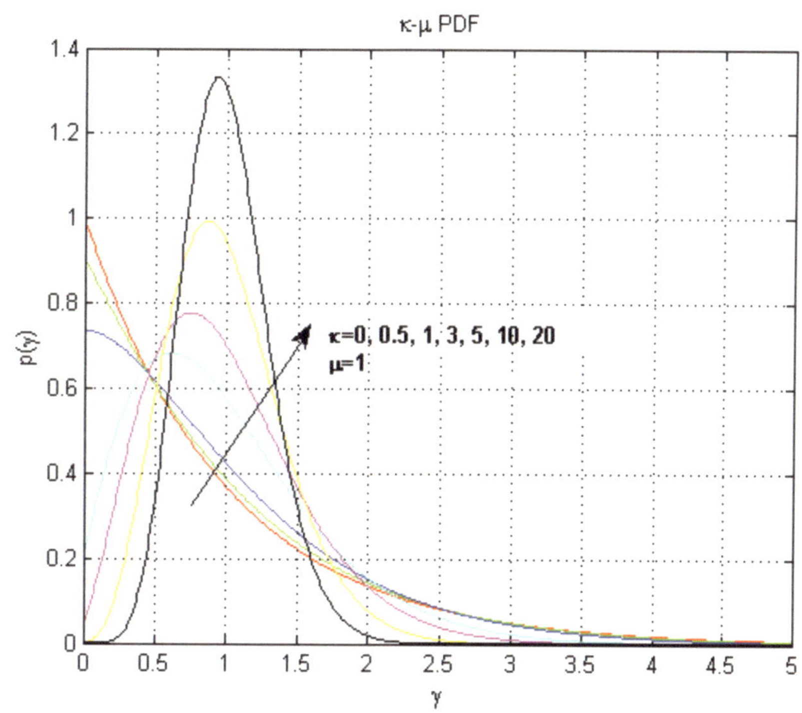

Fig. 2. PDF of SNR for μ=1 and various values of κ

2.2 The η-μ distribution

The η-μ distribution is a general fading distribution that can be used to better represent the small-scale variations of the fading signal in a No-Line-of-Sight (NLoS) conditions. The fading model for the η-μ distribution considers a signal composed of clusters of multipath waves, propagating in a nonhomogenous environment. Within single cluster, the phases of the scattered waves are random and have similar delay times, with delay-time spreads of different clusters being relatively large. The in-phase and quadrature components of the fading signal within each cluster are assumed to be independent from each other, and to have different powers. Envelope R, and instantaneous power P_R can be written in terms of the in-phase and quadrature components of the fading signal as:

$$R^2 = P_R = \sum_{i=1}^{n}\left(X_i^2 + Y_i^2\right) \tag{12}$$

, where X_i and Y_i are mutually independent Gaussian processes with $\overline{X_i} = \overline{Y_i} = 0$, $\overline{X_i^2} = \sigma_X^2$, $\overline{Y_i^2} = \sigma_Y^2$, and n is the number of clusters of multipath. Total power of the i-th cluster is $P_{R,i} = R_i^2 = X_i{}^2 + Y_i{}^2$. Since $P_{R,i}$ equals to the sum of two central Gaussian RVs with non-identical variances, its MGF yields in:

$$M_{P_{R,i}}(s) == \frac{\dfrac{\sqrt{h}}{\left(1+\eta^{-1}\right)\sigma_X^2}}{\sqrt{\left(s - \dfrac{h}{\left(1+\eta^{-1}\right)\sigma_X^2}\right)^2 - \left(\dfrac{H}{\left(1+\eta^{-1}\right)\sigma_X^2}\right)^2}} \tag{13}$$

, where $\eta = \dfrac{\sigma_X^2}{\sigma_Y^2}$ is the scattered-wave power ratio between the in-phase and quadrature components of each cluster of multipath, $h = \dfrac{2+\eta^{-1}+\eta}{4}$, $H = \dfrac{\eta^{-1}-\eta}{4}$. Following the same procedure as in the case of κ-μ distribution, the MGF of the $f_{P_R}(P_R)$, where $P_R = \sum_{i=1}^{n} P_{R,i}$, is found to be:

$$M_{P_R}(s) = \prod_{i=1}^{n} M_{P_{R,i}}(s) = \left[\frac{\dfrac{\sqrt{h}}{\left(1+\eta^{-1}\right)\sigma_X^2}}{\sqrt{\left(s + \dfrac{h}{\left(1+\eta^{-1}\right)\sigma_X^2}\right)^2 - \left(\dfrac{H}{\left(1+\eta^{-1}\right)\sigma_X^2}\right)^2}}\right]^n \tag{14}$$

The inverse of (14) is given by (Abramowitz and Stegun, 1972, eq. 29.3.60):

$$f_{P_R}(P_R) = \frac{\sqrt{\pi}\, n^{\frac{n+1}{2}} h^{\frac{n}{2}}}{(2H)^{\frac{n-1}{2}}\, \Gamma\!\left(\dfrac{n}{2}\right)} \cdot \frac{\left(P_R\right)^{\frac{n-1}{2}}}{\left(n\left(1+\eta^{-1}\right)\sigma_X^2\right)^{\frac{n+1}{2}}} \cdot \exp\left(-\frac{h \cdot P_R}{\left(1+\eta^{-1}\right)\sigma_X^2}\right) \cdot I_{\frac{n-1}{2}}\left(\frac{H \cdot P_R}{\left(1+\eta^{-1}\right)\sigma_X^2}\right) \tag{15}$$

, where $I_n(\bullet)$ stands for n-th order modified Bessel function of the first kind. It can be seen that $\Omega = \overline{R^2} = \overline{P_R} = n\left(1+\eta^{-1}\right)\sigma_X^2 = n\left(1+\eta\right)\sigma_Y^2$, and $\overline{R^4} = \overline{P_R^{\,2}} = \left[2n\left(1+\eta^2\right)+n^2\left(1+\eta^2\right)\right]\cdot\sigma_Y^4$. Thus,

$$\frac{\left(\overline{R^2}\right)^2}{\overline{R^4}-\left(\overline{R^2}\right)^2} = \frac{\left(\overline{P_R}\right)^2}{\overline{P_R^{\,2}}-\left(\overline{P_R}\right)^2} = \frac{n}{2}\cdot\frac{\left(1+\eta\right)^2}{\left(1+\eta^2\right)} \tag{16}$$

From (16), note that $n/2$ may be expressed in terms of physical parameters, such as mean-squared value of the power, the variance of the power, and the ratio of the total power of the dominant components and the total power of the scattered waves. Note also, that whereas these physical parameters are of a continuous nature, $n/2$ is of a discrete nature (integer multiple of $1/2$). It is plausible to presume that if these parameters are to be obtained by field measurements, their ratios, as defined in (16), will certainly lead to figures that may depart from the exact $n/2$. Several reasons exist for this. One of them, and probably the most meaningful, is that although the model proposed here is general, it is in fact an approximate solution to the so-called random phase problem, as are all the other well-known fading models approximate solutions to the random phase problem. The limitation of the model can be made less stringent by defining μ to be:

$$\mu = \frac{\left(\overline{R^2}\right)^2}{\overline{R^4}-\left(\overline{R^2}\right)^2}\cdot\frac{\left(1+\eta^2\right)}{\left(1+\eta\right)^2} = \frac{\left(\overline{P_R}\right)^2}{\overline{P_R^{\,2}}-\left(\overline{P_R}\right)^2}\cdot\frac{\left(1+\eta^2\right)}{\left(1+\eta\right)^2} = \frac{\left(\overline{P_R}\right)^2}{2\cdot\left(\overline{P_R^{\,2}}-\left(\overline{P_R}\right)^2\right)}\cdot\left(1+\left(\frac{H}{h}\right)^2\right) \tag{17}$$

, with μ being the real extension of $\frac{n}{2}$. Values of μ that differ from $\frac{n}{2}$ may account for:

- non-Gaussian nature of the in-phase and quadrature components of each cluster of the fading signal,
- non-zero correlation among the clusters of multipath components,
- non-zero correlation between in-phase and quadrature components within each cluster, etc.

So, the same analysis conduced for κ-μ applies to η-μ as well. Using the definitions for parameters η and μ, and the considerations as given above, the η-μ power PDF can be written from (15) as:

$$f_{P_R}(P_R) = \frac{2\sqrt{\pi}\cdot\mu^{\mu+0.5}\cdot h^{\mu}}{\Gamma(\mu)\cdot H^{\mu-0.5}\cdot\Omega^{\mu+0.5}}\cdot P_R^{\,\mu-0.5}\cdot\times\exp\left(-\frac{2\mu h\cdot P_R}{\Omega}\right)\cdot I_{\mu-0.5}\left(\frac{2\mu H\cdot P_R}{\Omega}\right) \tag{18}$$

Therefore, the η-μ envelope PDF can be obtained from (18) as:

$$f_R(R) = \frac{4\sqrt{\pi}\cdot\mu^{\mu+0,5}\cdot R^{2\mu}}{\Gamma(\mu)\cdot H^{\mu-0,5}\cdot\Omega^{\mu+0,5}}\cdot\exp\left[-\frac{2\mu h\cdot R^2}{\Omega}\right]\cdot I_{\mu-0,5}\left[\frac{2\mu H\cdot R^2}{\Omega}\right] \tag{19}$$

Instantaneous and average SNR are given by: $\gamma = \frac{P_R}{N_0} = \frac{R^2}{N_0}$, $\bar{\gamma} = \frac{\overline{P_R}}{N_0} = \frac{\Omega}{N_0}$, and therefore the η-μ SNR PDF can be obtained from (18) as:

$$f_\gamma(\gamma)=\frac{2\sqrt{\pi}\cdot\mu^{\mu+0.5}\cdot h^{\mu}\cdot\gamma^{\mu-0.5}}{\Gamma(\mu)\cdot H^{\mu-0.5}\cdot\left(\bar{\gamma}\right)^{\mu+0.5}}\cdot\exp\left(-\frac{2\mu\cdot h\cdot\gamma}{\bar{\gamma}}\right)\cdot I_{\mu-0.5}\left(\frac{2\mu\cdot H\cdot\gamma}{\bar{\gamma}}\right) \tag{20}$$

From (20) we can derive CDF of instantaneous SNR as:

$$F_\gamma(\gamma)=\int_0^\gamma f_\gamma(t)dt=1-\Theta_\mu\left[\frac{H}{h},\sqrt{\frac{2h\mu\cdot\gamma}{\bar{\gamma}}}\right] \tag{21}$$

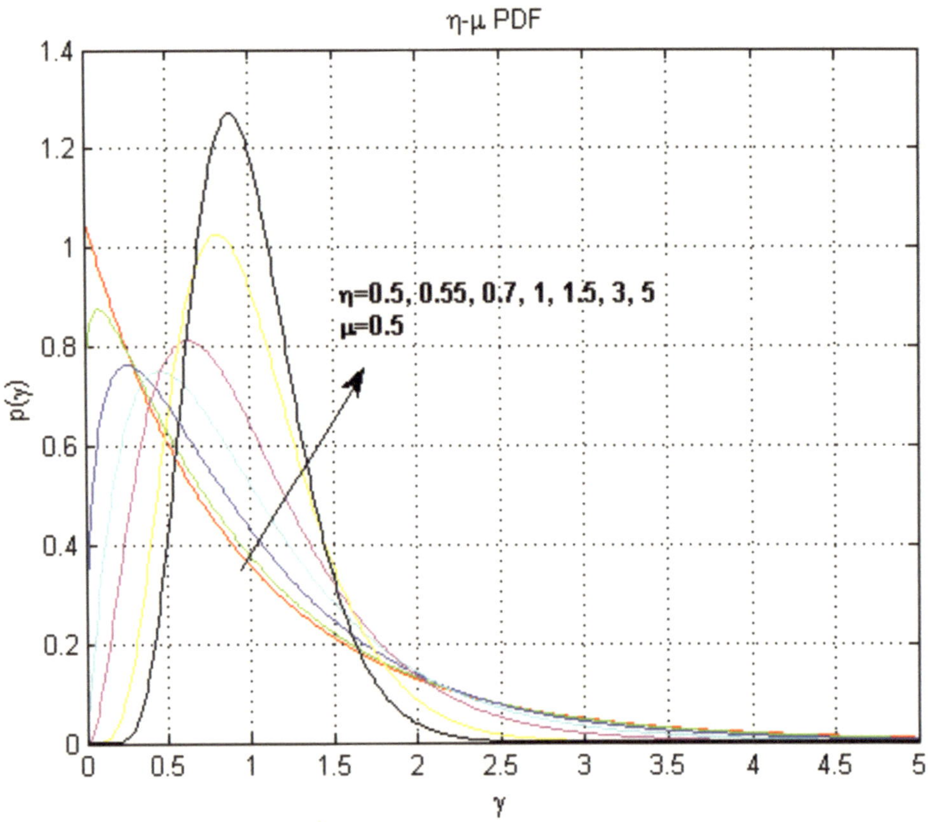

Fig. 3. PDF of SNR for μ=0,5 and various values of η

Since previous integral doesen't have closed-form solution, function Θ has been defined as:

$$\Theta_\nu\left(a,b\right)=\frac{2^{1,5-\nu}\cdot\sqrt{\pi}\cdot\left(1-a^2\right)^\nu}{a^{\nu-0,5}\cdot\Gamma(\nu)}\cdot\int_b^\infty t^{2\nu}\cdot\exp(-t^2)\cdot I_{\nu-0,5}\left(at^2\right)\cdot dt \tag{22}$$

Now we can obtain closed-form expression for the n-th order moment of RV γ as:

$$\overline{\gamma^n}=\frac{\Gamma(2\mu+n)}{\Gamma(2\mu)\cdot h^{\mu+n}}\cdot\left(\frac{\bar{\gamma}}{2\mu}\right)^n\cdot{}_2F_1\left(\mu+n+0,5,\mu+\frac{n}{2};\mu+0,5;\left(\frac{H}{h}\right)^2\right) \tag{23}$$

, where ${}_2F_1(\bullet,\bullet;\bullet;\bullet)$ represents Gauss hypergeometric function (Abramowitz and Stegun, 1972, eq. 15.1.1). The κ-μ and the η-μ distributions are indeed general fading distributions, which can be seen by a following special cases:

- μ=0,5 and κ=0, the κ-μ distribution becomes One-sided Gaussian distribution;
- μ=0,5 and η=0 (or η→∞), the η-μ distribution becomes One-sided Gaussian distribution;
- μ=1 and κ=0, the κ-μ distribution becomes Rayleigh distribution;
- μ=0,5 and η=1, the η-μ distribution becomes Rayleigh distribution;
- μ=1 and κ=K, the κ-μ distribution becomes Rice distribution, where K represents Rice K parameter;
- μ=0,5 and η=q^2, the η-μ distribution becomes Hoyt (Nakagami-q) distribution, where q represents Hoyt q parameter;
- μ=m and κ=0, the κ-μ distribution becomes Nakagami-m distribution, where m represents Nakagami m parameter;
- μ=m and η=0 (or η→∞), or μ=m/2 and η=1, the η-μ distribution becomes Nakagami-m distribution, where m represents Nakagami m parameter.

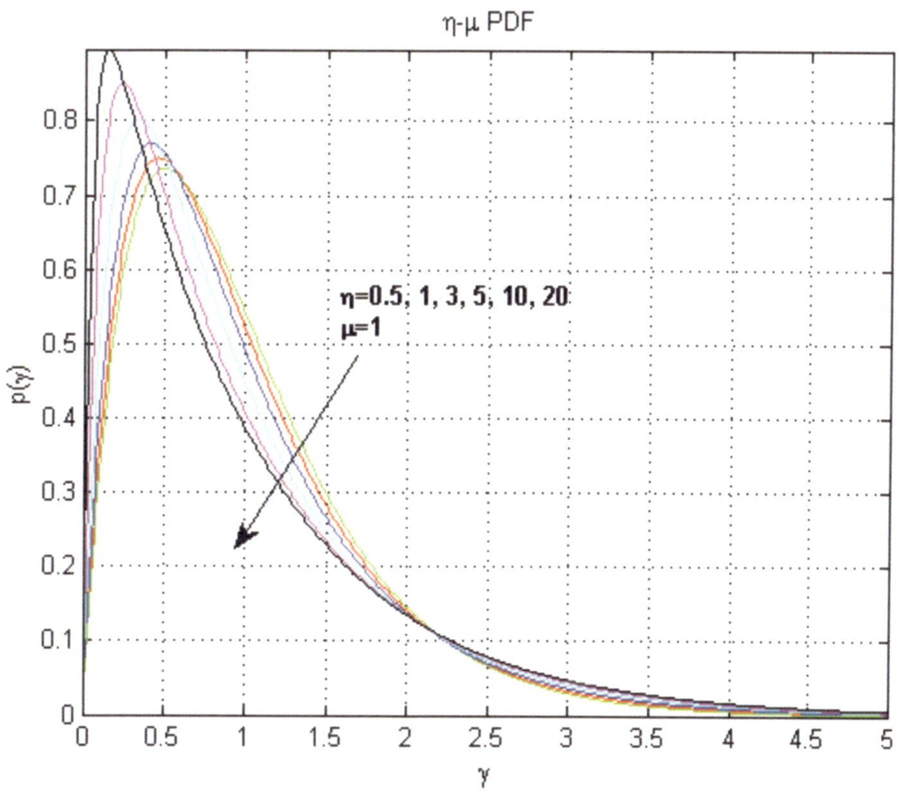

Fig. 4. PDF of SNR for η=0,5 and various values of μ

3. Maximal-ratio combining

There are four principal types of combining techniques (Simon & Alouini, 2005.) that depend essentially on the complexity restrictions put on the communication system and amount of channel state information (CSI) available at the receiver. As shown in (Simon & Alouini, 2005.), in the absence of interference, Maximal-Ratio Combining is the optimal combining scheme, regardless of fading statistics, but most complex since MRC requires knowledge of all channel fading parameters (amplitudes, phases and time delays). Since knowledge of channel fading amplitudes is needed for MRC, this scheme can be used in conjunction with unequal energy signals, such as M-QAM or any other amplitude/phase modulations. In this paper we will treat L-branch MRC receiver. As shown in (Simon & Alouini, 2005.) MRC receiver is the optimal multichannel receiver, regardless of fading statistics in various diversity branches since it results in a ML receiver. For equiprobable transmitted symbols, the total SNR per symbol at the output of the MRC is given by (Stuber, 1996.): $\gamma = \sum_{j=1}^{L} \gamma_j$, where γ_j is instantaneous SNR in j-th branch of L-branch MRC receiver.

3.1 Maximal-ratio combining in presence of κ-μ distributed fading

We will first analyze statistics of received power and SNR at MRC output in presence of κ-μ distributed fading, and then we will obtain expression for outage probability at MRC output. Repeating the same procedure as in the previous section, previous relation can be written as:

$$\gamma = \sum_{j=1}^{L} \gamma_j = \frac{1}{N_0} \cdot \sum_{j=1}^{L} P_{R,j} = \frac{1}{N_0} \cdot \sum_{j=1}^{L} \sum_{i=1}^{n} P_{R,j,i} \tag{24}$$

, where $P_{R,j,i}$ represents total received power of the i-th cluster manifested in the j-th branch of the MRC receiver. Using (1) one can obtain:

$$P_R = \sum_{j=1}^{L} \sum_{i=1}^{n} \left(X_{i,j} + p_{i,j}\right)^2 + \left(Y_{i,j} + q_{i,j}\right)^2 . \tag{25}$$

Repeating the same procedure as in the previous section, one can obtain MGF of the RV P_R:

$$M_{P_R}(s) = \prod_{j=1}^{L} M_{P_{R,j}}(s) = \frac{\exp\left(\dfrac{sd^2}{1-2s\sigma^2}\right)}{\left(1-2s\sigma^2\right)^{L \cdot n}} \tag{26}$$

, where $d^2 = \sum_{j=1}^{L} d_j^2$. Inverse of (26) yields in PDF of the RV P_R:

$$f_{P_R}(P_R) = \frac{L\mu(1+\kappa)^{\frac{L\mu+1}{2}}}{\kappa^{\frac{L\mu-1}{2}} \cdot \exp(L\mu\kappa) \cdot (L \cdot \Omega)^{\frac{L\mu+1}{2}}} \cdot P_R^{\frac{L\mu-1}{2}} \cdot \exp\left[-\frac{\mu(1+\kappa) \cdot P_R}{\Omega}\right] \cdot I_{L\mu-1}\left[2\mu\sqrt{L\frac{\kappa(1+\kappa) \cdot P_R}{\Omega}}\right] \tag{27}$$

From PDF of received signal power at MRC output, we can obtain PDF of instantaneous SNR at MRC output:

$$f_\gamma(\gamma)=\frac{L\mu(1+\kappa)^{\frac{L\mu+1}{2}}}{\kappa^{\frac{L\mu-1}{2}}\cdot\exp(L\mu\kappa)\cdot\left(L\cdot\bar{\gamma}\right)^{\frac{L\mu+1}{2}}}\cdot\gamma^{\frac{L\mu-1}{2}}\cdot\exp\left[-\frac{\mu(1+\kappa)\cdot\gamma}{\bar{\gamma}}\right]\cdot I_{L\mu-1}\left[2\mu\sqrt{L\frac{\kappa(1+\kappa)\cdot\gamma}{\bar{\gamma}}}\right] \quad (28)$$

Note that sum of L squares of the κ-μ distributed RVs is a square of κ-μ distributed RV, but with different parameters, which means SNR at the output of the MRC receiver subdue to the κ-μ distribution with parameters:

$$\mu_{MRC}=L\cdot\mu\,,\ \kappa_{MRC}=\kappa\,,\ \bar{\gamma}_{MRC}=L\cdot\bar{\gamma}\,.$$

Now, it is easy to obtain CDF of SNR at MRC output:

$$F_\gamma(\gamma)=\int_0^\gamma f_\gamma(t)\cdot dt=1-Q_{L\mu}\left[\sqrt{2L\kappa\mu},\sqrt{\frac{2(\kappa+1)\mu\cdot\gamma}{\bar{\gamma}}}\right] \quad (29)$$

For fixed SNR threshold γ_{th}, outage probability at MRC output is given by:

$$P_{out}(\gamma_{th})=F_\gamma(\gamma_{th})=\int_0^{\gamma_{th}} f_\gamma(t)\cdot dt=1-Q_{L\mu}\left[\sqrt{2L\kappa\mu},\sqrt{\frac{2(\kappa+1)\mu\cdot\gamma_{th}}{\bar{\gamma}}}\right] \quad (30)$$

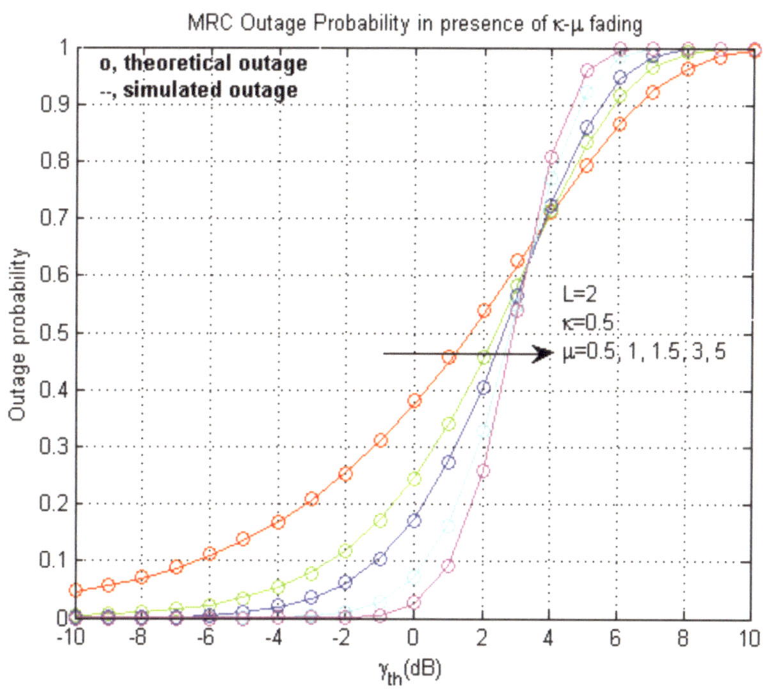

Fig. 5. Outage probability for dual-branch MRC (L=2), fixed κ, and various μ

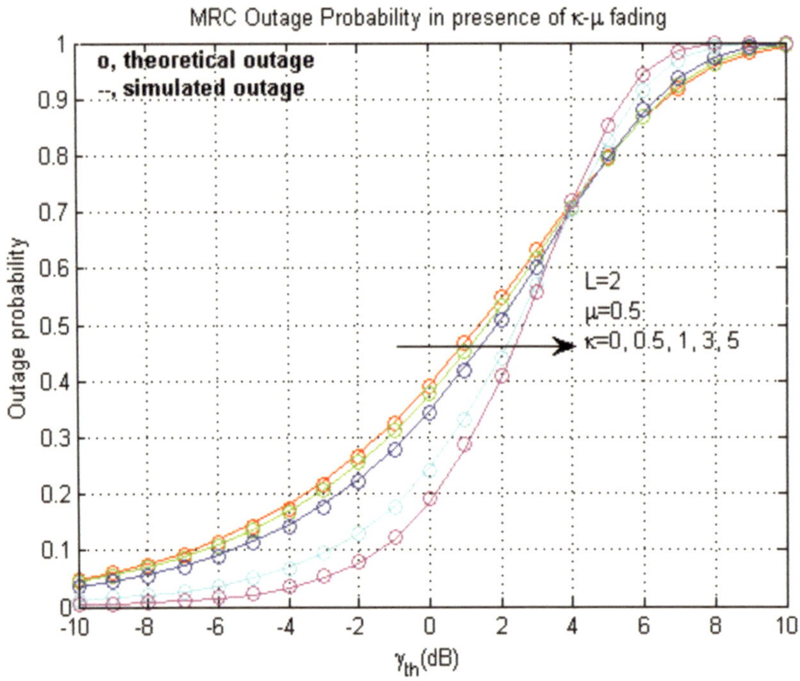

Fig. 6. Outage probability for dual-branch MRC (L=2), fixed μ, and various κ

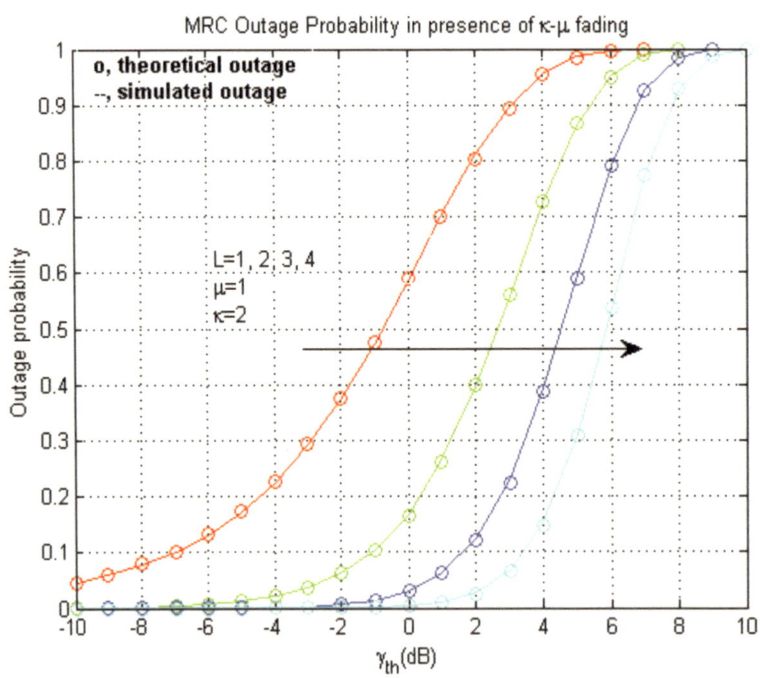

Fig. 7. Outage probability for fixed μ and κ, L=1, 2, 3 and 4

3.2 Maximal-ratio combining in presence of η-μ distributed fading

Now we analyze statistics of received power and SNR at MRC output in presence of η-μ distributed fading, and after that, we will obtain expression for outage probability at MRC output. Using (12) one can obtain:

$$P_R = \sum_{j=1}^{L}\sum_{i=1}^{n}\left(X_{i,j}^2 + Y_{i,j}^2\right) \tag{31}$$

Repeating the same procedure as in the previous section one can obtain MGF of the RV P_R:

$$M_{P_R}(s) = \prod_{j=1}^{L} M_{P_{R,j}}(s) = \left[\frac{\dfrac{\sqrt{h}}{\left(1+\eta^{-1}\right)\sigma_X^2}}{\sqrt{\left(s+\dfrac{h}{\left(1+\eta^{-1}\right)\sigma_X^2}\right)^2 - \left(\dfrac{H}{\left(1+\eta^{-1}\right)\sigma_X^2}\right)^2}}\right]^{L\cdot n} \tag{32}$$

Inverse of Equation (32) yields to PDF of the RV P_R:

$$f_{P_R}\left(P_R\right) = \frac{2\sqrt{\pi}\cdot\mu^{L\cdot\mu+0.5}\cdot h^{L\cdot\mu}}{\Gamma\left(L\cdot\mu\right)\cdot H^{L\cdot\mu-0.5}\cdot\Omega^{L\cdot\mu+0.5}}\cdot P_R^{\,L\cdot\mu-0.5}\cdot\exp\left(-\frac{2\mu h\cdot P_R}{\Omega}\right)\cdot I_{L\cdot\mu-0.5}\left(\frac{2\mu H\cdot P_R}{\Omega}\right) \tag{33}$$

From PDF of received signal power at MRC output, we can obtain PDF of instantaneous SNR at MRC output:

$$f_\gamma(\gamma) = \frac{2\sqrt{\pi}\cdot\mu^{L\cdot\mu+0.5}\cdot h^{L\cdot\mu}}{\Gamma\left(L\cdot\mu\right)\cdot H^{L\cdot\mu-0.5}\cdot\left(\bar{\gamma}\right)^{L\cdot\mu+0.5}}\cdot\gamma^{L\cdot\mu-0.5}\cdot\exp\left(-\frac{2\mu h\cdot\gamma}{\bar{\gamma}}\right)\cdot I_{L\cdot\mu-0.5}\left(\frac{2\mu H\cdot\gamma}{\bar{\gamma}}\right) \tag{34}$$

Note that sum of L squares of the η-μ distributed RVs is a square of η-μ distributed RV, but with different parameters, which means SNR at the output of the MRC receiver subdue to the η-μ distribution with parameters:

$$\mu_{MRC} = L\cdot\mu\,,\ \eta_{MRC} = \eta\,,\ \bar{\gamma}_{MRC} = L\cdot\bar{\gamma}\,.$$

Now, CDF of instantaneous SNR at MRC output is given by:

$$F_\gamma(\gamma) = \int_0^{\gamma} f_\gamma(t)\cdot dt = 1-\Theta_{L\mu}\left[\frac{H}{h},\sqrt{\frac{2h\mu\cdot\gamma}{\bar{\gamma}}}\right] \tag{35}$$

, where Θ is defined in (22). For fixed SNR threshold γ_{th}, outage probability at MRC output is given by:

$$P_{out}(\gamma_{th}) = F_\gamma(\gamma_{th}) = \int_0^{\gamma_{th}} f_\gamma(t)\cdot dt = 1-\Theta_{L\mu}\left[\frac{H}{h},\sqrt{\frac{2h\mu\cdot\gamma_{th}}{\bar{\gamma}}}\right] \tag{36}$$

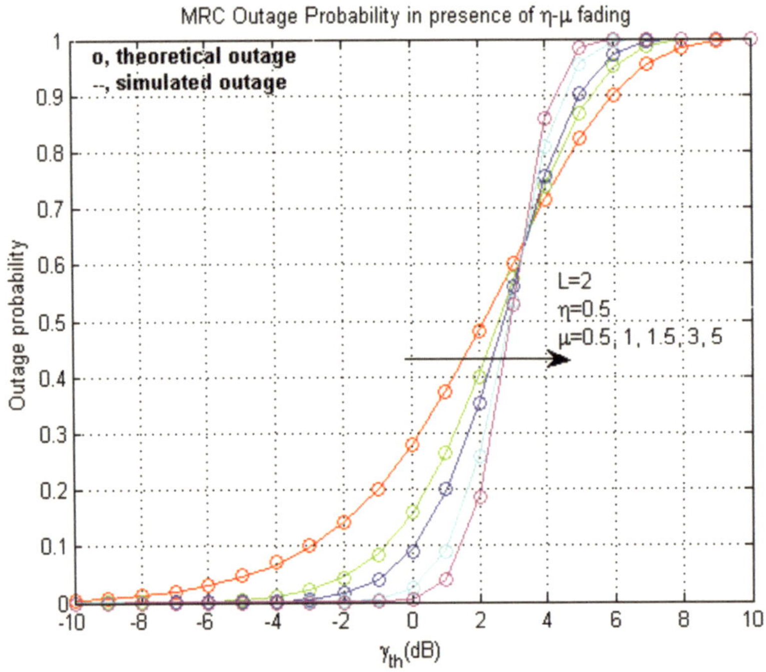

Fig. 8. Outage probability for dual-branch MRC (L=2), fixed η, and various μ

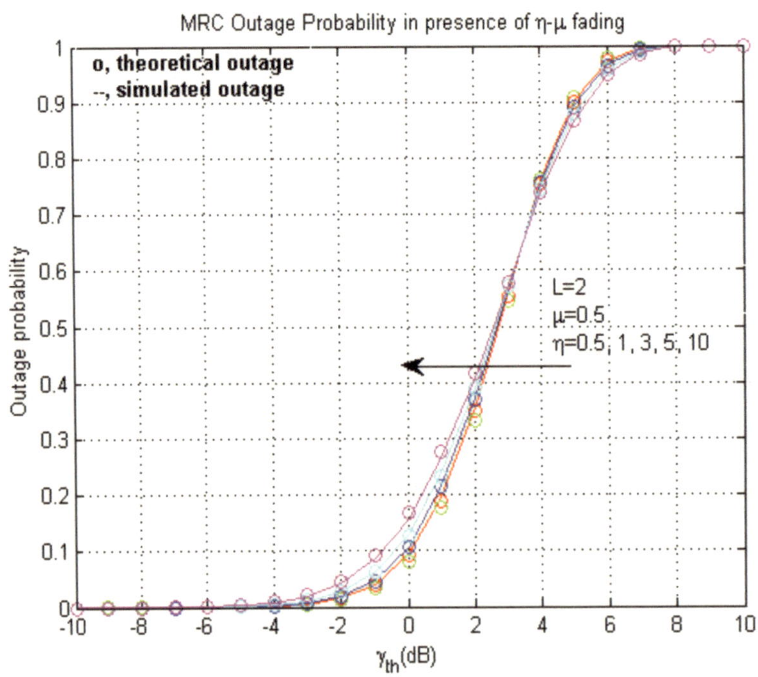

Fig. 9. Outage probability for dual-branch MRC (L=2), fixed μ, and various η

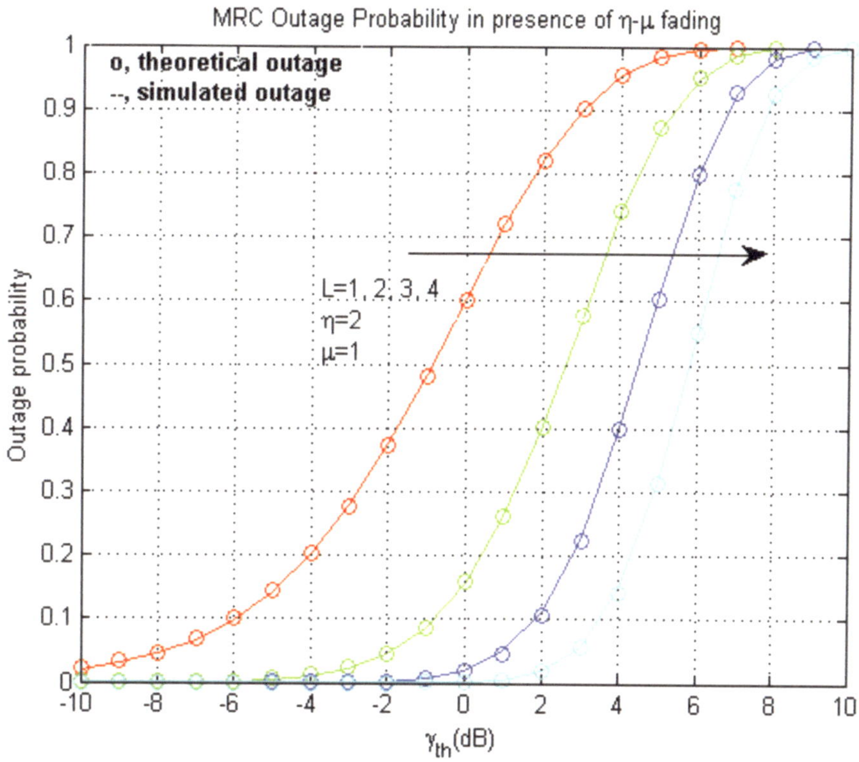

Fig. 10. Outage probability for fixed μ and η, L=1, 2, 3 and 4

4. Symbol error probability analysis

When we analyze symbol error probability (SEP), we must focus upon single modulation format because different modulations result in different SEPs. We must also consider type of detection (coherent or non-coherent). Namely, decision-block of coherent receiver performs decision based upon information gained in both In-phase and quadrature components, while decision-block of non-coherent receiver performs decision based only upon envelope of the received signal. Although coherent detection results in smaller SEP than corresponding non-coherent detection for the same SNR, sometimes it is suitable to perform non-coherent detection depending on receiver structure complexity.

Non-coherent detection

To obtain average SEP for non-coherent detection, we will use generic expression for instantaneous SEP: $SEP = a \cdot \exp(-b \cdot \gamma)$, where γ represents instantaneous received SNR, and non-negative parameters a and b depend on used modulation format.
Average SEP can be obtained by averaging expression for SEP with respect to γ:

$$ASEP = \int_0^{+\infty} SEP \cdot f_\gamma(\gamma) \cdot d\gamma = \int_0^{+\infty} a \cdot \exp(-b \cdot \gamma) \cdot f_\gamma(\gamma) \cdot d\gamma \qquad (38)$$

a \ b	0.5	1
0.5	BFSK	DBPSK
1	/	/
$\frac{M-1}{2}$	MFSK	/

Table 1. Values of *a* and *b* for some non-coherent modulations

Coherent detection

To obtain average SEP for coherent detection, we will use generic expression for instantaneous SEP: $SEP = a \cdot Q\left(\sqrt{b \cdot \gamma}\right)$, where $Q(\bullet)$ function is defined as:

$$Q(x) = \frac{1}{\sqrt{2\pi}} \int_{x}^{+\infty} \exp\left(\frac{-t^2}{2}\right) \cdot dt$$

, and non-negative parameters *a* and *b* depend on used modulation format.

a \ b	1	2	$2\sin^2\left(\frac{\pi}{M}\right)$	$\frac{3}{M-1}$
1	BFSK	BPSK	/	/
2	QPSK	DBPSK	MPSK	/
$4\frac{\sqrt{M}-1}{\sqrt{M}}$	/	/	/	Rect. QAM

Table 2. Values of *a* and *b* for some coherent modulations

Average SEP can be obtained by averaging expression for SEP with respect to γ:

$$ASEP = \int_{0}^{+\infty} SEP \cdot f_\gamma(\gamma) \cdot d\gamma = \int_{0}^{+\infty} a \cdot Q\left(\sqrt{b \cdot \gamma}\right) \cdot f_\gamma(\gamma) \cdot d\gamma \tag{39}$$

Nevertheless, it is sometimes impossible to find closed-form solution for (11). Because of that, we will present exact MGF-based solution, and solutions based on lower and upper bound of the *Q* function. To obtain MGF-based solution, first we have to rewrite *Q* function in more suitable form as in (Simon & Alouini, 2005.):

$$Q(x) = \frac{1}{\pi} \cdot \int_{0}^{\frac{\pi}{2}} \exp\left(\frac{-x^2}{2\sin^2\theta}\right) \cdot d\theta$$

Introducing alternate form of *Q* function in to the ASEP expression, we obtain MGF-based solution:

$$ASEP = a \cdot \int_0^{+\infty} Q\left(\sqrt{b \cdot \gamma}\right) \cdot f_\gamma(\gamma) \cdot d\gamma =$$

$$= \frac{a}{\pi} \cdot \int_0^{+\infty} \int_0^{\frac{\pi}{2}} \exp\left(\frac{-b \cdot \gamma}{2\sin^2\theta}\right) \cdot d\theta \cdot f_\gamma(\gamma) \cdot d\gamma$$

$$ASEP = \frac{a}{\pi} \cdot \int_0^{\frac{\pi}{2}} \left(\int_0^{+\infty} \exp\left(\frac{-b \cdot \gamma}{2\sin^2\theta}\right) \cdot f_\gamma(\gamma) \cdot d\gamma \right) \cdot d\theta = \frac{a}{\pi} \cdot \int_0^{\frac{\pi}{2}} M_\gamma\left(\frac{-b}{2\sin^2\theta}\right) \cdot d\theta \quad (40)$$

Now we seek solutions based on upper and lower bound of Q function. Q function can be bounded as:

$$\frac{\exp\left(\frac{-x^2}{2}\right)}{x\sqrt{2\pi}} \cdot \left(1 - \frac{1}{x^2}\right) < Q(x) < \frac{\exp\left(\frac{-x^2}{2}\right)}{x\sqrt{2\pi}} \quad (41)$$

Upper bound for ASEP can be obtained by introducing upper bound of the Q function in (39):

$$ASEP_{UB} = \int_0^{+\infty} \frac{a}{\sqrt{2\pi b \cdot \gamma}} \cdot \exp\left(\frac{-b \cdot \gamma}{2}\right) \cdot f_\gamma(\gamma) \cdot d\gamma \quad (42)$$

Lower bound for ASEP can be obtained by introducing lower bound of the Q function in (39):

$$\begin{aligned} ASEP_{LB} &= \int_0^{+\infty} \frac{a}{\sqrt{2\pi b \cdot \gamma}} \cdot \left(1 - \frac{1}{b \cdot \gamma}\right) \cdot \exp\left(\frac{-b \cdot \gamma}{2}\right) \cdot f_\gamma(\gamma) \cdot d\gamma = \\ &= ASEP_{UB} - \int_0^{+\infty} \frac{a}{\sqrt{2\pi (b \cdot \gamma)^3}} \cdot \exp\left(\frac{-b \cdot \gamma}{2}\right) \cdot f_\gamma(\gamma) \cdot d\gamma \end{aligned} \quad (43)$$

4.1 Symbol error probability analysis for maximal-ratio combiner in presence of κ-μ distributed fading

To obtain ASEP at MRC output for κ-μ fading for non-coherent detection, we introduce (28) in (38):

$$ASEP = a \cdot \int_0^{+\infty} \exp(-b \cdot \gamma) \cdot f_\gamma(\gamma) \cdot d\gamma =$$

$$= a \cdot \frac{L\mu(1+\kappa)^{\frac{L\mu+1}{2}}}{\kappa^{\frac{L\mu-1}{2}} \cdot \exp(L\mu\kappa) \cdot \left(L \cdot \bar{\gamma}\right)^{\frac{L\mu+1}{2}}} \cdot \int_0^{+\infty} \gamma^{\frac{L\mu-1}{2}} \cdot \exp\left[-\left(b + \frac{\mu(1+\kappa)}{\bar{\gamma}}\right) \cdot \gamma\right] \cdot I_{L\mu-1}\left[2\mu\sqrt{L\frac{\kappa(1+\kappa) \cdot \gamma}{\bar{\gamma}}}\right] \cdot d\gamma$$

Using (Prudnikov et al., 1992, eq. 5, page 318) we obtain closed-form expression for ASEP for non-coherent detection:

$$ASEP = a \cdot \left[\frac{\mu(1+\kappa)}{b \cdot \bar{\gamma} + \mu(1+\kappa)} \cdot \exp\left(\frac{-b\kappa \cdot \bar{\gamma}}{b \cdot \bar{\gamma} + \mu(1+\kappa)} \right) \right]^{L\mu} \quad (44)$$

Now we have to obtain ASEP at MRC output for κ-μ fading for coherent detection. To do so, first we have to manipulate (26) to obtain MGF for RV γ at MRC output:

$$M_{\gamma}(s) = \frac{\exp\left(\frac{\frac{L\kappa\bar{\gamma}}{1+\kappa} \cdot s}{1 - \frac{\bar{\gamma}}{\mu(1+\kappa)} \cdot s} \right)}{\left(1 - \frac{\bar{\gamma}}{\mu(1+\kappa)} \cdot s \right)^{L\mu}} \quad (45)$$

Introducing (45) in (40) we obtain:

$$ASEP = \frac{a}{\pi} \cdot \int_0^{\frac{\pi}{2}} M_{\gamma}\left(\frac{-b}{2\sin^2\theta} \right) \cdot d\theta = \frac{a}{\pi} \cdot \int_0^{\frac{\pi}{2}} \frac{\exp\left(\frac{\frac{L\kappa\bar{\gamma}}{1+\kappa} \cdot \frac{-b}{2\sin^2\theta}}{1 - \frac{\bar{\gamma}}{\mu(1+\kappa)} \cdot \frac{-b}{2\sin^2\theta}} \right)}{\left(1 - \frac{\bar{\gamma}}{\mu(1+\kappa)} \cdot \frac{-b}{2\sin^2\theta} \right)^{L\mu}} \cdot d\theta \quad (46)$$

Now we seek upper bound for ASEP for coherent detection by introducing (28) in (42):

$$ASEP_{UB} = \int_0^{+\infty} \frac{a}{\sqrt{2\pi b \cdot \gamma}} \cdot \exp\left(\frac{-b \cdot \gamma}{2} \right) \cdot f_{\gamma}(\gamma) \cdot d\gamma =$$

$$= \int_0^{+\infty} \frac{a}{\sqrt{2\pi b}} \cdot \frac{L\mu(1+\kappa)^{\frac{L\mu+1}{2}}}{\kappa^{\frac{L\mu-1}{2}} \cdot \exp(L\mu\kappa) \cdot \left(L \cdot \bar{\gamma} \right)^{\frac{L\mu+1}{2}}} \cdot \gamma^{\frac{L\mu-2}{2}} \exp\left[-\gamma\left(\frac{b}{2} + \frac{\mu(1+\kappa)}{\bar{\gamma}} \right) \right] \cdot I_{L\mu-1}\left[2\mu\sqrt{L \cdot \frac{\kappa(1+\kappa)\gamma}{\bar{\gamma}}} \right] \cdot d\gamma$$

Using (Prudnikov et al., 1992, eq. 5, page 318) we obtain closed-form expression for upper bound for average SEP for coherent detection:

$$ASEP_{UB} = \frac{a \cdot \Gamma(L\mu - 0.5)}{\sqrt{2\pi b} \cdot \Gamma(L\mu)} \cdot \left(\frac{\mu(1+\kappa)}{\exp(\kappa) \cdot \bar{\gamma}} \right)^{L\mu} \cdot \left(\frac{b}{2} + \frac{\mu(1+\kappa)}{\bar{\gamma}} \right)^{0.5 - L\mu} \cdot {}_1F_1\left(L\mu - 0.5; L\mu; \frac{\mu^2 L\kappa(1+\kappa)}{\frac{b \cdot \bar{\gamma}}{2} + \mu(1+\kappa)} \right) \quad (47)$$

, where ${}_1F_1(\bullet;\bullet;\bullet)$ is Kummer confluent hypergeometric function defined in (Wolfram, http://functions.wolfram.com/07.20.02.0001.01). Lower bound for ASEP can be obtained by introducing (28) in (43), and using the same solution as in the previous case:

$$ASEP_{LB} = \frac{a\cdot\Gamma(L\mu-0.5)}{\sqrt{2\pi b}\cdot\Gamma(L\mu)}\cdot\left(\frac{\mu(1+\kappa)}{\exp(\kappa)\cdot\bar{\gamma}}\right)^{L\mu}\cdot\left(\frac{b}{2}+\frac{\mu(1+\kappa)}{\bar{\gamma}}\right)^{0.5-L\mu}\cdot$$
$$\cdot{}_1F_1\left(L\mu-0.5;L\mu;\frac{\mu^2L\kappa(1+\kappa)}{\frac{b\cdot\bar{\gamma}}{2}+\mu(1+\kappa)}\right)-$$
$$-\frac{a\cdot\Gamma(L\mu-1.5)}{\sqrt{2\pi b^3}\cdot\Gamma(L\mu)}\cdot\left(\frac{\mu(1+\kappa)}{\exp(\kappa)\cdot\bar{\gamma}}\right)^{L\mu}\cdot\left(\frac{b}{2}+\frac{\mu(1+\kappa)}{\bar{\gamma}}\right)^{1.5-L\mu}\cdot$$
$$\cdot{}_1F_1\left(L\mu-1.5;L\mu;\frac{\mu^2L\kappa(1+\kappa)}{\frac{b\cdot\bar{\gamma}}{2}+\mu(1+\kappa)}\right) \tag{48}$$

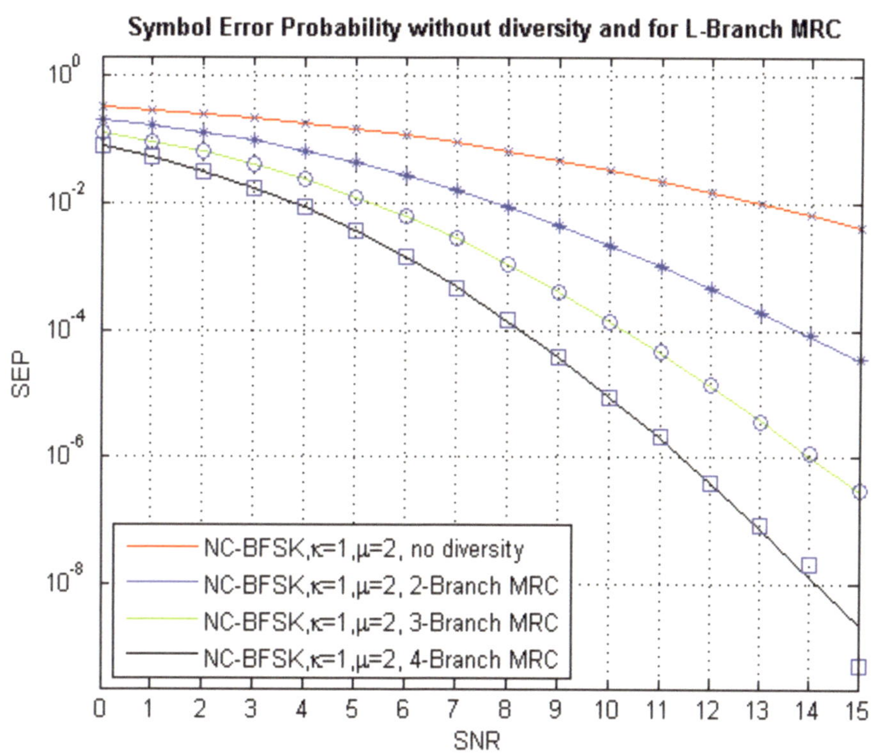

Fig. 11. Average symbol error probability for non-coherent BFSK, L=1, 2, 3 and 4

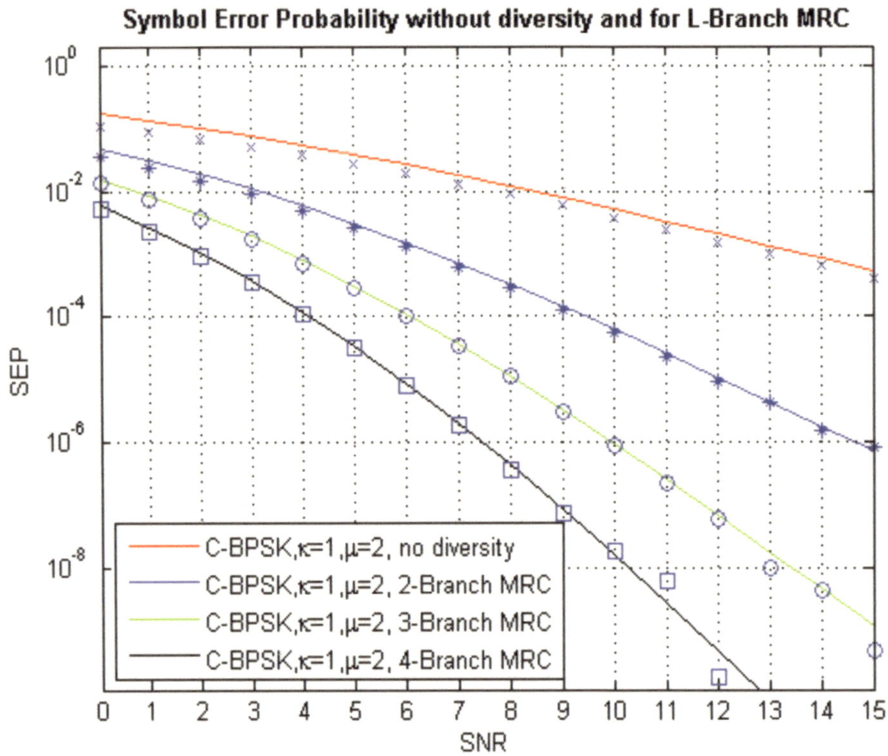

Fig. 12. Average symbol error probability for coherent BPSK, L=1, 2, 3 and 4

4.2 Symbol error probability analysis for maximal-ratio combiner in presence of η-μ distributed fading

To obtain ASEP at MRC output for η-μ fading for non-coherent detection, we introduce (34) in (38):

$$ASEP = a \cdot \int_0^{+\infty} \exp(-b \cdot \gamma) \cdot f_\gamma(\gamma) \cdot d\gamma =$$
$$= a \cdot \frac{2\sqrt{\pi} \cdot \mu^{L\cdot\mu+0.5} \cdot h^{L\cdot\mu}}{\Gamma(L\cdot\mu) \cdot H^{L\cdot\mu-0.5} \cdot (\bar{\gamma})^{L\cdot\mu+0.5}} \cdot \int_0^{+\infty} \gamma^{L\cdot\mu-0.5} \cdot \exp\left[-\left(b + \frac{2\mu h}{\bar{\gamma}}\right) \cdot \gamma\right] \cdot I_{L\cdot\mu-0.5}\left(\frac{2\mu H \cdot \gamma}{\bar{\gamma}}\right) \cdot d\gamma$$

Integration of previous expression will be carried out via Meijer-G functions, defined in (Wolfram, http://functions.wolfram.com/HypergeometricFunctions/MeijerG/). First we have to transform exponential and Bessel functions in Meijer-G functions in accordance to (Wolfram, http://functions.wolfram.com/07.34.03.0228.01) and (Wolfram, http://functions.wolfram.com/03.02.26.0009.01). Integration is performed with (Wolfram, http://functions.wolfram.com/07.34.21.0011.01). After some algebraic manipulations, and simplifications in accordance to (Wolfram, http://functions.wolfram.com/07.34.03.0734.01)

and (Wolfram, http://functions.wolfram.com/07.23.03.0079.01), we obtain closed-form expression for average SEP for non-coherent detection:

$$ASEP = a \cdot \left[\frac{4\mu^2 h}{\left(b \cdot \bar{\gamma} + 2\mu h\right)^2 - \left(2\mu H\right)^2} \right]^{L\mu} \tag{49}$$

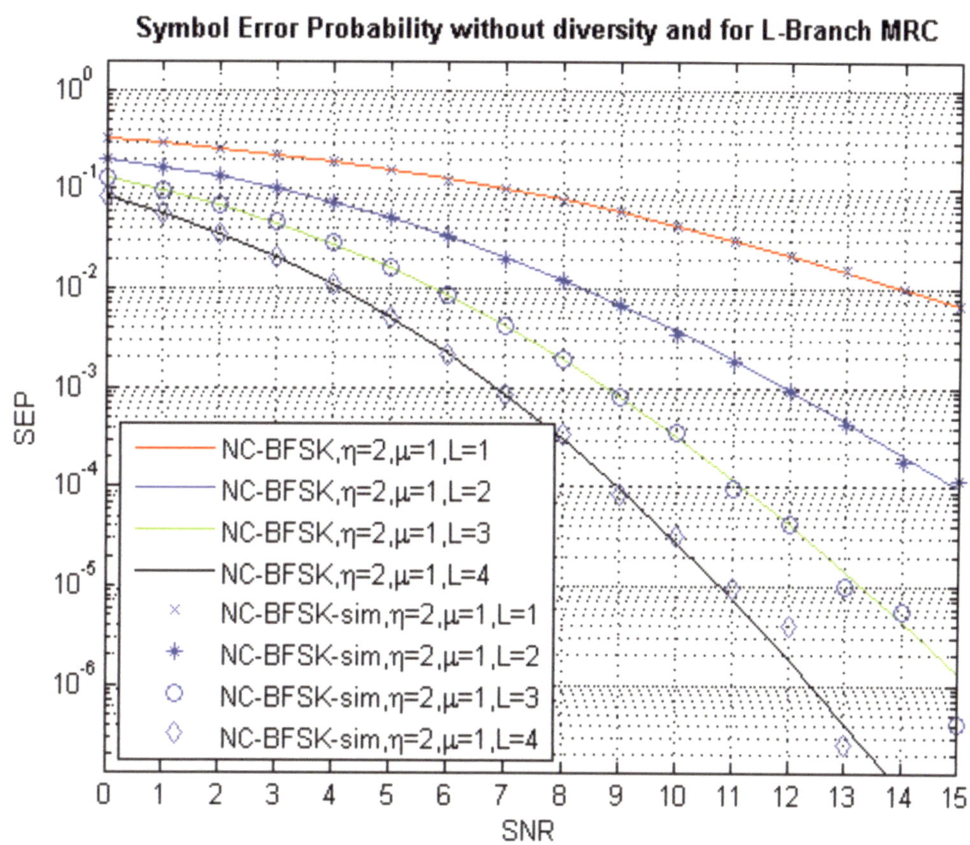

Fig. 13. Average symbol error probability for coherent BFSK, L=1, 2, 3 and 4

Now we have to obtain ASEP at MRC output for η-μ fading for coherent detection. First we manipulate (32) to obtain MGF for RV γ at MRC output:

$$M_\gamma(s) = \left[\frac{4\mu^2 h}{\left(s \cdot \bar{\gamma} - 2\mu h\right)^2 - \left(2\mu H\right)^2} \right]^{L\mu} \tag{50}$$

Introducing (50) in (40) we obtain:

$$ASEP = \frac{a}{\pi} \cdot \int_0^{\frac{\pi}{2}} M_\gamma \left(\frac{-b}{2\sin^2\theta} \right) \cdot d\theta = \frac{a}{\pi} \cdot \int_0^{\frac{\pi}{2}} \left[\frac{4\mu^2 h}{\left(\frac{b \cdot \bar{\gamma}}{2\sin^2\theta} + 2\mu h \right)^2 - (2\mu H)^2} \right]^{L\mu} \cdot d\theta \tag{51}$$

Now we seek upper bound for ASEP for coherent detection by introducing (34) in (42):

$$ASEP_{UB} = \frac{a}{\sqrt{2\pi b \cdot \bar{\gamma}}} \cdot \frac{(2\mu)^{2L\cdot\mu} \cdot h^{L\cdot\mu}}{\left(\frac{b \cdot \bar{\gamma}}{2} + 2\mu(h-H) \right)^{2L\cdot\mu - 0,5}} \cdot \frac{\Gamma(2L\cdot\mu - 0,5)}{\Gamma(2L\cdot\mu)} \cdot {}_2F_1\left(L\cdot\mu, 2L\cdot\mu - 0,5; 2L\cdot\mu; \frac{-4\mu H}{\frac{b\cdot\bar{\gamma}}{2} + 2\mu(h-H)} \right) \tag{52}$$

Lower bound for ASEP can be obtained by introducing (34) in (43):

$$\begin{aligned} ASEP_{LB} = {} & \frac{a}{\sqrt{2\pi b \cdot \bar{\gamma}}} \cdot \frac{(2\mu)^{2L\cdot\mu} \cdot h^{L\cdot\mu}}{\left(\frac{b \cdot \bar{\gamma}}{2} + 2\mu(h-H) \right)^{2L\cdot\mu - 0,5}} \cdot \frac{\Gamma(2L\cdot\mu - 0,5)}{\Gamma(2L\cdot\mu)} \cdot \\ & \cdot {}_2F_1\left(L\cdot\mu, 2L\cdot\mu - 0,5; 2L\cdot\mu; \frac{-4\mu H}{\frac{b\cdot\bar{\gamma}}{2} + 2\mu(h-H)} \right) - \\ & - \frac{a}{\sqrt{2\pi (b \cdot \bar{\gamma})^3}} \cdot \frac{(2\mu)^{2L\cdot\mu} \cdot h^{L\cdot\mu}}{\left(\frac{b \cdot \bar{\gamma}}{2} + 2\mu(h-H) \right)^{2L\cdot\mu - 1,5}} \cdot \frac{\Gamma(2L\cdot\mu - 1,5)}{\Gamma(2L\cdot\mu)} \cdot \\ & \cdot {}_2F_1\left(L\cdot\mu, 2L\cdot\mu - 1,5; 2L\cdot\mu; \frac{-4\mu H}{\frac{b\cdot\bar{\gamma}}{2} + 2\mu(h-H)} \right) \end{aligned} \tag{53}$$

5. Simulations and discussion of the results

For the purposes of simulations, in this section first we discuss ways for generation of κ-μ and η-μ RVs. Since we have $f_\gamma(\gamma)$, and since we can't obtain inverse $f_\gamma^{-1}(\gamma)$, we have to apply Accept-Reject method. So, our goal is to generate random numbers from a continuous κ-μ and η-μ distributions with probability distribution functions given by (9) and (20), respectively. Although this method begins with uniform random number generator (RNG), it requires additional RNG. Namely, we first generate a random number from a continuous distribution with probability distribution function $g_\gamma(\gamma)$, satisfying $f_\gamma(\gamma) \le C \cdot g_\gamma(\gamma)$, for some constant C and for all γ. A continuous Accept-Reject RNG proceeds as follows:

1. we choose $g_\gamma(\gamma)$;
2. we find a constant C such that $f_\gamma(\gamma) / g_\gamma(\gamma) \le C$ for all γ;
3. we generate a uniform random number U;
4. we generate a random number V from $g_\gamma(\gamma)$;
5. if $C \cdot U \le f_\gamma(V) / g_\gamma(V)$, we accept V;
6. else, we reject V and return to step 3.

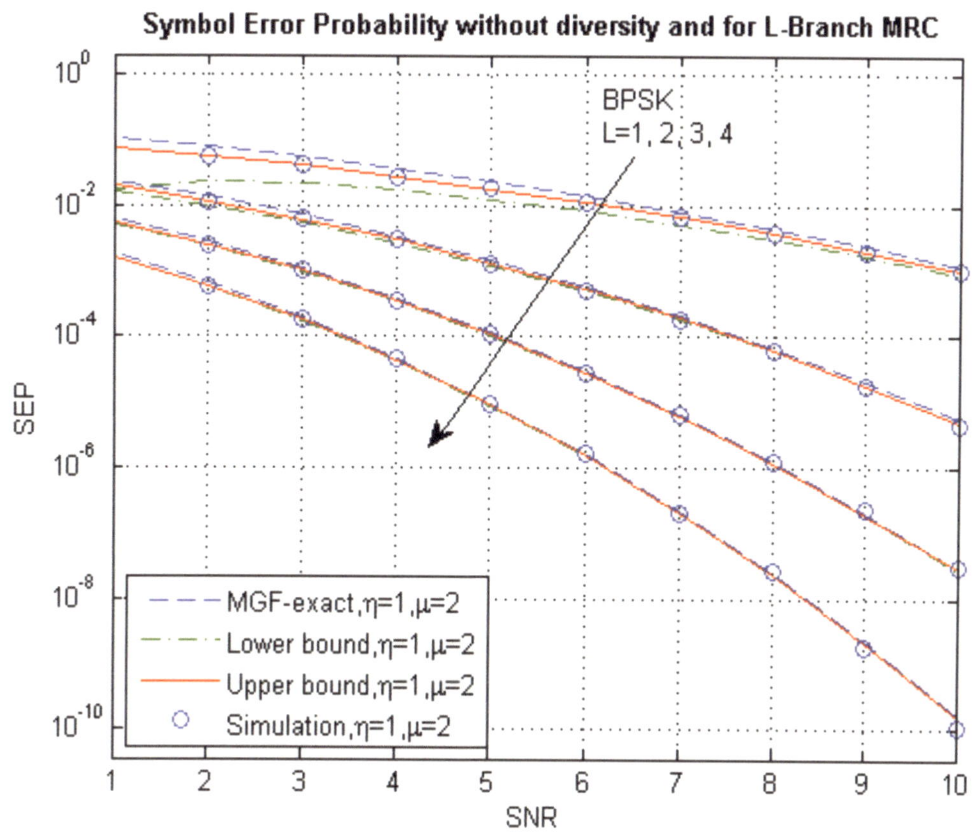

Fig. 14. Average symbol error probability for coherent BPSK, L=1, 2, 3 and 4

For efficiency of generation of random numbers V, we choose $g_\gamma(\gamma)$ as a exponential distribution. We find constant C so a condition $f_\gamma(\gamma) \le C \cdot g_\gamma(\gamma)$ is satisfied. There is another, more efficient method for generation of κ-μ and η-μ RVs. For κ-μ and η-μ distributions, in accordance to (1) and (12) respectively, if $\mu = 0,5 \cdot q$, where q is an integer number, then it is possible to obtain κ-μ and η-μ distributed random numbers as a sum of squares of q Gaussian random numbers generated from a generator with adequate parameters. We designed simulator of κ-μ and η-μ based on outlines given above. We used this simulator to generate samples of κ-μ and η-μ distributed instantaneous SNR. These samples are used to obtain outage probability as shown in Figs. 5-7 for κ-μ fading, and Figs. 8-10 for η-μ. As we can see from Figs. 7 and 10, there is not much need to increase number of combiner's branches beyond 4, because average SNR gained this way decreases for the same outage probability. On Figs. 11 and 13 ASEP for non-coherent BFSK has been depicted. Full lines represent theoretical ASEP curves given by (44) and (49), respectively. Markers on these figures represent values obtained by simulation. As we can see, theoretical and simulation results concur very well. Figs. 12 and 14 depict ASEP for coherent BPSK. On Fig. 12 we presented only simulation results (given by markers), and ASEP based on *Q* function upper-bound given by (47) (full lines). Here we can see some deviations between simulation results and theoretical expression. On Fig. 14 we presented 16 curves. Full lines represent curve of ASEP obtained by MGF (51); dashed curve represent ASEP based on *Q* function upper-bound given by (52); dot-dashed curve represent ASEP based on *Q* function lower-bound given by (53); markers represent curve obtained by simulation. We can see that simulation result concur with ASEP obtained by MGF (which was to be expected), while these two curves lay under upper-bound ASEP, and above lower-bound ASEP. Also, we can see that curves obtained by (52) and (53) are almost concurring with exact ASEP obtained by MGF.

6. Conclusion

Throughout this chapter we presented two general fading distributions, the κ-μ distribution and the η-μ distribution. Since we have placed accent on MRC in this chapter, we investigated properties of these distributions (we derived probability density functions for envelope, received power and instantaneous SNR; cumulative distribution function, n-th order moment and moment generating functions for instantaneous SNR), and derived relationships concerning distribution of SNR at MRC output (outage probability). Then we have analyzed average symbol error probability at MRC output in presence of κ-μ and η-μ distributed fading (we derived average symbol error probability for coherent and non-coherent detection; upper and lower bound of average symbol error probability for coherent). The results obtained clearly stated the obvious:

- for larger κ outage probability and symbol error probability were smaller for fixed μ, and fixed average SNR;
- for larger μ outage probability and symbol error probability were smaller for fixed, κ and fixed average SNR;
- for larger μ outage probability and symbol error probability were smaller for fixed, η and fixed average SNR;
- for η and $1/\eta$ we obtain the same results;
- for a greater number of MRC branches, outage probability and symbol error rate were smaller for fixed κ and μ, and for fixed η and μ.

Also, we gave some outlines for design of κ-μ and η-μ RNG.

Still, there is a lot of investigation in this field of engineering. Namely, scenarios for κ-μ and η-μ can be generalized in manner to assume that all clusters of multipath have dominant components with arbitrary powers and scattered components with different powers. Also, we can introduce nonlinearity in this fading model in the way Weibull did. Also, one should consider correlation among clusters of multipath. For suggested models, one should analyze combining performances: switched combining, equal-gain combining, maximal-ratio combining, general-switched combining, etc.

7. References

Abramowitz, M.; Stegun, I.A. (1972). *Handbook of Mathematical Functions*, US Dept. of Commerce, National Bureau of Standards, Washington, DC

Annamalai, W.A.; Tellambura, C. (2002). Analysis of hybrid selection/maximal-ratio diversity combiners with Gaussian errors, *IEEE Transactions on Wireless Communications*, Vol. 1, No. 3, July 2002, pp. 498 - 511

Asplund, H.; Molisch, A.F.; Steinbauer, M. & Mehta, N.B. (2002), Clustering of Scatterers in Mobile Radio Channels - Evaluation and Modeling in the COST259 Directional Channel Model, *IEEE Proceedings of* International Conference on Communications, April-May 2002

da Costa, D.B.; Yacoub, M.D., Fraidenraich, G. (2005). Second-order statistics for diversity-combining of non-identical, unbalanced, correlated Weibull signals, *SBMO/IEEE MTT-S Proceedings of International Conference on Microwave and Optoelectronics*, pp. 501 - 505, July 2005

Fraidenraich, G.; Santos Filho, J.C.S.; Yacoub, M.D. (2005). Second-order statistics of maximal-ratio and equal-gain combining in Hoyt fading, *IEEE Communications Letters*, Vol. 9, No. 1, January 2005, pp. 19 - 21

Fraidenraich, G.; Yacoub, M.D.; Santos Filho, J.C.S. (2005). Second-order statistics of maximal-ratio and equal-gain combining in Weibull fading, *IEEE Communications Letters*, Vol. 9, No. 6, Jun 2005, pp. 499 - 501

Kim, S.W.; Kim, Y.G. ; Simon, M.K. (2003). Generalized selection combining based on the log-likelihood ratio, *IEEE Proceedings of International Conference on Communications*, pp. 2789 - 2794, May 2003

Marcum, J.I. (1947). *A Statistical Theory of Target Detection by Pulsed Radar*, Project RAND, Douglas Aircraft Company, Inc.,RA-15061, December 1947.

Milišić , M.; Hamza, M.; Behlilović, N.; Hadžialić, M. (2009). Symbol Error Probability Analysis of L-Branch Maximal-Ratio Combiner for Generalized η-μ Fading, *IEEE Proceedings of International Conference on Vehicular Technology*, pp. 1-5, April 2009

Milišić , M.; Hamza, M.; Hadžialić, M. (2008). Outage and symbol error probability performance of L-branch Maximal-Ratio combiner for generalized κ-μ fading, *IEEE Proceedings of International Symposium on Electronics in Marine - ELMAR*, pp. 231-236, September 2008

Milišić , M.; Hamza, M.; Hadžialić, M. (2008). Outage Performance of L-branch Maximal-Ratio Combiner for Generalized κ-μ Fading, *IEEE Proceedings of International Conference on Vehicular Technology*, pp. 325-329, May 2008

Milišić , M.; Hamza, M.; Hadžialić, M. (2009) BEP/SEP and Outage Performance Analysis of L-Branch Maximal-Ratio Combiner for κ-μ Fading, *International Journal of Digital Multimedia Broadcasting*, Vol.2009, 2009, 8 pages

Prudnikov, A.P.; Brychkov, Yu.A.; Marichev, O.I. (1992). *Integrals and series : Direct Laplace Transforms*, Gordon and Breach Science Publishers

Simon, M.K.; Alouini, M-S (2005). *Digital Communications over Fading Channels*, second edition, Wiley

Stuber, G. L. (1996). *Principles of Mobile Communications*, Kluwer Academic Publishers, Norwell, MA.

Yacoub, M. D. (2007). The κ-μ Distribution and the η-μ Distribution, *IEEE Antennas and Propagation Magazine*, Vol. 49, No. 1, February 2007, pp. 68 - 81

20

Technological Issues in the Design of Cost-Efficient Electronic Toll Collection Systems

José Santa[1], Rafael Toledo-Moreo[2], Benito Úbeda[3], Miguel A. Zamora-Izquierdo[4] and Antonio F. Gómez-Skarmeta[5]
University of Murcia
Spain

1. Introduction

Nowadays, communications become essential in the information society. Everyone can get information anywhere, even in mobility environments, using different kinds of devices and communication technologies. In this frame the vehicle is another place where users stay for long periods. Thus, in addition to safety applications, considered as the most important services, other networked applications could bring an additional value for the comfort of drivers and passengers, as well as for driving efficiency, in terms of mobility, traffic fluency and environment preservation.

The payment methods for road usage have received a great attention during the past two decades. More recently, new advances in ICT (information and communication technologies) have encouraged researchers all around the world to develop automatic charging systems aiming at avoiding manual payments at toll plazas while enabling administrations to deploy charging schemes capable to reduce congestion and pollution. The recent application of Global Navigation Satellite Systems (GNSS) on these charging platforms can present important advances, and the research community in ITS (Intelligent Transportation Systems) is aware of this.

Although charging systems for road use have been called in many different names, the two most extended have been toll collection and Road User Charging (RUC), which were established considering the prime two reasons for deploying these systems (Rad, 2001). Firstly, toll collection was initially employed for charging the users of certain road infrastructures, with the aim of recovering the costs in construction, operation and maintenance. Many studies defend the application of this economic model to finance future road networks (Yan et al., 2009), instead of using public taxes or charging vehicle owners with a periodic fee (this is the case of Spain, for instance). On the other hand, road user charging has been the term used when the final aim of the system is not only to obtain revenue for road deployment expenses, but also to modify certain traffic behaviors in order to reduce pollution or congestion (among others) (Fields et al., 2009). The application of ICT to automate the charging process has introduced new terms, such as electronic tolling or electronic toll collection. In practice, many authors in the literature use all these terms indistinctively.

During the past years, dedicated short-range communications (DSRC) have been a key technology to automate the charging process on roads. By means of an on-board transceiver, the vehicle is detected when passing toll points. In real deployments there are usually

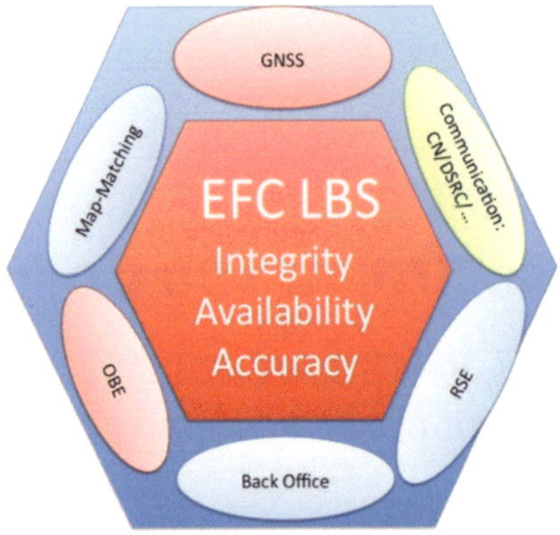

Fig. 1. Several elements comprise a GNSS-based electronic fee collection system.

speed limitations, since the communication channel between the on-board unit (OBU) and the roadside unit must be maintained for a while to allow the exchange of charging data. However, DSRC-based solutions present important problems, such as the cost of deploying roadside equipments when new roads want to be included in the system (a scalability problem) and a lack of flexibility for varying the set of road objects subject to charge. In this context, GNSS is lately considered as a good alternative. Essentially, GNSS-based RUC use geographic positions to locate vehicles in charging areas or roads, and this information is sent to the operator's back office to finally create the bill. The European Union is promoting the European Electronic Tolling Service (EETS) (Eur, 2009) as an interoperable system throughout Europe. This is based on a number of technologies, as it is shown in Fig. 1, although three of them are essential:

- Satellite positioning, GNSS;
- Mobile communications using cellular networks (CN);
- DSRC technology, using the microwave 5.8 GHz band.

Several standardization actions concerning electronic fee collection have been already considered by the European Commission, such as the security framework needed for an interoperable EETS, to enable trust between all stakeholders, and the definition of an examination framework for charging performing metrics.

Currently, some of the most important deployments of electronic RUC already use GNSS. In Switzerland, the LSVA system (also known as HVF for the English acronym of Heavy Vehicle Fee) complements a distance-based model that uses odometry and DSRC to check vehicle routes with GPS measurements. The role of GNSS in the German Toll Collect system is more remarkable, since GPS positions are used to identify road segments. Nevertheless, other extra mechanisms are used to assure vehicle charging in places where the GPS accuracy cannot guarantee the road identification. This problem has been analyzed for a potential deployment of a GNSS-based RUC in Denmark (Zabic, 2009), comparing the GPS performances obtained in 2003 and 2008. Although availability and accuracy problems had limited the usage of GNSS for RUC in the city of Copenhagen in 2003, more recent results showed that advances in receiver technology and updates in the GPS system made possible this application in 2008. This study supports this thesis primarily on the rise of the number of satellites in sight. In our opinion, these results must be taken with caution, since the experiments do not analyze how many satellites in view are affected by non-line-of-sight (NLOS) multi-path. In The Netherlands, the plans for creating a distance-based charging system for all vehicles on all roads also consider GNSS as the potential base technology (Eisses, 2008).

As it has been aforementioned, the accuracy of the position estimates is one of the main concerns towards the application of GNSS for RUC. It is necessary to provide a confidence level that assures that the estimate of the vehicle location is close enough to the real one with a certain high probability. This is the reason why the integrity concept is receiving a great attention in GNSS-based RUC these days (Pickford & Blythe, 2006). Per contra, the importance of the map-matching process is many times forgotten. When users are charged in accordance with the infrastructure used, the identification of charging objects (e.g. the road segment) is of key importance for the system. Even when the tariff scheme is not based on charging objects, the usage of additional digital cartography can be useful to improve the performance of the navigation system. Additionally, the communication subsystem, crucial for EFC, has not been properly attended in the literature so far. It is important to keep in mind that payment transactions cannot be completed if charging information does not arrive to credit and control centers. This chapter deals with these performance aspects regarding the navigation and communication subsystem and the map-matching algorithm used, all of these being key elements of EFC systems.

The rest of the chapter goes as follows. After presenting the concept of GNSS-based EFC in Section 2, more remarkable operation requirements and problems are analyzed in Section 3. The performance of the GNSS subsystem from the EFC perspective is then analyzed in more detail in Section 4. Next, some common methods for map-matching used in RUC are introduced in Section 5. Section 6 describes a proposal that further improves the performance of the navigation and map-matching subsystems, combining digital (and enhanced) maps with both GNSS and inertial sensors. Then, the performance of the communication subsystem when an on-board EFC unit is used is discussed in Section 7. Finally, Section 8 concludes the paper.

2. GNSS-based electronic fee collection

In GNSS-based RUC, information from the GNSS sensor is used to locate vehicles at charging places. The use of GNSS as the main positioning technology to charge users for the road usage, has several benefits related to flexibility and deployment costs:

- A minimum set of roadside units would be needed for enforcement purposes.
- OBU capabilities can be as simple as reporting GPS positions to a processing center, or as complex as calculating the charge and reporting payment transactions.
- A software-based OBU allows for software updates, reducing maintenance and system upgrade costs.
- GNSS sensors are cheaper and cheaper, and its performance is increasing.
- Cellular networks, which are the main communication technology considered, have a wide coverage, more than enough data rates for RUC, and decreasing costs that are also subject to agreements with operators.

Due to the flexibility of GNSS-based RUC, multitude of approaches can be designed to charge users. As main distinction factor, GNSS-based RUC solutions can be classified according to the tariff scheme used in the system. According to the literature (Cosmen-Schortmann et al., 2009; Grush et al., 2009), three tariff schemes can be distinguished:

Discrete charging In this case toll events are associated to the identification of road objects subject to be charged. This group includes single object charging (bridges, tunnels, etc.), closed road charging on certain motorway segments, discrete road links charging, cordon charging, or zone presence charging.

Continuous charging The tariff is calculated based on a cumulative value of time or distance. Distance-based charging and time-in-use charging are included in this group.

Mixed charging A combination of aforementioned approaches is used. An example of this tariff scheme is charging for cumulative distance or time considering a different price for each road segment.

3. Measuring the performance of GNSS-based RUC

A clear definition of the performance requirements for a road user charging system is needed for two main reasons. First of all, the industrial consortiums that apply for a deployment must be equally evaluated and the final choice must be based on the goodness of each solution according to some previously established performance needs. Secondly, the interests of users and authorities must be guaranteed.

Performance requirements must be described in such a way that any possible implementation that fulfills the needs may be under consideration and verifiable by means of field tests. Thus, requirements must be independent of the technology and internal calculations for charging. As the authors of Cosmen-Schortmann et al. (2009) claim, the issue of the positioning errors must be addressed by the proposed system, but not directly evaluated by the third part examiner that will evaluate all the proposals. The description of the performance requirements depends on the final charging scheme. Since it is likely that any final charging scheme is based on a combination of continuous and discrete ones, let us analyze briefly both cases here.

For a discrete charging scheme, there are only four possible cases: a correct detection (CD), a correct rejection (CR), a missed detection (MD) and a false detection (FD). Last two cases cause undercharging and overcharging respectively. Because the consequences of a MD and a FD are not the same, it is necessary to analyze these effects separately, and not by a single index of overall correct detection rate. Therefore, there must be two different performance requirements to avoid overcharges (for users) and to ensure revenues by avoiding undercharges (for authorities). Furthermore, it must be decided whether the requirements must be satisfied any time, for any trip in any scenario and under any circumstance, or it is enough if the average and some statistical parameters show that the overall errors of overcharge and undercharge are within desirable thresholds. The latter may lead to persistent errors in the bills of some users who repeatedly drive trajectories not well covered by the RUC system, due for example to bad satellite visibility conditions in the area. These special cases should be handled as exceptions, because it cannot be accepted that a system does not treat fairly every user.

Analogously, for continuous schemes two parameters are also needed to protect the interests of both users and service providers. Inspired by the notation of the navigation community (Santa, 2009), some authors introduce the concepts of charging availability and charging integrity (Cosmen-Schortmann et al., 2006). Charging availability can be defined as the probability that the charging error is within a desirable error interval. This parameter protects the interest of both the user and the toll charger, since it covers positive and negative errors (overcharges and undercharges respectively). Its main mission is to provide the toll charger with a level of warranty that the user will pay for the road infrastructure usage. On the contrary, charging integrity can be defined as the probability that the error is not over an upper limit; this is, that the user is not overcharged, and its value must be more restrictive than the charging availability (this is why we claim that the main objective of charging availability is to protect the interests of the authorities).

Since the charging integrity cannot be compromised, the developers must find a way to be aware of the reliability of every charge. In case of reasonable doubt, it is preferable not to charge, rather than to charge wrongly. For this reason, some integrity indexes must be calculated to verify the certainty of the charges. If integrity indexes inform of a possibly unsafe charge and the user is finally not charged, the probability associated to charging availability becomes smaller, but not the one linked to charging integrity. On the contrary, if the user is charged wrongly, both values of probabilities become smaller and the charging availability and integrity are compromised. The tuning of the integrity indexes must be done in such a way that it satisfies the needs regarding availability and integrity. If this tuning cannot be found, the system is incapable of providing the aimed level of reliability and it must be disregarded.

Although a good estimation of the integrity parameters is crucial for the developers, this aspect must neither appear in the definition of performance requirements, nor being tracked during the evaluations. It must be understood only as an internal parameter that eventually affects the charging availability and integrity.

Finally, one must bear in mind that the performance indexes coming from both discrete and continuous schemes must be transformed into a unique performance parameter, based for example on the impact of each error (discrete or continuous) on the eventual charge. This is necessary since despite the fact that the proposals coming from the industry could be based on different charging schemes, there must be a possible direct comparison for all of them and the final system must be seen as a sole charging system independent of the scheme particularities. Furthermore, the integration of continuous and discrete performance indexes turns into essential for mixed charging schemes.

4. GNSS performance issues

The main technological drawback of GNSS-based RUC is the performance of the GNSS sensor. The lack of availability of the GPS signals at places where there is no line of sight with satellites is a remarkable problem in urban canyons, tunnels or mountain roads, for instance. A research assignment demanded by the Dutch Ministry of Transport, Public Works and Water Management (Zijderhand et al., 2006) focuses on the accuracy and reliability of distance and position measurements by GNSS systems. The trials involved 19 vehicles during one month, and concluded that during the 13% of the traveling time there was no valid GPS position, although the overwhelming part of the unavailability was due to time to first fix (TTFF).

Highly related to this, the continuity of the GPS services is also dependent on military decisions of the US government, since GPS is not a pure-civil navigation system. Moreover, the accuracy of the position estimates, although it has been improved thanks to enhancements in the space segment and in the receiver technology, is still not fully reliable to decide whether or not a user must be charged for supposedly using a road. Although some performance problems can be compensated (satellite clock bias, signal propagation delay, etc.), others such as multi-path effects in the user plane are not yet modeled and degrade the accuracy in urban canyons above all. All these problems can reduce the performance of a liability critical service such as RUC. The analysis made in Zijderhand et al. (2006) for GNSS positioning accuracy shows that its 95% level is 37 m. Nevertheless, this number must be taken with caution when considering RUC applications, because many other factors apart from the

GPS inaccuracies themselves can affect this result, such as inaccuracies in digital maps or errors in the map-matching process. The consequences of the positioning errors in the system performance would not be so severe if current GNSS devices provide a fully meaningful value of the reliability of the positioning: its integrity. In this case, although the performance of the system may diminish, its integrity remains and users would be protected against overcharge. It is then up to the authorities to decide whether or not the expected performance is good enough to deploy the system, in other words, to ensure the revenue of the investment. However, current integrity values provided for GNSS devices are inappropriate.

An approximation to provide integrity in GNSS-based positioning is given by the Receiver Autonomous Integrity Monitoring (RAIM) algorithm. This technique, initially created for aerial navigation, is based on an over-determined solution to evaluate its consistency, and therefore it requires a minimum of five satellites to detect a satellite anomaly, and six or more to be able to reject it (Kaplan, 1996). Unfortunately, this cannot be assumed in usual road traffic situations, especially in cities (Verhoef & Mohring, 2009). In addition, the RAIM method assumes that only one failure appears at once, something feasible in the aerial field, but not in road scenarios: it is usual that several satellite signals are affected by simultaneous multi-path propagations in an urban area. Satellite Based Augmentation Systems (SBAS), such as EGNOS (European Geostationary Navigation Overlay Service) or WAAS (Wide Area Augmentation System), also offer integrity calculation. By means of the information about the GNSS operational state, broadcasted by GEO satellites, it is possible to compute a parameter of system integrity (Bacci et al., 2005; Toledo-Moreo et al., 2008). However, this approach does not consider local errors such as multi-path, which are of key importance in terrestrial navigation.

Due to these problems, in the last years some authors have suggested new paradigms to estimate the system integrity (Martinez-Olague & Cosmen-Schortmann, 2007; Yan et al., 2009). In concrete, the work described in Yan et al. (2009) shows an interesting approach for integrity provision based solely on GNSS that obtains interesting results. Fig. 2 illustrates the solutions provided by two different approaches (Santa, 2009; Toledo-Moreo et al., 2008; 2007) (among a number of the literature) for position integrity. The red line represents the HPL (Horizontal Protection Level) estimated by using the information provided by EGNOS. HPL does not include local errors at the user plane (such as multi-path) or the contribution of the aiding sensors. The green line shows the HIT (Horizontal Integrity Threshold) values along the trajectory. HIT represents the confidence on the horizontal position estimated by the filter that fuses the sensor data (this could be a particle filter or a Kalman filter, for instance). In Fig. 2, HIT does not show the peaks that appear in HPL caused by bad GNSS coverage, since HIT follows errors models that consider the vehicle and the aiding sensors. In this way, although HIT does not consider EGNOS integrity information for each satellite, it usually offers a better estimation of the real performance of the navigation system, since a multi-sensor approach (which supports periods of GNSS absence) is considered.

5. Map-matching for road user charging

In tariff schemes where the user is charged for driving along a road stretch or using a certain road infrastructure, the map-matching algorithm plays an essential role. However, as far as the authors know, there is not enough information in the literature about these algorithms applied to RUC, since current approximations are inside proprietary RUC solutions. This is identified as a problem towards standardization and calibration, apart from making more difficult the comparison between different algorithms.

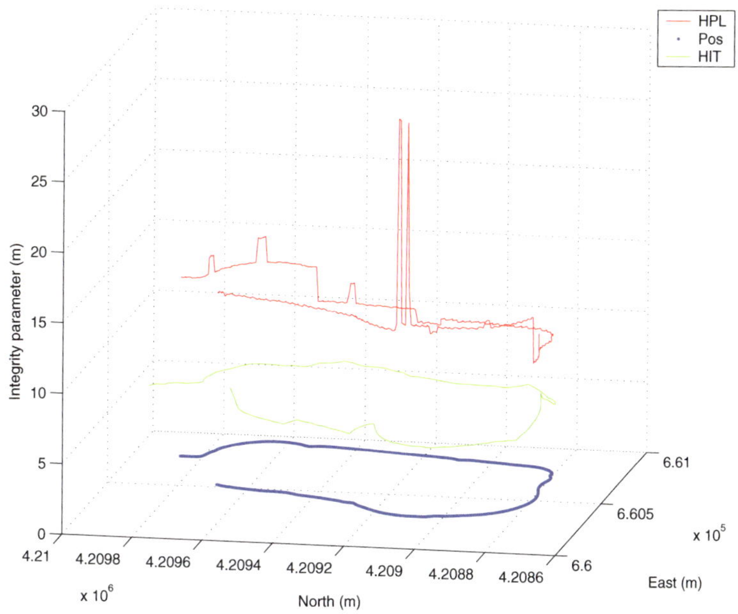

Fig. 2. Two different solutions for navigation integrity over the vehicle trajectory (in blue).

The most common algorithm used in map-matching is considering the distance between the vehicle location and the nearest road segment. In this way, apart from the GNSS sensor, digital information about the road network is necessary. Fig. 3 illustrates this algorithm, based on the point to segment distance. An ENU (East, North, Up) cartesian coordinate system is considered, and the computed fix for the vehicle at moment t_k is denoted as $P_{km} = (x_{km}, y_{km})$. The algorithm has three main steps:

1. Search for a road segment near the vehicle position, with coordinates $P_1 = (x_1, y_1)$ and $P_2 = (x_2, y_2)$.
2. Calculate the distance d_m between P_{km} and the segment.
3. If current segment is closer than previous segments to the position estimate, take it as a candidate.

An scenario which illustrates a correct operation of the previous algorithm is shown in Fig. 4. The vehicle is correctly detected at the entrance and exit points in the charing link, and the K road segments pertaining to the stretch are also identified. However, in real complex scenarios, GNSS performance problems can imply misdetection of road segments and overcharging or undercharging.

An extra problem appears when vehicles drive near a charge link but the real driving road is not present in the digital cartography. An umbral factor to detect roads can help to solve this problem. Fig. 5 illustrates this solution over a distance-based charging scheme. It considers a 57 km travel of a vehicle along a mix of charge and free roads. The last ones were selected from the available secondary roads which are parallel to the main highway. For this case, a threshold of 10 m was found useful to solve the misdetection problem. According to our large

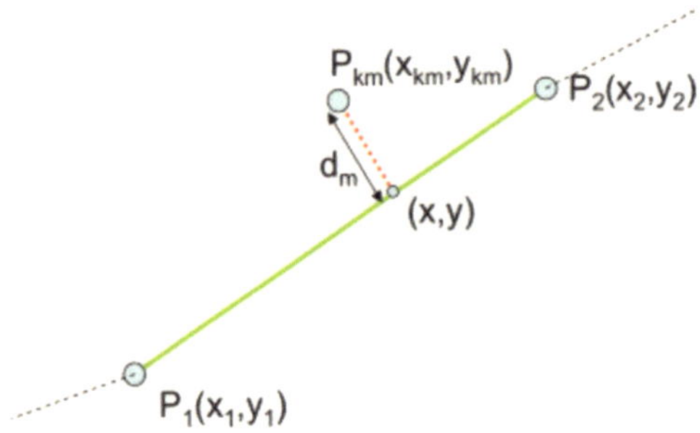

Fig. 3. Point-segment distance in map-matching.

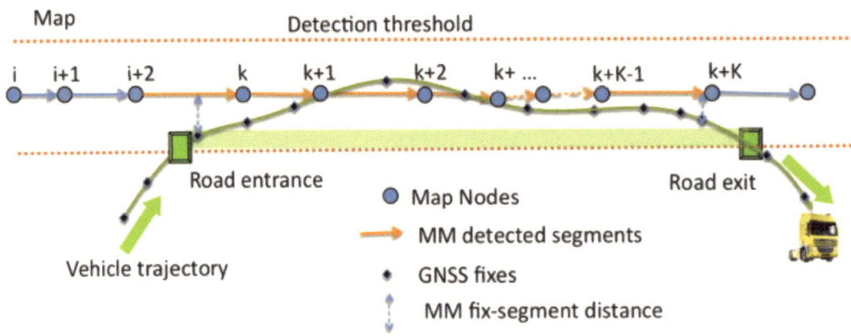

Fig. 4. Correct operation of map-matching using point-segment distance.

number of tests on Spanish roads, this technique and a suitable threshold can be useful to solve the problem of non-digitalized parallel secondary roads. However, further mechanisms are necessary to assure the correct identification of roads under potential GNSS performance errors and when more than one applicable road require a disambiguation decision.

6. Complementing GNSS in RUC

According to the current literature and our own tests, at its present form, the simplest approach for GNSS-based location for RUC based on single GPS positioning or GPS positioning map-matched to a standard digital cartography is not capable to ensure the demanded levels of performance availability and integrity. To enhance these results, standard positioning can be aided by different sensors in both the onboard equipment (OBE) and the road side equipment (RSE). We analyze in this section the main benefits of GNSS aided location with on-board sensors and maps for the purpose of RUC and its effect on the provision of performance integrity, with a especial emphasis on map aided road user charging.

6.1 Aiding positioning sensors

Many advanced positioning systems employ a minimum set of a GNSS receiver, an odometer for speed values and a gyroscope for heading estimates. This configuration presents a good balance between performance and budget. During GNSS outages, the dead-reckoning system

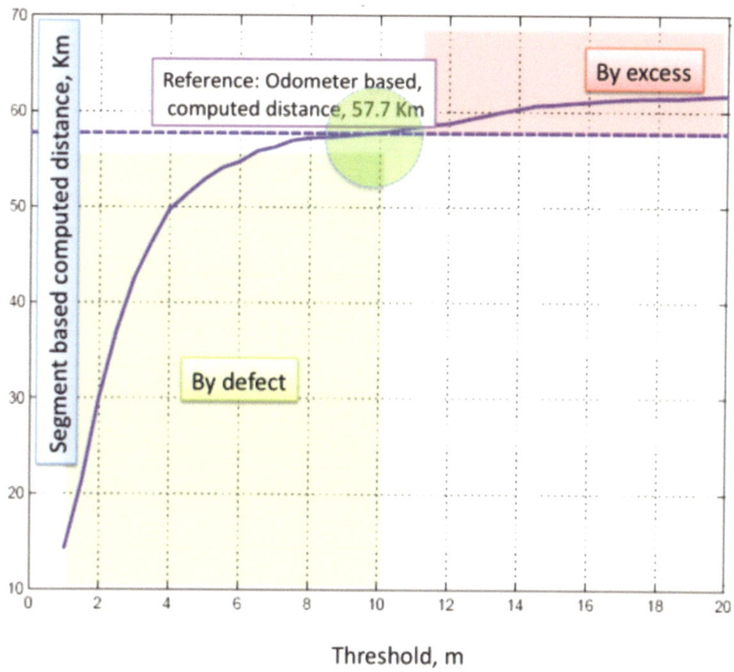

Fig. 5. Undercharging and overcharging and the selection of a threshold value in map-matching.

keeps estimating positions. The magnitude of position drifts depends primarily on the quality of the aiding sensors, but also on the skills of the algorithm used for sensor integration. Some interesting examples of GNSS-aided positioning in either loose or tight coupling modes can be found in Farrell & Barth (1998); Toledo-Moreo et al. (2007); Yang & Farrell (2003).

Aiding positioning supports RUC because as long as the quality of the position is kept and guaranteed, the road user charging system can stay available. Another advantage comes from the fact that hybridization algorithms smooth the noisy trajectories generated by the GNSS positions and represent more realistically the movements of vehicles, what can be useful to eliminate to some extent the overcharge accumulated in distance-based charging schemes that employ the GNSS positions to estimate the distance. It is also possible to compare the odometer distance and the GNSS-based one for enforcement purposes.

6.2 Exploiting enhanced maps for road user charging

Most Geographical Information Systems (GIS) represent roads with one or two polylines depending whether or not lanes with opposite driving directions are physically separated, being these polylines series of nodes and shape points, connected by segments. Apart from the global inaccuracy (from 5 m in urban areas up to 20 m in intercity roads) and the inaccuracy consequences of the local approximation of the road by series of linear segments, standard maps lack in contents. All these factors limit significantly the benefits of map-aided location for RUC.

The concept of enhanced maps (Emaps) was introduced with the objectives of reaching decimeter accuracy both globally and locally, respecting the shape of the road, and

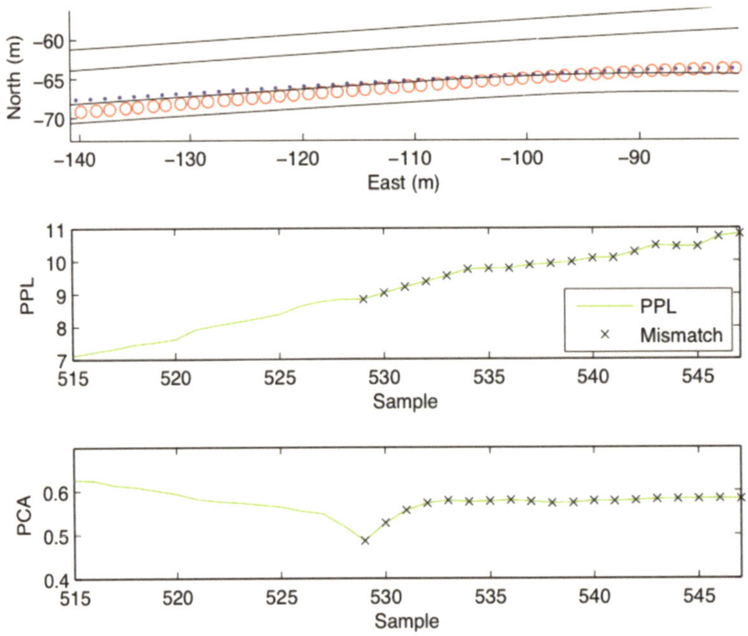

Fig. 6. (Top) Stretch of the trajectory during a period when the position estimates drifted as a consequence of a simulated GPS outage: solid black lines are the map; blue dots are the assumed true positions given by DGPS; red circles are the position estimates given by the PF. (Middle) PPL values during the period when the GNSS coverage is gone. The period of lane mismatch has been marked manually with black crosses. (Bottom) PCA during the same period.

representing all the lanes of the carriageway and their topological links. Our group collaborated with the Geolocalization research team of the Laboratoire Central des Ponts et Chaussées of Nantes, France, in the creation of a novel Emap introduced in the frame of the European Cooperative Vehicle Infrastructure Systems (CVIS) project (*Cooperative Vehicle-Infrastructure Systems (CVIS) project*, n.d.). This work proved to be useful for enhanced positioning and map-matching at the lane level (Bétaille et al., 2008). It is the authors' opinion that the benefits of Emaps can be exploited for RUC in the scenarios where standard maps fail.

Fig. 6 shows a stretch of one test carried out for a demonstration of the CVIS project. In the upper image the lanes are plotted from the data stored in the Emap. The high accuracy of the lanes allows one to distinguish at the first sight the drift in the position estimate that was caused by the simulation of GPS outage. The map building process, based on kinematic GPS integrated with inertial sensors and off-line processing, assured an error lower than five centimeters with respect to the driven middle-lane. Therefore, it can be claimed that the vehicle is actually on a lane if the confidence on the positioning is high enough. Blue dots represent the assumed true trajectory given by DGPS during a test. Middle and bottom images are to present the benefits of the confidence indicators, to be explained in next section.

6.3 Integrity provision

Independently of the charging scheme applicable for a road stretch or area, toll chargers need to know the level of reliability on the charge. For continuous schemes based on cumulative distance, this level can be represented by error-free positioning estimates, such as the one presented in Martinez-Olague & Cosmen-Schortmann (2007). However, as it has been previously stated, continuous charging schemes are likely to be completed with discrete ones, bringing the need of map-matching. It is the authors' opinion that when map-matching is needed for making the decision of whether or not a user should be charged, a single integrity indicator of the positioning error is not enough to represent the situation.

In the frame of the CVIS project, a double integrity indicator that represents the reliability of an algorithm for positioning and map-matching at the lane level was proposed in Toledo-Moreo et al. (2010). We believe that this paradigm can be exploited for road user charging purposes. To do it so, the proposed integrity parameters should represent the confidence on the positioning itself, as well as the confidence on the assignation of a position (or trajectory) to a road segment. The combination of both indexes may offer a representative idea of the positioning and map-matching process because they complement one another. The interest of the confidence on the position for continuous charging schemes is clear and it was discussed before. Nevertheless, there may be cases where the confidence on the position estimates is low due to bad GNSS coverage, but the map-matching problem is trivial, and the final confidence on the assignment can be high. On the contrary, even with a high confidence on the position estimates, if map-matching is difficult in a concrete scenario, the overall confidence should probably be low. In Toledo-Moreo et al. (2010) our paradigm of a double integrity index was proven to detect a significant number of wrong assignments at the lane level, improving the overall perception of the vehicle location. This can be exploited for road user charging purposes. The readers are welcome to follow this reference for further details.

The advantages of using these integrity indicators for the distinction of two adjacent lanes of the same carriageway is shown in Fig. 6. This situation could represent the RUC scenario of having contiguous lanes of a highway subject to different charges or the common case of two roads with different tariffs that go parallel and close along a certain distance. As it can be seen in the upper image of the figure, at one point of the trajectory the position estimates drift as a consequence of a simulated GPS gap (in this test the vehicle moves from the left to the right side of the image). The aiding sensor-set keeps the position estimates in good track for a while, but due to its low cost, eventually drifts and locates the vehicle in the contiguous lane. The increasing lack of confidence on the position is represented by the Positioning Protection Level (PPL) value and visible in the middle image of Fig. 6. PPL represents here a protection level based on the covariance of the positioning variables of the filter. However, even with lack of GPS coverage, the positioning and map-matching algorithm would not have allocated the vehicle in the wrong lane if both lanes would not be topologically connected: i.e., if due to physical or legal constraints the vehicle could not make a lane change to the left at that point. This is because the topological information stored in the Emap binds the vehicle location to the areas of feasible maneuvers. In that case, PPL values would still be high and the confidence on the position low, even though the vehicle is correctly assigned to the lane. The use of the Probability of Correct Allocation (PCA) indicator provides the information needed to distinguish between these two scenarios and deciding whether or not the vehicle can be charged. At the bottom of Fig. 6 it can be seen how PCA values become lower and lower, reaching the lowest value when the lane mismatch begins (the PCA value confirms the mismatch). Therefore, PCA enables the decision making of whether or not a rise of the PPL value corresponds to an incorrect lane allocation.

7. Performance of the communication link

There is another part of the OBU of key importance to perform payment transactions in GNSS-based EFC: the communication subsystem. Independently of the OBU capabilities to save the vehicle route or calculate the toll fee, a communication channel is necessary to send this information on time to a processing center. The performance of this link is also relevant, since payment transactions must be committed on the back office.

Among the different communication technologies used in vehicle telematics, the most considered for GNSS-based EFC are cellular networks. These are highly useful for EFC, since they offer vehicle to infrastructure communications by means of a wide deployment along road networks. Other communication technologies, such as DSRC or WiFi, are less suitable, since they are against the flexibility principle of GNSS road-pricing schemes, but they can be considered as complementary technologies. A set of communication performance parameters must be assessed in a GNSS-based EFC system, in order to include the necessary OBU capabilities to cache/calculate charging information or choose a telecom operator. Main performance issues of cellular networks include network coverage, network capacity, mobility conditions or vehicle speed (Santa et al., 2006).

Network availability is one of the main drawbacks of CN systems. Telecom operators do not offer the same service over the terrestrial surface. In urban environments, the CN coverage is excellent, due to the number of base stations where the mobile terminal can connect. In rural areas, however, the CN deployment is poor, or even null in some places. It is also important to differentiate between two important concepts regarding network availability in CN: coverage and capacity. Coverage can be understood as the possibility of the mobile terminal to use the network, because in this exact location, an operator has deployed (or not) the necessary infrastructure. However, even under good coverage conditions, the user can be rejected to establish a call or a data connection if the capacity of the network has been exceeded. Depending on several technological issues, such as modulation, frequency allocation and time slot scheduling, this effect has a different behavior. This way, the number of users who are concurrently using the network also restricts the CN availability. Since capacity or coverage problems can appear suddenly, the OBU must be ready to cache all charging information, which will be sent as soon as a data link is available again.

Apart from potential access to the network, some problems arise due to mobility effects. Handoffs between base stations are also important, due to potential decrease of performance in the process. If the mobile terminal is moving at locations far away from the base station without performing a handoff, poor latency and throughput results can be obtained (Alexiou et al., 2004). However, this effect is not remarkable in EFC environments these days, since most of the charged road networks are highways, where handoffs are less frequent. Nevertheless, the distance between two physical edges in the communication is not the only noticeable effect of mobility. Interference with other radio equipments and especially bad orography conditions could also cause communication problems in CN systems.

Finally, the speed is also a noticeable issue in cellular networks (Litjens, 2002). At the physical level, effects such as Doppler shift, Rayleigh fading and multipath propagation limit the bit rate allowable in CN at high speed. At link level, handoff issues must also be considered. Because the handoff process takes time (depending on the type), vehicles at high speed could have problems in places where the base station density is too high. For this reason, a high rate

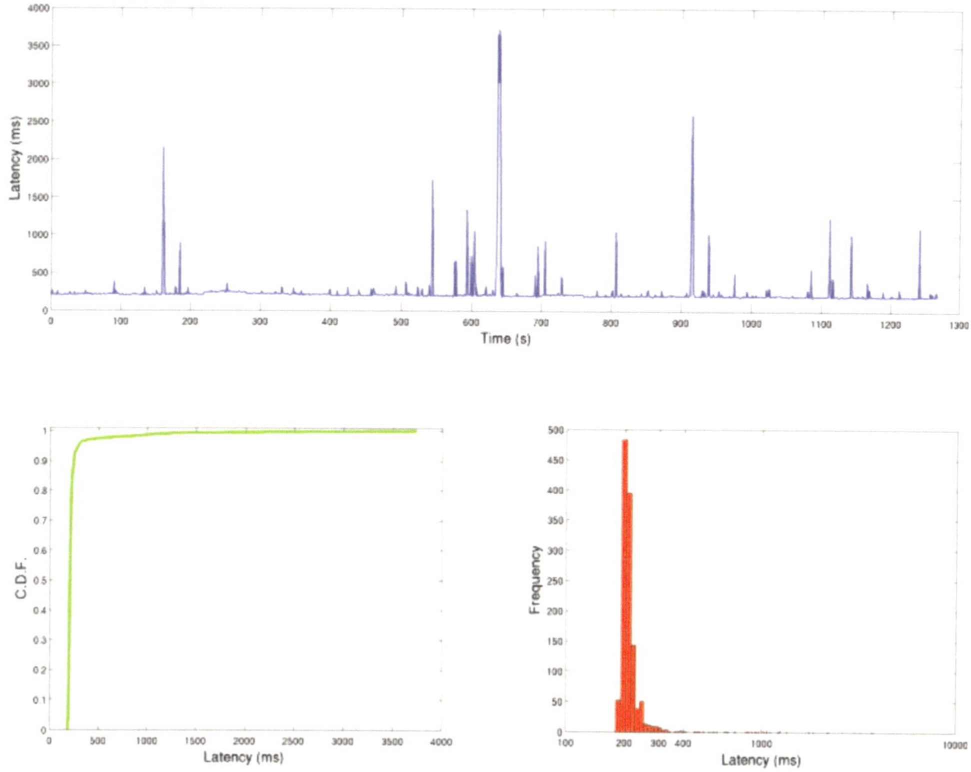

Fig. 7. An evaluation of a vehicle-to-infrastructure communication test using the UMTS cellular network.

of base stations at highways is favorable, apart from the obvious reduction in deployment costs. The tests shown in Fig. 7 illustrates some of the previous communication issues using a UMTS (Universal Mobile Telecommunications System) link. The on-board system considered in the field trials (Santa et al., 2010) uses the cellular network to send periodical messages to a server connected to Internet at the infrastructure side. The first graph shows the delay values for each message sent from the OBU. At a first glance, it is noticeable how mobility conditions cause continuous delay peaks.

There are three main problematic areas in the results shown in Fig. 7, which are noticeable in the groups of delay peaks observed in the first graph. The first one, between times 150 s and 200 s, comprises a road stretch where the car drives near a parking area covered with a metallic roof. Since radio signals are partially reflected, the network performance decreases. The second problematic area is more evident, between times 540 s and 700 s. The vehicle reaches the farthest position from the CN base station used for the connection and several buildings also block the line of sight with it, provoking a severe coverage problem. After leaving this area, the vehicle comes back towards the base station, where the network performance is good again. However, just before reaching the base station, the vehicle turns and goes across a third problematic zone between times 900 s and 950 s. At this location, a small hill between the vehicle and the base station decreases the channel quality. The graph that illustrates the cumulative distribution function (CDF) of the delay results shows that values between 180 and

240 ms comprise more than 90% of the messages. The rest of latency values are distributed in a quasi-logarithmic trend, since high latencies are less and less common. The last graph in Fig. 7 clearly illustrates this distribution of values, showing a histogram plot of the latency results with containers that are situated in a logarithmic scale on the axis of abscissas.

As it has been checked, mobile conditions in real scenarios cause performance variations when a cellular link is used to communicate with a node located in the wired Internet. Due to this, a good EFC system must be capable of caching enough charging information to commit payment transactions when the network is not available. Nonetheless, it is important to realize the good performance of new UMTS networks (a mean delay of 240 ms in the previous test) in places where telecom operators have deployed the necessary infrastructure. This should encourage researchers and engineers to consider this communication technology for telematic services such as road user charging in the next years.

8. Conclusion

The deployment of GNSS-based road pricing schemes needs to consider the impact of the different technologies involved in such a great system. The performances of the navigation and communication subsystems are found essential for a correct operation of a GNSS-based EFC but, as it has been described, the design of the map-matching algorithm is also a challenge. The analysis carried out in this chapter is considered as a first step prior to the assessment of high-level performance measurements in GNSS-based EFC, as charging reliability.

For today, it seems really difficult that any solution exclusively based on GNSS for positioning can ensure the high standards for charging reliability for road user charging. For this reason, it is the authors opinion that GNSS technology must be supported by aiding information coming from onboard sensors and even digital maps, following the line given in this chapter. Moreover, the calculation of integrity factors that indicate the position goodness at any time is considered a key need to measure and track the performance of the navigation system in liability critical services such as road charging, independently of the navigation system used. A proposal introduced in this chapter shows how the integration of multi-sensor navigation systems and next generation digital maps presents interesting advances to provide integrity and further map-matching capabilities.

Cellular networks have been identified as a good communication technology for GNSS-based EFC, due to their wide deployment and recent advances. However, several performance issues in real environments indicate that on-board units must be prepared to overcome eventual communication problems with backend systems. Among all these issues, the coverage and the capacity of the network in road segments are identified as the most important ones. In fact, these are being considered by telecom operators nowadays, installing more base stations or making the most of the available frequency spectrum. Only a reliable communication link can assure that payment transactions are committed at the back office of an electronic RUC system.

A deep study about all performance requirements in GNSS EFC is needed, or at least considering these three important elements: navigation subsystem, communication technology and map-matching technique. According to authors understanding, this preliminary analysis is needed for any GNSS-based road tolling system subject to be widely deployed in real environments, in order to guarantee both the user service and a fair charging scheme.

9. Acknowledgements

This work has been sponsored by the Spanish Ministry of Education and Science, through the SEISCIENTOS Project (TIN2008-06441-C02), and the Seneca Foundation, by means of the Excellence Researching Group Program (04552/GERM/06). In addition, the authors would like to thank the colleagues of the GNSS Metering Association for Road User Charging (GMAR), for the fruitful discussions about charging integrity, and also acknowledge the joint work carried out between University of Murcia and Laboratoire Central des Ponts et Chaussées of Nantes, in part of the designs and results briefly introduced in this chapter.

10. References

Alexiou, A., Bouras, C. & Igglesis, V. (2004). Performance evaluation of TCP over UMTS transport channels, *Proceedings of International Symposium on Communications Interworking*, Ottawa.

Bacci, G., Principe, F., Luise, M., Terzi, C. & Casucci, M. (2005). Soft-rec: A gps real-time software receiver with egnos augmentation, *Proceedings of Workshop on EGNOS performance and applications*, Gdynia.

Bétaille, D., Toledo-Moreo, R. & Laneurit, J. (2008). Making an enhanced map for lane location based services, *Proceedings of 11th International IEEE Conference on Intelligent Transportation Systems*, Beijing.

Cooperative Vehicle-Infrastructure Systems (CVIS) project (n.d.). http://www.cvisproject.org.

Cosmen-Schortmann, J., Azaola-Saenz, M., Martinez-Olague, M. & Toledo-Lopez, J. (2006). Integrity in urban and road environments and its use in liability critical applications, *Proceedings of 2008 IEEE/ION Position, Location and Navigation Symposium*, Monterey.

Cosmen-Schortmann, J., Grush, B., Hamilton, C., Azaola-Saenz, M. & Martinez-Olague, M. (2009). The need for standard performance definitions for GNSS road use metering, *Proceedings of 16th World Congress on Intelligent Transport Systems*, Stockholm.

Eisses, S. (2008). The distance-based charging project in the netherlands, *Proceedings of 7th European Congress and Exhibition on Intelligent Transport Systems and Services*, Geneva.

Eur (2009). *Decision 2009/750/EC of the European Commission on the definition of the European Electronic Toll Service and its technical elements*, document c(2009) 7547 edn.

Farrell, J. & Barth, M. (1998). *The Global Positioning System and Inertial Navigation : Theory and Practice*, McGraw-Hill, USA.

Fields, G., Hartgen, D., Moore, A. & Poole, R. (2009). Relieving congestion by adding road capacity and tolling, *International Journal of Sustainable Transportation* 3(5): 360–372.

Grush, B., Cosmen-Schrtmann, J., Hamilton, C. & Martinez-Olague, M. (2009). GMAR standard performance definitions for GNSS road use metering, *Proceedings of 16th World Congress on Intelligent Transport Systems*, Stockholm.

Kaplan, E. (1996). *Understanding GPS: Principles and Applications*, Artech House, USA.

Litjens, R. (2002). The impact of mobility on UMTS network planning, *Computer Networks* 38(4): 497–515.

Martinez-Olague, M. & Cosmen-Schortmann, J. (2007). Gnss based electronic toll collection system of guaranteed performances, *Proceedings of 14th World Congress on Intelligent Transport Systems*, Beijing.

Pickford, A. & Blythe, T. (2006). *Road User Charging and Electronic Toll Collection*, Artech House, USA.

Rad (2001). *Minimum Operational Performance Standards For Global Positioning System/Wide Area Augmentation System Airborne Equipment*, RTCA/DO-229C edn.

Santa, J. (2009). *Service Deployment Platform for Intelligent Transportation Systems*, Lambert, Germany.

Santa, J., Toledo-Moreo, R., Zamora-Izquierdo, M., Ubeda, B. & Gomez-Skarmeta, A. (2010). An analysis of communication and navigation issues in collision avoidance support systems, *Transportation Research Part C: Emerging Technologies* 18(3): 351–366.

Santa, J., Ubeda, B., R., T.-M. & Skarmeta, A. (2006). Monitoring the position integrity in road transport localization based services, *Proceedings of IEEE Vehicular Technology Conference Fall*, Montreal.

Toledo-Moreo, R., Bétaille, D. & François, P. (2010). Lane-level integrity provision for navigation and map matching with GNSS, dead reckoning, and enhanced maps, *IEEE Transactions on Intelligent Transportation Systems* 11(1): 100–112.

Toledo-Moreo, R., Santa, J., Zamora, M., Úbeda, B. & Skarmeta, A. (2008). A study of integrity indicators in outdoor navigation systems for modern road vehicle applications, *Proceedings of IEEE/RAS 2nd Workshop on Planning, Perception and Navigation for Intelligent Vehicles*, Nice.

Toledo-Moreo, R., Zamora-Izquierdo, M., Ubeda-Miñarro, B. & Gomez-Skarmeta, A. (2007). High-integrity IMM-EKF-based road vehicle navigation with low-cost GPS/SBAS/INS, *IEEE Transactions on Intelligent Transportation Systems* 8(3): 491–511.

Verhoef, E. & Mohring, H. (2009). Self-financing roads, *International Journal of Sustainable Transportation* 3(5): 293–311.

Yan, G., Chen, X. & Olariu, S. (2009). Providing VANET position integrity through filtering, *Proceedings of 12th IEEE Intelligent Transportation Systems Conference*, St Louis.

Yang, Y. & Farrell, J. (2003). Magnetometer and differential carrier phase GPS-aided INS for advanced vehicle control, *IEEE Transactions on Robotics and Automation* 2(2): 269–282.

Zabic, M. (2009). Road charging in copenhagen: A comparative study of the GPS performance, *Proceedings of 16th World Congress on Intelligent Transport Systems*, Stockholm.

Zijderhand, F., Van Nifterick, W. & Zwiers, A. (2006). Accuracy and reliability of distance and position measurements by GNSS systems, *Technical report*, ARS Consulting, The Hague.

21

Propagation Aspects in Vehicular Networks

Lorenzo Rubio[1], Juan Reig[1] and Herman Fernández[2]
[1]*Universidad Politécnica de Valencia*
[2]*Universidad Pedagógica y Tecnológica of Colombia*
[1]*Spain*
[2]*Colombia*

1. Introduction

Traffic accidents have become an important health and social problem due to the enormous number of fatalities and injuries. The total number of deaths and injuries in the European Union (EU), United States of America (USA) and Japan has been steadily reduced over the last decade. This reduction is mainly attributed to the implementation of a set of road safety measures, such as seat-belt use, vehicle crash protection, traffic-calming interventions and traffic law enforcement. However, the number of accidents has remained uniform due to the increasing number of vehicles and total distance driven (Peden et al., 2004).

In addition to passive vehicle safety systems, such as airbags, anti-lock braking system (ABS) and electronic stability control (ESP), new active safety systems have been introduced to improve vehicular safety. To this end, the last decade has witnessed the traffic management industry, engage and promote the integration of information and communications technology (wireless, computing and advanced sensor technologies) into both vehicles and the wider transport infrastructure. These proposals have led to the intelligent transportation system (ITS) concept. At present, different ITS applications have been introduced, such as variable message signs (VMS), located at strategic points (e.g., tunnels and merging highways) or spaced at given distances, to inform drivers about traffic and dangerous situations; automated toll collection systems for highways and parkings; and real-time traffic information broadcasted in the FM radio band.

Besides this, onboard ITS applications have improved the assistance and protection mechanisms for drivers: navigation systems, rear and front parking radars, and cameras are extensively used in the vehicle. Vehicles now incorporate sophisticated computing systems, with several sensors interconnected. However, short-range sensors employed in emergency systems, such as forward collision warning and lane keeping assist, are insufficient, specially when these sensors need to extend their communication horizon in emergency cases due to the limitation of their operating range to line-of-sight (LOS) conditions (Vlacic et al., 2001).

Therefore, there are safety applications for distance emergency situations, such as blind corners and traffic crossing, where large-range vehicular communication systems are required, e.g., operating rages of about 1000 meters, in both LOS and non-LOS (NLOS) conditions (Gallagher & Akatsuka, 2006). Wireless communications systems which can operate with these constraints are known as cooperative systems on the road. In the cooperative system concept, vehicles and infrastructure exchange safety messages to extend the distance horizon and provide more information in real time to drivers. Cooperative systems involve two capabilities: vehicle-to-infrastructure (V2I) and vehicle-to-vehicle (V2V) communications. V2I and V2V communications are also referred to in the literature as V2V communications.

Government radio management organizations have been allocated specific bands for ITS throughout the world. In USA, the Federal Communication Commission (FCC) assigned 75 MHz of licensed spectrum, from 5.85 GHz to 5.925 GHz, as part of ITS for dedicated short-range communications (DSRC) in 1999. The European Telecommunication Standard Institute (ETSI) allocated 70 MHz for ITS in the 5.9 GHz band, within the frequency range from 5.850 GHz to 5.925 GHz, in 2003. This band was divided in three subbands: (1) a band of 2x10 MHz, which is prioritized for critical road safety applications, (2) 30 MHz of spectrum for road safety related applications, and (3) 20 MHz of spectrum for non-safety related applications (ETSI TR 102 492-1 Part 1, 2006), (ETSI TR 102 492-2 Part 2, 2006). In Japan, a 10 MHz band from 715 MHz to 725 MHz was assigned for ITS using V2V communications in 2007 (Sai et al., 2009). Other unlicensed bands as the 2.4 GHz or 5.4 GHz bands can be also used for non-safety ITS applications.

The protocol stack standardized for DSRC systems is known as wireless access in vehicular environments (WAVE), whose physical layer (PHY) is the IEEE 802.11p (IEEE 802.11p, 2010) which is an adaptation of the IEEE 802.11a standard (IEEE 802.11, 2007) for DSRC systems.

In order to optimize the design and development of ITS applications based on wireless systems, it is necessary to understand the propagation channel characteristics and how these can affect the final performance of safety applications, where a low latency in the communication is vital. The V2V channel, where both the transmitter and receiver can be in motion with low elevation antennas, is a relatively novel research area in channel modeling. Differences between V2V channels and fixed-to-mobile (F2M) channels make that channel models developed for F2M cannot be applied to the performance evaluation and deployment of future ITS applications based on vehicular communications systems. In addition, V2I communications in open areas, such as intercity roads, where the fixed transmitting antennas are located along the roads at low height, have not yet been sufficiently analyzed in traditional wireless communication models. Also, particular characteristics of vehicular environments, such as traffic density and vehicle speeds, have a direct influence on the final performance of vehicular safety systems. Therefore, the main features of V2X channels must be taken into account in the channel modeling and its corresponding channel model.

In the context of V2X communications under the ITS concept and vehicular safety applications, this Chapter provides an overview of propagation aspects in vehicular networks based on the available literature, inspecting the approaches introduced in vehicular channel modeling, as well as the most important channel measurement campaigns carried out to validate and develop new and more accurate channel models. Narrowband and wideband characteristics of the vehicular propagation channel are also examined. The most important path loss models published in the literature, as well as Doppler spread models and fading statistics based on channel measured data are revised. Finally, general aspects related to the antennas for vehicular communications and future advances in channel modeling are also dealt with.

2. DSRC link specifications

In 2003, the DSRC specifications for ITS applications in the 5.9 GHz band were standardized in (ASTM E2213, 2003), where the spectrum allocated in USA was divided into seven channels of 10 MHz. The channel numbering and the maximum equivalent isotropically radiated power (EIRP) for DSRC devices are illustrated in Table 1. Channel 178 is the control channel (CCH), and it is used principally to broadcast communications related to safety applications, while non-safety data exchange is strictly limited in terms of transmission time and interval. The remaining six channels are referred to as service channels (SCH) and are dedicated to safety and non-safety applications. Channel 172 is used in V2V communications, and channel 184 is dedicated to V2I communications, especially in intersection applications. The rest of the service channels, i.e., 174, 176, 180 and 182, which may be combined in 2 channels of 20 MHz (channels 175 and 181), are used to share safety-related applications. A guard band of 5 MHz is located at the lower edge. In the EU, 20 MHz were allocated in the spectrum in the unlicensed band, from 5.795 to 5.815 GHz for short-range applications, such as electronic fee collection and parking band (ETSI EN 300 674, 1999).

Channel Number	Central frequency (MHz)	Bandwidth (MHz)	RSU EIRP max. (dBm) Pub./Priv	OBU EIRP max. (dBm) Pub./Priv
172	5860	10	33/33	33/33
174	5870	10	33/33	33/33
175	5875	20	23/33	23/33
176	5880	10	33/33	33/33
178	5890	10	44.8/33	44.8/33
180	5900	10	23/23	23/23
181	5905	20	23/23	23/23
182	5910	10	23/23	23/23
184	5920	10	40/33	40/33

Table 1. DSRC channel allocation in the 5.9 GHz ITS band

The maximum EIRP depends on the type of device. The radio transmitter onboard vehicles is called an on-board unit (OBU). The roadside unites (RSUs) are equipped within the infrastructure along the roadside. The EIRP is also limited depending on the application be it public or private. In the EU, the maximum EIRP in the ITS band was limited to 33 dBm (TR 102 492-1 Part 1, 2005). In the 5.8 GHz band, the EIRP of RSUs was also limited to 33 dBm and the maximum single side band EIRP was fixed on -21 dBm with backscattering using a passive OBU or tag (ETSI EN 300 674, 1999).

The radio specifications of ASTM E2213-03 standard are compiled in the IEEE 802.11p standard (IEEE 802.11p, 2010), which is an adaptation to vehicular radio communications of the IEEE 802.11a standard in the 5.9 GHz band (IEEE 802.11, 2007). Each channel of 10 MHz or 20 MHz is OFDM modulated using 52 subcarriers, with 4 pilots and the remaining 48 subcarriers being used for data transmission.

The main characteristics of the radio link are summarized in the Table 2. The media access control (MAC) protocol is the carrier sense multiple access with collision avoidance (CSMA/CA). Consequently, the channel is shared amongst the communication transmissions which simultaneously use this channel. The efficiency of the data rate, due to the MAC protocol, depends on the transmitter number of nodes, the transmission mode and the packet size (CEPT Report 20, 2007). Using a data rate of 6 Mbps and 100 nodes simultaneously transmitting, the efficiency varies from about 0.4 to 0.5 with a packet size of

256 and 1024 bytes, respectively, which correspond to a shared data rate of 2.4 Mbps and 3 Mbps, respectively.

Channel bandwidth	10 MHz	20 MHz
RF link range (m)	1000[1]	
Average LOS packet error rate (%)	10	
Average NLOS packet error rate (%)	Not defined	
Transmitted output power class (dBm)	0, 10, 20, 29 (Class A, B, C, D)	
Modulations	BPSK[2], QPSK[3], 16 QAM, 64 QAM	
Channel data rates (Mbps)	3, 4.5, 6[3], 9, 12, 18, 24, 27	6, 9, 12, 18, 24, 36, 48, 54
Receiver sensitivities (dBm)	-85 (3 Mbps), -84 (4.5 Mbps), -82 (6 Mbps), -80 (9 Mbps), -77 (12 Mbps), -73 (18 Mbps), -69 (24 Mbps), -68 (27 Mbps)	-82 (6 Mbps), -81 (9 Mbps), -79 (12 Mbps), -77 (18 Mbps), -74 (24 Mbps), -70 (36 Mbps), -66 (48 Mbps), -65 (54 Mbps)
Number of receiver antennas	1 (default)	
Maximum speed vehicle	200 km/h (120 mph)	
Packet size (bytes)	64, 1000[4]	
Maximum latency	From 20 ms to 1 s[5]	
OFDM symbol duration (μs)	8	4

Table 2. DSRC physical specifications (ASTM E2213-03, 2003), (IEEE 802.11p, 2010)

From channel measurements carried out in a highway environment under real traffic conditions, LOS ranges varying from 880 m to 1327 m in V2V and V2I links were achieved, respectively (Gallagher & Akatsuka, 2006). The measured packet error rate oscillates from 0.63% (V2V) to 1.21% (V2I). A coverage distance from 58 m to 230 m was obtained under NLOS conditions for V2V radio links.

3. Vehicular channel modeling

In any wireless communication, there are numerous interacting objects (reflectors and/or scatterers) between the transmitter and the receiver which condition the radio propagation, such as mountains, buildings, trees, vehicles, etc. that comprise the propagation environment. In the receiving antenna, multiple contributions or replicas of the transmitted signal will be received. These replicas will suffer different reflection, diffraction and scattering processes. The received signal corresponds to the coherent addition of these

---

[1] This range corresponds to the ASTM E2213-03 specification. The required range depends on the application. The range oscillates from 50 m in precrash sensing to 1000 m for approaching emergency vehicle warning or emergency vehicle signal preemption services (Almalag, 2009).

[2] The first OFDM symbol is always BPSK modulated.

[3] The control channel uses 10 MHz of bandwidth, with QPSK modulation and 6 Mbps data rate.

[4] These packet sizes correspond to ASTM E2213-03. However, the packet size depends on the application characteristics. The message length oscillates from 150 bytes to 1200 bytes (CEPT Report 20, 2007).

[5] The maximum allowed latency depends on the application. The maximum latency oscillates from 20 ms in precrash sensing and 1 s in various sign extension applications (Almalag, 2009).

replicas, in amplitude and phase. Such replicas have different attenuation and delay related to the travelled propagation path. Then, the received signal can experience time-dispersion (frequency-selectivity) due to the multipath propagation. In addition, due to the motion of the terminals, or the movement of any interacting object in the propagation environment, the received signal can experience time-selectivity (frequency-dispersion), as a consequence of changes in the phase relationship among the received replicas. In these conditions, the propagation environment will limit the final quality of any wireless communication system, obliging to the implementation of different transmission techniques to mitigate the impairments caused by the channel, such as diversity, equalization, power control and error correction, among others.

The term *channel modeling*, also referred to as *channel characterization*, is used to describe the approaches, models, and channel measurements conducted to understand how the propagation channel impairs and distorts the transmitted signal that propagates through it in a specific environment. The different propagation processes that can occur in wireless channels, i.e., free-space, diffraction, reflection, scattering and transmission through irregular objects, make difficult channel modeling. For practical purposes, it is necessary to adopt some approaches and simplifications of the propagation environment. Therefore, a *channel model* is a simplified representation of the propagation channel centered on those aspects of the propagation channel that can significantly affect the final performance of the wireless system (Michelson & Ghassemzadeh, 2009). In this sense, a channel model can be considered as a set of parameters which describes the propagation paths characteristics allowing one to analyze, simulate and design wireless communication systems. The knowledge of the propagation phenomenon and an accurate channel model are essential for a flexible and practical design of any wireless communications system under realistic propagation conditions.

Wireless channel investigation has decades of history encompassing the deployment of mobile and personal communication systems (cellular systems) in the early eighties. Nevertheless, the vehicular channel, in which both the transmitter and receiver can be in motion, is at present an important area in channel modeling (Matolak and Wu, 2009), (Wang et al., 2009) and (Molisch et al., 2009). The interest in vehicular channel modeling is mainly motivated by the ITS concept for dedicated short-range communications in vehicular environments, together with the appearance of new applications related to driving safety, that makes the knowledge of the V2V and V2I channels of paramount importance for the design and performance evaluation of vehicular communication systems.

The motion of terminals and the use of low elevation antennas make V2V systems differ from conventional F2M or cellular systems, where the base station is fixed with a height greater than the mobile terminal. Another difference with cellular systems is the high mobility that can be observed in V2V systems due to the relative velocities of the vehicular terminals. The high mobility makes vehicular channels exhibit greater temporal variability than other conventional channels, such as F2M channels, and can suffer more severe fast fading. On the other hand, the probability of link obstruction increases as a consequence of both the interacting objects displacement and the use of low elevation antennas. These differences, together with different frequency bands operation, make that the channel knowledge and channel models developed for cellular systems cannot be used in the performance evaluation and deployment of future V2V communication systems.

The high dynamics experienced in vehicular environments and multipath propagation cause vehicular channels to be both time- and frequency-selective. The time-selectivity refers to channel changes over time due to the motion of both terminals and scatterers. In the frequency domain, time-selectivity appears as frequency-dispersion. The frequency-selectivity refers to the variations of the channel in the frequency. In the time domain, frequency-selectivity reverts to time-dispersion.

The channel parameters describing the time- and frequency-dispersion are closely related to the environment where the propagation occurs. From a channel modeling point of view, the main features of the vehicular environment which are necessary to take into account are the following: (1) type of propagation link (V2V or V2I), (2) type of environment (urban, suburban, rural, expressway, highway, open areas, etc.), (3) vehicles speeds, (4) vehicular traffic density, and (5) direction of motion of the transmitter and receiver vehicles (the same or in opposite directions). Although some of the above features are overlapped, for example traffic densities are usually higher in urban environments and higher vehicle speeds are given in expressway and highway environments, these features increase the possible combinations of real propagation conditions thus making difficult the development of deterministic propagation models. Therefore, the characteristics of vehicular propagation channels are mainly determined either by means of (1) simulation models or (2) through measurement campaigns collected using a channel sounder in the time or frequency domain. Also, from a system design point of view, due to the large possible combinations of real propagation conditions, transmission techniques incorporated to reduce channel impairments will observe channels with different time (delay) and frequency (Doppler) dispersion.

Next, Subsection 3.1 is oriented towards the introduction of the most common parameters used in channel modeling to describe the time- and frequency-dispersive behavior, in a general wireless channel. A review of the most important V2V channel models and the approaches made to build them is presented in the Subsection 3.2.

3.1 Parameters for characterizing wireless channels

The most important parameters used to describe wireless channels, in a deterministic or statistical sense, can be derived from the channel impulse response (CIR). We will introduce the CIR of a time-variant wireless channel, as the basis to make a description of the time- and frequency-dispersive behavior of the channel.

In a time-variant environment characterized by multipath propagation, the received signal can be seen as the temporal superposition of the transmitted signal replicas (contributions from reflectors and scatterers) with different attenuation and delay. If the transmitted bandpass signal, denoted by $s(t)$, is

$$s(t) = \mathrm{Re}\left\{s_p(t)\exp(j2\pi f_c t)\right\}, \tag{1}$$

where $s_p(t)$ is the complex envelope and f_c being the carrier frequency, the bandpass received signal, denoted by $y(t)$, is given by

$$\begin{aligned} y(t) &= \mathrm{Re}\left\{\sum_{k=1}^{\infty}\sum_{i=1}^{\infty} a_{k,i} s_p(t-\tau_i)\exp\left(j2\pi\left(f_c + f_d\cos\theta_{k,i}\right)t + \phi_{k,i}\right)\right\} \\ &= \mathrm{Re}\left\{\left[\sum_{i=1}^{\infty} h(t,\tau_i)\delta(t-\tau_i)\otimes s_p(t)\right]\exp(j2\pi f_c t)\right\} = \mathrm{Re}\left\{\left[h(t,\tau)\otimes s_p(t)\right]\exp(j2\pi f_c t)\right\} \end{aligned} \tag{2}$$

where

$$h(t,\tau_i)=\sum_{k=0}^{\infty}a_{k,i}\exp\left(j2\pi f_d\cos\theta_{k,i}t+\phi_{k,i}\right), \tag{3}$$

and

$$h(t,\tau)=\sum_{i=0}^{\infty}h(t,\tau_i)\delta(\tau-\tau_i), \tag{4}$$

with $a_{k,i}$ and $\phi_{k,i}$ being the amplitude and phase of the i-th contribution that arrives to the receiver with an angle $\theta_{k,i}$ with respect to the direction of motion, and delay τ_i. The term f_d is the maximum Doppler frequency, also called Doppler shift, i.e., $f_d=v/\lambda_c$, where v refers to the receiver velocity, $\lambda_c=c_0/f_c$ is the wavelength associated to the carrier frequency f_c, and c_0 is the speed of light. The $\delta(\cdot)$ function is the Dirac delta, and $\otimes$ denotes convolution. Eq. (4) corresponds to the time-variant impulse response of the wireless radio channel. Specifically, $h(t,\tau)$ is the response of the lowpass equivalent channel at time t to a unit impulse generated τ seconds in the pass (Parsons, 2000). $h(t,\tau)$ is known as the *input delay-spread function*, and is one of the four system functions described by Bello (Bello, 1963), which can be used to fully characterize linear time-variant (LTV) radio channels. The term $h(t,\tau_i)$ is the time-dependent complex coefficient associated to a delay τ_i, and can be expressed as

$$h(t,\tau_i)=h_{iR}(t)+jh_{iQ}(t), \tag{5}$$

where

$$h_{iR}(t)=\sum_{k=0}^{\infty}a_{k,i}\cos\left(2\pi f_d\cos\theta_{k,i}t+\phi_{k,i}\right) \tag{6}$$

and

$$h_{iQ}(t)=\sum_{k=0}^{\infty}a_{k,i}\sin\left(2\pi f_d\cos\theta_{k,i}t+\phi_{k,i}\right) \tag{7}$$

are the in-phase and quadrature components, respectively.

In practice, many physical channels can be considered stationary over short periods of time, or equivalently over small spatial distances due to the transmitter/receiver or interacting objects displacement. Although these channels may not be necessarily stationary in a strict sense, they usually are considered wide sense stationary (WSS) channels. Also, a channel can exhibit uncorrelated scattering (US) in the time variable, i.e., contributions with different delays are uncorrelated. The combination of the WSS and US assumptions yields the WSSUS assumption, which has been very used in channel modeling for cellular systems.

Under the WSSUS assumption, the channel can be represented as a tapped delay line (TDL), where the CIR is written as

$$h(t,\tau)=\sum_{i=1}^{N}h_i(t)\delta(\tau-\tau_i), \tag{8}$$

where $h_i(t) \triangleq h(t,\tau_i)$ refers to the complex amplitude of the i-th tap. Using this representation of the CIR, and taking into account Eq. (3), each of the N taps corresponds to one group of closely delayed/spaced multipath components (MPCs). This representation is commonly used in the channel characterization theory because the time resolution of the receiver is not sufficient to resolve all MPCs in most practical cases. It is worth noting that the number of taps, N, and MPCs delay associated to the i-th tap, τ_i, remain constant during a short period of time, where the WSS assumption is valid. For this reason, N and τ_i are not dependent on the time variable in Eq. (8). Fig. (1) shows a graphical representation of the TDL channel model based on delay elements.

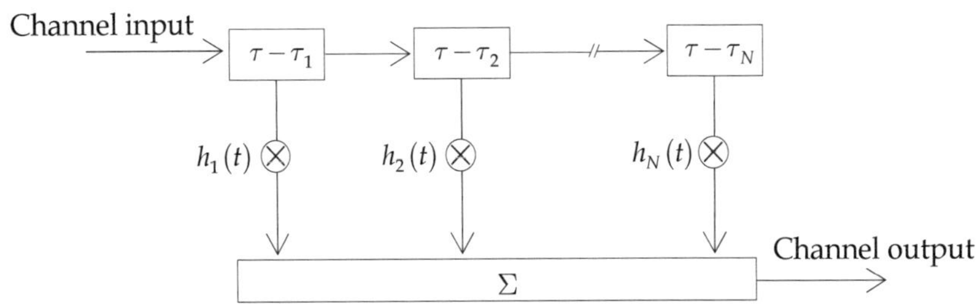

Fig. 1. TDL channel model description based on delay elements

The time variation of the taps complex amplitude is due to the MPCs relative phases that change in time for short displacements, in terms of the wavelength, of the transmitter/receiver and/or interacting objects. Thereby, $h_i(t)$ is referred to in the literature as the *fading complex envelope* of the i-th path. The time variation of $h_i(t)$ is model through the Rayleigh, Rice or Nakagami-m distributions, among others (for a mathematical description of these distributions, the reader is referred to the Section 5 of this Chapter).

As introduced previously, vehicular environments can be dynamic due to the movement of the terminals and/or the interacting objects. Therefore, vehicular channels may be non-stationary and the channel statistics and the CIR can change within a rather short period of time. In this situation, the WSSUS assumption is not applicable anymore. Sen and Matolak proposed to model the non-stationarities of vehicular channels through a *birth/death persistence process* in order to take into account the appearance and disappearance of taps in the CIR (Sen & Matolak, 2008). Then, the CIR can be rewritten as

$$h(t,\tau) = \sum_{i=1}^{N} h_i(t) z_i(t) \delta(\tau - \tau_i), \quad (9)$$

where $z_i(t)$ is introduced to model the birth/death persistence process. $z_i(t)$ takes the 0 and 1 values to model the disappearance and appearance of the i-th tap, respectively. As appointed in (Molish et al., 2009), this persistence process can provide a non-stationarity of the channel, but it does not consider the drift of scatterers into a different delay bin, and as result can lead to the appearance or disappearance of sudden MPCs. The non-stationarity

problem in V2V channel has been also addressed in (Bernardó et al., 2008) and (Renaudin et al., 2010) from channel measurement data.
The dispersive behavior in the time and frequency domains of any wireless channel conditions the transmission techniques designed to mitigate the channel impairments and limits the system performance. As an example, time-dispersion (or frequency-selectivity) obliges to implement equalization techniques, and frequency dispersion (or time selectivity) forces the use of diversity and adaptive equalization techniques. Orthogonal frequency division multiplexing (OFDM) has been suggested to be used in IEEE 802.11p. In a V2V system based on OFDM, time-dispersion fixes the minimum length of the cyclic prefix, and frequency-dispersion can lead to inter-carrier interference (ICI). In the following, the most important parameters used to describe the time- and frequency-dispersion behavior will be introduced.

3.1.1 Time-dispersion parameters

For a time-varying channel with multipath propagation, a description of its time-dispersion characteristics can be obtained by expressing the autocorrelation function of the channel output (the received signal) in terms of the autocorrelation function of the *input delay-spread function,* denoted by $R_h(t,s;\tau,\eta)$ and defined as

$$R_h(t,s;\tau,\eta) \triangleq E\{h(t,\tau)h^*(s,\eta)\}, \tag{10}$$

where $E\{\cdot\}$ is the expectation operator. In Eq. (10) t and s are time variables, whereas τ and η correspond to delay variables. If the WSSUS assumption is considered, the autocorrelation function of the channel output can be described by the *delay cross-power spectral density,* denoted by $P_h(\xi,\tau)$, where $\xi = s-t$ is a short time interval in which the channel can be considered wide-sense stationary. The WSSUS assumption implies a small-scale characterization of the channel where a local scattering can be observed, i.e., short periods of time or equivalently short displacements of the terminals. When $\xi \to 0$, $P_h(\xi,\tau)$ is simplified by $P_h(\tau)$, which is referred to as *power delay profile* (PDP) in the literature
It is very common in the wideband channel characterization theory to express the time- and frequency-dispersion of a wireless channel by means of the *delay-Doppler cross-power spectral density,* also called *scattering function*[6], denoted by $P_S(\tau,\nu)$, where the ν variable refers to the Doppler shift. The scattering function can be regarded as the Fourier transform of the autocorrelation function $R_h(t,t+\xi;\tau,\eta) \equiv R_h(\xi;\tau,\eta) = \delta(\eta-\tau)P_h(\xi,\tau)$ with respect to the ξ variable. For a complete understanding of a stochastic wideband channel characterization, the reader can see the Reference (Parsons, 2000), widely cited in channel modeling studies.
From $P_S(\tau,\nu)$, $P_h(\tau)$ can be also derived integrating the scattering function over the Doppler shift variable, i.e.

$$P_h(\tau) = \int_{-\infty}^{+\infty} P_S(\tau,\nu)\,d\nu\,. \tag{11}$$

[6] The term scattering function was incorporated in the wireless channels characterization due to the $P_S(\tau,\nu)$ function is identical to the scattering function $\sigma(\tau,\nu)$ of a radar target.

From channel measurements of the CIR in a particular environment, and assuming ergodicity, the $P_h(\tau)$ can also be estimated for practical purposes as

$$P_h(\tau) = E_t\left\{\left|h(t,\tau)\right|^2\right\}, \tag{12}$$

where $E_t\{\cdot\}$ denotes expectation in the time variable. From Eq. (12), the PDP can be seen as the squared magnitude of the CIR, averaged over short periods of time or small local areas around the receiver (small-scale effect).

The most important parameter to characterize the time-dispersion behavior of the channel is the *rms* (root mean square) delay spread, denoted by τ_{rms}, which corresponds to the second central moment of the PDP, expressed as

$$\tau_{rms} \triangleq \sqrt{\frac{\int_0^{\infty}(\tau-\bar{\tau})^2 P_h(\tau)d\tau}{\int_0^{\infty}P_h(\tau)d\tau}}, \tag{13}$$

where $\bar{\tau}$ is the average delay spread, or first central moment of the PDP, given by

$$\bar{\tau} \triangleq \frac{\int_0^{\infty}\tau P_h(\tau)d\tau}{\int_0^{\infty}P_h(\tau)d\tau}. \tag{14}$$

Other metrics to describe the delay spread of wireless channels are the maximum delay spread, the delay window and the delay interval (Parsons, 2000).

The frequency-selective behavior of wireless channels is described using the *time-frequency correlation function*, denoted by $R_T(\Omega,\xi)$ [7], which is the Fourier transform of the $P_h(\xi,\tau)$ function over the delay variable, i.e.

$$R_T(\Omega,\xi) = \int_{-\infty}^{+\infty} P_h(\xi,\tau)\exp\{-j2\pi\Omega\tau\}d\tau. \tag{15}$$

The Ω variable refers to a frequency interval, i.e., $\Omega = f_2 - f_1$. When $\xi = 0$, expression (15) is written as

$$R_T(\Omega) = \int_{-\infty}^{+\infty} P_h(\tau)\exp\{-j2\pi\Omega\tau\}d\tau, \tag{16}$$

where $R_T(\Omega)$ is known as the *frequency correlation function*. A metric to measure the frequency-selectivity of the channel is the coherence bandwidth, denoted by B_C. From Eq. (16), B_C is the smallest value of Ω for which the normalized frequency correlation function, denoted by $\tilde{R}_T(\Omega) \triangleq R_T(\Omega)/\max\{R_T(\Omega)\}$, where $\max\{R_T(\Omega)\} = R_T(\Omega=0)$, is equal to some suitable correlation coefficient, e.g. 0.5 or 0.9 are typical values. Physically, the coherence bandwidth represents the channel bandwidth in which the channel experiences approximately a flat frequency response behavior.

[7] $R_T(\Omega,\xi)$ corresponds to the autocorrelation function of the *time-variant transfer function* $T(f,t)$.

Since $P_h(\tau)$ is related to $R_T(\Omega)$ by the Fourier transform, there is an inverse relationship between the *rms* delay spread and the coherence bandwidth, i.e., $B_C \propto 1/\tau_{rms}$.

3.1.2 Frequency-dispersion parameters

When a channel is time-variant, the received signal suffers time-selective fading and as a result frequency-dispersion occurs. A description of the frequency-dispersion characteristics can be derived from the *Doppler cross-power spectral density*, denoted by $P_H(\Omega,\nu)$. In a WSSUS channel, $P_H(\Omega,\nu)$ is related to the autocorrelation of the *output Doppler-spread function*, denoted by $R_H(\Omega;\nu,\mu)$,

$$R_H(f,f+\Omega;\nu,\mu)=\delta(\nu-\mu)P_H(\Omega,\nu), \tag{17}$$

where the autocorrelation function is defined as

$$R_H(f,m;\nu,\mu)\triangleq E\{H(f,\nu)H^*(m,\mu)\}, \tag{18}$$

being $H(f,\nu)$ the *output Doppler-spread function*. In Eq. (18) f and m are frequency variables, whereas ν and μ correspond to Doppler shifts variables. From the relationships between the autocorrelation functions in a WSSUS channel (Parsons, 2000), the $P_H(\Omega,\nu)$ function can also be regarded as the Fourier transform of the scattering function with respect to the τ variable. When $\Omega \to 0$, $P_H(\Omega,\nu)$ is simplified by $P_H(\nu)$, which is referred to in the literature as *Doppler power density spectrum* (PDS). In a similar manner to the PDP, the Doppler PDS can also be derived integrating the scattering function over the delay variable, i.e.

$$P_H(\nu)=\int_{-\infty}^{+\infty} P_S(\tau,\nu)\,d\tau\,. \tag{19}$$

Now, from the Doppler PDS some parameters can be defined to describe the frequency-dispersive behavior of the channel. The most important parameter is the *rms* Doppler spread, denoted by ν_{rms}, given by

$$\nu_{rms} \triangleq \sqrt{\frac{\int_0^{\infty}(\nu-\overline{\nu})^2 P_H(\nu)\,d\nu}{\int_0^{\infty} P_H(\nu)\,d\nu}}, \tag{20}$$

where $\overline{\nu}$ is the average Doppler spread, or first central moment of the Doppler PDS, given by

$$\overline{\nu} \triangleq \frac{\int_0^{\infty} \nu P_H(\nu)\,d\nu}{\int_0^{\infty} P_H(\nu)\,d\nu}. \tag{21}$$

Time-selective fading refers to the variations of the received signal envelope due to the movement of the transmitter/receiver and/or the interacting objects in the environment. This displacement causes destructive interference of MPCs at the receiver, which arrive with different delays that change in time or space. This type of fading is observed on spatial scales in terms of the wavelength, and is referred to in the literature as *small-scale fading* in opposite to the *large-scale fading* (also referred to as *shadowing*) due to the obstruction or blockage effect of propagation paths. The variation of the received signal envelope can also be modeled in a statistical way using the common Rayleigh, Rice or Nakagami-*m* distributions.

In a similar manner to the coherence bandwidth, for a time-variant channel it is possible to define a parameter called *coherence time*, denoted by T_C, to refer to the time interval in which the channel can be considered stationary. From the *time correlation function of the channel*, denoted by $R_T(\xi) \equiv R_T(\Omega \to 0, \xi)$, T_C is the smallest value of ξ for which the normalized time correlation function, denoted by $\tilde{R}_T(\xi) \triangleq R_T(\xi) / \max\{R_T(\xi)\}$, where $\max\{R_T(\xi)\} = R_T(\xi = 0)$, is equal to some suitable correlation coefficient, e.g., 0.5 and 0.9 are typical values. There is an inverse relationship between the *rms* Doppler spread and the coherence time, i.e., $T_C \propto 1 / \nu_{rms}$.

3.2 V2V channel models

In the available literature of channel modeling, one can find different classifications of wireless channel models, e.g., narrowband or wideband models, non-physical (analytical) and physical (realistic) models, and two-dimensional or three-dimensional models, among others. Regardless of the type of classification, wireless channel models are mainly based on any of the three following approaches (Molish & Tufvesson, 2004):

- *deterministic* approach, which characterizes the physical propagation channel using a geographical description of the environment and *ray approximation*[8] techniques (ray-tracing/launching),
- *stochastic* approach, oriented to the channel parameters characterization in terms of probability density functions often based on large measurement campaigns, and
- *geometry-based stochastic* (GBS) approach, which assumes a stochastic distribution of interacting objects around the transmitter and the receiver and then performing a deterministic analysis.

In the following, some published V2V channels models based on these approaches will be briefly introduced. Also, the main advantages and drawbacks when the above approaches are applied to V2V channel models will be indicated.

3.2.1 Deterministic models

A deterministic channel model[9] characterizes the physical channel parameters in specific environments solving the Maxwell's equations in a deterministic way, or using analytical descriptions of basic propagation mechanisms (e.g., free-space propagation, diffraction, reflection and scattering process). These models require a geographical description of the environment where the propagation occurs, together with the electromagnetic properties of the interacting objects. It is worth noting that the term *deterministic* refers to the way in which the propagation mechanisms are described. Evidently, the structure of interacting objects, their electrical parameters (e.g., the conductivity and permittivity), and some parts

---

[8] Ray approximation techniques refer to high frequency approximations, where the electromagnetic waves are modeled as rays using the geometrical optic theory.

[9] Deterministic models are also referred to in the literature as geometric-based deterministic models (GBDMs).

of the environment are introduced in the model in a simple way by means of simplified or idealistic representations. The main drawbacks of deterministic models are the large computational load and the need of a geographical database with high resolution to achieve a good accuracy. Therefore, it is necessary to seek a balance between computational load and simplified representations of the environment elements. On the other hand, deterministic models have the advantage that computer simulations are easier to perform than extensive measurements campaigns, which require enormous effort.

The use of deterministic models based on ray-tracing techniques allows us to perform realistic simulations of V2V channels. Earlier Reference (Maurer et al., 2001) presents a realistic description of road traffic scenarios for V2V propagation modeling. A V2V channel model based on ray-tracing techniques is presented in (Maurer et al., 2004). The model takes into account the road traffic and the environment nearby to the road line. A good agreement between simulations results, derived from the model, and wideband measurements at 5.2 GHz was achieved. Nevertheless, characteristics of vehicular environments and the resulting large combinations of real propagation conditions, make difficult the development of V2V deterministic models with certain accuracy.

3.2.2 Stochastic models

Stochastic models[10] describe the channel parameters behavior in a stochastic way, without knowledge of the environment geometry, and are based on measured channel data.

For system simulations and design purposes, the TDL channel model has been adopted due to its low complexity. The parameters of the TDL channel model are described in a stochastic manner. Reference (Acosta-Marum & Ingram, 2007) provides six time- and frequency-selective empirical models for vehicular communications, three models for V2V and another three for V2I communications. In these models, the amplitude of taps variations are modeled in a statistical way through the Rayleigh and Rice distributions, with different types of Doppler PDS. The models have been derived from channel measurements at 5.9 GHz in different environments (expressway, urban canyon and suburban streets). In the references (Sen & Matolak, 2007), (Sen & Matolak, 2008) and (Wu et al., 2010), complete stochastic models based on the TDL concept are provided in the 5 GHz frequency band for several V2V settings: urban with antennas inside/outside the cars, small cities and open areas (highways) with either high or low traffic densities. These models introduce the Weibull distribution to model the amplitude of taps variations. The main drawback of V2V stochastic models based on the TDL representation is the non-stationary behavior of the vehicular channel. To overcome this problem, Sen and Matolak have proposed a birth/death (on/off) process to consider the non-stationarity persistence feature[11] of the taps, modeled using a two-state first-order Markov chain (Sen & Matolak, 2008).

3.2.3 Geometry-based stochastic models

The deterministic and stochastic approaches can be combined to enhance the efficiency of the channel model, resulting in a geometry-based stochastic model (GBSM) (Molisch, 2005). The philosophy of GBSMs is to apply a deterministic characterization assuming a stochastic (or randomly) distribution of the interacting objects around the transmitter and the receiver

[10] In the literature, stochastic models are also referred to as non-geometrical stochastic models (NGSMs)
[11] The persistence process is modeled by $z(t)$ in Eq. (9).

positions. To reduce computational load, simplified ray-tracing techniques can be incorporated, and to reduce the complexity of the model, it can be assumed that the interacting objects are distributing in regular shapes.
The earlier GBSM oriented to mobile-to-mobile (M2M) communications was proposed in (Akki & Haber, 1986). Akki and Haber extended the one-ring scattering model[12] resulting in a two-ring scattering model, i.e., one ring of scatterers around the transmitter and other ring around the receiver. Recent works (Wan & Cheng, 2005) and (Zajic & Stüber, 2008) consider the inclusion of deterministic multipath contributions (LOS or specular components) combined with single- and double-bounced scattering paths. Recently, in (Cheng et al., 2009) a combination of the two-ring and the ellipse scattering model is provided to cover a large variety of scenarios, for example those where the scattering can be considered non-isotropic. Fig.2 (a) shows a typical V2V urban environment, and its corresponding geometrical description, based on two-ring and one ellipse where the scatterers are placed, is illustrated in Fig.2 (b), intuitively. To take into account the scatterers around the transmitter or the receiver in expressway/highway with more lanes than in urban/suburban environments, several rings of scatterers can be considered around the transmitter/receiver in the geometrical description of the propagation environment, resulting in the so-called multi-ring scattering model.

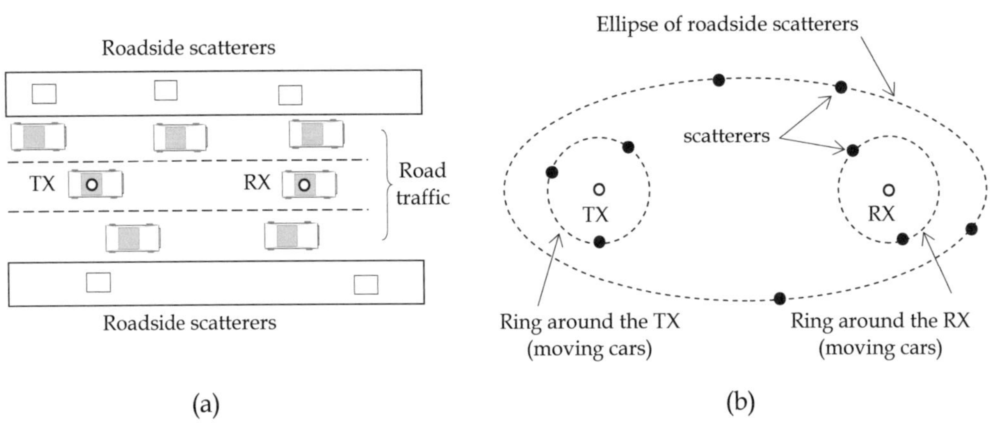

Fig. 2. Illustration of the concept of the two-ring and ellipse scattering model: (a) typical V2V urban environment with roadside scatterers along the route and road traffic (moving cars), and (b) its corresponding geometrical description to develop a GBSM

As pointed out previously, the channel parameters in vehicular environments are affected by traffic conditions. The effect of the vehicular traffic density (VTD) can be also incorporated in GBSMs. In the Reference (Cheng et al., 2009), a model that takes into account the impact of VTD on channel characteristics is presented. In Section 3.1, the problem of non-stionarity in vehicular environments due to high mobility of both the terminals and interacting objects was introduced. One advantage of GBSMs is that non-stationarities can

[12] The origin of the GBSMs goes back to the 1970s, with the introduction of antenna diversity techniques at the Base Stations in cellular systems. To evaluate the performance of diversity techniques, a set of scatterers distributed in a ring around the mobile terminal was considered.

be handled. In a very dynamic channel, as is the case of the V2V channel under high speeds and VTDs conditions, the WSSUS assumption cannot be accomplished. Results presented in (Karedal et al., 2009) show as MPCs can move through many delay bins during the terminal movement.

The manner in which GBSMs are built, permits a complete wideband channel description, as well as to derive closed-form expressions of the channel correlation functions. The latter is especially interesting in MIMO (Multiple-Input Multiple-Output) channel modeling, where the space-time correlation function[13] can be derived. The potential spectral efficiency increment of the system when MIMO techniques are introduced, together with the capability of placing multielement antennas in vehicles with large surfaces, makes MIMO techniques very attractive for V2V communications systems. The advantage of MIMO techniques, together with the MIMO channel modeling experience, explains that most of the emerging V2V GBSMs are oriented to MIMO communications. It is worth noting that MIMO techniques have generated a lot of interest and are an important part of modern wireless communications, as in the case of IEEE 802.11 standards.

The grade of accuracy in a GBSM can be increased introducing certain information or channel parameters derived from real channel data (e.g., path loss exponent and decay trend of the PDP). Reference (Karedal et al., 2009) provides a MIMO GBSM based on the results derived from an extensive MIMO measurements campaign carried out in highway and rural environments at 5.2 GHz. The model described in this reference introduces a generalization of the generic GBS approach for parameterizing it from measurements. Karedal et al. categorize the interacting objects in three types: mobile discrete scatterers (vehicles around the transmitter and receiver), static discrete scatterers (houses and road signs on and next to the road), and diffuse scatterers (smaller objects situated along the roadside).

Another important aspect to take into account in channel modeling is the three-dimensional (3D) propagation characteristics when geographical data are available. In channel modeling is frequent to distinguish between vertical and horizontal propagation. Vertical propagation takes into account the propagation mechanisms that take place in the vertical (elevation) plane, whereas horizontal propagation considers the propagation mechanisms that appear in the horizontal (azimuth) plane. The first models developed for cellular systems considered the propagation mechanisms in the vertical plane (e.g., Walfisch-Bertoni path loss model), resulting in the so-called two-dimensional (2D) models. These models were oriented to the narrowband channel characterization describing the path loss. Afterwards, the introduction of propagation mechanisms in the horizontal plane made possible a wideband characterization, resulting in 3D models. Although the V2V models cited in this section permit a wideband characterization, they only consider propagation mechanism in the horizontal plane. The assumption of horizontal propagation can be accomplished for vehicular communications in rural areas (Zajic & Stüber, 2009), whereas it can be questionable in urban environments, in which the height of the transmitting and the receiving antennas is lower than the surrounding buildings, or where the urban orography determines that the transmitter is at a different height than the receiver. For non-directional antennas in the vertical plane, the scattered/diffracted waves from the tops of buildings to the receiver located on the street are not necessarily in the horizontal plane. In this situation, a 3D propagation characterization can improve the accuracy of the channel model. The

---

[13] The space-time correlation function in MIMO theory can be use to compare the outage capacity of different arrays antenna geometries (i.e., linear, circular or spherical antenna array).

viability of 3D V2V GBSMs based on the two-cylinder model, as an extension of the one-cylinder model proposed by Aulin for F2M systems (Aulin, 1979), has been verified by Zajic et al. from channel measurements in urban and expressway environments (Zajic et al., 2009). The two-cylinder model can also be extended to a multi-cylinder in a similar way to the multi-ring scattering model.

4. Vehicular channel measurements

Channel measurements are essential to understand the propagation phenomenon in particular environments, and can be used to validate and improve the accuracy of existing channel models. A channel model can also take advantage of measured channel data, e.g., parameters estimated from channel measurements can be included in the channel model.

4.1 Channel measurement techniques and setups

The measurement setup used to measure the transfer function of a wireless channel, in either the time or frequency domain, is referred to as a *channel sounder*. The configuration and implementation of a channel sounder are related to the channel parameters to be measure. Thus, channel sounders may be classified as narrowband and wideband.

Narrowband channel sounders are used to make a narrowband channel characterization. Generally, the narrowband channel parameters explored are path loss, Doppler effect and fading statistics (small- and large-scale fading). The simplest narrowband channel sounder consists of a single carrier transmitter (RF transmitter) and a narrowband receiver (e.g., a specific narrowband power meter or a spectrum analyzer) to measure the received signal strength. Also, it is possible to use a vector signal analyzer (VSA) as a receiver. Since the channel response is measured at a single frequency, the time resolution of a narrowband channel sounder is infinity. This means that it is not possible to distinguish different replicas of the transmitted signal in the time delay domain.

When a wireless system experiences frequency-selectivity, or time-dispersion, a wideband characterization is necessary to understand the channel frequency-selective behavior. This is the case of the future DSRC system, which will use a minimum channel bandwidth of 10 MHz. To estimate the time-dispersion metrics defined in Section 3.1, a wideband channel sounder must be used. Wideband channel sounders can measure the channel response in either the frequency or time domain. In the frequency domain, a wideband channel sounder measures the channel frequency response at the t_0 instant, denoted by $T(f,t_0)$[14], and the CIR is estimated applying the inverse Fourier transform with respect to the frequency f variable, yielding $h(t_0,\tau)$. A vector network analyzer (VNA) can be use to estimate the channel frequency response from the S_{21} scattering parameter, where the DUT (dispositive under test) is the propagation channel and the transmitting and receiving antennas[15]. The main drawbacks of using a VNA are that the channel must be stationary during the acquisition time of the frequency response, i.e., the acquisition time must be lower than the

[14] $T(f,t)$ is the time variant transfer function of the propagation channel.

[15] When a VNA is used, the frequency response measured takes into account the channel responses and the frequency response of the transmitting and receiving antennas. A calibration process is necessary to extract the antennas effect.

coherence time of the channel, and the short transmitter-receiver separation distances since the transmitting and receiving antennas must be connected to the VNA. Due to these drawbacks, a VNA cannot be used in vehicular channel measurements. The VNA has been frequently used as a wideband channel sounder to measure the CIR over short distances, e.g., indoor scenarios. An alternative approach consists of transmitting a multicarrier signal, with known amplitudes and relative phases. As a receiver, a VSA can be used to estimate the measured complex frequency spectrum.

In the time domain, the CIR is measured directly. There are two possible implementations of wideband channel sounders operating in the time domain. The first implementation consists of using an impulse generator at the transmitter and a digital sampling oscilloscope (DSO) at the receiver, resulting in the so-called *impulse channel sounder*. The main drawback of an impulse channel sounder is that a probe antenna for the trigger pulse is necessary. Impulse channel sounders have been used in ultra-wideband (UWB) channel measurements. The second implementation of a wideband channel sounder in the time domain is based on the transmission of a wideband pseudo-random noise (PN) sequence[16]. The CIR is estimated as the cross-correlation between the received signal and the transmitted PN sequence, resulting in the so-called *correlative channel sounder*. Fig. 3 shows the operating principle of a correlative channel sounder. The correlative channel sounder is commonly used in wideband channel measurements, particularly in vehicular channel measurements. The time resolution of a correlative channel sounder and the maximum resolvable delay are related to the chip duration and the length of the PN sequence, respectively. In practice, the simplest correlative channel sounder consists of an arbitrary waveform generator (WG) as transmitter and a VSA as receiver. The WG transmits a PN sequence, the VSA collects the in-phase and quadrature components of the received signal, and then through post-processing the CIR is estimated.

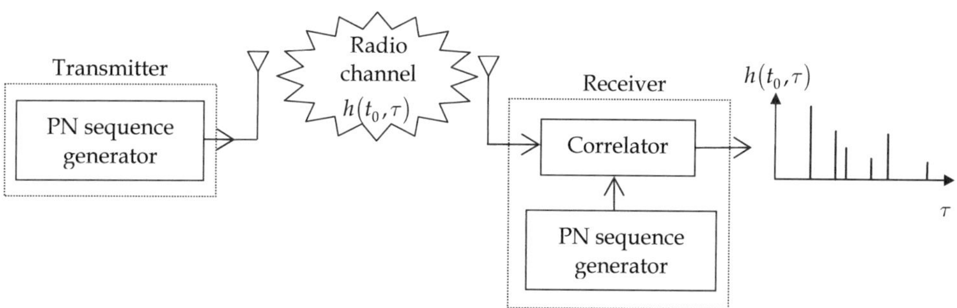

Fig. 3. Correlative channel sounder based on the PN signal principle

A wideband channel sounder can also use multiple antennas (array antennas) at the transmitter and the receiver to explore aspects related to the directional character of the propagation channel, e.g. correlation degree among the received signal at each antenna element used in spatial diversity and MIMO techniques. In MIMO channel modeling, a full characterization of the channel requires the knowledge of the direction-of-arrival (DOA) of

[16] In a PN sequence, the transmitted symbols are called *chips*. Also, the PN technique is referred to in the literature as direct sequence spread spectrum (DSSS) technique.

MPCs distribution at the receiver, and the direction-of-departure (DOD) distribution at the transmitter.
Independently of the measuring technique (i.e., narrowband or wideband, and time or frequency domain), the channel parameters estimated from measurements are influenced by the measurement setup, especially by the directional characteristics of the antennas. The influence of the measurement setup is not always easily separable from the CIR.

4.2 Vehicular channel measurement campaigns

Several vehicular channel measurement campaigns have been conducted to investigate the propagation channel characteristics at different environments and frequencies, mainly in the last five years. Table 3 summarizes the most representative measurement campaigns, indicating the frequency, the measuring technique (narrowband or wideband), the type of antennas (SISO[17] refers to a single antenna at the transmitter/receiver, and MIMO refers to multiple antennas at the transmitter/receiver), the propagation link measured (V2V, V2I or V2X) and the environment where the measures were collected. Five types of environments have been considered[18], i.e., urban, suburban, rural, expressway and highway.
The frequency bands correspond to the IEEE 801.11b/g band (2.4 GHz), the IEEE 801.11a band (5.2 GHz) and the DSRC band (5.9 GHz). A study about the channel parameters describing the path loss, as well as the time- and frequency-dispersion behavior will be carried out in Section 5 and 6 of this chapter.

Reference	Frequency	Channel sounder	System configuration	Propagation link	Environment
(Acosta et al., 2004)	2.4 GHz	WB	SISO	V2V	SU, E
(Zajic et al., 2009)	2.4 GHz	WB	MIMO	V2V	H
(Ito et al., 2007)	5.0 GHz	NW	SISO	V2V	U
(Sen & Matolak, 2008)	5.0 GHz	WB	SISO	V2V	U, SU, E, H
(Paier et al., 2007)	5.2 GHz	WB	MIMO	V2X	U, R, H
(Maurer et al., 2002)	5.2 GHz	NB	SISO	V2V	U, SU, E, H
(Renaudin et al., 2009)	5.3 GHz	WB	MIMO	V2V	U, SU, H
(Paschalidis et al., 2008)	5.7 GHz	WB	MIMO	V2V	U
(Acosta & Ingram, 2007)	5.9 GHz	WB	SISO	V2X	U, SU, E
(Cheng et al., 2007)	5.9 GHz	NB	SISO	V2V	SU
(Cheng et al., 2008a)	5.9 GHz	WB	SISO	V2V	SU, R, H
(Tan et al., 2008)	5.9 GHz	WB	SISO	V2X	U, R, H
(Kunisch & Pamp, 2008)	5.9 GHz	WB	SISO	V2V	U, R, H

Table 3. Vehicular channel measurements. NB: Narrowband, WB: Wideband, U: Urban, SU: Suburban, R: Rural, E: Expressway, H: Highway

[17] Single-Input Single-Output.
[18] This is a general classification to facilitate comparisons between empirical results. To know the specific characteristics of the environment where the measurement campaign was conducted, the reader can see the corresponding reference.

5. V2V narrowband channel characterization

Based on the available literature, the most important narrowband parameters of vehicular propagation channels are reviewed in this Section, i.e., path loss, narrowband fading statistics and Doppler spectrum.

5.1 Path loss modeling

Path loss is one of the most important parameters used in the link budget, being a measure of the channel quality. Path loss takes into account all propagation mechanisms that occur in the radio channel, such as free-space, reflection, diffraction and scattering, and is influenced by the propagation environment (e.g., urban, suburban or rural), the directional characteristic and height of the antennas, and the distance between the transmitter and receiver. Path loss is inversely related to the signal-to-noise ratio (SNR), i.e., the higher path loss the lower SNR, thus limiting the coverage area.

Under free space propagation conditions, for a transmitter to receiver separation distance d, the averaged received power in logarithm units (dBm), denoted by $P_R(d)$, is given by the Friis transmission formula as

$$P_R(d) = P_T + G_T + G_R - 10\log(4\pi d / \lambda_c)^2, \tag{22}$$

where P_T is the transmitted power (in dBm), G_T and G_R are the transmitter and receiver antenna gain (in dBi) in the direction of the propagation wave, and λ_c is the wavelength associated to the carrier frequency f_c. The last term in Eq. (22) represents the path loss for free space propagation conditions, $PL_{FS}(d)$, expressed in decibels (dB) as

$$PL_{FS}(d) = 10\log(4\pi d / \lambda_c)^2. \tag{23}$$

In different propagation conditions to free space, the path loss expressed in dB, denoted by $PL(d)$, can be expressed in a general way as

$$PL(d) = \overline{PL}(d) + X_\sigma, \tag{24}$$

where $\overline{PL}(d)$ is the average path loss (in dB), and X_σ is a Gaussian random variable with zero mean and standard deviation σ. The X_σ variable accounts for the large-scale fading or shadowing.

In many channel models, the average path loss is proportional to the logarithm of the distance, i.e., $\overline{PL}(d) \propto 10\gamma \log d$, being γ a path loss exponent extracted from channel measured data (from Eq. (23), $\gamma = 2$ in free space).

There are different path loss models proposed in the literature for V2I and V2V propagation links. Next, some path loss models which can be used to vehicular ad hoc networks (VANET) simulations will be presented. In the following, and unless otherwise indicated, path loss will be expressed in dB; distances, antenna heights and wavelength in meters; and frequencies in GHz.

5.1.1 V2I path loss models

Over the last few decades, intense efforts have been carried out to obtain accurate microcell path loss models in dense urban environments. In these models, the transmitting antenna is placed several meters over the streets floor. Nevertheless, only those models which consider the DSRC band can be used to estimate the path loss, as is the case of the microcell urban propagation model developed within the WINNER (Wireless World Initiative New Radio) European Project.

Microcell WINNER path loss model

The model reproduced here corresponds to the extension of the B1 microcell model (WINNER, 2007). This model is based on channel measurement results, considering both LOS and NLOS conditions. The validity frequency range of the model is from 2 to 6 GHz. The model consists of a dual-slope model with and effective height and a *breakpoint* or *critical distance*, denoted by d_c, estimated as

$$d_c = \frac{4h_T^{'}h_R^{'}}{\lambda_c}, \tag{25}$$

where $\lambda_c = c_0 / f_c$ is the wavelength associated to the carrier frequency f_c, $h_T^{'} = h_T - h_0$ and $h_R^{'} = h_R - h_0$, h_T and h_R are the height of the transmitter and receiver antennas, respectively, and h_0 is the effective height due to the presence of vehicles between the transmitter and receiver. h_0 is related to traffic conditions, varying from 0.5 (no or low traffic) to 1.5 (heavy traffic). For moderate traffic $h_0 = 1$. If d is the distance between the transmitter and receiver, the average path loss under LOS conditions is expressed as follows

$$\overline{PL_{LOS}}(d) = \begin{cases} 22.7\log d + 41 + 20\log(f_c/5), & d < d_c \\ 40\log d + 41 - 17.3\log d_c + 20\log(f_c/5), & d_c \geq d \end{cases} \tag{26}$$

The maximum range of the model is assumed as several kilometres. For the transmitter and receiver antennas height, the following ranges are proposed: $5\,m < h_T < 20\,m$ and $1.5\,m < h_R < 20\,m$. This model can be applied to V2I links making $h_T = h_{RSU}$ and $h_R = h_{OBU}$ where h_{RSU} and h_{OBU} are the RSU and OBU antenna heights, respectively.
Under NLOS conditions, the average path loss can be expressed as

$$\overline{PL_{NLOS}}(d_1, d_2) = \overline{PL_{LOS}}(d = d_1) + 20 - 12.5n_j + 10n_j\log(d_2), \tag{27}$$

where

$$n_j = \max\{2.8 - 0.0024d_1,\ 1.84\}. \tag{28}$$

The geometry for the microcell WINNER path loss model for NLOS conditions is shown in Fig. 4, where the distances d_1 and d_2 are also illustrated. Eq. (27) is valid for $d_2 > W_S / 2$, being W_S the street width. For $d_2 \leq W_S / 2$ the LOS model can be applied.
According to Eq. (24), the shadowing effect can be modeled by a standard deviation equal to 3 dB for LOS conditions and 4 dB for NLOS conditions.

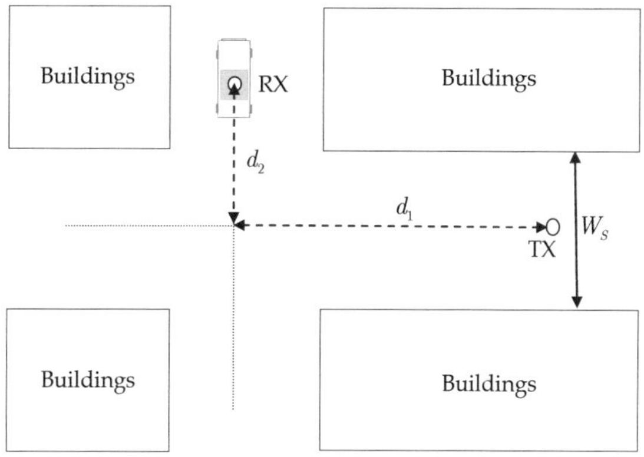

Fig. 4. Geometry for the NLOS WINNER microcell path loss model

5.1.2 V2V path loss models

Differences between V2V and F2M channels oblige the development of new models to estimate the average path loss. Next, we will describe the single- and dual-slope models, indicating the values of their parameters based on different narrowband and wideband channel measurements, as well as the typical two-ray model. Finally, a more elaborated path loss model developed in the University of Kangaku will also be presented.

Single- and dual-slope path loss models

In wireless channel propagation, the conventional single-slope path loss model assumes that the received power decreases with the logarithm of the separation distance between the transmitter and receiver. The average path loss can be estimated as

$$\overline{PL}(d) = \overline{PL}(d_0) + 10\gamma \log(d / d_0), \tag{29}$$

where d is the transmitter to receiver separation distance, $\overline{PL}(d_0)$ is the average path loss for a reference distance d_0, and γ is the well-known propagation path loss exponent. The value of γ takes into account the characteristics of the environment when the propagation occurs. In practice, linear regression techniques based on measured data are used to find the values of the path loss exponent.

However, there are environments where a dual-slope piecewise linear model is able to fit measured data more accurately. A dual-slope model is characterized by a path loss exponent γ_1 and a standard deviation σ_1 above a reference distance up to a breakpoint or critical distance d_c, and by a path loss exponent γ_2 and a standard deviation σ_2 for a distance higher than the critical distance. Using this model, the average path loss can be estimated as

$$\overline{PL}(d) = \begin{cases} \overline{PL}(d_0) + 10\gamma_1 \log(d / d_0), & d_0 \le d \le d_c \\ \overline{PL}(d_0) + 10\gamma_1 \log(d_c / d_0) + 10\gamma_2 \log(d / d_0), & d > d_c \end{cases} \tag{30}$$

In (Green & Hata, 1991), the critical distance d_c is based on the experience and is estimated as $d_c = 2\pi h_T h_R / \lambda_c$, where h_T and h_R are the heights of the transmitter and receiver antennas, respectively, and being λ_c the wavelength associated to the carrier frequency f_c. Nevertheless, Xia et al. proposed a different critical distance $d_c = 4h_T h_R / \lambda_c$, which is more accurate for V2V links (Xia et al., 1993). However, in practice the critical distance is related to the propagation characteristics environment, and of course, there are important differences among urban, suburban, highway and rural areas. In a similar way as the WINNER path loss model, where an effective height was defined to take into account traffic conditions, Emmelmann et al. have proposed a critical distance $d_c = 4h_T^{'} h_R^{'} / \lambda_c$, with $h_T^{'} = h_T - h_0$, $h_R^{'} = h_R - h_0$, and h_0 being an effective ground offset to model different propagation conditions (Emmelmann et al., 2010). From a measurement campaign at 5.9 GHz in an urban environment, with antennas heights $h_T = 1.51\ m$ and $h_R = 1.93\ m$ at the transmitter and receiver, respectively, a critical distance $d_c = 100\ m$ was derived, resulting in an effective ground offset $h_0 = 0.57\ m$ (Emmelmann et al., 2010).

Table 4 summarizes the values of the path loss exponent and the standard deviations of shadowing derived from channel measurement campaigns conducted in different vehicular environments.

Model	Parameter	Urban	Suburban	Highway	Rural
Single slope	$\gamma / \sigma(dB)$	1.61/3.4 (c)	2.32/7.1 (a) 2.75/5.5 (a)	1.85/3.2 (c)	1.79/3.3 (c)
Dual slope	$\gamma_1 / \sigma_1(dB)$	-	2/5.6 (a) 2.1/2.6 (a)	1.9/2.5 (b)	2.3/3.2 (b)
	$\gamma_2 / \sigma_2(dB)$	-	3.8/4.4 (a) 4/8.4 (a)	4/0.9 (b)	4.0/0.4 (b)
	$d_c(m)$	-	100 (a)	220 (b)	226 (b)

Table 4. Parameters for single- and dual-slope path loss models from different channel measured data: (a) (Cheng et al., 2007), (b) (Cheng et al., 2008b), and (c) (Kunisch & Pamp, 2008)

In addition, a three slope path loss model, with two breakpoints, has been proposed in (CEPT Report 101, 2007), where the first interval of distances considers free-space propagation.

Two-ray path loss model

The two-ray model is widely used in LOS conditions due to its simplicity. According to the classical definition of the two-ray propagation model, the direct path and the reflected wave by the ground can be taken into account to estimate the path loss when there is LOS between the transmitter and the receiver and there are no vehicles between them. The geometry for the two-ray propagation model is illustrated in Fig.5.

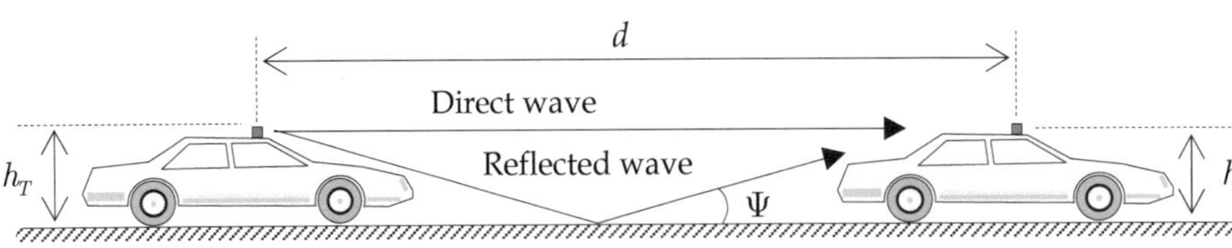

Fig. 5. Geometry for the two-ray propagation model

The superposition of both the direct and the reflected paths at the receiver antenna, results in a fieldstrength gain represented by the term $|E_T / E_D|$, where E_T and E_D are the total and the direct path fieldstrength at the receiver antenna, respectively. From the geometry given in Fig. 5, the term $|E_T / E_D|$ can be calculated as follows

$$|E_T / E_D| = \left| 1 + \rho_{\perp,\parallel}(\Psi) \frac{d}{d+\Delta d} \exp(-jk\Delta d) \sqrt{\frac{G_T(\theta_r,\varphi_r)}{G_T(\theta_d,\varphi_d)} \frac{G_R(\theta_r,\varphi_r)}{G_R(\theta_d,\varphi_d)}} \right|, \tag{31}$$

where $\rho_{\perp,\parallel}(\Psi)$ is the Fresnel reflection coefficient of the ground (the street floor in urban environments and the road in expressway and highway environments), for vertical ($\parallel$) or horizontal ($\perp$) polarization, associated to the angle of elevation Ψ; $k = 2\pi / \lambda_c$ is the wavenumber, and Δd is the different between the reflected path and the direct path lengths, that for $h_T, h_R \ll d$ can be estimated as

$$\Delta d \approx 2h_T h_R / d. \tag{32}$$

Eq. (31) takes also into account the gain pattern of both the transmitter and the receiver antennas, $G_T(\theta,\varphi)$ and $G_R(\theta,\varphi)$, respectively, with (θ_d,φ_d) and (θ_r,φ_r) being the angular directions associated to the direct and reflected paths, respectively.
Eq. (31) corresponds to a rigorous evaluation of the interference between the direct and the reflected paths, that must be taken into account for short-ranges, i.e., when the distance d is comparable with the antennas heights h_T and h_R. However, for large-ranges, i.e., when $d \gg h_T, h_R$, the following simplifications can be applied: (1) the elevation angle, Ψ in Fig.5, is close to 0 (grazing incidence), (2) $\rho_{\perp,\parallel}(\Psi \to 0) \approx -1$, and (3) $G_T(\theta_d,\varphi_d) \approx G_T(\theta_r,\varphi_r)$ and $G_R(\theta_d,\varphi_d) \approx G_R(\theta_r,\varphi_r)$. Thus, Eq. (31) can be simplified and rewritten as

$$|E_T / E_D| \approx |1 - \exp(-jk\Delta d)| = |2\sin(\pi \Delta d / \lambda_c)|_{\Delta d \approx 2h_T h_R / d} \approx \left| 2\sin\left(\frac{2\pi h_T h_R}{\lambda_c d}\right) \right|. \tag{33}$$

The term $|E_T / E_D|$ derived from Eq. (33) is shown in Fig.6 (a), for typical antennas heights in V2V communications ($h_T = h_R = 1.75$ m) and $\rho_{\parallel,\perp} = -1$.
For small distances between the transmitter and the receiver, the influence of the interference between the direct and the reflected paths is visible, and $|E_T / E_D|$ undergoes space fading until a breakpoint distance or critical distance given by $d_c = 4h_T h_R / \lambda_c$. It is worth noting that the maximum value (6 dB) of the term $|E_T / E_D|$ occurs for a distance where the interference of the reflected path is totally constructive; whereas the minimum value is obtained for a distance where the interference is totally destructive.
When the two-ray propagation model is considered, the Friis transmission formula given by Eq. (22), is reedited as follows

$$P_R(d) = P_T + G_T(\theta_d,\varphi_d) + G_R(\theta_d,\varphi_d) - PL_{FS}(d) + 10\log|E_T / E_D|^2, \tag{34}$$

being the total two-ray path loss of the link

$$PL(d) \triangleq 10\log(4\pi d / \lambda_c)^2 - 10\log|E_T / E_D|^2. \tag{35}$$

Fig.6 (b) shows the total path loss versus the transmitter to receiver separation distance, where $h_T = h_R = 1.75$ m and $\rho_{\|,\perp} = -1$ have been considered.

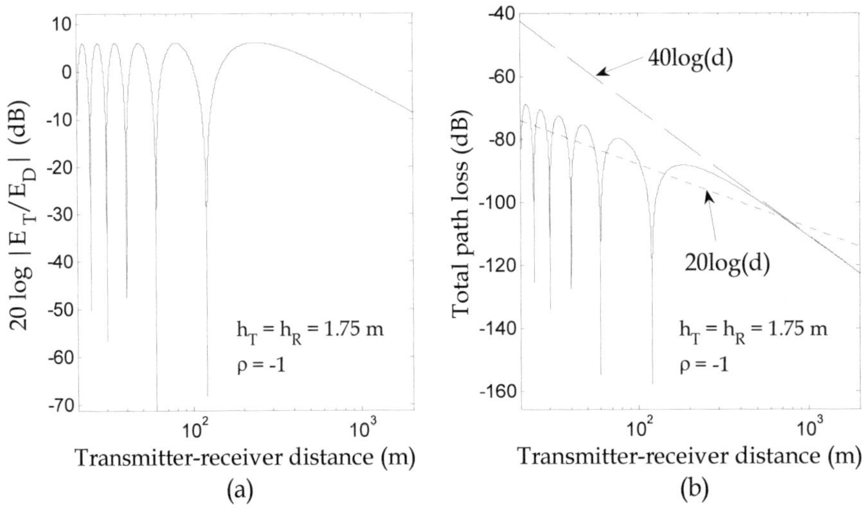

Fig. 6. Two-ray propagation model: (a) interference between the direct and the reflected paths, and (b) total path loss

Measurement results reported in the Reference (Kunisch & Pamp, 2008), show that the two-ray propagation model can be used in highway and rural environments. Due to the direct path can be affected by the surrounding environment, Kunisch and Pamp suggest an average path loss, denoted here by $\overline{PL}_{Two-ray}(d)$, in the way of

$$\overline{PL}_{Two-ray}(d) = PL_0 + 10\gamma \log d - 10\log\left|E_T / E_D\right|^2, \tag{36}$$

where γ is a path loss exponent different from 2 (LOS conditions), and PL_0 is a constant. Both γ and PL_0 can be extracted from measured data using linear regression techniques. When $\Delta d \ll \lambda_c$, Eq. (33) can be simplified once again and rewritten as

$$\left|E_T / E_D\right| \approx 2\frac{2\pi h_T h_R}{\lambda_c d}. \tag{37}$$

Then, the total average path loss is given by

$$PL(d) \approx 10\log\left(\frac{4\pi d}{\lambda_c}\right)^2 - 10\log\left(\frac{4\pi h_T h_R}{\lambda_c d}\right)^2 = 10\log\left(\frac{d^2}{h_T h_R}\right)^2, \tag{38}$$

yielding to the well-known d^4 power law, i.e., the total path loss is proportional to the fourth power of the transmitter to receiver separation distance.

Kangaku University model

Based on channel measurements, a prediction formula to estimate the average path loss in urban environments, with both LOS and NLOS conditions, was proposed in (Ito et al., 2007).

The average path loss under LOS conditions is estimated as follows

$$\overline{PL}_{LOS} = \left[10.4 + 1.3\log\left(\frac{h_T h_R}{\lambda_c}\right)\right]\log d + 24.6\log\left(1 + \frac{d}{d_c}\right) + 19.4\log f_c + 3.9\log W_S + 33\,, \quad (39)$$

where f_c is the carrier frequency; d_c is a critical distance estimated by $d_c \approx 8h_T h_R / \lambda_c$; and W_s is the street width. The geometry of the path loss model for LOS conditions is shown in Fig. 7. Table 5 summarizes the validity range of Eq. (39).

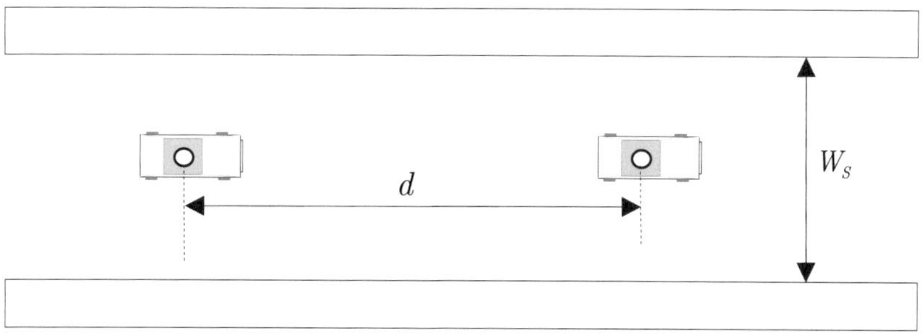

Fig. 7. Geometry of the Kangaku University path loss model for LOS conditions (Ito et al., 2007)

Parameter	Urban environment
$d(m)$	2-1000
$f_c(GHz)$	0.4-6
$W_S(m)$	8, 20, 40, 60
$h_T, h_R(m)$	0.5-3.5

Table 5. Applicable range of the Kangaku University LOS model

For NLOS conditions, the model considers the geometry and parameters illustrated in Fig. 8. The average path loss is estimated as follows

$$\overline{PL}_{NLOS} = \begin{cases} \overline{PL}_{LOS}, & D \leq d_{EL} \\ \min\left(\overline{PL}_{NLOS\text{-}1}, \overline{PL}_{NLOS\text{-}2}\right), & D > d_{EL} \end{cases} \quad (40)$$

where

$$D = d_1 + W_{S1} + W_{S2} + d_2\,, \quad (41)$$

$$d_{EL} = d_1 + W_{S1} + W_{S2} + \left(W_{S1}W_{S2} / d_1\right), \quad (42)$$

and

$$\overline{PL}_{NLOS\text{-}1} = \left\{\left(3.2 - 0.033W_1 - 0.022W_2\right)d_1 + 39.4\right\}\left\{\log(D) - \log(d_{EL})\right\} + L_{LOS}\left(d_{EL}\right), \quad (43)$$

$$\overline{PL}_{NLOS\text{-}2} = \left\{-6.7 + 11.2\log\left(\frac{h_T h_R}{\lambda_c}\right)\right\}\log(D) + \left\{25.9 + 10.1\log\left(\frac{d_1}{\lambda_c}\right)\right\}\log\left(1+\frac{D}{d_c}\right) + 19.8\log(f_c) - 3.8\log(W_1 W_2) + 57.7 \quad (44)$$

with $d_c = 4h_T h_R / \lambda_c$ being the critical distance.
However, a new modification which improves the prediction of path loss has been proposed in (Sai et al., 2009), for both LOS and NLOS conditions, where the average path loss given by Eqs. (39) and (44) are substituted by the following expressions

$$\overline{PL}_{LOS} = \left[7.2 + 7.1\log\left(\frac{h_T h_R}{\lambda_c}\right)\right]\log(D') + 28.3\log\left(1+\frac{D'}{d_c}\right) - 1.2\log(f_c) - 19.6\log(W_s) + 65.9 \quad (45)$$

$$\overline{PL}_{NLOS-2} = \left\{47.6 + 6.6\log\left(\frac{h_T h_R}{\lambda_c}\right)\right\}\log(D') + \left\{89.1 - 33\log\left(\frac{d_1}{\lambda_c}\right)\right\}\log\left(1+\frac{D'}{d_c}\right) + 19.9\log(f_c) - 11.3\log(W_1 W_2) + 2.8 \quad (46)$$

with $D' = d_1 + d_2$. Eqs. (45) and (46) were adjusted from measurement results under real traffic conditions.

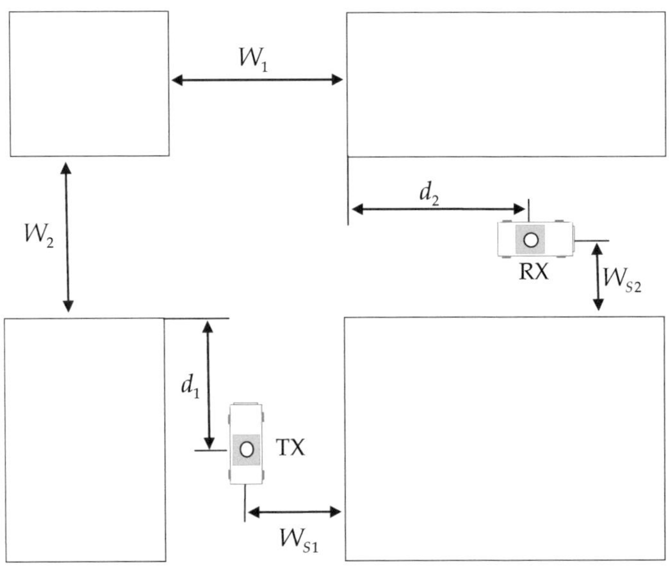

Fig. 8. Geometry of the Kangaku University path loss model for NLOS conditions (Sai et al., 2009)

5.2 Narrowband fading statistics

In a time-variant channel, the multipath propagation causes signal fading due to the phase differences among the MPCs. This signal fading, referred to as small-scale fading, can be modeled by the classical Rayleigh and Rice distributions. Rayleigh fading occurs when the number of MPCs is high having similar amplitudes. From the central limit theorem (CLT), the time variation of the received signal enveloped, denoted by $r(t)$, in a small local area around the receiver follows a Rayleigh distribution, with probability density function (pdf) given by

$$pdf_r(r) = \frac{r}{\sigma^2}\exp\left(-\frac{r^2}{2\sigma^2}\right), \tag{47}$$

where $E\{r^2\} = 2\sigma^2$ is the mean square value of the distribution, equivalent to the average power of the received signal. Rayleigh fading occurs generally in NLOS conditions. Whereas, in LOS conditions where a dominant contribution with a power A^2 exists, the received envelop follows a Rice distribution, given by

$$pdf_r(r) = \frac{r}{\sigma^2}\exp\left(-\frac{r^2 + A^2}{2\sigma^2}\right)I_0\left(\frac{rA}{\sigma^2}\right), \tag{48}$$

where $I_0(\cdot)$ is the zeroth-order modified Bessel function of the first kind (Gradsthteyn & Ryzhik [8.406], 2007). The mean square value of the Rice distribution is $E\{r^2\} = 2\sigma^2 + A^2$. To model the severity of fading, it is common to use the so-called Rice K factor defined as the relation between the power of the dominant contribution and the power associated to the diffuse MPCs, i.e.

$$K \triangleq \frac{A^2}{2\sigma^2}. \tag{49}$$

Note that when $K \to 0$, the Rice distribution tends to the Rayleigh distribution (Eq. (48) $\to$ Eq. (47)). Ricean fading occurs generally in LOS conditions. Based on measured data, other distributions have been applied to model the small-scale fading, such as the Nakagami-m and Weibull distributions. In Nakagami-m fading, the pdf of the received signal envelope is given by

$$pdf_r(r) = \frac{2}{\Gamma(m)}\left(\frac{m}{\Omega}\right)^m r^{2m-1}\exp\left(-\frac{m}{\Omega}r^2\right), \tag{50}$$

where $\Omega = E\{r^2\}$ and m is the so-called *shape factor* (fading parameter). When $m = 1$, the Nakagami-m distribution yields the Rayleigh distribution (note that $\Omega = 2\sigma^2$), and if $m \gg 1$ the Nakagami-m distribution can approximate the Rice distribution, with (Molisch, 2005)

$$\Omega = 2\sigma^2 + A^2 \tag{51}$$

$$m\big|_{m\gg 1} \approx (K+1)^2 / (2K+1). \tag{52}$$

In Weibull fading, the pdf of the received signal envelope is given by

$$pdf_r(r) = \frac{\beta}{\Omega}r^{\beta-1}\exp\left(-\frac{r^\beta}{\Omega}\right), \tag{53}$$

where $\beta > 0$ is the fading parameter. When $\beta = 2$, the Weibull distribution yields the Rayleigh distribution, and if $\beta \gg 2$ the Weibull distribution can also approximate the Rice distribution.

Measured results in urban and highway environments reported in (Maurer et al., 2002) show that small-scale fading can be approximated very well by the Rice and Nakagami-*m* distributions. Also, in (Cheng et al., 2007) the observed trend is a good match with the Rice distribution for short transmitter-receiver separation distances, whereas for large distances the small-scale fading is better modeled using the Rayleigh distribution. Nevertheless, there are situations of large distances, where the Nakagami-*m* distribution offers a better fit to the measured data, caused by an intermittent loss of the dominant MPC at large distances, e.g. when the transmitter and receiver are separated around corners. In (Cheng et al., 2007) based on two data sets, high values of m between 3 and 4 have been estimated from short transmitter to receiver distances (approximately less than 4.5 m), whereas for large distances $m < 1$ (e.g. $m = 0.45$ for distances ranging from 71 m to 177 m), indicating fading more severe than Rayleigh.

Regarding to the large-scale fading or shadowing, produced by the obstruction of MPCs, measured results reported in (Maurer et al., 2002) show that it can be well approximated by a log-normal distribution. Nevertheless, there are situations in which an intermittent blockage of the MPCs exists, causing that the random variable X_σ in Eq.(24) not follow a Gaussian distribution. In this latter case, based on empirical results Cheng et al. proposed to jointly model the small- and large-scale fading using the Nakagami-*m* distribution (Cheng et al., 2007).

5.3 Doppler spectrum

The channel variability over time and the multipath propagation effect cause frequency dispersion or Doppler spread. The movement of both the transmitter and receiver terminals, as well as the interacting objects, e.g. other vehicles between the terminals, make that the Doppler spread to be different to the conventional F2M systems. Theoretical Doppler spectrum models and experimental results are presented in this Subsection.

5.3.1 Theoretical models

The first theoretical approach of the Doppler spectrum in a V2V channel was based mainly on two important assumptions: (1) both the angle-of-departure (AOD) and the angle-of-arrival (AOA) of MPCs are uniformly distributed over (0, 2π[, and (2) the CIR has a uniform power profile. The above assumptions yield the isotropic scattering assumption, and thus the Doppler PDS, $P_H(\nu)$, is given by (Akki & Haber, 1986)

$$P_H(\nu) = \begin{cases} \dfrac{1}{\pi^2 \nu_R \sqrt{a}} K\left[\dfrac{(1+a)}{2\sqrt{a}}\sqrt{1-\left(\dfrac{\nu}{(1+a)\nu_R}\right)^2}\right] & -(\nu_T+\nu_R) < \nu < \nu_T + \nu_R \\ 0 & \text{otherwise} \end{cases}, \qquad (54)$$

where $a = \nu_T / \nu_R$; $\nu_T = v_T / \lambda_c$ and $\nu_R = v_R / \lambda_c$ are the maximum Doppler frequencies of the transmitter and receiver, respectively, being v_T and v_R the transmitter and receiver speed, respectively, and λ_c the wavelength; and $K[\cdot]$ is the complete elliptic integral of the first kind (Gradsthteyn & Ryzhik [8.112 1], 2007). Fig. 9 shows the behavior of the Doppler PDS as a function of the frequency, for different values of a.

When $a = 0$, the maximum Doppler frequency of the transmitter is null and therefore, the transmitter is static, which corresponds to a V2I channel. In this case, Eq. (54) can be simplified an rewritten as follows

$$P_H(\nu) = \begin{cases} \dfrac{1}{\pi \nu_R \sqrt{1 - \left(\dfrac{\nu}{\nu_R}\right)^2}} & -\nu_R < \nu < \nu_R \\ 0 & \text{otherwise} \end{cases}, \tag{55}$$

which is the classical Jakes Doppler PDS (Clark, 1968), (Jakes, 1974). From Eq. (54), the *rms* Doppler spread can be calculated as

$$\nu_{rms} = \left(\frac{1}{\lambda\sqrt{2}}\right) v_{eff} = \left(\frac{1}{\lambda}\right)\sqrt{\frac{v_T^2 + v_R^2}{2}}, \tag{56}$$

where $v_{eff} \triangleq \sqrt{v_T^2 + v_R^2}$ is defined as the effective speed (Cheng et al., 2007). For non-isotropic scattering channels, i.e. non uniform distribution of both AOD and AOA, a general closed form for the $P_H(\nu)$ has not been obtained. However, simulation results of the PDS show that in a non-isotropic channel the Doppler PDS exhibits one-sided Doppler shifts (Zheng, 2006).

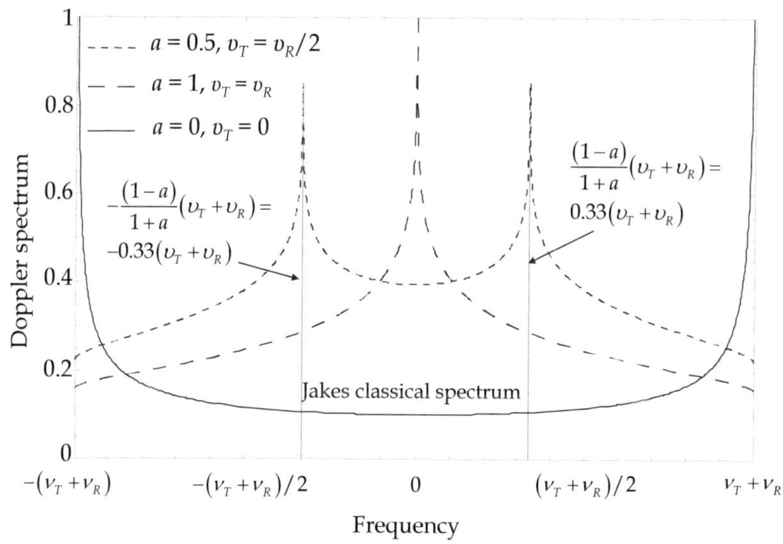

Fig. 9. Normalized Doppler PDS for the V2V channel (Akki & Habber, 1986)

Nevertheless, for a V2V specific scenario in an urban canyon, where there are two dense reflectors located along one side of the road as shown in the Fig. 10(a), the Doppler PDS can be estimated as (Jiang et al., 2010)

$$P_H(\xi)=\begin{cases}\frac{1}{A\nu_d}\frac{1}{\sqrt{1-\xi^2}}\times\left\{\left[\frac{d_2}{\sqrt{1-\xi^2}}+\sqrt{\left(d_0-\frac{d_2}{\xi}\sqrt{1-\xi^2}\right)^2+d_1^2}\right]^{-\gamma}+\left[\frac{d_2'}{\sqrt{1-\xi^2}}+\sqrt{\left(d_0-\frac{d_2'}{\xi}\sqrt{1-\xi^2}\right)^2+d_1'^2}\right]^{-\gamma}\right\} & -1\le\xi\le1, \\ 0 & \text{otherwise}\end{cases} \tag{57}$$

where $\xi=\nu/\nu_d$; $\nu_d=v_d/\lambda_c$, being v_d the relative speed between vehicles, d_0 is the distance between the transmitter and receiver along the road; d_1' and d_2' are the distances illustrated in Fig. 10(a); γ is the path loss exponent; and

$$A=\int_{-\pi}^{0}\frac{1}{(r')^{\gamma}}d\beta+\int_{0}^{\pi}\frac{1}{r^{\gamma}}d\alpha, \tag{58}$$

with

$$r=r_1+r_2=\sqrt{\left(d_0-\frac{d_2\cos\alpha}{\sin\alpha}\right)^2+d_1^2}+\frac{d_2}{\sin\alpha}, \tag{59}$$

$$r'=r_1'+r_2'=\sqrt{\left(d_0-\frac{d_2'\cos\beta}{\sin\beta}\right)^2+d_1'^2}+\frac{d_2'}{|\sin\beta|}, \tag{60}$$

where $|\cdot|$ denotes absolute value.

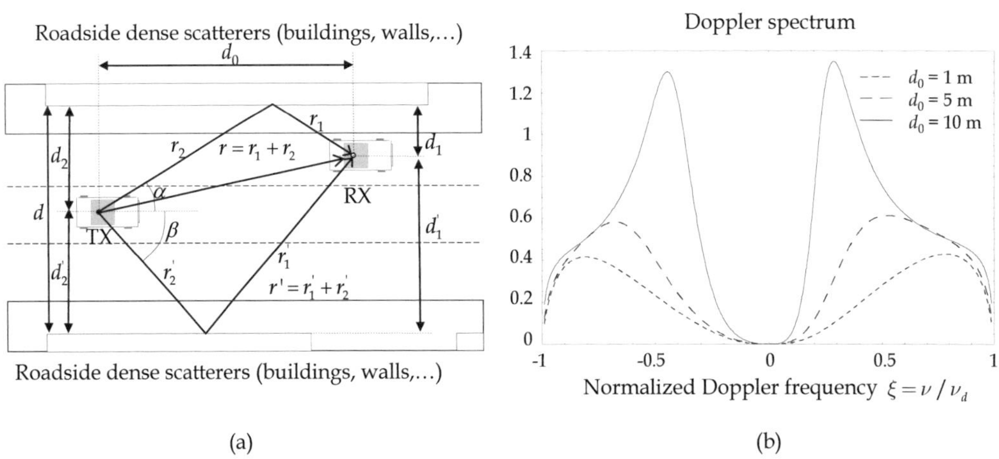

Fig. 10. (a) Illustration of a typical urban canyon environment and (b) its corresponding Doppler PDS, when $d_1=d_2'=3\ m$, $d_2=d_1'=5\ m$ and $\gamma=3$ (Jiang et al., 2010)

For distances along the road, d_0, with similar order of magnitude to the distances d_1, d_2, d_1' and d_2' shown in Fig. 10(a), the curves are smoothed and the asymptotes of Fig. 10(b) do not appear. Otherwise, for asymmetrical positions of vehicles, i.e. $d_1 \neq d_2'$ and $d_2 \neq d_1'$, the asymmetry of the Doppler PDS increases around the central Doppler frequency.

5.3.2 Experimental results

The *rms* Doppler spread derived from different measurement campaigns, (Cheng et al., 2007), (Cheng et al., 2008c) can be approximated as a function of the effective speed, v_{eff}, by

$$\nu_{rms} = \frac{s}{\lambda_c\sqrt{2}} v_{eff} + \delta, \tag{61}$$

where $s / \lambda_c\sqrt{2}$ is the slope and δ is an offset. Note that according to Eq. (56), $s=1$ and $\delta = 0$ for isotropic scattering. Table 6 summarizes the values of s and δ for highway, rural and suburban V2V environments (Cheng et al., 2008c).

Parameter	Suburban	Highway	Rural
Offset, δ	11.2	0.2	0.5
s	0.428	0.414	0.420

Table 6. Parameters measured of *rms* Doppler spread

It should be emphasized that s is significantly less than 1 due to the non-isotropically oriented scattering objects in real scenarios.

Likewise, it can be seen that the Doppler spread usually increases if the transmitter to receiver distance grows. Obviously, this spreading is the result of the tendency for the vehicle speeds to be higher at larger distances. In situations of platoon, since the average relative velocity between the vehicles is close to zero, it is comprehensible that the average Doppler spread, $\bar{\nu}$, is approximately 0.

Table 7 summarizes the results of Doppler parameters for a platoon of vehicles, with a distance between transmitter and receiver ranging from 5 to 300 m, under realistic traffic conditions.

Parameter	Urban	Highway
$\bar{\nu}$	2.03 Hz	0.93 Hz
ν_{rms}	85.6 Hz	119.98 Hz

Table 7. Doppler spectrum parameters measured under platoon conditions, from (Maurer et al., 2002)

5.3.3 Coherence time

The coherence time has an inverse relation to the *rms* Doppler spread. The coherence time can be related to the *rms* Doppler spread in terms of a proportionality *k*-constant

$$T_C \approx k / \nu_{rms}, \tag{62}$$

where k represents the time correlation degree. In a V2I channel, k can be analytically derived. For a time correlation degree greater than 50%, $k_{>50\%} \approx 9/\left(16\sqrt{2}\right)\pi$ (Sklar, 1997).

According to (Cheng et al., 2007), in a V2V channel an estimation of k for a correlation degree equal to 90% is $k_{90\%} \approx 0.3$.
From Eq. (61) and Eq. (62), the coherence time has an inverse relationship to the effective speed. Also, results reported in (Cheng et al., 2007) show that the coherence time decreases as the distance between transmitter and receiver grows.

6. V2V wideband channel characterization

This Section provides a brief wideband channel description based on channel data collected in the measurement campaigns summarized in the Table 3. It is worth noting that the measurement results published in the literature have shown that the propagation characteristics are closely related to the type of environment, and in similar environments with different frequencies can also vary significantly. For this reason, the metric values reported here will be applicable for environments with similar characteristics as those where the measurements were collected.

6.1 V2V TDL channel model

In channel modeling, the most popular wideband model is the TDL channel model. The CIR of a TDL channel model of N taps is written as

$$h(t,\tau) = \sum_{i=1}^{N} h_i(t)\delta(\tau - i\Delta\tau), \tag{63}$$

where $\Delta\tau = 1/B$, being B the channel bandwidth[19]. A first approximation to the total number of taps can be derived from the *rms* delay spread as

$$N \simeq \left\lceil \frac{\tau_{rms}}{\Delta\tau} \right\rceil + 1, \tag{64}$$

where $\lceil \cdot \rceil$ represents the nearest upper integer function. A TDL channel model should describe the number of taps, the relative taps power, the fading amplitude statistics and the Doppler PDS for each tap. Some V2V TDL channel models have been published in the literature (Acosta & Ingram, 2007), (Sen & Matolak et al., 2008), (Renaudin et al., 2009) and (Wu et al., 2010). The reader is referred to those references in order to have a better knowledge of how the models have been built.
The tap fading amplitude statistics can be described by the well-known Rayleigh or Rice distributions, as is the case of the model proposed in (Acosta & Ingram, 2007). Nevertheless, other researchers have found a better fit to the measured results through the Nakagami-m and Weibull distributions. Different small-scale fading conditions have been observed for a given tap, ranging from Rician fading (equivalent to fading parameters $m > 1$ and $\beta > 2$ for the Nakagami-m and Weibull distributions, respectively) to more severe than Rayleigh fading (equivalent to $m < 1$ and $\beta < 2$). In (Renaudin et al., 2009), the fading parameter β takes values from 1.56 to 3.75. In (Sen & Matolak. 2008) the best fit to the measurements is performed using the Weibull distribution, where the values of β oscillates from 1.29 to 5.15.

---

[19] When a correlative channel sounder is use to estimate the CIR, B is equal to the inverse of the time resolution.

In (Wu et al., 2010), as an extension work of (Sen & Matolak. 2008), additional TDL channel models are provided in terms of the channel bandwidth, where the tap fading statistics is also modeled by the Weibull distribution, with β ranging from 1.59 to 4.97 for a channel bandwidth of 20 MHz.

6.2 RMS delay spread

Statistical values of the *rms* delay spread and maximum delays have been reported in the literature based on measurement campaigns. In (Kunisch et al., 2008) values of *rms* delay spread equal to 52 ns, 41 ns and 47 ns were estimated for rural, highway and urban areas, respectively, for a maximum transmitter-receiver separation of about 300 m. Whereas, in (Tan et al., 2008) higher values have been estimated for the same transmitter-receiver separation. Table 8 summarizes the *rms* delay spread values reported in (Tan et al., 2008) for different environments, vehicles distances and propagation conditions (LOS and NLOS).

	LOS	NLOS
Rural (100 m)	21.6	-
Highway (300 m)	156.8	-
Highway (400 m)	141.8	398
Urban (200 m	157.5	295
Urban (400 m)	320.6	-
Urban (600 m)	286.6	-

Table 8. RMS delay spread vales, in ns, for different environments and transmitter-receiver separation distances, from (Tan et al., 2008)

Values of *rms* delay spread reported in (Sen & Matolack, 2008) take into account the location position of the antenna and the VTD. Thus, 256 ns has been estimated in urban environments with the antenna outside the car, and 125.8 ns with the antenna inside the car. For open areas, the values reported are 53.2 ns and 126.8 ns for low and high traffic conditions respectively.

In (Cheng et al., 2008a), a comparative study of the time- and frequency-dispersion behavior was performed. Values of maximum delay spread of 0.6 μs, 1.4 μs and 1.5 μs was estimated for suburban, highway and rural areas, respectively. These values correspond to a 90% probability, i.e., values which can be observed in the 90% of times or situations. For a correlation degree equal to 90%, the following coherence bandwidths have been estimated: 730 kHz, 520 kHz and 460 kHz for suburban, rural and highway areas, respectively.

In general, vehicular environments are mainly characterized by the presence of local interacting objects and the longest delay spread appears in canyon scenarios under NLOS conditions, i.e., street canyons in urban areas and highway environments when big vehicles exist in the transmitter and receiver proximity. This is a further difference with respect to cellular systems, where the longest time dispersion occurs within open and suburban areas.

6.3 RMS Doppler spread

As started in the previous Section, in V2V channels the Doppler PDS can differ to the classical U-shaped spectrum due to the AOA of MPCs has a non-uniform distribution. Values of the *rms* Doppler spread reported in the literature are clearly dependent on the vehicular speed used in the measurement campaign. Values of *rms* Doppler spread less than

1 kHz have been measured. These values are a small fraction of the inter-carrier frequency defined in IEEE 802.11p (156.25 kHz), indicating that the Doppler spread may not cause significant inter-carrier interference (ICI). Nevertheless, the coherence time in vehicular environments can introduce impairments when long symbol packets are transmitted.
Based on channel measurements, in (Acosta & Ingram, 2007) different Doppler PDS shapes for each tap of the TDL channel model are provided: flat, round, classic 3 dB and classic 6 dB.

7. Some aspects related to the antennas for vehicular communications

Both the infrastructure and the vehicle antennas in the ITS 5.9 GHz band are typically directive in the vertical plane and omni-directional in the azimuth plane. The antennas in vehicles are usually mounted either on the roof or inside the vehicle (e.g. under the windshield, in rear view mirror, on a seat or near a dashboard) although some vehicles can have additional antennas for radar applications in the bumper. For instance, the antennas are usually mounted under the windshield of the vehicle, inside the device known as tag or on-board equipment (OBE), for toll collection applications and parking control working in the CEN DSRC 5.8 GHz band (CEN EN 12253, 2004).
The effect of the antenna placement in vehicles is significant (Kaul et al., 2007). The differences in the cumulative link packet error rates are until 25-30% depending on the antenna locations under LOS conditions. Gain patterns of omni-directional antennas in the azimuth plane become asymmetric in certain positions of the vehicle, showing only an accurate approximation to the omni-directional pattern if the antenna is mounted on the center of the vehicle roof.
The antenna can be designed as either wideband dipole antenna (Chen et al., 2005) or double-fed printed monopole (Jensen et al., 2007) for OBU, and double-looped monopole (Kim et al., 2003) for RSU, operating in WLAN 2.4-2.485 GHz, 5.47-5.725 GHz and ITS 5.9 GHz bands.
The antenna radiation pattern in the vertical plane can be approximated using the ITU-R F.1336 Recommendation (CEPT Report 101, 2007), (ITU-R F.1336, 2007), where the gain, expressed in dBi, for an elevation angle θ_1 is given by

$$G(\theta)=\max\left[G_1(\theta),G_2(\theta)\right], \tag{65}$$

with

$$G_1(\theta)=G_0-12\left(\frac{\left|\theta-\theta_1\right|}{\Delta\theta_{-3dB}}\right)^2, \tag{66}$$

$$G_2(\theta)=G_0-15+10\log\left[\left(\max\left\{\frac{\left|\theta-\theta_1\right|}{\Delta\theta_{-3dB}},1\right\}\right)^{-1.5}+k\right], \tag{67}$$

where θ_1 is the tilt of the gain elevation pattern; $\Delta\theta_{-3dB}$ is the -3 dB beamwidth in the vertical plane $\Delta\theta_{-3dB}=107.6\times10^{-0.1\,G_0}$, being G_0 the maximum gain (in dBi); and $k=0$ is the sidelobe factor for the frequency range from 3 GHz to 70 GHz. All angles in Eqs.(65)-(67) are expressed in degrees.

Fig. 11(a) shows an example of the approximated gain pattern in the vertical plane for a vehicle antenna using Eqs.(65)-(67). Maximum gains of 5 and 8 dBi have been considered. In Fig. 11(b), the main parameters of the RSU and OBU antennas are illustrated. The height and the tilt of the RSU antenna depend on the range of the transmission. Regardless RSU antenna, the height is limited to 15 m above the road surface. For applications in CEN DSRC 5.8 GHz band (CEN EN 12253, 2004), the antenna height is around 5.5 m, with a usual antenna tilt of 55°; and for larger-range applications in the ITS 5.9 GHz band the typical RSU antenna elevation is about 6 m with an antenna tilt equal to 45°.

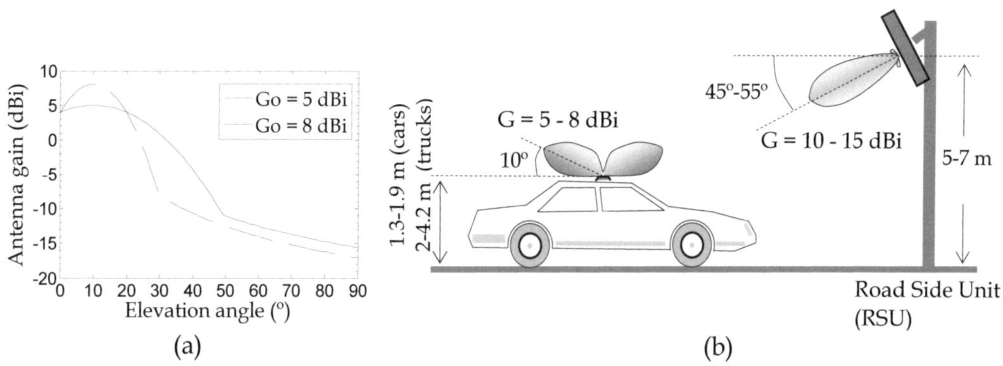

Fig. 11. (a) Gain pattern in the vertical plane for the ITS 5.9 GHz band, with maximum gains of 5 and 8 dBi, and (b) antenna installation in the vehicle and in the RSU for ITS applications at 5.9 GHz

8. Future advances in channel modeling for vehicular communications

Although advances have been made within V2V channel modeling, there are still many remaining unanswered questions. Further empirical studies should be conducted to improve the accuracy of channel models, considering the possible combinations of real propagation conditions. For this purpose, more measurement campaigns covering a greater variety of scatterers density, vehicles speed, LOS and NLOS conditions, are necessary. In this sense, the future challenges should be oriented to channel measurements and the development of parameterized channel models, where the main channel parameters estimated from real data could be introduced in the channel model.

The Doppler PDS is an important parameter which characterizes the frequency-dispersion behavior of vehicular environments. In order to develop more accurate channel models based on the TDL concept, further studies of non-isotropic scattering based on measurements should be conducted. In addition, the transmitter and receiver direction of motion affect the channel parameters. Few measurement campaigns have considered the transmitter and receiver moving in opposite directions, especially in expressways and highways, where the vehicles speed is high. In this sense, more measurement campaign should be conducted when the transmitter and receiver are moving in opposite directions and for various vehicular traffic densities in different environments.

The measurement campaigns performed consider short separation distances between the transmitter and the receiver, with a maximum distance of about 400 m. Although there are still few applications involving vehicles separated by large distances, measurement campaigns are required to explore the propagation characteristics for large ranges, e.g., larger than 1 km.

In the measurement campaigns, the antennas are placed in elevated positions of the vehicles. Since the antenna position in the vehicle can affect the packet error rate, other vehicles positions, near to real implementations, should be considered. Also, the typical vehicles involved in the measurements are vans. More measurements are necessary considering a greater variety of vehicles, since the propagation mechanisms, especially the obstruction effects, are related to the height of the antennas.

Exploring the available literature, little attention has been devoted to V2I measurement campaigns compared with the V2V case. Since some safety applications involve RSUs, more measurements of the V2I channel are required, developing more realistic channel models. It is also interesting to obtain V2I path loss models for highway and rural areas taking into account the effective height of the terrain, as well as the road curves and other important interacting objects that comprise the propagation environment.

Other frequency bands, as the low part of the UHF band may be also allocated to V2V communications. In Japan, a 10 MHz band from 715 MHz to 725 MHz has been already assigned for ITS applications to prevent oncoming traffic collisions (Sai et al., 2009). It is interesting to explore the propagation characteristics at these frequencies, because significant differences to the 5.9 GHz band exist.

Finally, since large differences exist among cities, expressways and highways around the world, to have a better knowledge of the vehicular channel propagation characteristics in specific environments more measurement campaigns must be conducted in different places.

9. Conclusions

The ITS concept for DSRC systems, together with new applications related to driving safety and mobile ad hoc networks, have triggered great interest in vehicular channel modeling during the last decade, and specially in the five last years. Differences between vehicular and traditional cellular channels require new channel models and measurement campaigns to understand the signal impairments introduced by the time- and frequency-dispersive behaviour of the vehicular channel. The main characteristics of vehicular environments, make that the vehicular channel have great incidence to the final system performance. As mentioned, an accurate knowledge of the vehicular channel is vital to implement safety applications.

In this Chapter, principal propagation aspects in vehicular networks have been analyzed. Based on the available literature, an overview of V2V and V2I channel modeling and channel measurements has been reported. A narrowband characterization of the V2V and V2I channel has been made, where the most important path loss and Doppler PDS models have been described. Also, information of wideband channel parameters has been provided. Finally, future advances in channel modeling for vehicular communications have been identified.

10. References

Acosta-Marum, G. & Ingram, M. A. (2007). Six time- and frequency-selective empirical channel models for vehicular wireless LANs. *IEEE Vehicular Technology Magazine*, Vol. 2, No. 4, pp. 4-11.

Acosta-Marum, G.; Tokuda, K. & Ingram, M. A. (2004). Measured joint Doppler-delay power profiles for vehicle-to-vehicle communications at 2.4 GHz. *IEEE Global Telecommunications Conference*, Vol. 6, pp. 3813-3817, Atlanta, GA, Dec. 2004.

Akki, A. S. & Haber, F. (1986). A statistical model for mobile-to-mobile land communication channel. *IEEE Transactions on Vehicular Technology*, Vol. 35, No. 1, pp. 2-7.

Akki, A. S. (1994). Statistical properties of mobile-to-mobile land communication. *IEEE Transactions on Vehicular Technology*, Vol. 43, No. 4, pp. 826-836.

Almalag, M. S. (2009). *Vehicular Networks : from Theory to Practice,* Olariu, S. & Weigle, M. C. , Ed., pp. 5.1-5.26, Chapman & Hall/CRC Press, Boca Raton, FL, USA.

Almers, P.; Bonek, E.; Burr, A.; Czink, N.; Debbah, M.; Degli-Esposti, V.; Hofstetter, H.; Kyösty, P.; Laurenson, D.; Matz, G.; Molisch, A. F.; Oestges, C. & Özcelik, H. (2007). Survey of channel and radio propagation models for wireless MIMO systems. *EURASIP Journal on Wireless Communications and Networking*, Article ID 19070, Vol. 2007, pp. 1-19.

ASTM E2213-03. (2003). Standard Specification for Telecommunications and Information Exchange Between Roadside and Vehicle Systems — 5 GHz Band Dedicated Short Range Communications (DSRC) Medium Access Control (MAC) and Physical Layer (PHY) Specifications, American Society for Testing Materials (ASTM), West Conshohocken, PA, USA.

Aulin, T. (1979). A modified model for the fading signal at a mobile radio channel. *IEEE Transactions on Vehicular Technology*, Vol. VT-28, No. 3, pp. 182-203.

Bello, P. A. (1963). Characterization of randomly time-variant linear channels. *IEEE Transactions on Communications Systems*, Vol. CM-11, No. 4, pp. 360-393.

Bernadó, L.; Zemen, T.; Paier, A.; Matz, G.; Karedal, J.; Cnik, N.; Dumard, C.; Tufvesson, F.; Hagenauer, M. ; Molisch, A.F. & Mecklenbräuker C. (2008). Non-WSSUS vehicular channel characterization at 5.2 GHz - spectral divergence and time-variant coherence parameters. *XXIX General Assembly of the International Union of Radio Science (URSI)*, Chicago, IL, Aug. 2008.

CEPT Report 101. (2007). *Compatibility Studies in the Band 5855- 5925 MHz between Intelligent Transport Systems (ITS) and other Systems,* European Union Electronic Communications Committee (ECC), within the Conférence Européenne des administrations des Postes et des Télécommunications (CEPT), Bern, Switzeland.

CEPT Report 20. (2007). *The Harmonised Radio Spectrum Use for Safety Critical Applications of Intelligent Transport Systems (ITS)*, European Union Electronic Communications Committee (ECC), within the Conférence Européenne des administrations des Postes et des Télécommunications (CEPT), Bern, Switzeland.

Chen, G. Y.; Huang, S. Y.; Sun, J. S. & Chen, S. Y. (2005). The multi-band dipole antenna. *2005 International Symposium on Communications, Kaohsiung*, Taiwan, Nov. 2005.

Cheng, L.; Henty, B. E.; Stancil, D. D.; Bai, F. & Mudalige. P. (2007). Mobile vehicle-to-vehicle narrowband channel measurement and characterization of the 5.9 GHz dedicated short range communication (DSRC) frequency band. *IEEE Journal on Selected Areas in Communications*, Vol. 25, No. 8, pp. 1501-1526.

Cheng, L.; Henty, B.; Cooper, R.; Stancil, D. D. & Bai, F. (2008a). Multi-path propagation measurements for vehicular networks at 5.9 GHz. *IEEE Wireless Communications and Networking Conference*, pp. 1239-1244, Las Vegas, NV, Ap. 2008.

Cheng, L.; Henty, B. E.; Stancil, D. D.; Bai, F. & Mudalige, P. (2008b). Highway and rural propagation channel modeling for vehicle-to-vehicle communications at 5.9 GHz. *IEEE Antennas and Propagation Society International Symposium*, San Diego, CA, Jul. 2008.

Cheng, L.; Henty, B. E.; Bai, F. & Stancil, D. D. (2008c). Doppler spread and coherence time of rural and highway vehicle-to-vehicle channels at 5.9 GHz. *IEEE Global Telecommunications Conference*, New Orleans, LO, Dec. 2008.

Cheng, X.; Wang, C. X.; Laurenson, D. I.; Salous, S. & Vasilakos, A. V. (2009). An adaptive geometry-based stochastic model for non-isotropic MIMO mobile-to-mobile channels. *IEEE Transactions on Wireless Communications*, Vol. 8, No. 9, pp. 4824-4835.

Clark; R. H. (1968). A statistical theory of mobile radio reception. *Bell Syst. Tech. J.*, Vol. 47, pp. 957-1000.

Emmelmann, M; Bochow B, & Kellum C. (2010). *Vehicular Networking. Automotive Applications and Beyond*, p. 52, Communications Systems for Car-2-X Networks, Wiley.

ETSI EN 300 674. (1999). *Technical Characteristics and Test Methods for Dedicated Short Range Communication (DSRC) Transmission Equipment (500 kbit/s / 250 kbit/s) Operating in the 5.8 GHz Industrial, Scientific and Medical (ISM) Band*, European Telecommunications Standard Institute (ETSI) , Technical Report, Sophia Antipolis, France.

ETSI TR 102 492-1 Part 1. (2006). *Technical Characteristics for Pan European Harmonized Communications Equipment Operating in the 5 Ghz Frequency Range and Intended for Critical Road Safety Applications*, European Telecommunications Standard Institute (ETSI), Technical Report, Sophia Antipolis, France.

ETSI TR 102 492-2 Part 2. (2006). *Technical Characteristics for Pan European Harmonized Communications Equipment Operating in the 5 Ghz Frequency Range Intended for Road Safety and Traffic Management, and for Non-Safety Related ITS Applications*, European Telecommunications Standard Institute (ETSI) , Technical Report, Sophia Antipolis, France.

Gallagher, B. & Akatsuka, H. (2006). Wireless communications for vehicle safety: radio link performance and wireless connectivity methods. *IEEE Vehicular Technonolgy Magazine*, Vol. 1, No. 4, pp. 4-16, 24.

Gradsthteyn, I. S. & Ryzhik I. M. (2007). *Table of Integrals, Series and Products*. San Diego, CA, Academic, 7th ed.

Green, E. & Hata, M. (1991). Microcellular propagation measurements in an urban environment. *IEEE Personal, Indoor and Mobile Radio Communications International Symposium*, pp. 324-328, London, UK, Sep. 1991.

IEEE 802.11. (2007). *Wireless LAN Medium Access Control (MAC) and Physical Layer (PHY) Specifications*, Institute of Electrical and Electronic Engineers (IEEE), New York, USA.

IEEE 802.11p. (2010). Wireless LAN Medium Access Control (MAC) and Physical Layer (PHY) Specifications Amendment 6: Wireless Access in Vehicular Environments, Institute of Electrical and Electronic Engineers (IEEE), New York, USA.

Ito, Y.; Taga, T.; Muramatsu, J. & Susuki, N. (2007). Prediction of line-of-sight propagation loss in inter-vehicle communication environments. *IEEE International Symposium on Personal, Indoor and Mobile Radio Communications*, Athens, Sep. 2007.

Jakes, C. W. (1974). *Microwave Mobile Communications*. Wiley, New York.

Jensen, I. & Halpe Gamage, J. K. (2007). CVIS vehicle rooftop antenna unit, *6th ITS in Europe Congress & Exhibition*, Aalborg, Denmark, Jun. 2007.

Jiang, T.; Chen, H.; Wu, H. & Yi, Y. (2010). Channel modeling and inter-carrier interference analysis for V2V communication systems in frequency-dispersive channels. *Mobile Networks and Applications*, Vol. 15, No. 1, pp. 4-12.

Karedal, J.; Tufvesson, F.; Czink, N.; Paier, A.; Dumard, C.; Zemen, T.; Mecklenbräuker, C. F. & Molisch, A. F. (2009). A geometry-based stochastic MIMO model for vehicle-to-vehicle communications. *IEEE Transactions on Wireless Communications*, Vol. 8, No. 7, pp. 3646-3657.

Kaul, S.; Ramachandran, K.; Shankar, P.; Oh, S.; Gruteser, M.; Seskar, I. & Nadeem, T. (2007). Effect of antenna placement and diversity on vehicular network communications. *IEEE Sensor, Mesh and Ad Hoc Communications and Networks Conference*, pp. 112-121, San Diego, CA, Jun. 2007.

Kim. Y; Song, C.; Koo, I; Choi, H. & Lee S. (2003). Design of a double-looped monopole array antenna for a DSRC system roadside base station. *Microwave and Optical Technology Letters*, Wiley, Vol. 37, No. 1, pp. 74-77.

Kunisch, J. & Pamp, J. (2008). Wideband car-to-car radio channel measurements and model at 5.9 GHz. *IEEE Vehicular Technology Conference*, pp. 1-5, Calgary, BC, Sep. 2008.

Matolak, D. W. (2008). Channel modeling for vehicle-to-vehicle communications. *IEEE Communications Magazine*, Vol. 46 , No. 5, pp. 76-83.

Matolak, D. W. & Wu, K. (2009). Vehicle-to-vehicle channels: Are we done yet?. *IEEE GLOBECOM*, pp. 1-6, Honolulu, HI, Dec. 2009.

Maurer, J.; Fügen, T. & Wiesbeck, W. (2001). A realistic description of the environment for inter-vehicle wave propagation modelling. *IEEE Vehicular Technology Conference*, Vol. 3, pp. 1437-1441, Atlantic City, NJ, Oct. 2001.

Maurer, J.; Fügen, T. & Wiesbeck, W. (2002). Narrow-band measurement and analysis of the inter-vehicle transmission channel at 5.2 GHz. *IEEE Vehicular Technology Conference*, Vol. 3, pp. 1274-1278, Birmingham, AL, May 2002.

Maurer, J.; Fügen, T.; Schäfer, T. & Wiesbeck, W. (2004). A new inter-vehicle communications (IVC) channel model. *IEEE Vehicular Technology Conference*, Vol. 1, pp. 9-13, Los Angeles, CA, Sep. 2004.

Michelson, D. G. & Ghassemzadeh, S. S. (2009). *New Directions in Wireless Communications Research*. Springer Science+Business Media (Chapter 1).

Molisch, A. F. & Tufvesson, F. (2004). Multipath propagation models for broadband wireless systems. *CRC Handbook of signal processing for wireless communications*, Chapter 2, 2004.

Molisch, A. F. (2005). *Wireless Communications*, IEEE Press-Wiley.

Molisch, A. F.; Tufvesson, F.; Karedal, J. & Mecklenbräuker, C. (2009). A survey on vehicle-to-vehicle propagation channels. *IEEE Wireless Communications*, Vol. 16, No. 6, pp. 12-22.

Paier, A.; Karedak, J; Czink, N.; Hofstetter, H.; Dumand, C.; Zemen, T.; Tufvesson, F.; Mecklenbräuker, C. F. & Molisch, A. F. (2007). First results from car-to-car and car-to-car infrastructure radio channel measurements at 5.2 GHz. *IEEE International Symposium on Personal, Indoor and Mobile Radio Communications*, Athens, Sep. 2007.

Parsons, J. D. (2000). *The Mobile Radio Propagation Channel*, 2nd ed., Wiley, New York.

Paschalidis, P.; Wisotzki, M.; Kortke, A.; Keusgen, W. & Peter, M. (2008). A wideband channel sounder for car-to-car radio channel measurements at 5.7 GHz and results for an urban scenario. *IEEE Vehicular Technology Conference*, Calgary, BC, Sep. 2008.

Peden, M.; Scurfield, R.; Sleet, D. ; Mohan, D. ; Hyder, A. A. ; Jarawan, E. & Mathers, C. (2004). *World Report on Road Traffic Injury Prevention*, pp. 31-66, World Health Organization, Geneva, Switzerland.

Renaudin, O.; Kolmonen, V-M.; Vainikainen, P. & Oestges, C. (2009). Car-to-car channel models based on wideband MIMO measurements at 5.3 GHz. *European Conference on Antennas and Propagation*, pp. 635-639, Berlin, Germany, Mar. 2009.

Renaudin, O.; Kolmonen, V-M.; Vainikainen, P. & Oestges, C. (2010). Non-stationary narrowband MIMO inter-vehicle channel characterization in the 5-GHz band. *IEEE Transactions on Vehicular Technology*, Vol. 59, No. 4, pp. 2007-2015.

Sai, S.; Niwa, E.; Mase, K.; Nishibori, M.; Inoue J.; Obuchi, M.; Harada, T.; Ito, H.; Mizutani, K. & Kizu, M. (2009). Field evaluation of UHF radio propagation for an ITS safety system in an urban environment, *IEEE Communications Magazine*, Vol. 47, No. 11, pp. 120-127, Nov. 2009.

Sen, I. & Matolak, D. W. (2007). V2V channels and performance of multiuser spread spectrum modulation. *IEEE Vehicular Technology Magazine*, Vo. 2, No. 4, pp. 19-25.

Sen, I. & Matolak, D. W. (2008). Vehicle-vehicle channel models for the 5-GHz band. *IEEE Transactions on Intelligent Transportation Systems*, Vol. 9, No. 2, pp. 235-245.

Sklar, B. (1997). Rayleigh fading channels in mobile digital communication systems part i: characterization. *IEEE Communications Magazine*, Vol. 35, No. 7, pp. 90-100.

Tan, I.; Tang, W.; Laberteaux, K. & Bahai, A. (2008) Measurement and analysis of wireless channel impairments in DSRC vehicular communications. *IEEE International Conference on Communications*, pp. 4882-4888, Beijing, China, May 2008.

Vlacic, L; Parent, M. & Harashima, F. (2001). *Intelligent Vehicle Technologies. Theory and Applications*, pp. 87-188, Butterworth-Heinemann, Oxford, UK.

Wan, R. & Cox, D. (2002). Channel modeling for ad hoc mobile wireless networks. *IEEE Vehicular Technology Conference*, Vol. 1, pp. 21-25, Birmingham, AL, May 2002.

Wan, L. C. & Cheng, Y. H. Cheng. (2005). A statistical mobile-to-mobile Rician fading channel model. *IEEE Vehicular Technology Conference*, Vol. 1, pp. 63-67, Stockholm, Sweden, May 2005.

Wang, C. X.; Cheng X. & Laurenson D. I. (2009). Vehicle-to-vehicle channel modeling and measurements: recent advances and future challenges. *IEEE Communications Magazine*, Vol. 47, No. 11, pp. 96-103.

Wu, Q.; Matolak, D. W. & Sen. I. (2010). 5-GHz-band vehicle-to-vehicle channels: models for multiple values of channel bandwidth. *IEEE Transactions on Vehicular Technology*, Vol. 59, No. 5, pp. 2620-2625.

Xia, H. H.; Bertoni, H. L.; Maciel, L. R.; Lindsay-Stewart, A. & Rowe, R. (1993). Radio propagation characteristics for line-of-sight microcellular and personal communications. *IEEE Transactions on Antennas and Propagation*, Vol. 41, No. 10, pp. 1439-1447.

WINNER European Project. (2007), D1.1.2. WINNER II Channel models. Part II Channel measurement and analysis results.
[Online] Available: http://www.ist-winner.org/deliverables.html

Zajic, A. & Stüber, G. L. (2008). Space-time correlated mobile-to-mobile channels: modelling and simulation. *IEEE Transactions on Vehicular Technology*, Vol. 57, No. 2, pp. 715-726.

Zajic, A. & Stüber, G. L. (2009). Wideband MIMO mobile-to-mobile channels: geometry-based statistical modeling with experimental verification. *IEEE Transactions on Vehicular Technology*, Vol. 58, No. 2, pp. 517-534.

Zheng, Y.R. (2006). A non-isotropic model for mobile-to-mobile fading channel simulations. *IEEE Military Communications Conference*, Washington, Oct. 2006.

22

Propagation Path Loss Modelling in Container Terminal Environment

Slawomir J. Ambroziak, Ryszard J. Katulski,
Jaroslaw Sadowski and Jacek Stefanski
Gdansk University of Technology
Poland

1. Introduction

Container port area should be treated as a very difficult radio waves propagation environment, because lots of containers made of steel are causing very strong multipath effect and there is time-varying container arrangement in stacks of different height. Path loss modelling for such area is still complex task and has not yet been considered in scientific research. But as the total amount of cargo carried yearly in containers by land and sea increases, the only effective way of controlling such huge number of containers is to build efficient electronic container supervision systems. Nowadays almost all the major container ports have some kind of radio monitoring of containers, based on available radio communication standards (GSM/GPRS, UMTS, TETRA, WiFi, WiMAX, ZigBee, Bluetooth, many different RFID systems or other solutions in unlicensed frequency band) working in frequency range from about 0.4GHz to 5GHz. It should be noted that ITU-R did not present any special recommendation for propagation path loss prediction for radio link in container terminal environment. Differences in spatial arrangement and structure between container stacks and typical urban or industry area can cause relevant path loss prediction errors in case of use inadequate path loss model, so the special survey of propagation phenomenon in container terminal area becomes crucial.

At the outset of the chapter, radio links are characterized in terms of transmission loss and its components. Then authors discuss the requirements concerning measuring equipment, its calibration process, measurement methodology, as well as the processing and presentation of their results (Ambroziak, 2010).

The main part of the chapter presents new analytical approach to path loss modelling in case of propagation in container port environment, based on empirical results from measurement campaign in Gdynia Container Terminal (Poland). Upon the results of almost 5 thousands propagation path measurements in real container terminal environment, a novel analytical model was developed. Additionally, authors present mobile measuring equipment used to research in DCT Gdansk Container Terminal (Poland) and planned results of the analysis of nearly 290 thousand of propagation cases which were collected. It is an introduction to generalization of the propagation model for container terminal environments (Katulski et al., 2009).

2. Normative requirements

The propagation medium is a factor that causes many difficulties in designing wireless networks, because of large diversity of propagation environments, which includes rural, urban, industrialized, marine and mountainous environments. The radio wave attenuation in each environment is determined by many variables phenomena and factors. It is essential to determine the radio wave attenuation (so-called transmission loss) to a specified accuracy. Knowledge of transmission loss is necessary to meet energy requirements in radio links designing (Katulski, 2009).

Therefore, there is a need to create empirical propagation models for different environments, based on measuring research results. So far a number of such models has been developed, mainly for urban and indoor environments. However, the environments in these groups may also differ within. Because of this, the issue of radio wave propagation measuring research is still a current topic, especially for designing the radio networks in specific environments.

At present, the Department of Radiocommunication Systems and Networks in the Gdansk University of Technology is carrying out the wide research on radio wave propagation. Very important are normative requirements - as described in literature, such as ITU-R Recommendations - that have to be met during research on radio wave propagation.

In this subsection a radio link is characterized in scope of transmission loss and its components. Then the next to be discussed are requirements concerning measuring equipment, its calibration process, measurement methodology, as well as the processing and presentation of results.

2.1 Description of the measuring radio link

As known, power of signal transmitted in the radio link is significantly attenuated. The effect of this is the large difference between signal power at the output of transmitter and power of the same signal available at the input of receiver. This difference depends on many factors, mainly transmission loss of propagation medium, as well as the power losses in the transmission feeder lines, the losses due to measuring devices, the antenna losses due to the impedances or polarization mismatch, etc.

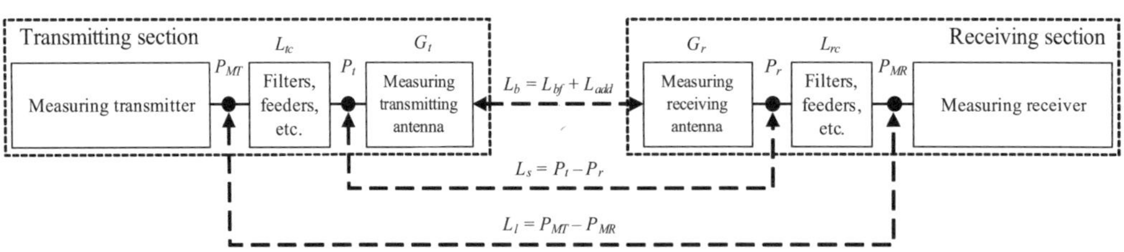

Fig. 1. Graphical presentation of terms used in the measuring transmission loss concept

Therefore, there is a necessity to systematize terminology and symbols used in analyzing the transmission loss and its components. It may be presented using a graphical depiction of terms used in the measuring transmission loss concept, shown in Fig. 1 (Ambroziak, 2010), which considered all essential factors affecting the energy level in radio link, such as:

- total loss of a measuring radio link between transmitter output and receiver input,

- system loss between input of the transmitting antenna and output of the receiving antenna,
- basic transmission loss of the radio link,
- free-space basic transmission loss, that is a basic component of transmission loss.

The total loss of a measuring radio link (symbol: L_l [dB]) is defined as the difference between power P_{MT} [dBW] supplied by the measuring transmitter and power P_{MR} [dBW] available at the input of the measuring receiver in real installation, propagation and operational conditions (ITU-R P.341-5, 1999). The total loss may be expressed by:

$$L_l[dB] = P_{MT}[dBW] - P_{MR}[dBW] = 10\log\left(\frac{p_{MT}[W]}{p_{MR}[W]}\right), \tag{1}$$

where lowercase letters, i.e. p_{MT} and p_{MR}, are power at the output of measuring transmitter and power at the input of measuring receiver, respectively. They can be expressed in absolute values, such as [W], or in relative values, such as [dBW], in that case they are written as uppercase letters, P_{MT} and P_{MR}, respectively. Total loss includes all factors affecting the power of received signal, i.e. basic transmission loss of propagation medium, gains of antennas, loss in feeder lines, etc. Knowledge of the total loss components is necessary to correctly determine the value of the basic transmission loss.

The system loss (symbol: L_s [dB]) is defined as the difference between power P_t [dBW] supplied at the terminals of measuring transmitting antenna and power P_r [dBW] available at the terminals of measuring receiving antenna (ITU-R P.341-5, 1999). By analogy with equation (1), it may be written as follows:

$$L_s[dB] = P_t[dBW] - P_r[dBW] = 10\log\left(\frac{p_t[W]}{p_r[W]}\right). \tag{2}$$

In addition to basic transmission loss, the system loss also includes influence of circuits associated with the measuring antennas, such as ground losses, dielectric losses, antenna loading coil losses and terminating resistor losses. But on the other hand, the system loss excludes losses in feeder lines, both in the transmitting section (L_{tc} [dB]) and in the receiving section (L_{rc} [dB]). Considering Fig. 1, it can be written as follow:

$$L_l[dB] = L_s[dB] - L_{tc}[dB] - L_{rc}[dB]. \tag{3}$$

The basic transmission loss (symbol: L_b [dB]) consists of free-space basic transmission loss L_{bf} [dB] and additional loss L_{add} [dB], resulting from the real conditions of propagation environment, different from ideal free space. From this point of view, the basic transmission loss may be expressed by:

$$L_b[dB] = L_{bf}[dB] + L_{add}[dB]. \tag{4}$$

The additional loss L_{add} includes phenomena occurring in real propagation environments. In terms of measurement procedures, the most important are:

- loss dependent on path clearance,
- diffraction fading,
- attenuation due to rain, other precipitation and fog,

- fading due to multipath.

Equation (4) is a case of isotropic radiation, i.e. it excludes characteristics of real antennas, especially its directional characteristics and power efficiency, which are described by power gain. Taking into consideration the link power budget, in case of free-space environment, the basic transmission loss may be expressed by:

$$L_b[dB] = P_t[dBW] - P_r[dBW] + G_t[dBi] + G_r[dBi], \qquad (5)$$

where G_t and G_r (in [dBi]) are the isotropic (absolute) gains of the transmitting and receiving antennas, respectively, in the direction of propagation. Table 1 gives the power gains for typical reference antennas (ITU-R P.341-5, 1999).

Reference antenna	g	G = 10 log g [dBi]
Isotropic in free space	1	0
Hertzian dipole in free space	1.5	1.75
Half-wave dipole in free space	1.65	2.15
Hertzian dipole, or a short vertical monopole on a perfectly conducting ground	3	4.8
Quarter wave monopole on a perfectly conducting ground	3.3	5.2

Table 1. The power gains for typical reference antennas

As known, free space is an ideal case of propagation environment, open and without any propagation obstacles. It is a perfectly dielectric, homogenous and unlimited environment, characterized by a lack of influence of Earth surface on radio wave propagation and non-absorbing the energy of the electromagnetic field (Katulski, 2009).

Assuming free-space propagation environment and distance (d [m]) between antennas of the measuring radio link much larger than wavelength (λ [m]) of test signal, the free-space basic transmission loss (symbol: L_{bf} [dB]) may be expressed by a well-known equation (ITU-R PN.525-2, 1994):

$$L_{bf}[dB] = 20\log\left(\frac{4\pi \cdot d[m]}{\lambda[m]}\right). \qquad (6)$$

2.2 Standardization of measuring apparatus

In order to ensure accurate measurement results in frequency range 9 kHz to 3 GHz and above (up to 40 GHz), the ITU-R recommends (in SM.378-7) the method of installation and calibration of measuring systems. The document also determines the accuracy, that are required in field-strength measurements, assuming no noise of receiver, atmospheric noise or external interference. Taking these assumptions into account, the expected accuracy of measurements should be:

- for frequency band 9kHz to 30MHz: ± 2dB,
- for frequency band 30MHz to 3GHz: ± 3dB.

If recommended values are not obtainable (for various reasons, such as limitation of the measuring receiver, interference, instability of the test signal, etc.), nevertheless the accuracy specified above should be taken into consideration (ITU-R SM.378-7, 2007).

Depending on the electrical parameters, which the receiving antenna and the measuring receiver were calibrated for, the measuring receiver may measure the following quantities:

- signal power at the receiver input, resulting from the power flux density of electromagnetic wave at the point of reception (the point of the receiving antenna placement),
- voltage at the receiver input, resulting from the electric field intensity at the point of reception,
- current at the receiver input, resulting from the magnetic field intensity at the point of reception.

And so, for the receiving antenna which was calibrated for power flux density of electromagnetic wave, at the receiver input the power P_{MR} is available and measured (Fig. 1). This power is the basis for determining of basic transmission loss L_b, according to equation (8). Similar equations may be written for the case of the receiving antennas, calibrated for electric or magnetic component of electromagnetic field.

Type of receiving antenna may affect the type of measuring receiver - the electrical signal, measured by the measuring receiver should correspond with electrical signal (which the antenna was calibrated for) available at output terminals of the receiving antenna. For example, for short monopole antenna of a specified length, the receiver should measure voltage of test signal, and for the inverted cone type vertical antenna the receiver should measure power of test signal.

Recommendation SM.378-7 contains examples of antennas for different frequency ranges. For frequencies below 30MHz it is recommended to use vertical or loop antennas. In case of the vertical antenna, the monopole antenna shorter than one-quarter of a wavelength may be used with a RF ground system, built of radial conductors at least twice the length of the antenna and spaced 30° or less. Instead of radial conductors, an equivalent RF ground screen may be used. There is also a possibility to use an inverted cone type vertical antenna with similar construction of RF ground system. It allows to obtain a greater power gain of measuring antenna than the quarter wave monopole antenna.

For frequency range 30MHz to 1GHz it is recommended to use a short monopole antennas, half-wave dipoles or high-gain directional antennas, but it is essential to ensure the same polarization of receiving antenna as the transmitting antenna. For field-strength measurements at frequencies above 1GHz it is recommended to use directional antennas with matched polarization.

It should be noted that the height of antenna installation has a significant influence on the measurement results, especially when the height is electrically small (Barclay, 2003). And so, if antennas are installed in close proximity to the ground, the electromagnetic waves take the form of surface waves, which takes effect to the wave depolarization, consequently there is wave attenuation resulting to the polarization mismatch in the radio links. In addition, the radio wave attenuation increases due to losses related to the penetration of radio waves into the propagation ground (Katulski, 2009). To minimize influence of the Earth surface on test signal, transmitting antenna has to be installed at a height that enables space waves propagation (Barclay, 2003). Therefore, the ITU-R recommends that for frequency range 30MHz to 1GHz, the installation of the transmitting antenna should be at least 10 meters high (ITU-R SM.378-7, 2007). The recommended height of the receiving antenna is 1.5 up to 3 meters (ITU-R SM.1708, 2005).

The measuring receiver primarily should have stable parameters (inter alia: gain, frequency, bandwidth), that have an influence on the accuracy of test signal measurement (its voltage, current or power). Local oscillators should have low phase noise, the operating dynamic range should be greater than 60dB and the bandwidth should be wide enough to allow reception of essential parts of the test signal spectrum. Type of detector depends on the bandwidth and the modulation mode of test signal. The required bandwidth and detector functions for various signal types are compiled in Table 2 (ITU-R SM.1708, 2005).

Example of signal types	Minimal bandwidth (kHz)	Detector function
AM DSB	9 or 10	Linear average
AM SSB	2.4	Peak
FM broadcast signal	170 or greater	Linear average (or log)
TV carrier	200 or greater	Peak
GSM signal	300	r.m.s.
DAB signal	1 500	
DVB-T signal Systems: 6 MHz 7 MHz 8 MHz	 6 000 7 000 8 000	
TETRA signal	30	
UMTS signal	3 840	
Narrow-band FM radio Channel spacing: 12.5kHz 20kHz 25kHz	 7.5 12 12	Linear average (or log)

Table 2. The required bandwidth and detector functions for various signal types

Properly configured spectrum analyzer may be used as the measuring receiver, whose work may also be automated. The measuring receiver, with remainder of the receiving section, may be mounted on a vehicle or a hand-cart, that enables mobile measurements in the area of propagation research.

Each of measuring devices and circuits (feeder lines, filters, etc.), that affect total loss of a measuring radio link, are usually calibrated in accordance with certain standards as one of the stages of their production. Nevertheless it is recommended to calibrate transmitting and receiving section as a single entities (ITU-R SM.378-7, 2007). The above allows to take into account the influence of all elements of the measuring radio link, including attenuation due to the ground, masts, etc.

The calibration procedures, presented below, deal with the case of basic transmission loss calculation based on power measurement. Calibration of the transmitting section concerns set-up of power P_{MT} value, in order to obtain required value of P_t at the input terminals of transmitting antenna. Calibration process of the receiving section deals with calculation of the difference (in logarithmic scale) between power of test signal available at the output terminals of receiving antenna and the power of test signal at the input of the measuring receiver. After taking into account the power gain of receiving antenna, it is possible to calculate a correction factor (F_c [dB]) as follows (see Fig. 1):

$$F_c[dB] = L_{rc}[dB] - G_r[dBi] = P_r[dBW] - P_{MR}[dBW] - G_r[dBi] . \quad (7)$$

Considering equation (5) and Fig. 1, which implies that $P_r = P_{MR} + L_{rc}$, after simple transformation, the basic transmission loss may be calculated using following equation:

$$L_b[dB] - P_t[dBW] + G_t[dBi] - P_{MR}[dBW] - F_c[dB] . \quad (8)$$

The equation (8) is very important in measuring research and calculation of the basic transmission loss on the basis of power P_{MR} measurements at input of the measuring receiver. To calculate basic transmission loss it is necessary to know the following values:

- the power gain G_t of the transmitting antenna,
- the power P_t on input of the transmitting antenna - set during calibration process of the transmitting section,
- the correction factor F_c - calculated during calibration process of the receiving section.

Measuring apparatus should be recalibrated at least once a year or every time after change any of its parts (ITU-R SM.378-7, 2007).

2.3 Standardization of measuring procedures

There may be many various reasons for measuring research on radio wave propagation, inter alia: to create empirical propagation models or to estimate coverage of radio networks. This information may be useful in increasing efficiency of radio resources management or for controlling proper use of this resources by particular entities, and so on. Considering the above-cited, ITU-R recommends to unify methodology of measuring procedures and presentation of its results.

The measurement results should include information about slow and fast changes of the power flux density of electromagnetic field (slow and fast fading, respectively). So it is recommended to choose the measurement points in an appropriate manner. Measurements points should be spaced every 0.8λ along a route of radio waves propagation. It is recommended that the results should be averaged every 40λ (Lee, 1993).

Measurements may also be done automatically when the measuring receiver is mobile, but speed V [km/h] of the receiver is not arbitrary. It depends on the frequency f [MHz] of test signal and minimum time t_r [s] given by the receiver specifications to revisit a single frequency. It may be expressed by following equation (ITU-R SM.1708, 2005):

$$V[km/h] \leq \frac{864}{f[MHz] \cdot t_r[s]} . \quad (9)$$

In order to find relations between the basic transmission loss and the distance from the transmitting antenna, the result of each measurement should be correlated to the place of its execution. For this reason, the positioning system should be used for reading current position of measuring receiver. It is recommended to use one of three systems specified in the ITU-R Recommendation SM.1708.

The GPS is a preferred positioning system, although its accuracy is limited in tunnels, narrow streets or valleys. Accuracy in position determining should be a few meters, which in most cases can be provided by GPS.

If unable to determine the position using GPS system, it is recommended to use dead reckoning system. Position is determined basing on information about starting point, direction of movement and distance covered by the receiver. It is also possible to use the complex navigation system, which is the combination of the above-mentioned systems.

Due to the large instability of propagation environment, the result of single measurement is not reliable or repeatable. Therefore, the measurement results should be classified in terms of probability of exceeding a particular value by the power of received signal. This probability may be in range of 1-99%, but typical values for this parameter are as follows: 1%, 10%, 50%, 90% and 99%. During research on radio wave propagation, the median value is recommended (ITU-R SM.1708, 2005), i.e. the value from an ordered subset of measurement results, which is exceeded by 50% of the other values from this subset.

In practice, for each *i*-th subset of measurement data, it is necessary to calculate the median of test signal power P_{MR}^{i} at the receiver input. Each subset of data is created on the basis of *n* measurement results, collected along the route at 40λ spacing in accordance with the following equation:

$$P_{MR}^{i} = \begin{cases} P_{MR,\frac{n+1}{2}}^{i} & ,if\ n\ is\ odd, \\ \frac{1}{2}\left(P_{MR,\frac{n}{2}}^{i} + P_{MR,\frac{n}{2}+1}^{i}\right) & ,if\ n\ is\ even, \end{cases} \tag{10}$$

where $P_{MR,1}^{i} \le P_{MR,2}^{i} \le P_{MR,3}^{i} \le \ldots \le P_{MR,n-1}^{i} \le P_{MR,n}^{i}$ is the subset of measurement results in non-descending order, and *n* is the number of measurement results taken on the *i*-th ($i = 1,2,3 \ldots$) section of the radio waves propagation route.

Calculation of median values of the test signal power may be done in real time during the measuring research, but only calculated median values are recorded. It is also possible to record all the results and calculate median values after measuring research. Results obtained using both methods may be used to basic transmission loss modeling or estimating coverage of radio networks in the area under research.

There are three, recommended by ITU-R (SM.1708), methods of the measurement results presentation. The first one and the easiest is a table containing results of all measurements before calculating the median values. The advantage of this method is an access to information about local fading of test signal. However, there is a large number of data to analyze. In addition, it is hard to interpret a single result.

The second possibility is graphical representation of the pre-processed median values - as a function of distance - in the Cartesian coordinates. This way of data presentation helps to illustrate changes of the basic transmission loss in dependence on the distance from transmitting antenna.

The third one is a digital map of the area under research with marked colored points, that are representing a range of measured values of test signal power at input of the measuring receiver, assuming a known value of the equivalent isotropic radiated power (EIRP), which

is equal to the signal power P_t [dBW] supplied to the terminals of transmitting antenna plus power gain G_t [dBi] of this antenna. Map scale is dependent on the area where research is carried out. The advantage of this results presentation method is a simultaneous view on the value of received signal power and localization of each measurement. It is also possible to interpolate the results of measurements in order to estimate the radio coverage in the area.
It should be noted, that on the basis of power measurements at the receiver input and using equation (8) it is easy to calculate the basic transmission loss in given propagation environment.

3. A novel empirical path loss model for container terminal

This subsection presents new analytical approach to path loss modeling in case of propagation in container port environment, based on empirical results from measurement campaign in Gdynia Container Terminal (Poland). Precise classification of propagation environment and selection of parameters which influence the propagation mechanism in essential way, allowed to define adequate multivariate error function for multidimensional regression analysis. As a result of this research, new analytical relation between propagation path parameters and path loss in container terminal scenario is proposed.

3.1 Measuring equipment

Block diagram of primary equipment set used in propagation measurements in container terminal scenario is presented in Fig. 2 (Katulski et al., 2008).

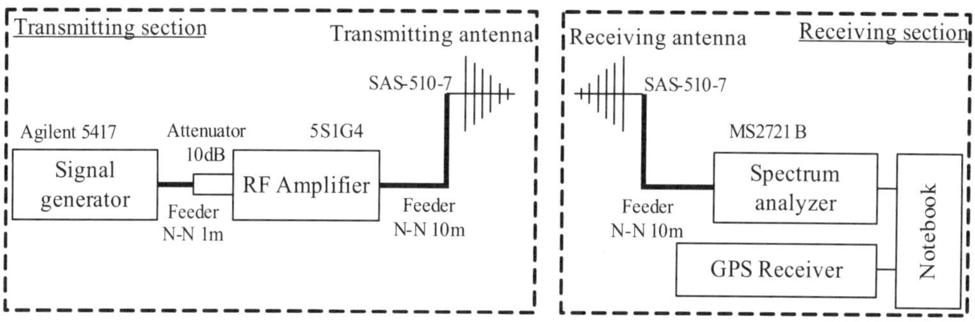

Fig. 2. Block diagram of primary measuring equipment set

Propagation path loss measuring equipment concept was based on fixed reference signal transmitter and mobile receiver equipment placed in many different positions in the area of container terminal. Harmonic signal without modulation, with frequency in range 0.5GHz to 4GHz, was emitted by transmitting antenna situated in various places in port. Power RF amplifier input was protected by precise 10dB attenuator. The receiving section was made of handheld signal spectrum analyzer working as a sensitive received signal power meter, GPS receiver and notebook with special software. All the receiver section components were battery powered. Log-periodic directional wideband antennas of the same type were used in both transmitter and receiver side. These antennas were calibrated by producer and have precise parameters in whole frequency range of interest.
Firstly the measurement plan assumed four reference signal frequencies: 1, 2, 3 and 4GHz, but during the measurement campaign additional frequency of 0.5GHz was also put into

investigation. Because the power amplifier used in transmitting section works properly only in frequency range 800MHz to 4.2GHz, schematic diagram of transmitting section in case of measurement at frequency 500MHz was slightly modified: additional attenuator and power amplifier had to be removed and the output of signal generator (with power level set to maximum value equal +9.6dBm) was directly connected to transmitting antenna via 10m long feeder.

3.2 Calibration procedure

In order to precisely compute the propagation path loss from power level of signal detected by handheld spectrum analyzer, radio link power budget equation have to include parameters of all the components from Fig. 2.

Because in the container terminal scenario, path loss of over 100dB should be expected, relatively high power test signal should be connected to the transmitting antenna. For the frequencies of 1GHz and above, constant power level +30dBm at the input of transmitting antenna was chosen. As the antenna's power gain at all the frequencies of interest is known (measured by manufacturer) and transmitter power level is being kept constant, equivalent isotropic radiated power (EIRP) can be simply computed for every frequency.

To ensure that accuracy of measurements doesn't vary with frequency, the transmitting and receiving section was calibrated in the Gdansk University of Technology laboratory. Firstly, the attenuation of transmitting section feeders at all the frequencies of interest was measured using vector network analyzer. The results are compared in Table 3.

Frequency [GHz]	**1**	**2**	**3**	**4**
Feeder loss between generator and additional attenuator [dB]	0.25	0.59	1.25	0.70
Feeder loss between amplifier and antenna [dB]	3.27	4.89	6.15	7.10

Table 3. Attenuation of transmitter section feeders

Although the power amplifier has smooth gain adjustment, authors decided to set the amplification to fixed value of 38dB (amplifier setting, real amplification value was not measured) and determine the signal generator output power that is necessary to achieve signal power at the input of transmitting antenna equal +30dBm. In laboratory conditions, spectrum analyzer from receiver section together with precise attenuator 20dB was used instead of antenna as a power meter (Fig. 3).

Fig. 3. Transmitting section calibration set schematic diagram

Evaluated generator output power values, which provided signal level of +10dBm at the input of spectrum analyzer (+30dBm at the input of transmitting antenna), are presented in Table 4. As the receiver antenna gain in whole band of interest was precisely measured by producer, the only part of receiver section from primary block diagram (Fig. 2) with unknown parameters is the feeder between antenna and handheld spectrum analyzer.

Frequency [GHz]	1	2	3	4
Signal generator output power [dBm]	4.5	7.4	7.2	7.8

Table 4. Output power level of signal generator for particular measurement frequencies

The MS2721B spectrum analyzer is able to measure and present received signal power level directly in [dBm]. Using the same device during calibration phase and in final measurement campaign should compensate eventual received signal power measurement errors. The receiving section feeder attenuation values are presented in Table 5.

Frequency [GHz]	1	2	3	4
Feeder loss between antenna and spectrum analyzer [dB]	3.24	4.86	6.25	7.7

Table 5. Attenuation of receiver section feeder

Obviously, similar but not the same calibration procedure was repeated at frequency 0.5GHz after measurement campaign to obtain the power level at the input of transmitting antenna and attenuation of feeders for this specified frequency.

Because the receiving antenna has directional spatial characteristic, path loss measurement procedure required pointing the antenna in direction of transmitter in case of line of sight (LOS) condition or in direction of maximum received signal power in case of non-line of sight (NLOS) for every position of receiving section. To simplify the search of maximum signal direction, both transmit and receive antennas were fastened to movable masts with tripods, which allow to change azimuth of reception while height of antenna above terrain remained unchanged.

As the maximum transmitter output power was set to +30dBm and the transmitting antenna gain did not exceed 8dBi, the value of EIRP was far below 15W limit. According to Polish law, electromagnetic radiation sources with EIRP less than 15W are objects that do not affect environment or human, so nobody from the measurement team was exposed to harmful electromagnetic radiation.

To improve measurement speed and accuracy, data from spectrum analyzer (received signal power) and GPS receiver (geographic coordinates and time of each measurement) data were collected by notebook. Special software running on computer with Linux operating system allowed to define the time between successive measurements, frequency and bandwidth of received signal, type of applied power detector, additional averaging of results etc. It is also possible to record signal spectrum in each measurement point.

3.3 Path loss measurements in the Gdynia Container Terminal

With the help from administration of the Gdynia Container Terminal, complex survey of propagation aspect in container port was made in term from June to September 2007. Almost 5000 data sets were collected during measurement campaigns, which means about thousand measurement points for each analyzed frequency. The analyses were made in different weather conditions - sunny, cloudy and rainy days with temperature from 5°C to 20°C.

Exemplary results of propagation path loss measurements in area of container terminal are shown on map in Fig. 4, where blue rectangles symbolize stacks of containers, dots symbolize location of successive measurement points and colour of each dot indicates basic transmission loss in [dB].

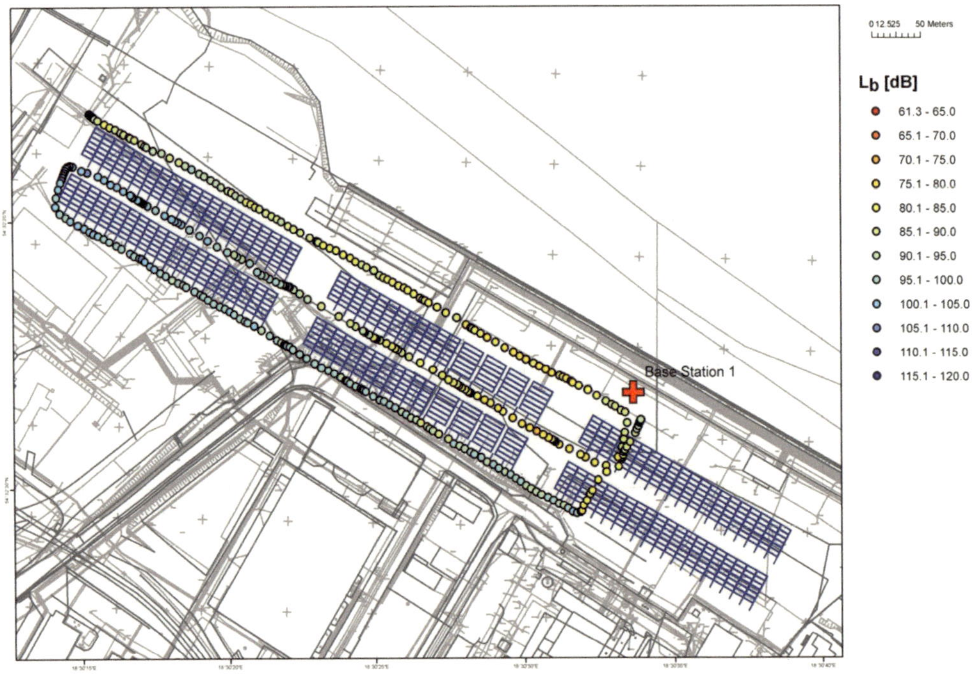

Fig. 4. Propagation path loss measurement results at 2GHz in the Gdynia Container Terminal

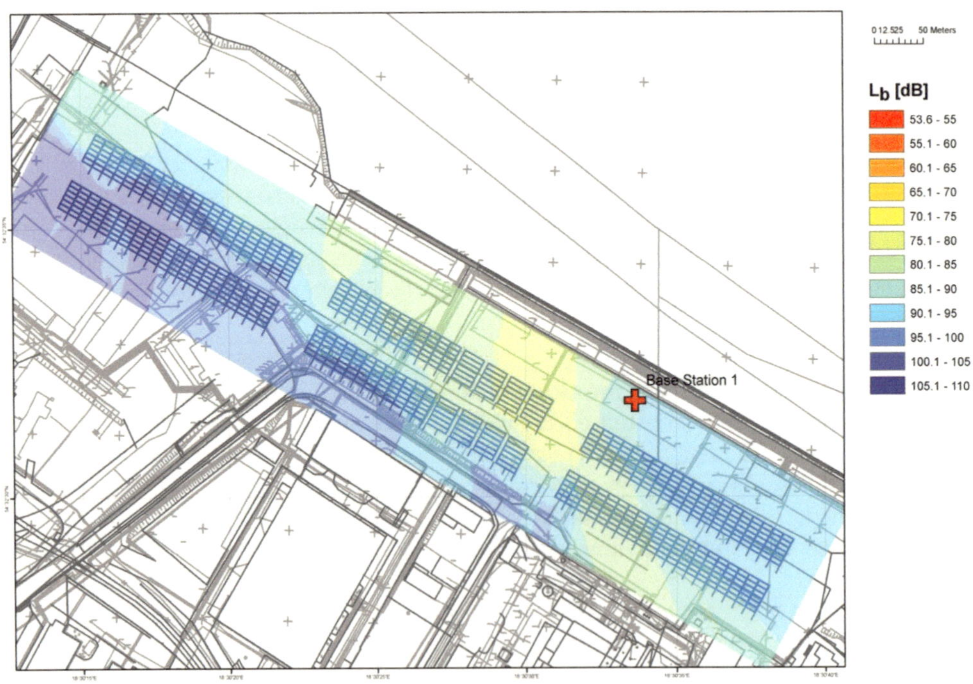

Fig. 5. Spatial interpolation of measurement results from Fig. 4

The next figure (Fig. 5) was created using spatial interpolation. This interpolation should be understood as a prediction of expected basic transmission loss in places where measurements were not possible. There are many different methods of spatial data interpolation. In this example inverse distance weighting (IDW) interpolation was used, which is based on weighted averaging of value from nearby measurement points. This method assumes, that the greatest impact on value in point of interpolation have these points of real measurements which are the closest, so the weight used in averaging process should be the inverse of distance between interpolation point and measurement point.
The value in the interpolation point is calculated by equation:

$$Z = \frac{\sum_{i=1}^{n} \frac{Z_i}{d_i^p}}{\sum_{i=1}^{n} \frac{1}{d_i^p}}, \tag{11}$$

where:

Z - value in interpolation point,
d_i - distance between i-th measurement point and interpolation point,
Z_i - value in i-th measurement point,
p - exponent (in our case $p=2$),
n - number of nearby measurement points used to interpolate values between points (in our case $n=12$).

The value of exponent p determines how the measurement points impact the interpolated value: the bigger value of p, the smaller impact to the interpolated value has the measurement points located farther from interpolation point. Therefore as the value of p is bigger, local variations of measured values are more visible in the results of interpolation.
The IDW method, briefly described above, was chosen because of spatial nature of collected data: measurement points were located closely (distance between neighboring point varies from one to several meters) so the IDW method allowed to distinguish local variation in path loss value. All the spatial analysis and interpolations were made using ArcView 9.2 geographic information system (GIS) software.

3.4 A novel multivariate empirical path loss model

Upon the results of almost 5 thousands propagation path measurements in real container terminal environment, a novel analytical model was developed using multidimensional linear regression analysis with multiple independent variables. For the sake of this analysis a multivariate error function was defined (Katulski & Kiedrowski, 2005). The following parameters, which should affect the value of propagation path loss in port area, were chosen as independent variables in error function:

- frequency f,
- propagation path length d,
- path type qualification: line of sight or non line of sight condition,
- difference between transmitter antenna height h_T above terrain level and average height h_{av} of container stack, but two possible cases should be investigated separately: $h_T \geq h_{av}$ and $h_{av} > h_T$.

Because the container terminal, in which all the measurements were made, was permanently used for container transportation, safety restrictions forced authors to limit the height h_R of

receiver antenna to fixed value equal 2m. Due to fixed value of receiver antenna height, proposed propagation models do not include this height as a variable parameter.
As a result of defined error function analysis, regression coefficients for respective propagation cases were computed. Based on this, analytical formulas of propagation path loss in container terminal area can be presented (Katulski et al., 2008).
Propagation path loss in [dB] in line of sight scenario:
a) in case, when $h_T \geq h_{av}$ (LOS1):

$$L_{LOS1} = 55.2 + 20\log f + 5.8\log d - 22.1\log(h_T - h_{av}) , \tag{12}$$

b) otherwise, when $h_{av} > h_T$ (LOS2):

$$L_{LOS2} = 41.9 + 20\log f + 25.9\log d + 4.2\log(h_{av} - h_T) . \tag{13}$$

Propagation path loss in non line of sight scenario:
a) in case, when $h_T \geq h_{av}$ (NLOS1):

$$L_{NLOS1} = 32.6 + 20\log f + 7.9\log d + 0.8\log(h_T - h_{av}) , \tag{14}$$

b) otherwise, when $h_{av} > h_T$ (NLOS2):

$$L_{NLOS2} = 38.6 + 20\log f + 13\log d + 5.9\log(h_{av} - h_T) . \tag{15}$$

The frequency f in equations (12) – (15) should be in [MHz], propagation distance d in [km], height of transmit antenna and average height of container stack in [m].
Mean error (ME) and mean square error (MSE) are commonly being used to verify accuracy of path loss models. These errors are defined by expression (16) and (17) respectively (Katulski & Kiedrowski, 2006):

$$ME = \frac{1}{N}\sum_{i=1}^{N}(L_{meas,i} - L_{reg,i}) , \tag{16}$$

$$MSE = \sqrt{\frac{1}{N}\sum_{i=1}^{N}(L_{meas,i} - L_{reg,i})^2} , \tag{17}$$

where $L_{meas,i}$ is the value of measured path loss in i-th position of receiver equipment (i=1,...,N), $L_{reg,i}$ is the path loss value computed using equations (12) to (15) for i-th position, and N is the total number of considered results. Mean error value reflect the expected average difference between path loss values obtained using proposed model and real path loss measurement results, while mean square error is the ratio of dispersion of measured path loss values and describes how good the propagation model matches experimental data.
Mean errors and mean square errors for all the considered propagation path variants separately (different height of transmitter antenna, line of sight condition) and summary for all measurement results together, are presented in Table 6.
The propagation path loss calculated using proposed analytical model fits very well to the results from measurement campaign for all propagation path variants, which is confirmed by very low values of mean errors and acceptably low values of mean square errors.

LOS				*NLOS*				*SUMMARY*	
LOS1		*LOS2*		*NLOS1*		*NLOS2*			
ME	*MSE*	*ME*	*MSE*	*ME*	*MSE*	*ME*	*MSE*	*ME*	*MSE*
0.00	8.51	0.01	6.02	0.00	6.73	0.00	6.28	0.00	6.82
ME=0.01, MSE=7.22				ME=0.00, MSE=6.49					

Table 6. Mean errors and mean square errors for proposed propagation model

4. Future research in the DCT Gdansk Container Terminal

In order to generalize the path loss model for container terminal environments there was a necessity to carry out a wider research in other type of container terminal. It has been made in the DCT Gdansk Container Terminal (Poland). New multipurpose mobile equipment for propagation measurements allowed to carry out the research in accordance with described normative requirements.

Measurement equipment consists of two parts: immobile transmitting section (Fig. 6) and mobile receiving section (Fig. 7). These block diagrams exclude descriptions of devices types and feeders lengths.

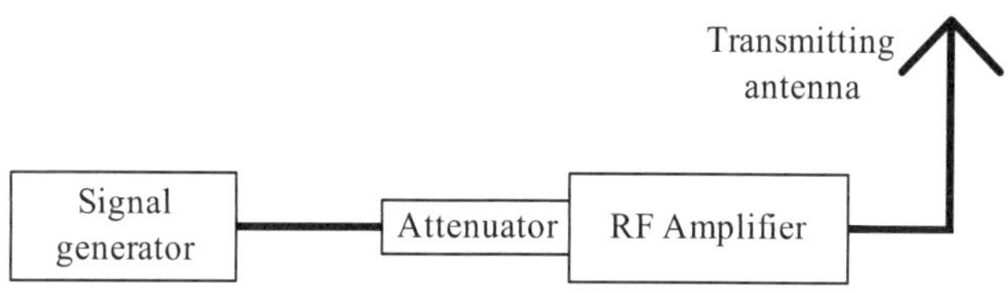

Fig. 6. Simplified block diagram of the immobile transmitting section

The transmitting section of the equipment for propagation measurements consists of signal generator connected to transmitting antenna through the RF amplifier and the attenuator. The generator is a source of the test signal, that is going to be investigated. The attenuator protects the amplifier from damage caused by high level signals. The signal generator and the amplifier are a source of test signal with power of P_{MT} (see Fig. 1), which is supplied to input terminals of transmitting antenna. The transmitting antenna is a monopole vertical antenna with electrical length of one-quarter of a wavelength. It has been developed and implemented in a manner, that allows to change its linear length, so it may be used for research on various frequencies. During the research antenna was installed on various heights, but always higher than 10 meters above a ground level (ITU-R SM.378-7, 2007), to minimize the influence of the Earth surface on the test signal attenuation.

In order to prepare the transmitting section for tests, the calibrating spectrum analyzer should be connected in place of the transmitting antenna in the same way as presented in Fig. 3. The desirable value P_t of test signal should be set by changing settings of the generator and RF amplifier and taking into account attenuation of the attenuator. It should be noted that the calibrated equipment should not be changed during the tests.

In mobile receiving section (Fig. 7), the spectrum analyzer is used as the measuring receiver. It is also equipped with a GPS receiver, which allows to determine the test vehicle position and assign it to appropriate measurement result. The receiving antenna is the same type as the transmitting antenna. During the research the receiving antenna was installed at a height of 2 meters above ground level. The receiving section is carried by test vehicle (in our case a hand-cart). It is moving along a route of radio wave propagation with velocity not exceeding the value resulting from equation (9).

The rotary encoder is used to determine the distance from starting point and to determine the point where the measurement should be triggered. This encoder is connected to the test wheel and the encoder controller. For every distance of 0.8λ, the encoder controller sends an impulse to the industrial computer, which triggers next measurement of signal power P_{MR} at the receiver input. The industrial computer is responsible for the spectrum analyzer configuring, measurements triggering and recording its results. The LCD display shows the following data: current measurement result, distance from starting point, current velocity of test vehicle. Whole receiving section is powered by battery with sufficient capacity.

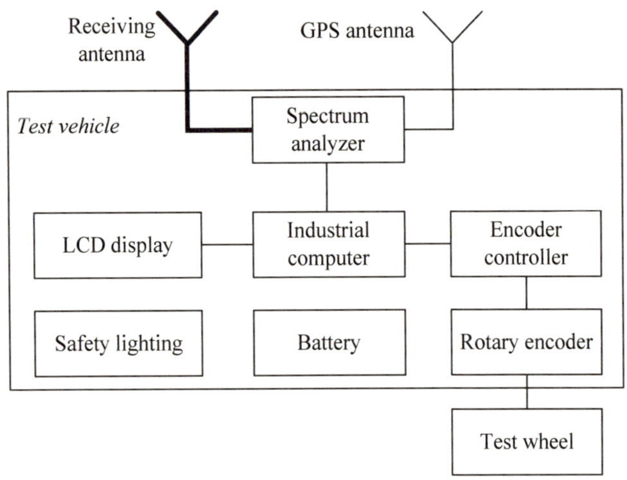

Fig. 7. Simplified block diagram of the mobile receiving section (Ambroziak, 2010)

Fig. 8. Measurement team during research in the DCT Gdansk Container Terminal

It is very important to ensure safety of measurement team during research in container environment. The DCT Port internal safety procedures required that all the objects moving between container stacks has to be clearly visible in all conditions. For this reason, the receiving section is equipped with pulsing safety lighting to make the test vehicle more visible. This lighting is mounted in the highest point of the test vehicle, at the antenna mast. Fig. 8 presents measurement team from the Department of Radiocommunication Systems and Networks (Gdansk University of Technology) during measuring research on radio wave propagation in the DCT Gdansk Container Terminal (Poland).

5. Conclusion

The chapter presents the normative requirements concerning the methodology of measurement research on radio wave propagation and the measuring apparatus. This requirements are in accordance with current ITU-R Recommendations. On the basis of these recommendations there were carried out the propagation research in the container terminals in Gdynia and Gdansk. Radio propagation analysis in container terminal scenario, presented in this chapter, was the first such measurement in Poland and unique in the worldwide area of radio communication research.
Upon the analysis of path loss measurement data collected in the Gdynia Container Terminal, the novel container port area propagation model was proposed. This model has been verified in real propagation conditions in wide frequency range from 0.5GHz to 4GHz and can be used to predict propagation path loss in case of designing radio communication systems for container ports or even other related propagation environments.
During the research in the DCT Gdansk Container Terminal the data about nearly 290 thousand of propagation cases was collected. These cases concern the propagation routes with various lengths, various frequencies and various heights of transmitting antenna. The results of these research will be used for verification, extending and generalize new-elaborated propagation model for container terminals.

6. References

Ambroziak, S.J. (2010). Measuring research on radio wave propagation - normative requirements (in Polish), *Telecommunication review and telecommunication news*, No. 2-3/2010, pp.84-89, ISSN: 1230-3496

Barclay, L. (2003). *Propagation of radiowaves*, The Institution of Engineering and Technology, ISBN: 978-0-85296-102-5, London, United Kingdom

ITU-R P.341-5 (1999). *The concept of transmission loss for radio links*

ITU-R PN.525-2 (1994). *Calculation of free-space attenuation*

ITU-R SM.1708 (2005). *Field-strength measurements along a rout with geographical coordinate registration*

ITU-R SM.378-7 (2007). *Field-strength measurements at monitoring stations*

Katulski, R.J. & Kiedrowski, A. (2005). *Empirical formulas for determination of the propagation loss in urban radio access links*, Procedings of IEEE 62nd Vehicular Technology Conference, pp. 1742-1746, ISBN: 0-7803-9152-7, Dallas 25-28 September 2005, USA

Katulski, R.J. & Kiedrowski, A. (2006). Propagation path loss modeling for fixed radio networks in urban scenario (in Polish), *Electronics and Telecommunications Quarterly*, vol. 52 no. 2, pp. 193-210, ISSN : 0867-6747

Katulski, R.J.; Sadowski, J. & Stefanski, J. (2008). *Propagation Path Modeling In Container*

Terminal Environment, Proceedings of VTC 2008-Fall: IEEE 68th Vehicular Technology Conference, pp. 1-4, ISBN: 978-1-4244-1721-6, Calgary 21-24 September 2008, IEEE, Canada

Katulski, R.J. (2009). *The radio wave propagation in wireless telecommunication* (in Polish), Transport and Communication Publishers, ISBN: 978-83-206-1749-8, Warsaw, Poland

Katulski, R.J. et al. (2009). *Description of the system for cargo containers monitoring in a container terminal* (in Polish), Gdansk University of Technology, Faculty of Electronics, Telecommunications and Informatics, Department of Radiocommunication Systems and Networks, Scientific Report 5/2009

Lee, William C.Y. (1993). *Mobile communications design fundamentals,* John Wiley & Sons, ISBN: 0-471-57446-5 , New York, USA

23

Link Budgets: How Much Energy is Really Received

Aarne Mämmelä[1], Adrian Kotelba[1],
Marko Höyhtyä[1] and Desmond P. Taylor[2]
[1]VTT Technical Research Centre of Finland
[2]University of Canterbury
[1]Finland
[2]New Zealand

1. Introduction

In this chapter we consider various factors affecting link power or energy budgets, which are introduced in some detail for mobile and satellite communications in (Saunders, 1999, pp. 97-99; Sklar, 2001, pp. 242–303). We use the link budget as presented in (Saunders, 1999) for mobile cellular systems as a starting point. Some loss factors may be random. If we do not know the distributions of those random variables, it is reasonable to use the worst case analysis (French, 1986, p. 36), which is useful but in general highly pessimistic. We discuss the alternatives based on decision theory. In some cases it is reasonable to compute link budgets for the average performance instead of worst case performance. We can also use outage probabilities in the worst case analysis.

The goal of a link budget is to estimate the received signal-to-noise ratio (SNR) per bit at the input of the combiner/equalizer in the receiver (Saunders, 1999; Sklar, 2001), as in Fig. 1. Normally the combiner/equalizer includes some kind of correlator, which tries to use the received signal in the best possible way (Proakis, 2001).

It has been known at least since the mid-1960's that the transmitter power or energy is a basic system resource (Lucky et al., 1968, pp. 51-58; Schwartz et al., 1966, pp. 555-558). The different loss factors are assumed to be multiplicative in the linear domain and additive in the logarithmic domain, i.e., in decibels. In addition, the additive noise level is estimated by measuring the receiver noise floor and certain link margins are added. Most of the additive noise actually arises in the receiver front end. If there are several antennas, the sum of the powers is used. With a comprehensive link budget the necessary design trade-offs can be made in a systematic way before the system is actually built.

Link budgets do not include phenomena in or preceding the power amplifier, for example its finite efficiency. We do not discuss transmitter and receiver power budgets, i.e., how much energy or power is actually taken from the battery or electricity network. There are also other link budgets such as delay budgets that we do not consider.

Many loss factors depend on the frequency range (Saunders, 1999; Chu & Greenstein, 1999). The wavelength is $\lambda = c/f$ where c is the propagation velocity of the radio waves and f is the frequency. Radio frequencies range from 3 kHz to 300 GHz with wavelengths from 100 km down to 1 mm. The use of frequencies below 3 kHz is in general impractical and above 300

GHz we have infrared frequencies. In many modern mobile communication systems the frequencies employed are between 1 GHz and 10 GHz, and the wavelengths are then between 30 cm and 3 cm. We will assume that the antennas are used only in the far field, which implies that the dimensions of all antennas are small compared to the distance between transmitter and receiver antennas. Thus we will not consider any evanescent or near field effects.

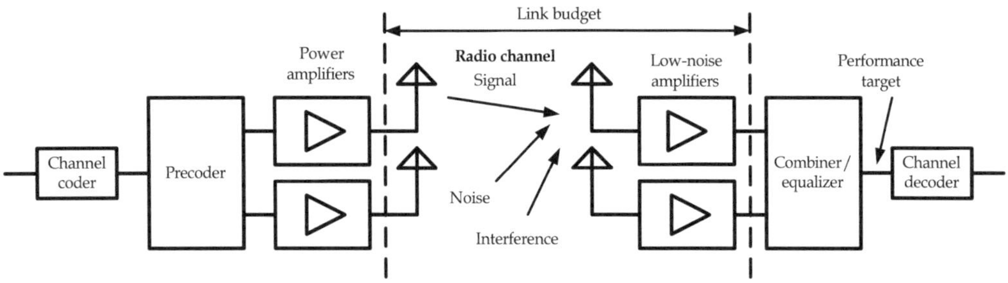

Fig. 1. Definition of a link budget

It is well known that energy is conserved in all physical processes. In communications, we are not especially interested in transmitting energy, but without energy, communication is not possible, and the performance of the receiver depends in part on the ratio of the received energy per bit to the noise power spectral density. Furthermore, it is not possible to observe the state of the channel without transmitting energy.

The physical channel is normally assumed to be linear (Bello, 1963; Proakis, 2001), but it may distort the transmitted signal because of temporal, frequency, and spatial selectivity. The noise is assumed to be additive white and Gaussian (AWGN). Some interfering signals from other users may also be received. Usually the transmitted signal is divided into blocks called frames, each containing multiple symbols, and each symbol includes one or more bits. The modulation and coding methods are taken into account in the received SNR required to attain the specified bit error rate or block error rate. Under worst case conditions, the raw bit error rate before channel decoding should be between 10^{-2} and 10^{-3}. Then, with a good channel coding scheme, it is possible to reduce the bit error rate to less than say 10^{-6}.

In some power-controlled systems the transmitted energy is zero when the channel is too weak. We must divide the systems into power-limited and energy-limited. *Power-limited systems* include, for example, base stations that are connected to the power-distribution network. *Energy-limited systems* include mobile terminals based on batteries. In such systems a fundamental question is how many bits can we reliably transmit for the expenditure of a given transmitted energy, which normally forms a major part of the total energy used in the transmitter. Thus the average transmitted energy per bit is a critical parameter.

In power-limited systems the power is averaged over all time, but in energy-limited systems the energy is averaged only when the transmitted energy is nonzero (Mämmelä et al., 2010). We refer to the latter kind of averaging as averaging with *expurgation*. Because of discontinuous transmission in a fading channel, the number of received symbols is a random variable and analysis becomes more difficult for energy-limited systems than in the power-limited case. If transmission is continuous, expurgation has no effect and there is no difference between power-limited and energy-limited systems.

Normally the received symbol energy is optimized to yield sufficiently good performance. In adaptive systems the changes in the channel must be slow compared to the symbol rate (Monsen, 1980; Proakis, 2001), which is valid if the coherence time of the channel is much larger than the symbol interval. This is a reasonable assumption when the bit rate is high and the symbol interval is short. In practice, because of the use of channel estimators whose observation interval is several symbol intervals, the channel should be essentially invariant during each block of symbols.

In this chapter we briefly review the major loss factors included in link budgets. We want to provide an intuitive explanation for these factors and introduce a new one. More specifically, we introduce a factor arising from the correlation between two loss factors. Such a situation is common in power-controlled systems where the transmitted energy and the channel energy gain $G = E_r/E_t$ are correlated where E_r is the received energy and E_t is the transmitted energy. If $E_t = 0$, we define $G = 0$. The correlation is usually ignored in link budgets although it has a significant effect, either positive or negative, depending on the power control rule. The analytical background for this is explained in (Mämmelä et al., 2010). There can be also some other nonlinear interactions between the gains.

This chapter is organized as follows. We first summarize the required decision-theoretic tools to properly formulate link budgets. Next we discuss path loss, shadowing, and multipath fading of the channel and the corresponding fade margins. Other loss factors are briefly mentioned. Some channels are reciprocal and the requirements for this are discussed. Next we discuss the effect of correlation between different loss factors. Various averaging methods are used in the literature, but not all of them are appropriate for our purposes. Normalization of the channel model may be problematic since some models have an infinite peak energy gain, which obviously is inconsistent with the energy conservation law and may sometimes lead to confusing results as explained in (Xiang & Pietrobon, 2003).

2. Decision-theoretic tools

Alternative approaches for formulating link budgets are based on rational decision-making under strict uncertainty (French, 1986, pp. 33-60). The first approach is called the *max-min approach approach*, which is sometimes called Wald's rule (French, 1986, p. 36). We use this approach when the distributions of random variables are unknown and we maximize the worst case performance. This approach is used in conventional link budgets, which are based in various margins such as fade and interference margin. Another approach called *Laplace's principle of insufficient reason* (French, 1986, p. 38) can also be used. Laplace argues in his "Essai philosophique des probabilités" that knowing nothing at all about the true state of nature is equivalent to all states having equal probability (French, 1986, p. 38). Following Laplace's line of reasoning leads to the assumption that all random variables whose probability distributions are unknown to the system designer are uniformly distributed random variables. Under the uniform distribution assumption, maximization of average performance is well justified (French, 1986, p. 49). The third approach is called the *safety first approach*. There are three schemes in this approach. The schemes are summarized in (Kotelba & Mämmelä, 2008) where the original references can be found. The first scheme was devised by Roy. Here we minimize the probability of outage α for a given threshold c set for the received SNR γ_r, i.e., $\alpha = \min \Pr(\gamma_r < c)$. This leads to the search for a probability density function $p(\gamma_r)$ such that the ratio $(\mu_\gamma - c)/\sigma_\gamma$ is maximized where μ_γ is the mean of γ_r and σ_γ is its standard deviation (Roy 1952, p. 434). The second scheme was derived by Kataoka. Here we set the probability of outage at a certain level and maximize the threshold for the received SNR, thus $\max_c \Pr(\gamma_r < c) = \alpha$. The third scheme was published by Telser. In this scheme we maximize the mean of the received SNR for a given outage probability, i.e., max $\{E\{\gamma_r\}, \Pr(\gamma_r < c) = \alpha\}$. The approaches lead to the conclusion that as an alternative to the worst case analysis, we can also propose average analysis based on an outage probability.

3. Path loss of the channel

According to the energy conservation law, the energy gain of a channel is always equal to or less than unity. The loss is the inverse of the gain, and it is usually expressed in decibels. To simplify our discussion we often use the term gain although in general the gain is smaller than unity. The gain has temporal, frequency, and spatial selectivity. Therefore the average gain depends on the distribution of the transmitted signal in these domains (Mämmelä et al., 2010). The gain is not only a property of the channel, but depends on how well the transmitted signal is "matched" to the channel, which may be randomly changing. In general, the gain is larger if the transmitted signal is using those parts of the channel in time, frequency, and space where the gain is the largest.

The path loss is the average loss at a given distance. It consists of the free-space loss, the absorption and diffraction loss because of the intervening medium, and the loss because of the reception on many multipath components due to reflections and refractions (Rappaport, 2002). In some specific scenarios the path loss can be smaller than the free-space loss because of the additional energy received from multipath components as in a waveguide, corridor, or duct.

3.1 Free-space loss

The concept of free-space loss is based on a model of an isotropic antenna in a vacuum. The free-space gain of an isotropic antenna is $G_{fsi} = P_r/P_t = A_i/A_{ro} = (\lambda/4\pi r)^2$ (Saunders, 1999, p. 68). It depends on the wavelength λ and thus on the frequency $f = c/\lambda$. Here $A_i = \lambda^2/(4\pi)$ is the effective aperture of the isotropic antenna, and $A_{ro} = 4\pi r^2$ is the area of the surrounding sphere of radius r. Intuitively there should be no dependence on wavelength. The energy is distributed in free space in the same way in all directions at all frequencies. The size of the antenna is assumed to be very small compared to the wavelength so that the antenna has an isotropic antenna pattern. The effective aperture of an isotropic antenna depends on the wavelength. Thus the smaller the wavelength and the higher the frequency, the smaller the effective aperture. We can imagine a situation where the transmitter antenna in a vacuum is surrounded by $\lfloor 1/G_{fsi} \rfloor$ receiver antennas in such a way that almost all the transmitted energy is collected. In that case the loss does not depend on the wavelength in free space, and in fact there is no free-space loss.

Physical antennas have an effective aperture $A_e = G_a \cdot \lambda^2/(4\pi)$ where G_a is the gain of the antenna. The gain of a practical antenna depends on direction because of its finite size, and we usually refer to the highest gain. The gain is also frequency-dependent. The free-space gain between two antennas is specified as

$$G_{fs} = G_t\, G_r\, (\lambda/4\pi r)^2 \quad (1)$$

where G_t and G_r are the gains of the transmitter and receiver antennas. This is called Friis's equation (Saunders, 1999, pp. 69, 93-94). We have assumed that the gain is in the direction of the receiver at the transmitter and in the direction of the transmitter at the receiver, as otherwise there is an additional loss. Free-space loss can be approximated in practice when several Fresnel's zones defined by the transmitter and receiver are free around the antennas (Saunders, 1999, p. 49-50).

3.2 Absorption and diffraction loss

In any medium, part of the transmitted energy is absorbed and diffracted, for example due to different material objects (Saunders, 1999, pp. 45-55, 127-135). In addition, at certain frequencies the radio waves are highly attenuated because of gas molecules and rain. Energy is often changed to heat in absorption. Energy is absorbed because the electrical particles, for example electrons, start to move in the medium, and the energy is partially changed to heat that cannot be received by an antenna but is distributed throughout the environment. In general the absorption loss of radio waves is increased when the wavelength is made shorter. Material objects cause also another phenomenon called diffraction. Thus part of the radio waves may be bent near the objects. Diffraction is stronger at lower frequencies. Absorption and diffraction losses are difficult to predict analytically especially if the form of the objects is irregular. Such losses are usually estimated by measurements (Pahlavan & Levesque, 1995; Rappaport, 2002).

4. Shadowing and multipath fading

Fading is caused by shadowing and multipath propagation. The time-variant part of the absorption and diffraction loss is called shadowing. In general it is a much slower phenomenon than multipath fading. When the receiver is moved at a given distance from the transmitter, the medium seen by the radio waves is changed. In urban areas the coherence distance is in the order of ten meters (Gudmunson, 1991), which corresponds to tens or hundreds of wavelengths. In multipath fading the coherence distance may be less than a wavelength.

Under the worst case analysis, fading is normally included in link budgets as a margin. We are interested in the received SNR in the worst case in a slowly fading channel. As will be seen, typical fading models are such that the gain can approach zero. In practice we must accept a certain confidence level, say 90-99 %, with which the required received SNR is attained. If the SNR becomes lower, such a situation is called an *outage*. The corresponding outage probabilities are between 10 % and 1 %. In mobile communications, typical edge-of-cell margins are such that the confidence level is 90 % (Saunders, 1999, p. 183).

4.1 Shadowing

Shadowing is normally assumed to be frequency-nonselective, but in practice it is frequency-selective (Coulson et al., 1998). In a lossy medium the frequency has some effect because usually it is the thickness of the material in terms of the number of wavelengths and the electrical properties of the material that cause the loss (Pahlavan & Levesque, 1995). We need a database where the materials are classified according to their electrical properties and the losses are normalized to wavelength. The loss caused for example by a wall can then be estimated by measuring its thickness in wavelengths. Not all the multipath components propagate directly to the receiver. Different multipath components see a different absorption and diffraction loss, and therefore shadowing is frequency-selective.

Shadowing can be divided into global and local shadowing (Salo, 2006, pp. 33-35). Local shadowing refers to changes in the absorption and diffraction loss when the receiver moves in a given environment without changing its distance to the transmitter. A model for local shadowing should be used in link simulations. Based on the central limit theorem we usually assume that local shadowing is lognormal. The logarithm of the product of several independent gain factors $G_k = \log[\Pi_i (G_i)] = \Sigma_i \log(G_i)$ is normally distributed if the assumptions of the theorem are valid. The measured standard deviations are usually in the order of 3-6 dB.

Global shadowing refers to the changes in the absorption loss globally at a given distance. Those changes are much larger than in local shadowing since there can be large differences in the media. In a link budget global shadowing is pessimistic and is based on the assumption that we do not know much about the intervening medium. If we know the materials comprising the medium and their location, we can improve the estimate. Global shadowing is often assumed to be lognormal, but strictly speaking it cannot be lognormal if local shadowing is lognormal. The average of K lognormal random variables $G = (1/K) \Sigma_k G_k$ is not in general lognormal, and in fact, according to the central limit theorem it tends to be normal when K is large. However, there is a more or less general consensus that the average of independent lognormal random variables can be approximated by another lognormal random variable with appropriately chosen parameters (Stüber, 2001, pp. 129-139). The measured standard deviations are in the order of 5-12 dB (Stüber, 2001, p. 21).

4.2 Multipath fading

Multipath propagation is caused by reflection and scattering of radio waves in flat and rough zones between two media (Sounders, 1999). Multipath propagation can also be caused by deterministic or random changes in the refractive index of the medium such as the atmosphere. The fading caused by multipath propagation is best explained by using a sinusoidal signal. A physical distance d causes a delay $\tau = d/c$ and shifts the phase by $\varphi = -2\pi f \tau$. Multipath signals can have path-length differences of several kilometres (Turin, 1980; Pahlavan & Levesque, 1995). For example, a path-length difference of 6 km corresponds to a delay difference of 20 μs. Thus usually the path length difference Δd is in the order of several wavelengths, and the magnitude of the phase shift difference is $|\Delta\varphi| >> 2\pi$ rad. Therefore, the signals are added together in the receiver antenna either constructively when the phase difference $|\Delta\varphi| \approx 2n\pi$ where n is an integer or destructively when $|\Delta\varphi| \approx (2n + 1)\pi$, depending on the phase relationship. The phase shift φ is usually frequency-dependent, and thus multipath fading is frequency-selective. If $|\Delta\varphi| << 2\pi$ rad, the multipath fading would disappear, but this is not normally the case because usually $\lambda << |\Delta d|$. The fading is also time-selective since the delays are constantly changing due to terminal mobility.

The amplitude distribution of multipath fading is usually assumed to be Rayleigh or Rician (Proakis, 2001). These distributions arise due to the central limit theorem. The received power has then either a central or noncentral chi-square distribution. Sometimes also the Nagakami-m distribution also used for the amplitude distribution. The latter corresponds to a generalized Rayleigh distribution (Proakis, 2001, p. 45-48).

Energy does not disappear in multipath fading. The energy is spatially distributed and in different locations the receiver antenna sees a different amount of energy. The channel is thus also spatially selective. In theory all that energy could be collected if there were no absorption.

4.3 Beam forming and antenna diversity

Often transmitters and receivers use more than one antenna element. Such systems may be classified as beam forming or diversity systems. Beam forming can be seen to be more general than antenna diversity. In beam forming significant aliasing in the spatial or directional domain is avoided by using the sampling theorem in space. The maximum distance between antenna elements is half a wavelength. The antenna pattern can be modified by complex weighting. The finite size of the antenna array causes side lobes similar to the Gibbs phenomenon in filters (Hansen, 1966). Thus there is some aliasing because of the side lobes.

In beam forming we process correlated signals to or from certain directions. In receiver beam forming we make the effective aperture of the receiver larger and thus we receive more energy. In transmitter beam forming the spreading loss is made smaller because the same energy is focused in a given direction. If the number of antenna elements is N, the maximum antenna array gain is $10 \log_{10} N$ (Parker & Zimmermann, 2002). For example, if $N = 16$, the gain is 12 dB. Directional antennas can be used to reduce multipath components.

In multiple-input multiple-output (MIMO) diversity systems the distance between antenna elements is so large that we ideally process uncorrelated signals coming from all directions. This is called rich scattering. The distance is in the order of wavelengths. Antenna diversity is a special case of beam forming in the sense that some of the complex weights and the antenna elements are removed and aliasing is introduced. In the transmitter the signals are made orthogonal by using special coding. In the receiver the signals may be combined coherently as in maximal ratio combining. In antenna diversity the antenna array has an antenna pattern that looks random and it is constantly changing when the multipath propagation is changing. Because of aliasing we cannot freely modify the antenna pattern. One direction corresponds to many directions and the antenna pattern has some periodicity in different directions.

If we add the number of transmitter antennas in antenna diversity, the received energy is not in general increased since it must be divided among the antennas and the transmitted energy is the basic resource (Schwartz et al., 1966, pp. 555-558). The antenna pattern is approximately omnidirectional although random, and there is no array gain on average. In diversity systems, there is typically no array gain, but there is usually element gain. In receiver diversity the situation is different because the effective aperture is increased by using two or more antennas.

5. Other loss factors and margins and reciprocity

Antenna efficiency is defined as the total radiated energy divided by the total input energy. The efficiency depends on frequency. Normally the antenna radiates with the highest efficiency at frequencies where the wavelength is of the same order as the physical dimensions of the antenna. A physical antenna thus acts as a band-pass filter.

Additional loss factors include for example base station feeder loss, body and matching loss, and implementation loss (Saunders, 1999; Sklar, 2001). Some loss factors are not easy to estimate and thus it is common to use specific margins, for example interference margin (Saunders, 1999). Interference margin is especially important for cell edge performance in cellular systems.

Reciprocity is usually assumed in time-division duplex systems, particularly when the channel is changing sufficiently slowly. Antenna elements are in general reciprocal between the transmitter and receiver (Saunders, 1999, pp. 66-68). This means that the impulse response of the channel does not depend on direction. If we use a set of antenna elements, the situation may be different since we may weight the signals in different ways in the transmitter and receiver. Also the noise and interference depend on the environment and they are not in general reciprocal. Thus the SNR is not in general reciprocal. Electronic components such as power amplifiers are usually not reciprocal.

6. Correlation between gain factors

The gain factors are assumed to be multiplicative. Thus the logarithm of the total gain of the form $G = \Pi_i (G_i)$ is $\log(G) = \log[\Pi_i (G_i)] = \Sigma_i \log(G_i)$ where the gains G_i are deterministic. If the gains are correlated random variables, this approach is no longer valid. For brevity, we consider two gains G_1 and G_2, which are correlated with the covariance $\mathrm{Cov}\{G_1\, G_2\} = E\{[G_1 - E\{G_1\}][G_2 - E\{G_2\}]\} = E\{G_1\, G_1\} - E\{G_1\}E\{G_2\}$, where E refers to the expected value. The total instantaneous gain is $G = G_1\, G_2$, but the average gain is

$$E\{G\} = E\{G_1\, G_2\} = E\{G_1\}\, E\{G_2\} + \mathrm{Cov}\{G_1\, ,\, G_2\}. \tag{2}$$

The covariance in (2) can be positive, negative, or zero.

A good example is the covariance between the transmitted energy and the channel energy gain in a power-controlled transmitter. We have shown in (Mämmelä et al., 2010) that the covariance has a significant effect on the link budgets. The basic power control rules are water-filling, which maximizes capacity if the bit rate is optimally adjusted at the same time, and truncated channel inversion, which minimizes the outage probability for a given transmitter power constraint and bit error rate (Biglieri et al., 1998, pp. 2627-2628). Under water-filling we transmit more energy when the channel is good, but when the channel is sufficiently bad, we do not transmit at all. In this way the transmitted signal is well matched to the channel, and we actually receive a good fraction of the transmitted energy. The method is optimum only for isolated links due to possible interference to other users. In truncated channel inversion we do the opposite: we transmit more energy when the channel is bad, but if the channel is too bad, we do not transmit at all. In full channel inversion we transmit all the time. These alternatives can be approximated by recursive algorithms (Gilhousen et al., 1991; Caire et al., 1999; Yu et al., 2004; Höyhtyä & Mämmelä, 2007).

In practice it is easier to track shadowing which is much slower than multipath fading. However, this depends on the system parameters. There is no fundamental reason why we cannot track multipath fading if it is slow enough compared to the symbol rate of the system.

The covariance can be included in the link budgets as an additional gain or loss because the average received energy can be expressed as

$$E\{E_r\} = E\{E_t\, G\} = E\{E_t\}\, E\{G\} + \mathrm{Cov}\{E_t, G\}. \tag{3}$$

In practice, a more informative figure of merit is the *efficiency of power transmission* defined as

$$\eta = E\{E_r\}/(E\{E_t\}\, E\{G\}) = 1 + \mathrm{Cov}\{E_t\, ,\, G\}/(E\{E_t\}\, E\{G\}). \tag{4}$$

Here is a short interpretation of (3) and (4). The covariance $\mathrm{Cov}\{E_t, G\}$ captures the effect of adaptive power control on the average received signal energy, and therefore, on the link energy budget. The efficiency parameter η of (4) describes whether there is an additional gain or loss compared to the traditionally used approach where possible correlation between the transmitted energy and channel gain is not taken into account.

We illustrate the effect of the covariance at two extremes. The covariance can be either maximized or minimized with respect to some additional constraints on the moments of distribution of transmitted energy (Kotelba & Mämmelä, 2010b). For given average transmitted energy $E\{E_t\}$ and representative energy gain $E\{G\}$, the maximization of the covariance value is equivalent to maximization of the average received energy or maximization of transmission efficiency. The *representative energy gain* is the average energy gain of a signal whose energy is uniformly distributed in all the domains (Mämmelä et al., 2010). Similarly, minimization of the covariance value is equivalent to minimization of the average received energy or minimization of transmission efficiency. We demonstrated in (Kotelba & Mämmelä, 2010b) that to maximize the average received energy $E\{E_r\}$ the instantaneous transmitted energy E_t should be proportional to the channel gain G. On the other hand, if the instantaneous transmitted energy E_t is inversely proportional to the square of the channel gain G, then the average received energy $E\{E_r\}$ is minimized. In general there is some gain for water-filling and some loss for truncated channel inversion, compared to the case where the transmitted energy is constant (Mämmelä et al., 2010). The difference can be several decibels depending on the transmitted signal-to-noise ratio. A good way to include the effect of the covariance in the average link budget is to use the transmitted SNR instead of the received SNR in bit error rate measurements. The transmitted SNR γ_t is defined to be the ratio of the transmitted energy per bit and the received noise spectral density (Mämmelä et al., 2010).

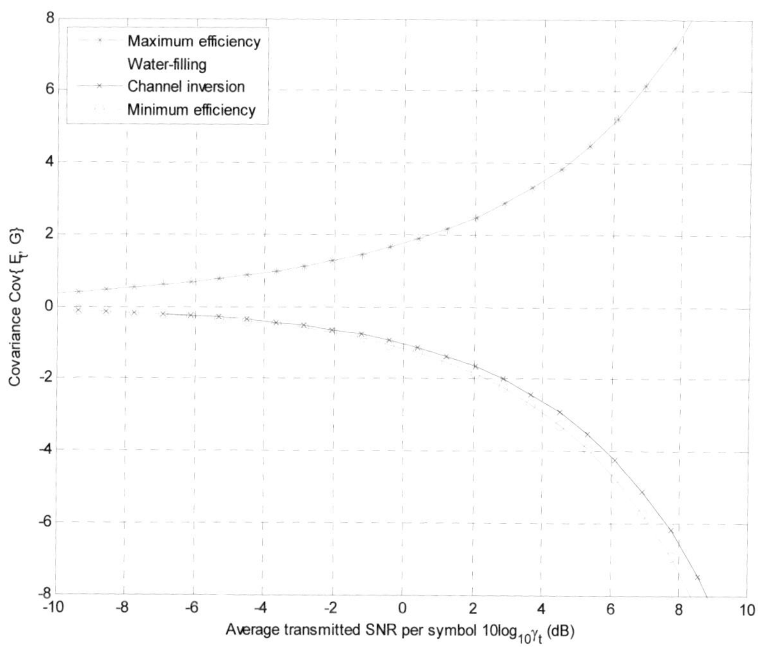

Fig. 2. Covariance between instantaneous transmitted energy and channel gain for various adaptive power control schemes

In the numerical examples in Figs. 3 through 5 we did not use expurgation, which implies that the results are valid for power-limited systems. In nonexpurgated systems the analysis is tractable and we are able to illustrate the concepts. In our numerical results, shown in Figs. 2 and 3, we used a composite gamma/lognormal fading channel model (Simon & Alouini, 2000, pp. 24-25) with multipath fading parameter m = 4, shadowing channel gain μ

= 0 dB, and shadowing standard deviation σ = 4 dB. For truncated channel inversion, the cut-off level was set to zero, that is, we considered full channel inversion.

The results shown in Fig. 3 can be used in average link budgets. In the worst case analysis we must define an outage probability and determine how far below the mean the SNR can be. The cumulative distribution function (cdf) $F(\gamma_r)$ of the received SNR γ_r is shown in Fig. 4. Three power control rules are used in a composite multipath/shadowing channel. The average transmitted SNR is constant for all power control rules and equal to 5 dB. The cutoff value in truncated channel inversion μ_{tci} was set to 0.223489 to meet the outage probability requirement. The respective averages are marked with an asterisk. Water-filling gives the largest average received SNR (positive covariance), followed by no power control (zero covariance), and truncated channel inversion (negative covariance). For a target outage probability of 10 %, the truncated channel inversion and water-filling require the smallest and the largest fade margin, respectively.

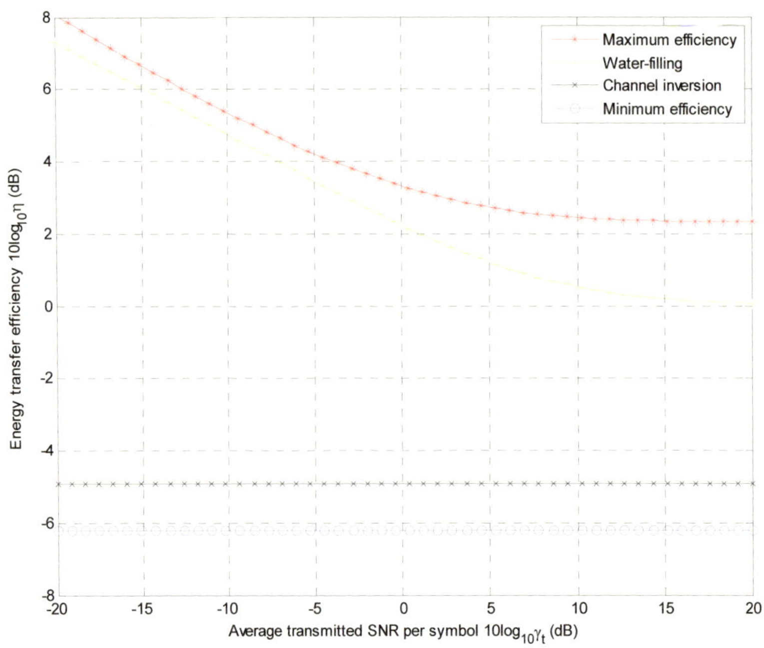

Fig. 3. Efficiency of power transmission in a composite multipath/shadowing channel

The corresponding fade margins are shown in Fig. 5, again for a target outage probability of 10 %. The results in Fig. 5 suggest that under the water-filling power control rule, the maximum outage probability requirement of 10 % can be satisfied only for sufficiently large average transmitted SNR. Below a certain value of the average transmitted SNR, the outage probability associated with a nonzero cutoff value μ_{wf} exceeds the target probability of 10 %. Since the water-filling cutoff value μ_{wf} is unique for a given average transmitted SNR, it is not possible to meet the maximum outage probability requirement and the required fade margin tends to infinity. This effect is illustrated in Fig. 5. The results in Fig. 5 can be used in the worst case analysis when we add the efficiency of power transmission shown in Fig. 3

and the fade margin shown in Fig. 5. In conventional link budgets the fade margin for no power control is used. Thus we have provided some tools to include power control in both the average and worst case analyses.

7. Averaging

In averaging we must consider possible commutability problems. We must know in which domain we must perform averaging. The expectation operation commutes over linear and affine transformations of the form $\mathbf{y} = \mathbf{Ax} + \mathbf{c}$, i.e., $E\{\mathbf{y}\} = \mathbf{A}\, E\{\mathbf{x}\} + \mathbf{c}$ where $\mathbf{x}$ is a random column vector, $\mathbf{A}$ is a constant matrix, and $\mathbf{c}$ is a constant column vector (Kay, 1993, pp. 149-150, 349-350, 390). If $\mathbf{c} = \mathbf{0}$, the transformation is *linear*, and the superposition theorem is valid, and if $\mathbf{c} \neq \mathbf{0}$, the transformation is *affine*. The property for the expectation operation does not in general carry over to nonlinear transformations, for example to those including logarithms and inversions. The problem is discussed in (Kay, 1993, pp. 173-177, 185) for common estimators.

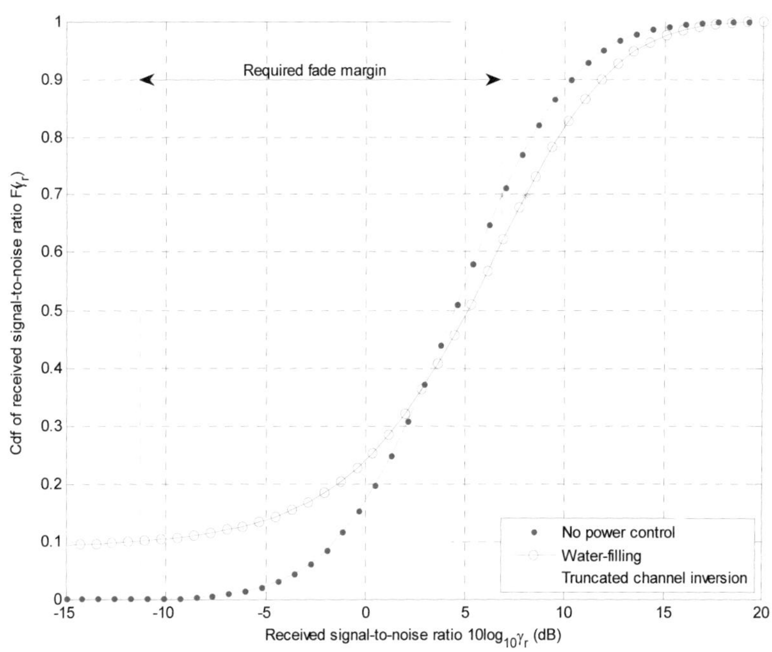

Fig. 4. Cdf's of the received SNR for no power control, water-filling, and truncated channel inversion

When we measure the attenuation of the channel, the received power or energy must be averaged and compared with the transmitted power or energy. By definition, averaging should be done in the linear domain, i.e., in watts or joules. However, sometimes it is more practical to average in the logarithmic domain using decibels since most link budgets are computed in decibels. The two averages are not in general identical even if expressed in the same domain (Pahlavan & Levesque, 1995), and thus averaging of decibels may be misleading. By using Jensen's inequality (Feller, 1971, p. 153) we have $E\{\log X\} \leq \log E\{X\}$. In general logarithmic averaging leads to pessimistic results and the estimated average gain is lower than the actual average gain.

In the literature the local shadowing is usually averaged in the linear domain and the global shadowing is averaged in the logarithmic domain (Rappaport 2002, pp. 139-141). If we assume that the lognormal approximation is good enough, the global average is actually the median in the linear domain, not the true linear average. If we do not use the average itself but the confidence limits in the worst case analysis, there is no problem with this approach.

In a fading channel we must often average ratios of the form X/Y where X and Y are random variables and $Y \neq 0$. Now $E\{X/Y\} = E\{X \cdot 1/Y\} = E\{X\}\, E\{1/Y\} + \mathrm{Cov}\{X, 1/Y\}$. If X and $1/Y$ were uncorrelated, we would have $E\{X/Y\} = E\{X\} \cdot E\{1/Y\}$. In general $E\{X/Y\} \neq E\{X\}/E\{Y\}$ where $E\{Y\} \neq 0$. Such ratios must be considered for example when the received energy $X = E_r$ and transmitted energy $Y = E_t$ are random in a power-controlled system, and we must estimate the average energy gain of the channel. The variables X and $1/Y$ are also correlated in this case. Another example is when the transmitted energy per block is $X = E_t$ and the number of bits in a block $Y = w$ is a random variable and we must estimate the average transmitted energy per bit $E\{E_t/w\}$ (Mämmelä et al., 2010). In such cases we obtain different results with different averaging methods $E\{X/Y\}$ or $E\{X\}/E\{Y\}$ and we must decide what we really want. In general, the observation interval should be sufficiently large that the random variable Y in the denominator can be assumed to be essentially constant. In such cases $E\{X/Y\} \approx E\{X\}/E\{Y\}$.

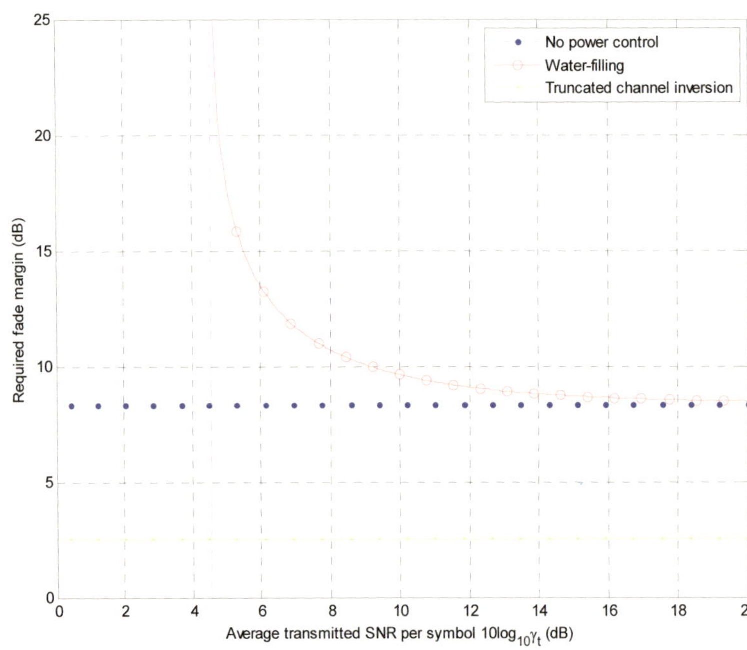

Fig. 5. Fade margins for no power control, water-filling, and truncated channel inversion

In a fading channel transmission may not be continuous. In such cases we must distinguish between power-limited and energy-limited systems. In energy-limited systems expurgation is used to measure the average transmitted energy and the corresponding SNR. In power-limited systems no expurgation is used for the SNR. In all systems expurgation is used when measuring the bit error rate since we are not interested in bits that are not actually transmitted. When using expurgation, we must know when the actual bits are received.

8. Normalization of the channel

The average energy gain of the channel depends on the signal structure: how the energy is distributed in the time, frequency and spatial domains and thus how well the transmitted signal is matched to the channel (Mämmelä et al., 2010). In link budgets the channel is usually normalized according to the so-called *representative energy gain*. For example, if the channel model has the form $Y = aX$, the representative energy gain is $E\{|a|^2\}$. Some channel models are such that their instantaneous energy gain is infinite although the average energy gain would be finite. Normally this does not matter since the path loss is rather large, and it is quite improbable that the instantaneous gain would be larger than the path loss. Strictly speaking our models should be truncated. If our aim is to compare the effect of different distortions, it may be reasonable to normalize the models according to the peak energy gain (Xiang & Pietrobon, 2003).

If in an antenna diversity system we add new receiver antennas at the same distance from the transmitter, the received energy would be increased linearly. In principle we can increase the effective aperture up to the whole sphere surrounding the transmitter antenna. We could have a suitably formed effective aperture that fits together like a jigsaw puzzle. We receive only part of the transmitted energy since part of the energy is absorbed and transformed to other types of energy. The received energy would saturate to that maximum and no longer increases linearly with the number of antenna elements.

The measured path loss includes all the multipath signal components, not just the direct path component. If directive antennas were to be used, some multipath components would be rejected. By including the multipath components we are able to clarify the normalization problem we have. In fair comparisons in energy-limited systems the starting point is the basic resource or the average transmitted energy per bit. If we have different channel models, we have the problem of how we should normalize them. Intuitively peak normalization would be fair. However, the theoretical peak is not always attained. For example, in a multipath channel the multipath components experience certain delays. In theory the peak would be such that all the multipath components are added constructively. This is not always possible if we cannot select the delays arbitrarily. A certain delay corresponds to a certain phase shift. For example, there can be additional phase shifts due to reflections. In general, the peak of the square of the magnitude of the transfer function of a slowly fading channel is less than or equal to the theoretical maximum.

Many channel models cannot be peak-normalized since the peak energy gain is unlimited. Such models include for example Rayleigh, Rician and lognormal fading channels. In such channels surprisingly there appears a St. Petersburg-like paradox (Kotelba & Mämmelä, 2010a). It can be resolved by using suitable performance metrics and fading models. Normalization is further discussed in (Loyka & Levin, 2009).

9. Conclusion

Link power or energy budgets are usually used to estimate the received SNR and thus the expected performance of the receiver. In a link budget we evaluate the different loss factors in a communication system. Usually the link budget is computed in the logarithmic domain and the different factors are added together in decibels. Normally a worst case analysis is made. We have described the factors in detail and have shown the uncertainties we have, and how to possibly reduce those uncertainties. We have also provided tools describing how to include power control in the worst case analysis or in the alternative average analysis.

A general assumption is that there are no nonlinear interactions between the different factors. When power control is used there is a strong correlation between the energy gain of the channel and the transmitted energy. Numerical results were presented to show the effect of this correlation. The chapter has concentrated mainly on single isolated links. An extension to multiple links would be useful, but so far no standard practices are available other than the interference margins.

We have also shown the limitations of the common fading channel models leading to some pathological phenomena that never occur in practical channels. We discussed the normalization of the channel models and averaging of the power and energy in power-limited and energy-limited systems. The emphasis was in phenomena that are not very well known in the scientific society.

10. References

Bello P. A. (1963). Characterization of randomly time-variant linear channels, *IEEE Transactions on Communications Systems*, Vol. CS-11, No. 4, pp. 360 - 393.

Biglieri E., Proakis J., and Shamai S. (1998). Fading channels: Information-theoretic and communications aspects, *IEEE Transactions on Information Theory,* Vol. 44, No. 6, pp. 2619 - 2692.

Caire G., Taricco G., and Biglieri E. (1999). Optimum power control over fading channels, *IEEE Transactions on Information Theory*, Vol. 45, No. 5, pp. 1468-1489.

Chu T.-S. and Greenstein L. J. (1999). A quantification of link budget differences between the cellular and PCS bands, *IEEE Transactions on Vehicular Technology*, Vol. 48, No. 1, pp. 60 – 65.

Coulson A. J., Williamson A. G., and Vaughan R. G. (1998). A statistical basis for lognormal shadowing effects in multipath fading channels, *IEEE Transactions on Communications*, Vol. 46, No. 4, pp. 494-502.

Feller W. (1971). *An Introduction to Probability Theory and Its Applications*, Vol. 2, 2nd ed., John Wiley & Sons, New York.

French S. (1986). *Decision Theory: An Introduction to the Mathematics of Rationality*. Halsted Press, New York.

Gilhousen K. S., Jacobs I. M., Padovani R., Viterbi A. J., Weaver L. A., Jr., and Wheatley C. E., III (1991). On the capacity of a cellular CDMA system, *IEEE Transactions on Vehicular Technology*, Vol. 40, No. 2, pp. 303–312.

Gudmundson M. (1991). Correlation model for shadow fading in mobile radio systems, *Electronics Letters*, vol. 27, no. 23, pp. 2145 - 2146.

Hansen R. C. (1966). *Microwave Scanning Antennas*, John Wiley & Sons, New York.

Höyhtyä M. and Mämmelä A. (2007). Adaptive inverse power control using an FxLMS algorithm, *Proceedings of the IEEE Vehicular Technology Conference (VTC) Spring*, Dublin, Ireland, pp. 3021-3025.

Kay, S. M. (1993). *Fundamentals of Statistical Signal Processing: Estimation Theory*. Prentice-Hall, Englewood Cliffs, NJ.

Kotelba A. and Mämmelä A. (2008). Application of financial risk-reward theory to adaptive transmission, *Proceedings of the IEEE Vehicular Technology Conference (VTC) Spring*, Singapore, pp. 1756 - 1760.

Kotelba A. and Mämmelä A. (2010a). St. Petersburg paradoxes in performance analysis of adaptive wireless systems, *Proceedings of the IEEE Vehicular Technology Conference (VTC) Spring*, Taipei, Taiwan.

Kotelba A. and Mämmelä A. (2010b). Efficiency of power transmission in adaptive communication systems, unpublished.

Loyka S. and Levin G. (2009). On physically-based normalization of MIMO channel matrices, *IEEE Transactions on Wireless Communications*, Vol. 8, No. 3, pp. 1107-1112.

Lucky R. W., Saltz J., and Weldon E. J., Jr. (1968). *Principles of Data Communication*. McGraw-Hill, New York.

Mämmelä A., Kotelba A., Höyhtyä M., and Taylor D. (2010). Relationship of average transmitted and received energies in adaptive transmission, *IEEE Transactions on Vehicular Technology*, Vol. 59, No. 3, pp. 1257-1268.

Monsen P. (1980). Fading channel communications, *IEEE Communications Magazine*, Vol. 18, No. 1, pp. 16-25.

Pahlavan K. and Levesque A. H. (1995). *Wireless Information Networks*. John Wiley & Sons, New York.

Parker D. and Zimmermann D. C. (2002). Phased arrays–Part I: Theory and architectures, *IEEE Transactions on Microwave Theory and Techniques*, Vol. 50, No. 3, pp. 678-687.

Proakis J. G. (2001). *Digital Communications*, 4rd ed., McGraw-Hill, New York.

Rappaport T. S. (2002). *Wireless Communications: Principles and Practice*, 2nd ed., Prentice-Hall, Englewood Cliffs, NJ.

Roy A. D. (1952). Safety first and the holding of assets, *Econometrica*, Vol. 20, No. 3, pp. 431 - 449.

Salo, J. (2006). Statistical analysis of the wireless propagation channel and its mutual information, Doctoral thesis, Helsinki University of Technology, Espoo, Finland.

Saunders S. R. (1999). *Antennas and Propagation for Wireless Communication Systems*. John Wiley & Sons, West Sussex, UK.

Schwartz M., Bennett W. R., and Stein S. (1966). *Communication Systems and Techniques*. McGraw-Hill, New York.

Simon M. K. and Alouini M.-S. (2000). *Digital Communication Over Fading Channels: A Unified Approach to Performance Analysis*. John Wiley & Sons, New York.

Sklar B. (2001). *Digital Communications: Fundamentals and Applications*, 2nd ed., Prentice-Hall, Upper Saddle River, NJ.

Stüber G. L. (2001). *Principles of Mobile Communication*, 2nd ed., Kluwer Academic Publishers, Norwell, MA.

Turin G. L. (1980). Introduction to spread-spectrum antimultipath techniques and their application to urban digital radio, *Proceedings of the IEEE*, Vol. 68, No. 3, pp. 328 - 353.

Xiang W. and Pietrobon S. S. (2003). On the capacity and normalization of ISI channels, *IEEE Transactions on Information Theory*, Vol. 49, No. 9, pp. 2263–2268.

Yu W., Rhee W., Boyd S., and Cioffi J. M. (2004). Iterative water-filling for Gaussian vector multiple-access channels, *IEEE Transactions on Information Theory*, Vol. 50, No. 1, pp. 145-152.